Total Quality Process Control for Injection Molding

WILEY SERIES IN PLASTICS ENGINEERING AND TECHNOLOGY
Series Editor: Richard F. Grossman

Handbook of Vinyl Formulating / Edited by Richard F. Grossman
Total Quality Process Control for Injection Molding, Second Edition / M. Joseph Gordon, Jr.

Total Quality Process Control for Injection Molding

Second Edition

M. Joseph Gordon, Jr.

WILEY

A John Wiley & Sons, Inc., Publication

Published by John Wiley & Sons, Inc., Hoboken, New Jersey
Published simultaneously in Canada

ISBN 978-0-470-22963-7

Library of Congress Cataloging-in-Publication Data is available.

10 9 8 7 6 5 4 3 2 1

Contents

Preface

Total quality process control (TQPC) for injection molding is the process for the repeatable manufacture of a product that consistently meets the customer's requirements. Senior management is responsible for providing the assets, direction, and support to ensure TQPC is implemented, maintained, and practiced daily throughout all company business and manufacturing operations.

Quality begins with senior management implementing a policy for excellence and an attitude that it is achievable. An example of a successful company's quality policy is as follows:

> We, as employees of "COMPANY," are dedicated to the delivery of quality product and technical services contributing to the success of our customers throughout the world. We believe high ethical standards are essential to achievement of our individual and organizational goals.

How a company achieves this or its own specific quality policy and goals is through the use of proven quality management, operations, and methods (e.g., ISO 9001:2008, Total Quality Management, Six Sigma) and other proven quality methods. Process control, with statistical process control (SPC), is just one section of this national standard that requires the company to develop quality methodology to ensure a quality operation is built to provide continuous quality product and services to its customers in a repeatable process. Quality is not the standard; it is the *only* standard for successful business operations.

This book focuses all the personnel and resources of a company toward a plan to implement total quality process control procedures for the production of plastic parts.

The focus is on management's desire and direction to implement the program by providing the assets, guidance, and information to manufacture plastic parts "right the first time."

The quality process begins with sales and continues through the company's different departments, be they large or small, including finance, purchasing, design, tooling, manufacturing, assembly, decorating, and shipping. All personnel have a responsibility and effect on the success of their total quality process control program.

The book explores in detail the methods and procedures that have obtained solid positive results in satisfying their customers' quality part requirements. These techniques have reduced cost, improved product performance, and increased customer satisfaction and profitability for both themselves and their customers.

Each chapter explores in detail different ways to improve part design, processability, and total manufacturing and part quality. Also included are material and process control procedures with control charting in real time to monitor quality through the entire manufacturing system. By adherence to these methods, the tooling for part production and the manufacturing equipment will always be capable of producing product to meet the customer's quality requirements.

Problem analysis techniques and troubleshooting procedures are also presented to improve a company's process control system and solve manufacturing problems with a minimum of time and expense to maintain production schedules and delivery requirements.

Any company, large or small, cannot afford not to adapt all or at least a major portion of the total quality process control procedures to be discussed. Competition is always knocking on our customers' doors, and the only way to counter their threat is to provide a high-quality product within a realistic time schedule and at a fair market price.

ACKNOWLEDGMENTS

I want to extend my appreciation for the love and support I received form my family and especially my wife Joyce during the years of writing this technical book. I also want to thank Dean Wakefield, Carolina Jacobson, Ron Smith of Cooper Industries, and Kermit Lawson of Black and Decker for reviewing the text, adding information, and offering suggestions. Many thanks to my friend and typist, Michelle Jenkins, for her loyalty and timely meeting of deadlines. This book has been a labor of love intended to help improve the quality of the plastics' injection molding industry and the parts it supplies to its customers.

The updating of quality methods for today and beyond was necessary to keep the information current with industry standards.

June 2008 M. JOSEPH GORDON, JR.
September 1992

1

Total Quality Process Control

Total quality process control (TQPC) for injection molding is an operation and quality analysis of the entire injection molding process. TQPC begins with customer involvement and continues through customer satisfaction. It is involved with all the major and minor equipment systems, material requirements, and operation and quality control requirements for repeatably producing good products in "real time," cycle to cycle, to meet customer requirements.

The injection molding process is composed of a multitude of business and manufacturing networking support systems. The analysis begins by developing and understanding all the business variables operating in concert with the manufacturing variables, which include all the design and equipment variables that operate at the same time and that are necessary to produce a quality product. Combined with material handling systems, secondary assembly, and decorating operations (welding, electroplating, and printing) the product supplier must coordinate design and manufacture requirements with material, multiple machine operations, and support equipment and trained personnel for the process to produce a quality product for their customers.

All company operations begin with a well-designed quality program and process system that will encompass all the product and quality requirements necessary to produce a quality product in a repeatable operation. To support this task, the plastics industry is following the most current ISO 9000:2008 and automotive (section specific) ISO/TS16949:2009 quality standard system for

Total Quality Process Control for Injection Molding, by M. Joseph Gordon

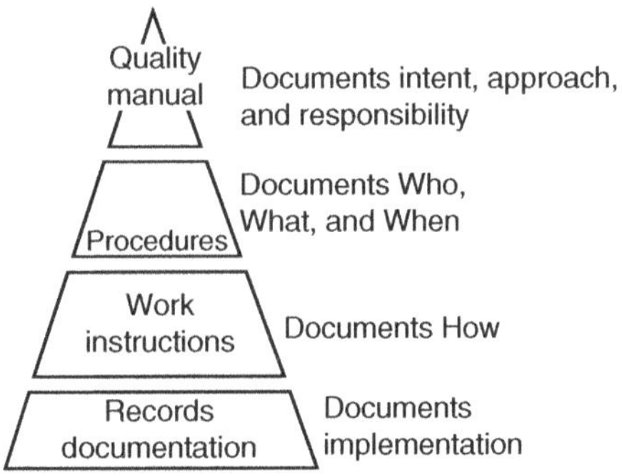

FIGURE 1.1. *The ISO triangle of documentation.*

meeting their quality goals and customer requirements. A survey conducted by the Independent Association of Accredited Registrars[1] listed the main reason for ISO accreditation as follows:

- 29% customer mandate
- 17% competitive pressure or advantage
- 16% continuous improvement based on customer requirements
- 14% improve own quality

To achieve good quality requires dedicated personnel, an executable quality program with management support, and good documentation and communication between employees and the customer by communicating what you will do, doing it, and documenting it. This requires that all personnel work together as a highly motivated quality and manufacturing team to achieve TQPC results.

ISO 9001

The implementation of a good quality program begins with quality documentation as shown in The ISO Triangle of Documentation (Figure 1.1) for ISO 9001:2008. A good quality program, ISO 9001:2008, begins with a quality manual. The ISO accreditation program has additional requirements, which include six procedures for specific documentation on how to handle control of the following:

1. Documents
2. Records
3. Nonconforming items

[1]Smith, L., "The Hidden Cost of Cheap Certification," *Quality Digest* May 2007: 32–35.

4. Audits
5. Corrective action
6. Preventative action

Plus, the company can, if it deems it necessary, add any special and/or specific business and manufacturing operation procedures and operation-specific instructions to its system. Automotive, consumer, and aerospace companies have required their product suppliers to be in compliance and to be registered with ISO/TS 16949:2009 or AS9100, which demands more company quality documentation.

It is the responsibility of the company's senior management to develop a quality program to assure customers that quality is their goal and that only products meeting their customer's specifications will be shipped.

Even if a company does not become ISO certified, the company can use it as a guide in establishing a quality system. The quality manual is typically 30 to 35 pages with detailed, streamlined procedures and instructions for specific operations. Standardized templates are available on the Internet to be used as guides for all the documentation; a procedure example (Template) called "Control of Documents" is available in Appendix A.

I recommend a company document the individual and/or specific information and instructions for their equipment and process operations as individual instructions. The company can then record all data from its business and manufacturing operations into an established company program and project documentation and record storage and retrieval system. Such a system is called the molding data record sheet. Information on company operations is stored in this system. Documentation and operations data and records can be recorded at machine side for the individual injection molding machines in a molding data record sheet (Figure 1.2) and/or stored electronically in the file memory of the process control equipment setup instruction, which is downloaded into the configuration management system (CMS). Electronic storage is preferred as it will then be accessible at all stations with a computer operating with the CMS storage and retrieval system.

Documentation is necessary for each job setup because each mold and molding machine setup is specific and independent of all other setups that occur daily in a manufacturing environment. The molding data record sheet is a record of the specific settings used and of the process information on how the product was manufactured for the customer. It is based on the customer's specifications as well as on the manufacturing setup instructions and records for how the product was produced. A copy of this information should remain as a record of the molding operation with a copy of the molds operation put in the program file. A lot of redundant information is filed, but it is necessary for a complete record of each item in the manufacturing operation. Remember, the next time the mold is run, it may be scheduled on another molding machine and set up by different technicians. These records assist in ensuring that the customer will receive the same product quality.

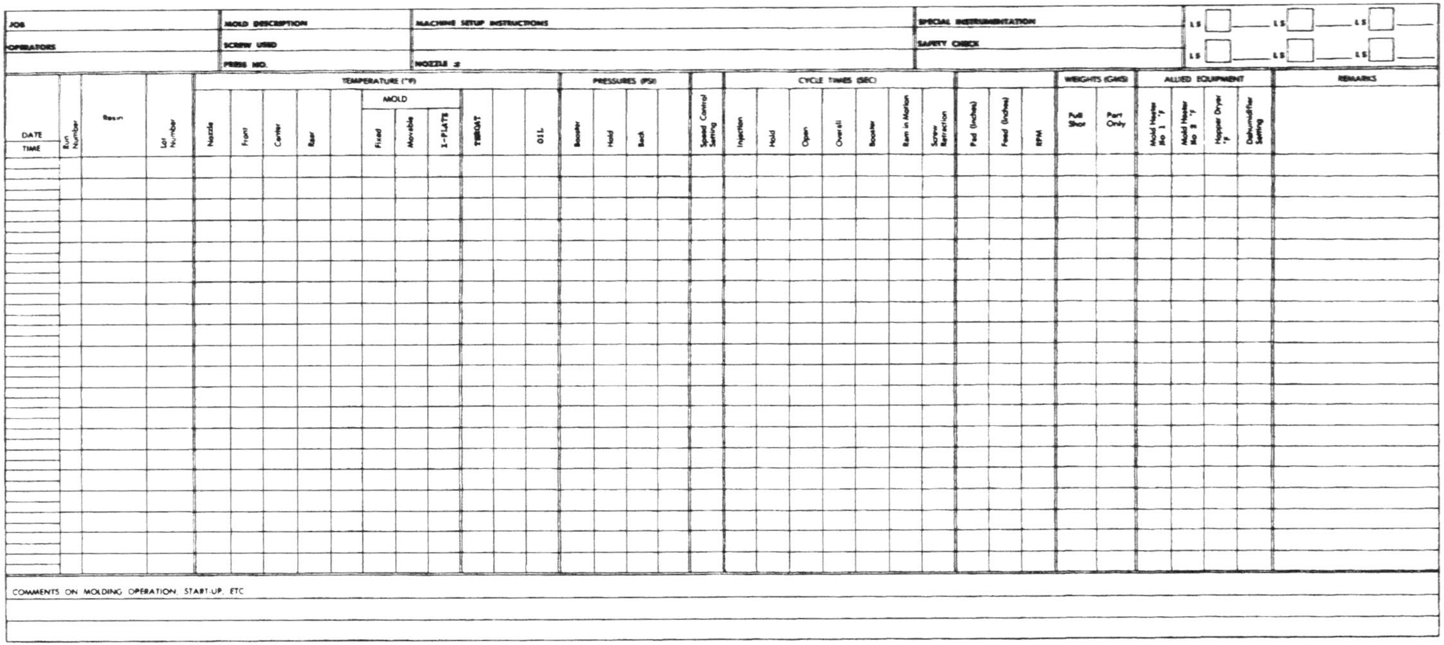

FIGURE 1.2. *Molding data record sheet.*

DOCUMENTATION

The quality program's documentation process begins with the necessary company information and documentation, which is written as procedures and necessary instructions. These may be selected operations of the business, beginning with program initiation, design and development, manufacture, and service for the products provided to their customers. These instructions can be used as the basis for a company training program for new-hires and for training operators in performing additional and new functions. Keep documentation simple, to the point, and in a separate and easily accessible section of the configuration management system.

Information should flow from main documents, the quality manual and specific procedures, with any updating and revisions on the lower level documents as with your daily operating instructions and documentation. Machine setup and startup instructions can be laminated and located at machine side as an operation guide, in addition to any checklists and molding record sheet information.

Customer and program documentation also include information as meeting notes, verbal discussions, communications, and records produced during the customer's program discussions and negotiations. Also, as the program progresses, the design, manufacturing information, and data are filed, respectively, in the CMS storage system.

Remember, the information and instructions not documented are quickly forgotten and may result in later problems requiring corrective action. Injection molding is one of the more variable intense manufacturing operations for producing a single product. Problems can occur quickly if a key variable is forgotten. And when a key person leaves, he or she can take information with them that was never documented on how a specific operation was conducted.

Process control is involved with determining, knowing, controlling, and documenting these variables as a record of the operation for the entire manufacturing process, step by step, from product design to shipping. This should also include all supplier information and support provided for product design and prototype assistance, if within the supplier's capability level.

ESTABLISHING PROCESS OWNERSHIP

For any process to be successful, ownership must be assigned, accepted, and implemented within the organization. Ownership is defined as belonging to the one most to benefit from a successful program or well-running process. To determine who this, not always obvious, person is the following questions should be answered first. Who is the person with the most of the following qualities:

1. Ability to affect change
2. Resources (e.g., people, systems, and budget)
3. Problems (customer complaints, critiques, and endless defects)
4. Time available/necessary to make changes
5. Credit to gain when all works well
6. Actual or potential credit

The owner, as defined by this list of questions, should have the most to gain from these planned improvements. They should also have delegated authority to act, essentially, anywhere within the defined system, and even out of the supposed system operating area. Because the root cause of a problem may not always be in their direct line of authority, the leader must have senior management's authority for the entire process. Responsible actions should always be coordinated through the managing authority in the other area if cause is found for the process problem originating from their actions.

I helped to solve a problem, at the request of the Vice President (VP) of Operations that was discovered at the final test point of their major product line. The solution involved an analysis of the product's design, which involved multiple molds, assembly operations, and final testing. This problem had been occurring frequently for more than three years without a satisfactory and lasting solution. The final solution involved four departments and retraining assembly and test personnel after determining the multiple solutions that solved the problem. This problem was not in one person's area of responsibility, but as in most cases, there was one person with the most to gain, in this case, the VP of Operations.

The business process owner should be given authority to operate at a level high enough to do the following:

1. Identify effects of any new business directions on the process
2. Influence changes in procedures and/or policies on the process
3. Plan and implement process changes as appropriate
4. Monitor the effects on the process for efficiency and effectiveness[2]

The next set of criteria for effective process improvement involves the leader's ability to lead. The team leader should possess leadership characteristics such as follows:

1. Recognition as a creditable leader in the company
2. Ability to direct and lead a group
3. Ability to keep the team on schedule
4. Ability to obtain the assets needed for support of the team
5. Ability to provide encouragement and direction for the team
6. Ability to induce change and have it accepted

7. Ability to deal and work with senior management
8. Reputation as a skilled negotiator
9. Ability to push aside roadblocks
10. Ability to live up to commitments
11. Ability to change poor performance into acceptable performance[2]

It is best if the owner knows and understands the process. He or she does not have to be a member of management, but he or she is in many situations. The solution of a problem begins with a team selected for assistance. The process with the problem is then presented on a diagram or flowchart for the team to improve understanding of all the involved operations. It is then advised to run a failure mode and effects analysis (FMEA) with a fishbone in-depth analysis to uncover all variables that act on the entire process.

The FMEA is a step-by-step analysis of a process that lists all potential failure or problem points in the process and the results if not corrected or controlled. The fishbone analysis is a detailed analysis of a situation that lists all known variables that act on the situation. More in-depth information on the workings of these two quality methods will be discussed later.

Once all the available information of the process is known, analysis begins by making corrections, monitoring, and implementing preventive actions with the operation put back in service, corrected, and in perfect operation.

Five steps for achieving the TQPC goal are as follows:

1. ***Standard selection.*** Select the quality standard for the organization based on customer requirements and future business potential.
2. ***Management support.*** Management establishes the business goals, policy, and objectives and provides the ongoing assets and support.
3. ***Corrective and Preventative Actions.*** User satisfaction is first with the "root causer" of problems eliminated in all areas of the company.
4. ***Continual improvement.*** The quality management system (QMS) is continually reviewed, improved, and updated for quality performance.
5. ***Know your system's capability.*** Maintaining your system's equipment to a known standard is essential for repeatable manufacture.

The methods to achieve the quality required are not easy, inexpensive, or quick. Considerable time, money, and hard work are involved, which initially do not show a return on investment as quickly as management would like to achieve. Therefore, plan your quality improvement program well (checklists), use the information in this text as a guide develop your implementation plan in stages with check points and milestones for review of progress, and train

[2]Harrington, H.S., Performance Improvement "Who Owns the Process?" May 2007: 16.

yourself and employees in the methods and practices of achieving and retaining a quality operation.

Work to ensure every employee can be the best he or she can be and provide the assets to have it happen. Have employees strive for repeatability of operations, with improvements as needed to reduce problems and cost, plus provide incentives for continual improvements in forms that are achievable by your personnel. Provide employees with the tools to do this, such as checklists, operation guides, instructions, procedures, and so on.

Review the classic quality methods for inclusion and consideration of use at your company. They may be old, but most are still active at progressive companies. Quality leaders have expressed their views that the Six Sigma advances were made using these "tried and true" quality methods listed in Table 1.1.[3]

To add some order to the quality area as far as methodology, what you see today is not really new; it is just presented in a different box. Quality essentially started with control charting and progressed to what it is today. New names have been applied to proven methods.

Armand V. Feigenbaum's *Quality Control: Principles, Practice, and Administration* (McGraw-Hill & Co., 1951) set the standard in 1951. His definition of total quality control (TQC) included the following plus many others:

- Design of experiments
- Quality cost
- Design review
- Statistical process control
- Process certification
- Involvement by top management
- Supplier controls
- Trained, certified quality engineers
- Reliability engineers
- Employee training

The next major change, which was implemented in approximately 1975, occurred with total quality management (TQM) and included the following requirements:

- All of TQC
- ISO 9001
- Benchmarking
- Team problem solving
- Five S

[3]Six Sigma. Available at: http://en.wikipedia.org/wiki/Six_Sigma.

TABLE 1.1. Quality Improvement Methods.

Quality Methodology Understood:

Program Name	Worker Involvement	Specialist Orientec	Group	Individual	Procedure	Work Methods	Quality	Prod Design	Morale Enhancement	Motivation
Quality Circles	X		X		X	X	X		X	X
Zero Defect	X			X			X			
Employe Suggest	X			X		X	X	X	X	X
Work Simplify	X		X		X	X				X
Qual of Work life	X		X			X	X	X	X	X
Scanion Plan	X		X			X				X
VE/VA		X	X		X			X		
IE Work Study		X		X	X	X				
QA/QC		X		X	X		X			
Org Developmt	X	X	X	X	X				X	X
Fish Bone	X		X	X	X	X	X	X	X	X
SPC	X			X	X		X			X
DOE	X		X	X		X	X	X	X	X
CP/CpK		X		X	X		X		X	X
FMEA		X		X	X	X	X	X	X	
PAP	X		X		X	X	X	X		X
PPAP	X		X		X	X	X	X		X
QFD		X		X	X	X	X		X	X

TABLE 1.1. (*Continued*)
Quality Methodology Understood:

Program Name	Worker Involvement	Specialist Oriented	Group	Individual	Procedure	Work Methods	Quality	Prod Design	Morale Enhancement	Motivation
GMP	X		X		X	X	X		X	X
Kaizen	X	X	X		X		X		X	X
ISO 9000	X	X		X	X	X	X		X	X
TS16949	X	X		X	X	X	X		X	X
CEA	X		X		X	X	X		X	X
8-D	X		X		X	X	X		X	X
Poka-yoke	X		X		X	X	X	X	X	X
VSM (value Stream mapping)	X		X	X	X	X	X		X	X
CTQ	X			X	X		X	X	X	X
VOC	X		X		X		X		X	X
TPS (Toyota)	X	X	X	X	X	X	X	X	X	X
FEA		X		X	X	X	X	X		
TQM	X	X	X		X	X	X	X	X	X
Lean	X		X	X	X	X	X		X	X
JIT	X		X	X	X	X			X	X
5S	X		X	X	X	X	X			X
C&A	X		X	X	X	X	X		X	X
Triz		X	X	X				X		X

- Toyota production system
- Strategic quality plans
- Lean
- Process focus

The TQM mantra is as follows: "Do it right the first and every time, no level of defects is acceptable."

In 1984, the new program was business process improvement (BPI), which attacked the core of current white-collar problems by focusing on waste and bureaucracy. Quality output was the foundation with organizations simplifying and streamlining operations. The main objectives of BPI were to ensure the organization has the following business processes that:

- Eliminate waste
- Eliminate errors
- Eliminate delays
- Maximize use of assets
- Promote understanding
- Are easy to use
- Adapt to customers' needs
- Provide a competitive advantage
- Reduce excess head count

Then in 1986, Motorola developed Six Sigma and focused on business improvement as consisting of the following:

- Understanding and managing customer requirements
- Aligning key business processes to achieve those requirements
- Using rigorous data analysis to minimize variation in those processes
- Driving rapid and sustainable improvement to business processes

The heart of the Six Sigma system is the methodology called "DMAIC" (define, measure, analyze, improve, and control process improvement). Six Sigma included the following:

- Selected TQM tools
- Selected BPI tools
- Full-time problem solvers called Black/Green Belts
- Expanded statistical training for a selected group of problem solvers

Tying all of the latest quality information together leads us to the current "Total Six Sigma" system. This came from the 1987 improvements of Six

Standards & Practices of Plastics Molders

Material
Acrylonitrile Butadiene Styrene (ABS)

Note: The *Commercial* values shown below represent common production tolerances at the most economical level. The *Fine* values represent closer tolerances that can be held but at a greater cost. Any addition of fillers will compromise physical properties and alter dimensional stability. Please consult the manufacturer.

Drawing Code:
A = Diameter (See note #1)
B = Depth (See note #3)
C = Height (See note #3)

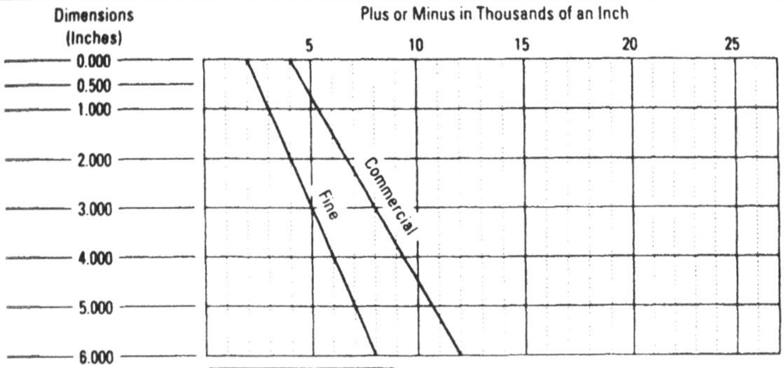

Drawing Code	Dimensions (Inches)	Comm. ±	Fine ±
	6.000 to 12.000 for each additional inch add (inches)	0.003	0.002
D = Bottom Wall	(See note #3)	0.004	0.002
E = Side Wall	(See note #4)	0.002	0.003
F = Hole Size Diameter (See note #1)	0.000 to 0.125	0.002	0.001
	0.126 to 0.250	0.002	0.001
	0.251 to 0.500	0.003	0.002
	0.501 & over	0.004	0.002
G = Hole Size Depth (See note #5)	0.000 to 0.250	0.003	0.002
	0.251 to 0.5000	0.004	0.002
	0.501–1.000	0.005	0.003
H = Corners, Ribs, Fillets	(See note #6)	0.027	0.017
Flatness	0.000 to 3.000	0.015	0.010
(See note #4)	3.001 to 6.000	0.030	0.020
Thread Size (Class)	Internal	1	2
	External	1	2
Concentricity	(See note #4) (F.I.M.)	0.009	0.005
Draft Allowance per side	(See note #5)	2.0°	1.0°
Surface finish	(See note #7)		
Color Stability	(See note #7)		

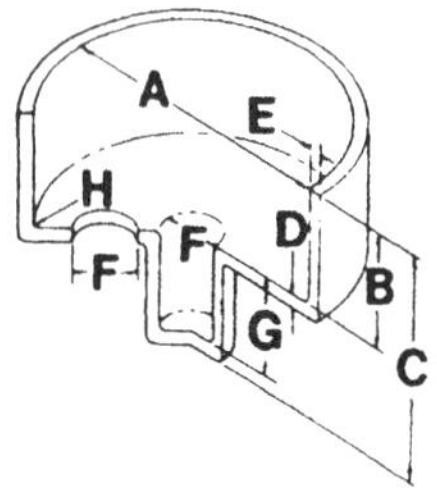

REFERENCE NOTES

1. These tolerances do not include allowance for aging characteristics of material
2. Tolerances are based on 0.125 inch wall section.
3. Parting line must be taken into consideration.
4. Part design should maintain a wall thickness as nearly constant as possible. Complete uniformity in this dimension is sometimes impossible to achieve. Walls of non-uniform thickness should be gradually blended from thick to thin.
5. Care must be taken that the ratio of the depth of a cored hole to its diameter does not reach a point that will result in excessive pin damage.
6. These values should be increased whenever compatible with desired design and good molding techniques.
7. Customer-Molder understanding is necessary prior to tooling.

Revised 1991
15

FIGURE 1.3. *Cavity hold tolerances, dimensionally.*

Sigma, lean Six Sigma, and Total Improvement Management. The common bonds between these are the following:

- Top management leadership
- Process focus
- Similar problem-solving approaches
- Measurements of dollars saved
- Customer focus

The prime use of these methods is to ensure they are all applied correctly, never poorly.

When you begin a quality improvement program, research it so well you can explain it to your peers. Study the benefits that could be achieved and the time and cost of each method you may consider implementing. The Internet[4] has a lot of free information on these methods that will give you a brief overview as to what they can accomplish when applied correctly. I have used several that returned considerable quality benefits when implemented. I believe in using statistical process control (SPC), fishbone analysis, quality circles, FMEA, checklists, equipment and process procedures, and instructions. The Lean and Six Sigma methods are discussed and have considerable merit when correctly applied by a trained implementer.

Total quality process control is composed of a QMS, trained personnel, and management support systems to ensure all customers' specifications (within injection molding capabilities) are achieved. This means that metal working tolerances are not used for plastic parts. Tolerances, both fine and commercial, for the manufacture of injection molded plastic products, in this case, for the unfilled plastic material acrylonitrile butadiene styrene (ABS) as documented per the Society of the Plastics Industry, Inc. (SPI), are shown in Figure 1.3. Each generic plastic has its corresponding tolerance value variance figure available from the SPI.

The tighter the tolerance requirement, the greater the cost of the product because the manufacturer will have to hold tighter tolerances in a variety of molding areas from the choice of designing the part, material, mold design, molding parameters, post cure, part assembly, and handling methods.

IDEAS AND METHODS

When the ideas and concepts for creating a TQPC program are accepted by all levels of an organization, the result will be profitable products for the customer. The TQPC program effectively completes the customer–supplier design and manufacturing cycle by focusing on development of a

[4]http://www.statsoft.com/textbook/stquacon.html#process.

quality-conscious organization for product development that covers design, material selection, tool design, and manufacturing through assembly and decoration, to the final shipment of the product to the customer.

It is best to use statistical process control methods to supervise the manufacturing of plastic parts. Unlike earlier statistical part checking methods, TQPC does not rely on inspection to separate the good from the bad parts. Rather, from the start, it focuses on all the variables that can influence plastic part manufacture. Success is achieved through a combination of good design principles, the use of capable manufacturing equipment, and appropriate selection of part tolerances, materials, and tooling. Finally, the manufacturing process must be controlled to meet customer requirements.

In no-nonsense terms, TQPC explains tried-and-true methods that work and ways to motivate the organization to accomplish the common goal of product quality. The plastics injection molding industry has long needed this type of information, which ties all the many product and manufacturing variables together in an organized and readable format.

Many companies already using these methods are reaping the rewards by becoming preferred suppliers. As a result, they are continuing to grow in a very competitive marketplace. In fact, most companies, from large original equipment manufacturers (OEMs) to small part suppliers, which now use these principles, can with a little more effort and practice become even better quality-product suppliers and more competitive in the marketplace.

Readers who apply TQPC methods will find them easier to implement than had been thought earlier and, through a good program, can achieve even greater returns at minimum cost while expanding their customer bases.

2

Implementing Total Quality Process Control (TQPC)

TQPC uses the quality methods developed by the quality leaders including Juran, Deming, Taguchi, Feigenbaum, and others to develop a system where the best quality methods are used for control of the design, development, and manufacturing processes. Based on today's quality leaders who suggest that the lean style of manufacturing is best, non-batch style of production, TQPC strives to meet this type of production. If it is not capable of meeting production for the batch style of manufacture for injection molding, then it will be productive in later operations as during the decorating, assembly, and final testing of finished products in the original equipment manufacturers (OEM) plant.

Products produced today are not allowed to have an acceptable amount of defects, as with the acceptable quality level method of manufacture and quality inspection, which is illustrated in Table 2.1. Today, management wants all their parts to be in the acceptable category, without any defects. This is possible with TQPC when all variables remain in control and instructions are followed. This zero-defect type of manufacture ensures all variables are in control and are kept there during the entire production run. This is not easy to do but is a goal to achieve.

A "quality improvement plan" with step-by-step instructions lists the steps for the implementation of improving quality with minimum effort. Quality can always be improved when the quality team "accepts the challenge."

TABLE 2.1. Master Sampling Table.

Lot Size	Sample Size	Acceptable Quality Level 0.25		0.5		0.75		1		1.5		2		3		4		5		6		7		8		9		10		12	
		A	R	A	R	A	R	A	R	A	R	A	R	A	R	A	R	A	R	A	R	A	R	A	R	A	R	A	R	A	R
499 or less	40	↓		↓				0	2	0	2	0	3	1	4	1	4	1	6	2	6	2	7	3	7	3	8	4	9	4	9
	50	↓		↓				0	2	0	3	1	3	1	4	2	5	2	6	3	7	3	9	4	9	4	9	5	10	5	11
	60	↓		↓				0	3	1	3	1	3	2	5	2	6	3	7	4	8	4	9	5	10	5	11	7	12	7	13
	70	↓		↓				1	3	1	3	1	4	2	5	3	6	4	8	5	9	5	9	6	10	7	11	8	13	8	12
	80	↓		↓				2	3	3	4	3	4	4	5	5	6	7	8	8	9	8	9	9	10	10	11	12	13	12	13
500 to 799	40	↓		↓		*	1	*	2	0	3	0	3	0	4	1	5	1	5	1	6	1	7	2	8	2	8	2	9	4	10
	60	↓		↓		*	1	0	2	0	3	1	4	1	5	2	6	2	7	3	8	3	9	4	10	5	11	5	11	6	12
	80	↓		↓		0	2	1	3	1	4	1	5	1	6	3	7	3	8	5	10	5	11	6	12	7	13	8	14	9	15
	100	↓		↓		0	2	1	3	2	4	2	5	2	6	4	8	5	9	6	11	7	13	8	14	9	16	10	17	12	13
	120	↓		↓		1	2	2	3	3	4	4	5	5	6	7	8	8	9	10	11	12	13	13	14	15	16	16	17	13	19
800 to 1,299	40	↓		*	1	*	1	*	2	*	3	0	3	0	4	0	5	0	6	1	6	1	7	1	8	2	8	2	9	2	10
	60	↓		*	1	0	2	0	2	0	3	0	4	1	5	1	6	2	7	2	8	3	9	3	10	4	11	4	12	5	12
	80	↓		0	2	0	2	0	3	1	4	1	5	2	6	2	7	3	8	4	10	5	11	5	12	6	13	8	15	8	15
	100	↓		0	2	0	2	0	3	1	4	1	5	2	6	3	8	5	10	5	11	7	13	7	14	9	15	10	17	10	18
	120	↓		0	2	1	3	1	3	2	5	2	6	3	7	5	9	6	11	7	13	8	14	9	16	11	18	12	19	13	21
	160	↓		1	2	2	3	3	4	4	5	5	6	7	8	9	10	10	11	13	14	15	16	16	17	18	19	19	20	22	23
1,300 to 3,199	50	*	1	*	1	*	2	*	3	*	3	0	4	0	4	0	5	0	6	1	7	1	8	2	9	2	10	3	10	3	11
	75	*	1	*	1	0	2	0	3	0	4	0	5	1	5	2	7	2	8	3	9	4	10	4	12	5	12	6	14	6	15
	100	*	1	0	2	0	2	1	4	1	4	1	5	2	6	3	8	4	9	5	11	6	12	6	14	8	15	9	17	10	18
	125	0	2	0	2	1	3	1	4	2	5	2	6	3	7	4	9	5	11	7	13	8	15	9	16	11	18	12	20	13	21
	150	0	2	0	2	1	3	2	5	2	5	3	7	4	8	6	10	7	13	9	15	10	17	11	19	14	21	15	23	16	25
	200	1	2	1	2	2	3	4	5	5	6	6	7	8	9	10	11	13	14	17	18	17	18	20	21	22	23	25	26	27	28

A—Acceptance number; R—Rejection number; *No acceptance at this sample size.

Arrows: When there is an arrow under a given AQL, use the first sampling data below the arrow. (Form larger lots if possible.)

Adapted from Ref. [1].

QUALITY IMPROVEMENT PLAN

1. Company management commits to improving quality.
2. Management team appoints a "leader" to undertake quality improvement, with accountability.
3. Leader forms a quality team to determine the degree of quality improvement and where to begin.
4. Determine needs by monitoring the areas of the operation that need improvement from problems.
5. Select the quality system/accreditation and methods to use for determining the system to select.
6. Examine system for "root cause" of each problem detected.
7. Document all the results of the problem definition and analyze the problem for root cause and repeatability.
8. Discuss the requirements with personnel involved for each problem and document all possible solutions.
9. Develop possible solutions for problems with confirmation that the solution is correctly implemented, monitored, and proven to eliminate the problem without causing a new problem.
10. Write new update existing manufacturing/service procedures and instructions, train personnel, and implement them.
11. Conduct quality failure mode and effects analysis (FMEA) operations for monitoring the business and manufacturing operation.
12. Develop and implement procedures and monitor them in real time for operation.
13. Develop checklists for all operations to establish repeatability of operations.
14. Implement quality training for personnel.
15. Monitor operations for quality business and manufacturing operations.
16. Use quality function deployment (QFD) methods with the customer to develop better information to meet their requirements.
17. Implement ISO 9001:2008 or beyond for operations and customer quality.
18. Monitor and maintain manufacturing equipment for compliance with manufacturer specifications of operations.
19. Monitoring operations for data accuracy to meet customer and internal quality requirements.

These are the starting methods to use when performing quality improvement, no matter what quality system is used in your company.

TQPC focuses on the "total manufacturing system," not just the molding machine. In an analysis, the injection molding machine is just one of many variable-producing mechanisms that the manufacturing system must keep in control. When all the major variables are considered, such as material, control systems and auxiliary equipment, mold, plant environment, and maintenance services, there are a sizable amount of variables that need to be controlled.

There are, then, the secondary operations to consider for the part, such as assembly, decoration, information transfer, and any other operations required of the parts end-use service. An example of this is shown in Figure 2.1 in a partial fishbone diagram of some molding variables.

Note that as one main element is identified, there are support variables that contribute to the main elements actions on the total process or item, as selected for analysis. Always take each element to its basic factors so no single item is ever left unidentified in the analysis process. If necessary, take it to the supplier of an item, as the problem could have originated in their system, and was not told to your personnel.

Unfortunately, some suppliers do not inform their customers of all changes they may make in their products. They often assume the change is so minor it will not matter, or it is proprietary and need not disclose any changes as long as the end use is not affected. But this is not always the case. Therefore, create a trust with your suppliers to ensure, if they ever modify their product, to disclose this information and require that they send a trial sample for evaluation before the change is actually made. This will give you time to evaluate the modification in your product for processing and end-use performance.

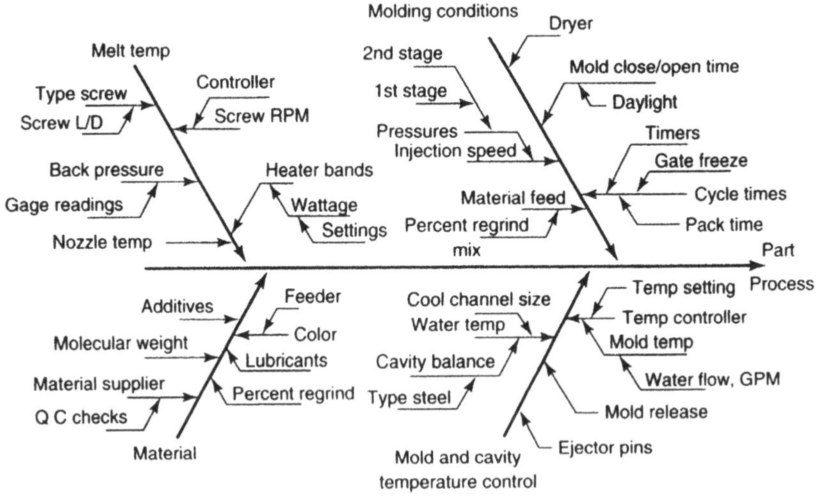

FIGURE 2.1. Fishbone diagram.

STATISTICAL PROCESS CONTROL (SPC)

SPC is used to gather statistical control data for your operations. An example includes the data gathered to measure the degree of control [capability (Cp)]; thus, a machine, process, and/or operation can reach and then maintain this level of productivity during manufacture. Because injection molding at a custom molder is a typical short-term program, the parts required for the customer must be produced, in a specific time period, and shipped. Then, change the molds and begin the next program, possibly with a new material and different customer quality requirements for the product. The important item to remember is that the quality program does not change; only the mold, machine settings, and material at the machine will change.

Many injection molding machines have the SPC "closed loop, continuous feedback machine controls," which continuously measure the variance in machine process and/or product during each machine or molding cycle. If a variance is noted, depending on how the control system is set up, it automatically attempts to adjust the affecting variable to keep the process in control, using pre-assigned control limit values of adjustment, to reestablish the specific control parameter, sensed, out of tolerance.

If a variable seems to change the process, the SPC control system attempts to correct for the variance. If the variance is too great or continues out of control and crosses over the established upper specification limit (USL) and/or lower specification limit (LSL), an alarm will sound calling the machine operator's attention to adjust the machine manually or determine what has changed to cause the dramatic out-of-control problem. The change could be minor or major depending on the "root cause" of the problem.

Today, statistics are used to determine and control those areas of strength in a manufacturing system that can be used to improve the total system for manufacturing plastic products. Statistics are a very useful tool for controlling the manufacture and monitoring the quality of plastic products.

CONTROLLING THE PROCESS

In any manufacturing process, it is extremely important to maintain the highest degree of process control. Injection molding is dependent on a cycle-to-cycle repeatability in process control. It is also very important to know whether the process equipment can maintain the type of control required to produce good parts repeatably.

It is a statistical fact that when the Capability Index (Cp) of a process or system is 1.00 or greater, the variables in the process being monitored are in control during the period of time they were monitored. Therefore, monitoring Cp is one of the key quality operations that show whether the system and all the supporting branches are in control for that time period.

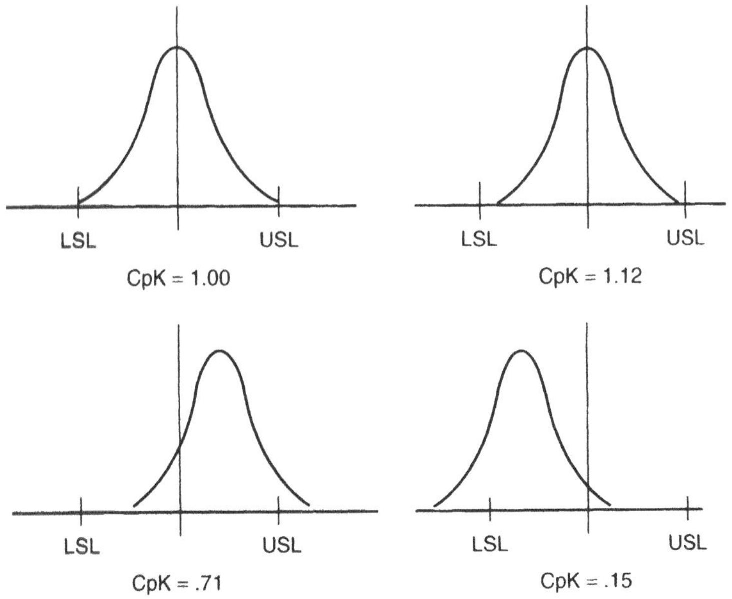

FIGURE 2.2. *Cpk is a measure of spread and centeredness; the higher the Cpk value, the more in control is the process.*

The centeredness of the curve indicates the degree of control of the machine and/or process when the data are plotted as shown in Figure 2.2. Equipment and software systems are available to perform the data collection, analysis, and plotting automatically to show the degree of control within the monitored system. The spread of the ends of the curve indicates the degree of centeredness or control of the system. As the curved ends of the data spread beyond the USL/LSL, the result is a direct reflection on the control of the process, either high or low.

CP THE CONTROL OF OPERATIONS

It is recommended that each time a mold and injection molding machine combination is used, a Cp system analysis is conducted. This will show when the operation reaches equilibrium with the system's operating variables. As monitoring continues, it will validate the control the process is capable of obtaining to produce good parts, with the data recorded for process control. If needed the data can be given to the customer showing the degree of control achieved during their product run. This is explained in greater detail in the author's book from J. Wiley & Sons, *Industrial Design of Plastic Products*.

The processes cycle's Cp index is used for determining the capability of the system for continued repeatability of the manufacturing operation. It is also used to determine how tight the processing tolerance must be held so acceptable parts are achieved on every cycle. Monitoring the cycle and process variables in real time is critical to ensure the parts stay within acceptable process parameters. Typically, the injection molding machines' main variables, pressure, temperature, and timer settings are monitored for cycle consistency, which results in the machines Cp value. A capability analysis will also provide management with a good analysis of the quality of their "preventative" maintenance program, or if one is necessary. The Cp is generated on operation data and analyzed during the startup and continued molding of the product for consistent repeatability.

To assist in determining the Cp value of the machine, most injection molding machines and support equipment come with the option of having real-time process control systems installed on the equipment. The machine manufacturer provides the options of what is installed on the machine, often at the buyer's suggestion or selection. The better the control system, the better the output of the machine.

TQPC is involved in maintaining the highest degree of process equipment capability by monitoring the machine and system's index of capability, either Cp, Cpk, or Ppk. (Cp is the ability of a process to produce consistent results, Cpk is a capability index for how well a system can meet specification limits, and Ppk is an index of longer term process performance for how well a system is meeting specifications.)

Cp is the ratio between the permissible and the actual spread of a process. Permissible spread is the difference between the USL and the LSL of acceptability or the total tolerance, where the actual spread is the difference between the upper and lower $3 \times \sigma$ deviations from the mean value (representing 99.7 percent of the normal distribution). The formula is $Cp = (USL - LSL)/(6 \times \sigma)$. Note: In some cases, the term "specification" is replaced with "control", I have used specification here.

In statistics, sigma (the lowercase Greek letter σ) is defined to represent, the standard deviation (a measure of variation, http://en.wikipedia.org/wiki/Standard_deviation) of a population based on a sample. Its units of measurement are dependent on the selected sample, which is defined as the square root of the variance. In a capability/study, sigma refers to the number of standard deviations between the process mean and the nearest specification limit as shown in Figure 2.3, with the mean at 0 and the specification limits at ±6 sigma.

To understand standard deviation, remember the variance is the average of the squared differences between the data points and the mean. Variance is tabulated in units squared. Standard deviation is then the square root of that quantity that measures the spread of data about the mean, measured in the same units as the data.

As an example, in a population of (4, 8), the mean is 6 and the deviations from the mean are (−2, +2). These deviations squared are (4, 4), the average of which (the variance) is 4. Therefore, the standard deviation is 2. In this case,

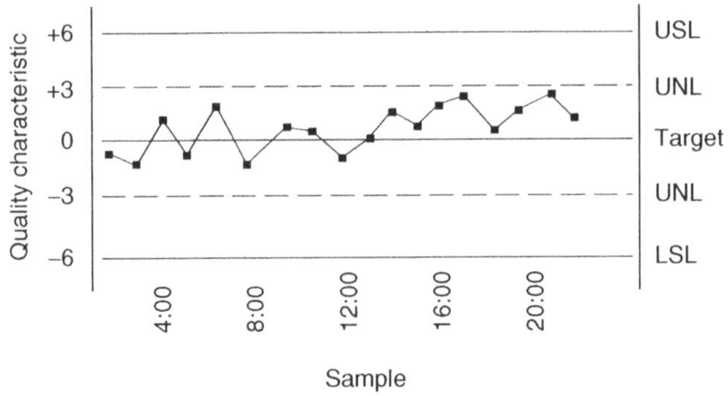

FIGURE 2.3. *A run chart depicting a +1.5σ drift in a 6σ process. The upper natural tolerance limit (UNL) and the lowernational tolerance limit (LNL) of normal cycle variance during operations are shown.*

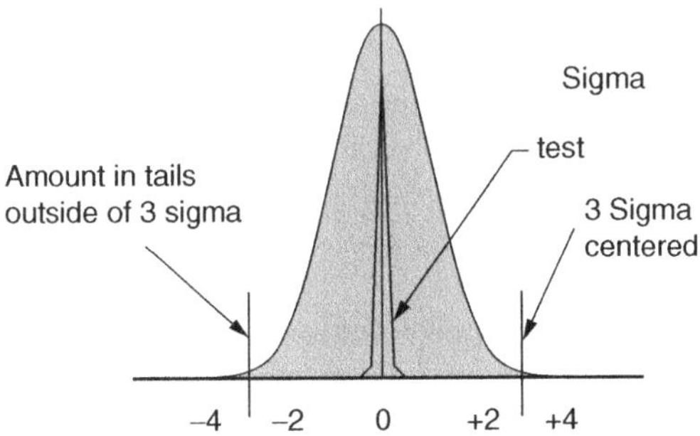

FIGURE 2.4. *Six Sigma process. (Adapted from Ref. [2].)*

100 percent of the values in the population (4, 8) are at one standard deviation (2) from the mean.

Formally stated, the standard deviation is the root mean square (RMS) deviation of values from their arithmetic mean. Cp (process capability) can be thought of in the following ways:

- Cp measures the capability of a process to meet its specification limits. It is the ratio between the required and the actual variability
- Mathematically, the Cp is expressed as Cp = (USL − LSL)/(6 × sigma). This is the spread of a normal curve. Capability statistics are basically a ratio between the allowable process spread (the width of the specification limits) and the actual process spread (6 sigma)
- Graphically, as shown in Figure 2.4, a normal curve is centered between the specifications. Notice the tail-end areas that exceed the specification limits. The smaller the area outside the specifications, the larger the Cp. This is similar to looking at a parts per million (PPM) value for the number of items that exceed the specification.

CPK-CENTERED PROCESS CONTROL

Cpk or the process capability index is a measure of the off-centeredness of a Cp-centered process producing a similar level of defects—the ratio between permissible deviation, which is measured from the mean value to the nearest specific limit of acceptability, and the actual one-sided 3 × sigma spread of the process. As a formula, Cpk = either [(USL – Mean)/(3 × sigma)] or [(Mean – LSL)/(3 × sigma)], whichever is smaller (i.e., depending on whether the shift is up or down). Note that this ignores the vanishing small probability of defects at the opposite end of the tolerance range. A Cpk of at least 1.33 or greater is the desired value.

Note: Do not connect the term "Six Sigma Process" as the same as (6 × sigma), which is the process control charting of Cp and Cpk. They are not the same![1]

Process control includes the following:

- Documentation—Documenting what you say you will do, how you will do it, and how you will ensure or enforce it being done each and every time.
- Training—Provided to ensure it is always done correctly.
- Process monitoring—Real-time monitoring and instant feedback of process variables and machine status, along with access to real-time process data uploaded to remote computer terminals with alarms for out-of-tolerance conditions.
- Data entry—Operators enter downtime reasons and update work order status, part production, and scrap information in real time for production control at the press.
- Automated graphical reporting—Machine uptime and production reports to reduce burden on resources and provide timely access to information with graphic run charts provided for tolerance control capability and either operator/computer determining or maintaining machines and system at optimum Cp efficiency.
- Instant notification—E-mail and paging notification immediately alert decision makers of machine, plant, and equipment status and of molding process variations.
- Diagnostic tools—Tools available for determining root cause of problems associated with equipment and system control and operation when out of tolerance conditions occur.
- Advanced features—Depending on software selected, most allow modular architecture to add modules as their needs grow. Optional modules enable

[1]Search for "Process Capability Cp" on the Internet or go to http://en.wikipedia.org/wiki/Six_Sigma for additional information.

production tracking, advanced planning and scheduling, as well as statistical process control.

The use of pure statistics, however, will not impart quality to a product, but it can identify where problems exist and quantify the type and frequency of occurrence. Separating the acceptable from the nonconforming is costly, but in some situations, to get the product to the customer, suppliers have resorted to this method of manufacture. When this occurs, identify the defects and their percentages with Parato charts (Figure 2.5), which report on the frequency and type of nonconforming items produced. This will give emphasis to providing a solution to this problem. When this is recognized and acknowledged, then corrective action can be employed.

Likewise, employee quality groups, such as quality circles, will not prevent poor designs and tooling in the manufacture of products, and zero-defect commitments cannot solve machine capability problems that produce bad parts. What is needed is management commitment to providing the quality resources, equipment, personnel, methods, training, and to ensure motivated and trained personnel properly apply these assets. This commitment to produce only quality products must start with senior management and continue down the lines of authority in the company. A quality circle group, newly implemented, at a major Japanese automotive company saved more than $75,000.00 by implementing preventative problem solving solutions in their department in one year.

Quality improvement must be the goal of all employees, from senior management personnel to the shipping clerk. All operations of a company affect the quality of the final product and service to the customer. Customers

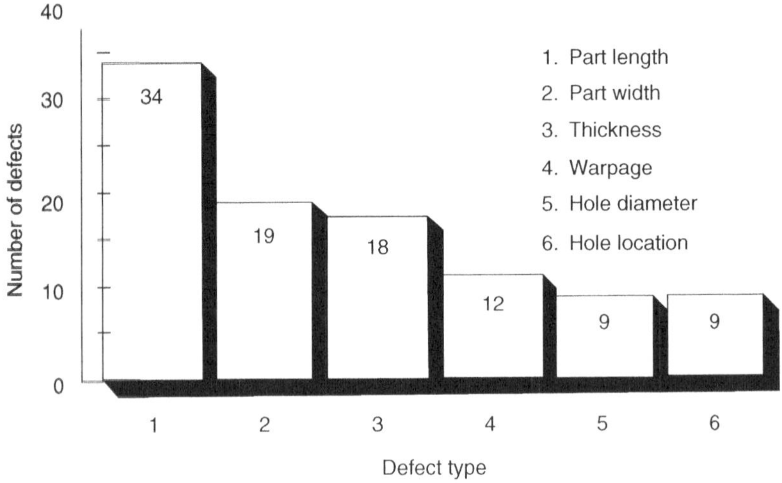

FIGURE 2.5. *Parato chart of defect types.*

have come to anticipate that only quality products are shipped to their receiving dock, which can go directly to their assembly line, often without inspection. This puts the responsibility where it belongs, on the supplier of the product.

This is the "Do it right the first time" mantra that management must send to their employees with the support and training they need to do it. Customers are reducing their supplier base and relying on their proven and quality-minded suppliers. Customer attitude is not based solely on unit price also but depends on their cost of handling the unit once it has been received at their plant and a problem is found. The cost is now almost doubled because of their required incoming inspection cost plus the loss of the unit and their time loss waiting for a replacement.

ESTABLISHING COMPANY QUALITY OBJECTIVES

Quality objectives should be a reach for a company. The objectives must be attainable within the scope of the company policy, yet a goal that is not easily obtained. Management must be kept alert and always searching for new methods for improvement, including how to do it better while providing more value, for the customer, for the price charged. Quality is always defined as "customer satisfaction." What is important is how to satisfy customer requirements while justifying costs and earning a profit. The "best" in relation to quality control means satisfying the customers needs and wants within part requirements and cost structure.

For a company to commit to TQPC, it must first ensure it has the internal structure on which to build the quality system. Second, it must take the time to write down, first their short-term and then long-term, business, financial, and quality objectives. The company should believe it will be capable of meeting these requirements; then, it should implement the structure to accom plish this requirement. The objectives should meet the needs and expectations of company management and customers. Objectives should be straightforward and to the point.

1. Ensure that the guide for establishing the company's goals is the company plan, with departments selecting their individual yearly objectives for meeting their goals and making their operations more error and problem free.
2. Separate management and quality for independent operation. Ensure joint agreement on the supplying of assets for customer satisfaction.
3. Ensure all employees are aware of their customer's product requirements. Sales using QFD (discussed in Chapter 3) will develop information for determining and establishing the customer's wants and needs beyond the current product.

4. Ensure the products' design, manufacture, and end-use requirements are known and established by the customer when design is involved for the product. Use checklists specifically designed to gather all the information required for the product and its manufacture.
5. Establish preproduction reviews with all involved departments to ensure all details, specifications, and requirements are established and questions are answered.
6. Establish the manufacturing requirements, equipment dedicated, and suppliers of materials selected and approved with manufacturing instructions written and an FMEA conducted to ensure all variables and potential problem areas have been considered and evaluated.
7. Select suppliers who can provide the products and services necessary and within the specifications and price required for the products. Both response time and customer service are critical for injection molding because of the variety of products and materials used daily.
8. Daily process control measures for maintenance of equipment is mandatory, as equipment is often used in a variety of manufacturing conditions and its maintenance and cleanup after each job is critical to avoid the cross contamination of materials. Vacuum up material; never blow it!
9. All data generated must be used in real time. Data collected during a process reaching equilibrium is historic data. Only when the system has reached "temperature equilibrium" should adjustments to the cycle be made.
10. Always observe the cycle and give it time for an adjustment to be incorporated into the operation. Too many hasty adjustments can create cycle instability and have it go out of control. It is very critical for control of the process to maintain control of the entire operation. Be sure the data collected are meaningful and are analyzed right away.
11. Train personnel to use checklists, process sheets, and instructions during their daily work operations and to follow established procedures. Maintain operation sheets and run/log books with mold and machine operation conditions. Keep equipment maintenance records at the machine for reference and to know the items on the machine, screw and nozzle type, age of heater bands, and so on.
12. Recognize quality as a price necessary to pay for the product, not as a negative cost. Quality is essential for product and process and must be instilled in personnel as a necessity, not as a requirement. A goal is to have the price of quality less than 2 percent of sales. Keep quality as a positive company and department quantity!
13. Make corrective action a thing of the past and inspire preventive actions to identify, correct, and eliminate problems from the business and work area. Be proactive in daily maintenance and quality operations.

Success will result when you manage your area of responsibility as if you own it. You take the responsibility to ensure all is correct and processes are in control and remain within specifications.

CUSTOMER QUALITY

Customer quality requirements should not vary within the organization. Each customer's product is special and is manufactured using the same methods as any other product. When quality procedures are written, all jobs will require the same degree of supplier quality and with a well-established and managed system in place, all job objectives and requirements will be successful. The only recognizable feature will be that some jobs may have tighter specifications; your company will be capable of meeting these specifications on a daily schedule.

To consider having a tiered quality system is wrong. The company should only produce products to their best capability. The only difference is what the customer requires for their product's finished state. This is determined when the program is contracted and quality is discussed with the customer. No variance in quality operations should ever be allowed. List on the program setup sheets what the customer requires, not what the customer will accept!

3

Managing for Success, Commitment to Quality

Management must commit to producing a quality product! Without this commitment, it will not happen. This statement of quality excellence must be included and attainable in the company's policy statement and communicated to all of the employees.

OBJECTIVES FOR MANAGING A QUALITY SYSTEM

A well-organized and documented total quality process control (TQPC) system must meet the following objectives:

1. Positive customer orientation
2. Well-defined and specific quality policies and objectives
3. Departments and personnel oriented to achieving these objectives and carrying out the policies
4. Specific vendor control policies
5. Complete and identified part and process quality requirements
6. Full documentation of work instructions for operator use
7. Trained personnel with motivated and strong quality knowledge
8. Proactive preventative problem analysis program

Total Quality Process Control for Injection Molding, by M. Joseph Gordon

9. Continuous real-time process control with closed-loop, self-adjusting control of process parameters, if not, operators capable for control
10. Periodic audits of process systems for compliance to specifications

Quality encompasses all departments of the company. Sales and marketing promote and obtain sales of the company's products and/or services. They are the first contact your customers have with your company. Your customers' first impressions of your company are key along with ethical control of your business dealings with their company. Honesty, integrity, and quality make up a trio of reasons for conducting business. Honesty in providing the services at a fair price, integrity in providing the service as contracted, and quality in the provided service that meets with the customer's satisfaction. In between these actions are a multiple of required actions that will make the business relationship a success.

Management has responsibility for 85 percent control of the quality system, but it manages only 15 percent of the process. Management must be made aware of the assets needed and provide them in a form usable in their operations. Once a quality system is established and operating, they must support it and ensure it provides the services necessary to execute the actions needed for providing quality products and services to their customer base.

The first principle management must be aware of is: "Quality is never your problem, it is the solution to your problems." The price of ignoring quality has cost major corporations their loyal customers. Rival companies are waiting to compete and provide the products customers want with the quality consumers have been wanting, so the customer can purchase the product. Customers are willing to pay for quality when it is in the product and/or service.

The cost of ignoring quality has brought new competitors into their markets. Many have proven to themselves that a staggering 20 to 25 percent of a company's operating budget is spent fixing problems that should never have occurred.

PROACTIVE PREVENTIVE ACTION

Learning how to identify potential problems is the key to proactive preventative action. Identifying a potential problem before it occurs is one of the correct ways to spend quality assets. TQPC is dedicated to this means of identification of preventative problems. Will all of them be detected before manufacturing begins? Probably not all, but most will when the methods described here are implemented, practiced, and used daily. These methods are not difficult, but they must be followed and used to achieve the best results.

Providing the right working environment, equipment processes, assets, and people is the key to a successful TQPC program. Management must be held accountable and its performance measured by how well the company provides the support and assets necessary to achieve the quality goals. Management must be involved in more than name. The management quality objectives should be made known to and judged by all employees to determine whether they are committed and serious about obtaining their stated objectives.

Quality must become a way of work ethics. It must start with senior management and proceed down to all levels of employment in the company. Plans must be developed as to how this is to be done. All programs must have a good plan to succeed. It is also important to keep it as simple and as practical (worker friendly involved) as possible. It is interesting to note that when quality-control circles were first developed in Japan, management believed they were a waste of time and were initially reluctant to implement them on the factory floor.

Today, typical areas explored by the quality circle volunteers, usually up to ten plus their leader, are ways of improving safety, product design, and the manufacturing process. Quality circles have the advantage of continuity; the circle remains intact from project to project. Savings can be great, up to $100,000.00 and greater when correctly applied in the work place.

In their plan, management personnel must establish objectives for each type of service and product it wants to provide to their customers. These products and services include the assets, machinery, and equipment to make the product; the people to design and manufacture the product; and the sales team to solicit and service accounts, as shown in the ladder of operations (Figure 3.1). This progression of operations and specific actions must occur for the program to precede to completion. The manner in which these operations are performed includes the objectives of the TQPC plan. If one area in the plan should be faulty, then there needs to be a method of immediately making the correction to ensure continuation of the process.

TOTAL QUALITY PROCESS CONTROL

Attitude

All quality programs require a positive attitude toward accepting change in the organization. Just because it was always done one way for years, does not mean it cannot be improved. A positive searching attitude of new ways to do and improve the business is healthy for a business to instill in their employees. Remember, most quality methods were developed in the last 50 years by employees of very successful companies (e.g., Western Electric, Motorola, General Electric, Ford Motor Company, Toyota Motor Sales, and others). These companies fostered a growth in quality methods and improvements by changing the manufacture and quality of their operations.

Management provides the driving force for these operations to happen. They must also practice and support these quality operations even if the return

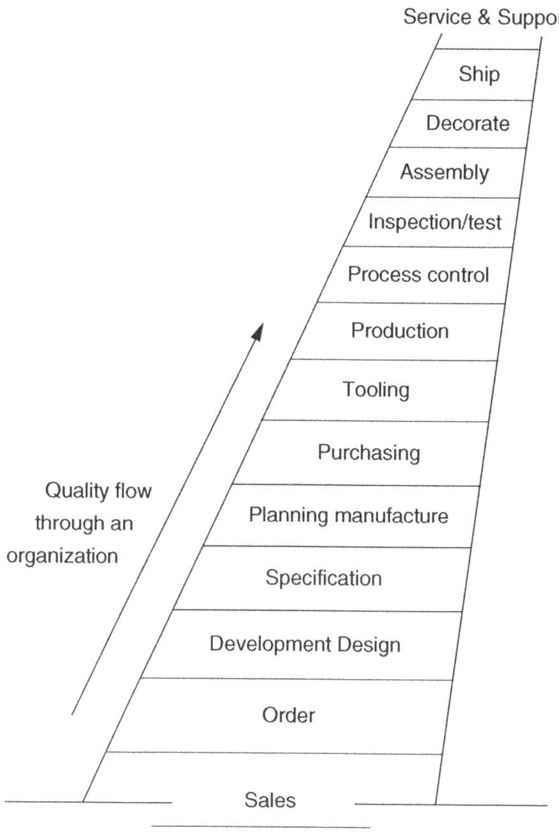

FIGURE 3.1. *Ladder of operations. The lean method of manufacture works in a similar fashion, which will be discussed later in the text.*

is not always as great as anticipated. Remember, the Six Sigma[1] quality programs were only initiated if the potential savings were identified as being $175,000 or greater and then required senior management approval. A designated management champion was then assigned to spearhead the program and ensure it had all the assets necessary for a positive result. In the beginning, that was not always achieved. Within Motorola, which is the developer of Six Sigma, the program leader, to become a designated Black Belt, had to manage a successful program of about $175,000 of savings to earn the title. As we know, not all programs can yield this amount of savings. Therefore, as the system spread, the monetary requirements were lowered to create more Black Belt quality experts within all sizes of companies. Unfortunately, this has turned into a money-making industry, as one organization for Six Sigma training advertises that two separate weeks of training and one project sandwiched

[1]Six Sigma. Available at: http://en.wikipedia.org/wiki/Six_Sigma.

in between equals the Black Belt. This format is not exactly what the initial program had in mind!

There are now many Green, Black, and Master Black Belt quality professionals within all sizes of companies with multiproduct diversity who are making major quality contributions to the success of these companies. Are their programs capable of attaining the same size savings and rewards? Probably not, but they are improving their companies' quality operations.

Control of Change

Improvement requires change to occur within an organization. The intent is to have a positive result, not to change because everyone else is doing it, as it is not always successful. Change can be as minor and simple as improving the lighting in the business and manufacturing areas. Light improvements have shown to increase worker output by over 35 percent in some companies. The work area must be as pleasant as possible, and it is a quality item for future consideration.

Improvements should be explained to the employees when changes are going to be required. It is helpful to explain why they are being made, the expected result, the benefits to them and the company, and what time, effort, and their involvement may be necessary. Whether any employees are to be moved, retrained, transferred, and so on has to be explained to alleviate fears of loss of jobs and smooth the transition when employees are moved within the company.

When a Kaizen, a fast work area improvement quality method, is performed, some employees may no longer be required to perform operations that were combined or even eliminated. They are not fired or laid off but are used in other areas of the operation, especially if long term, loyal, and knowledgeable in operations. Keep the pain of change low and the achievements to be gained from improvements high.

Also, investigate the anticipated effects of improvements even before they are made to ensure their effect will not cause a problem after the change. Plan the improvement, analyze the changes to be made, make the change, perform a failure mode and effects analysis (FMEA) if possible, and analyze the results. Last, ensure management agrees with the changes and will lead the improvement program, providing the incentive, support, and assets for it to be successful. Training is essential for all personnel especially if new operations are implemented. A trained employee pool is essential for a successful program starting up with minimum difficulties. It is also essential that management listen to their employees as they may have some positive input into the planned changes that will dramatically affect the success of the program. The use of a quality circle type of analysis is helpful in planning the changes, anticipating what problems may occur, and reviewing the amount of training and new instructions that are required to ensure the program is successful and has a positive startup.

Improvement with Control of Change

To implement change accurately, planning is essential. The requirements, equipment, their installation, training of personnel, and trial startup runs must be written down in detail to be sure the results are attained with good product output rate. There are numerous analysis results to be evaluated for all situations, and fine tuning may be necessary to obtain the best results.

An example may be a decision to move to quick mold change by preheating the molds. This involves an analysis of the following:

1. Justification (savings in time and product gain) of going to quick mold change
2. Molds and what machines they fit
3. Machine platens and molds modified for quick change
4. Availability of preheating equipment
5. Platen insulated from mold
6. Procedure and checklists to heat, install, and start up the mold
7. Access of lifting/installation equipment to install mold
8. Instructions written and trialed at press
9. Setup team trained in quick change methods
10. Molding machine considerations, material staging, clean out, and so on.

When a new idea finally becomes reality, considerable planning and work is done to ensure that if the change is made, it is justified and can be accomplished.

Documentation is the key, and recording of all events is necessary. The startup of a new molding cell, operation, or machine requires verification of required operations using, in my experience, a checklist of items that have to be accomplished so the operation would be successful. Even the omission of one item could cause the results to not be positive. I have been involved in troubleshooting multiple problem areas that were never solved until an analysis of the data and a set of in-detail instructions led to the final solution of the problem.

Keep good records and review the recorded information generated from the operation in real time—not an hour later but immediately after it was recorded so it can be used in the control of the process. If a problem should persist and a solution is not be possible, then shut down the operation, review the data, make a calculated analysis, or decide to run a "design of experiments" (DOE) (see Appendix B for an example) to determine the variable(s) that are the main contributor of the problem. See the Engineering Statistics Handbook (http://www.itl.nist.gov/div898/handbook/index.htm).

Quality Decisions

Decisions are made on the best information available at the time, and often a decision should not be made until additional information is obtained, which is a decision. The gathering of information begins with the sales department and spirals upward in the organization processing through each department that is affected by the operation. Each department requires a specific type of information to assist in its decision-making process of the order, and it affects the quality of the operation/product and the profitability of the operation.

During all phases of operations, the customer is continually evaluated for attaining satisfaction, which is the goal of quality operations. Flexibility is often required in design or manufacture, and the individual departments must share their knowledge and experience to attain the best possible results. Using measurable results and feedback, the system must adjust to new factors as they occur. You want to avoid fragmentation within the quality organization and to keep all departments working toward the common goal of quality.

PRINCIPLES FOR QUALITY SYSTEMS ENGINEERING

The principles that relate to quality systems engineering are as follows:

1. Relate quality technology to quality requirements through hardware, procedures, and plans to meet customer needs.
2. Relate quality technology to quality requirements by evaluating new and changing systems. Balance technology with these requirements, thereby guiding the introduction of practical improvements in the quality system.
3. Consider the total range of relevant human information and equipment factors needed for these procedures and controls. Integrate hardware–human–equipment–information factors as a functional system.
4. Using feedback, measure and fully evaluate the quality system in operation. Establish measurements to grade the system.
5. Quality systems engineering should structure the quality system objectively and provide for audits of the system.
6. Provide for the ongoing control of the quality system by combining quality systems engineering and management.

OBJECTIVES FOR MANAGING A QUALITY SYSTEM

A strongly engineered and well-managed total quality control system must meet the following objectives:

1. Positive customer orientation.
2. Well-defined and specific quality policies and objectives.

3. Departments oriented to achieving these objectives and carrying out the policies.
4. Integration of company departments to produce quality products.
5. Clearly defined personnel assigned to achieve quality.
6. Specific vendor control activities.
7. Complete and identified part and process quality requirements.
8. Defined and effective quality information records, flow processing, and control for manufacture.
9. Well-trained company personnel who are motivated and strongly quality minded.
10. Know quality costs and establish measurements and standards for quality performance analysis.
11. Positive corrective action procedures that will be effective.
12. Continuous control of the system with feedback and flow of information so that the analysis of results can be compared with present standards.
13. Periodic audit and checking of systems activities.

No efforts should be spared to produce a new part or evaluate an existing part prior to production. No new job should be accepted without an extensive evaluation of all these parameters. But, in many cases, for parts with existing tooling (the common industry term referring to the mold base and part cavity), if the tool is transferred to a new molder or part supplier, this is never done. As a result, the existing part and tooling problems for the old supplier become the same problems for the new supplier.

A lower piece-part price is not always the driving reason for tools to be moved to a new molder. Usually, the decision is based on a quality problem, which relates to parts that do not meet customer requirements and would result in late deliveries and increased part cost. The reasons for any tools transfer should be communicated to the new molder at the time of transfer. If, after transfer and review of the problems, the new molder accepts the tool anyway, then provisions should be made to provide the assets to fix the problems. All company departments should be involved in evaluating the transferred tooling before accepting it for production. If, after evaluation, the tool is deemed not capable of producing good parts, the job should be refused.

Once the company's quality objectives are defined, it is the responsibility of the sales department to solicit new business. It is also the responsibility of the other company departments to support the sales function, guided by management quality objectives, in obtaining the kinds of customers the company wants to cultivate.

Sales must sell the company's capabilities and its commitment to providing a quality product. There are four basic types of quality agreements a company can provide to meet customer requirements. Because all customers will not require the same quality for their products, the necessary degree of quality must be known and determined at the start of any new program. This degree of quality should never be lower than the company's own quality objectives. It will relate only to the quality level of the specific program. In short, the company must always provide the same high overall quality standards, but they should be adapted through job requirements to suit each specific customer product. To vary the company objectives would be to sabotage the whole total quality process control system.

CUSTOMER-SUPPLIER QUALITY AGREEMENTS

Captive Part Quality

The captive part quality method uses the first article out of the tool that the customer judges as acceptable in a form, fit, and function reference. This first article is used to define the minimum quality values acceptable for the part. Thereafter, quality reference is judged against this part with no critical divergence tolerated. Quality is based on the minimum, or low side, of the part, and value judgments are constantly being made against this standard. Color, clarity, no-flash, warpage, etc., may be the only standards the part must pass. This makes value judgments more acceptable by more people but, in disputes, the customer is the final decision maker. This is an example of quality set up for nonfunctioning or mainly high volume, low cost, aesthetic parts in the less expensive plastics. The part is either accepted or rejected with no middle ground. Documentation is minimal and no attempt to improve or evaluate part quality is expected or anticipated. These items are usually one-time use and throwaway items, or of a quality that should it fail are of little concern.

PRODUCT QUALITY DETERMINATION

Parts to Print

Quality by "parts to print" relies on the customer providing specifications that in turn become requirements for the product, on acceptance of a contract, by the supplier. These part drawing specifications become the standard against which the product is judged for acceptance. These specifications were determined by the designer to have the part meet end-use product functions. In many situations, the tolerances specified are for metal parts that do not take into effect the behavior of plastics. The designer may tolerate all dimensions per the drawing metal tolerance reference table that is part of the title block, but it is all wrong.

The part designer needs to know or determine what dimensions are actually required, what tolerance is acceptable, and referring to the Society of the

Plastics Industry, Inc. (SPI), material tolerance chart (Figure 1.3), to specify the part dimension tolerance accordingly. The tool builder and molder must also inject their comments on the design and dimensioning. The mold builder will use appropriate cavity dimensions and tolerances to meet the designer's dimensional requirements. The cavity dimensions are based on the material selected, the designer part dimensions specified, the estimated number of mold cavities to achieve dimensions, and the location and number of the cavity gate(s), the opening size, and the balanced cavity and runner system that feeds the part cavity. This is shown in Figure 3.2 for a balanced, unbalanced and family mold cavity layout. When the cavity pressures are not equal, some cavities will be overpacked and others will be underpacked, depending on the timing of the molding machines operations. The goal is to have all gates freeze off at the same time or within a second of each other. If not, the product may not meet the requirements of the designer, even though it may still function as required. Decisions on product tolerances, number of cavities in the mold, and other mold requirements must be made now, not later after the mold is built and production has started.

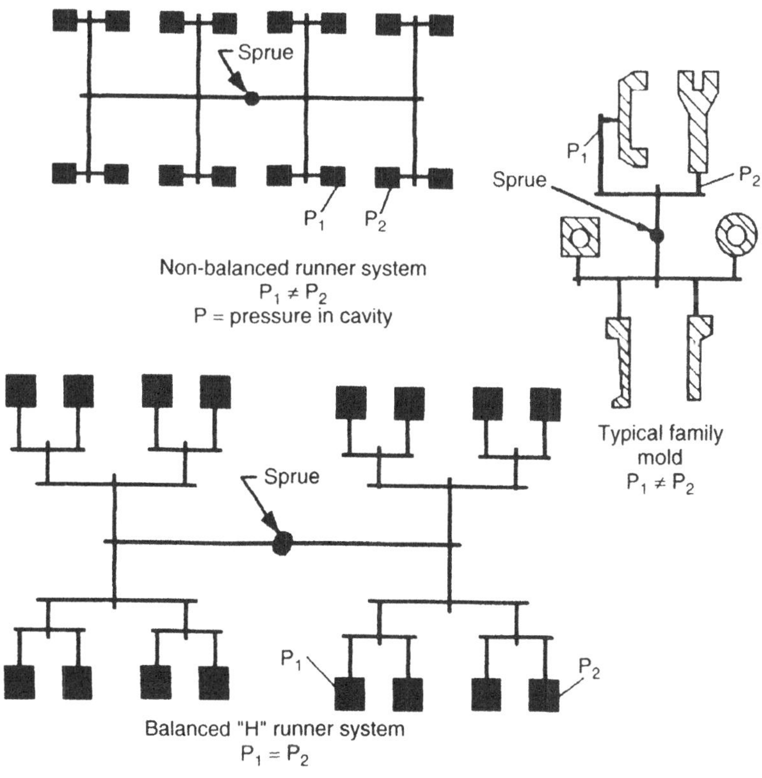

FIGURE 3.2. *Mold cavity layout.*

Quality by part to print requires that procedures and/or instructions are developed and followed for the manufacture of the product based on the stated specifications. These instructions specify exactly what is required of the product, as well as the tolerances, and dimensions critical to the function of the part. This must be known so the mold can be designed to meet the part requirements. The tighter the specifications, the fewer the number of mold cavities permitted and the more critical are the mold tolerances, gate size, location, and the runner feed system. This also includes the cooling requirements for dimensional control, material type, and even the material source in some situations.

It is very difficult for an injection molder to hold metal manufacturing tolerances on a plastic part. At best, one or two specific tolerances can be held to metal-like tolerances. Plastic materials that are reinforced and/or filled can be held to tighter, like metal, tolerances because of the addition of filler and reinforcing mediums (see Figure 3.3 for a microscopic view of the fibers). Fillers and reinforcement cause lower in-mold shrinkage of the matrix resin because they take up a respective amount of resin volume. In filled resins, the filler material does not chemically or physically attach itself to the base resin, acting only as an inert filler adding a higher degree of stiffness to the part but lower elongation and toughness. The reinforced material is chemically and physically attached to the fibers, and it binds itself to the resin and increases the part's physical properties. The reinforced materials (e.g., short or long glass fibers) will also experience differential shrinkage because the fibers line

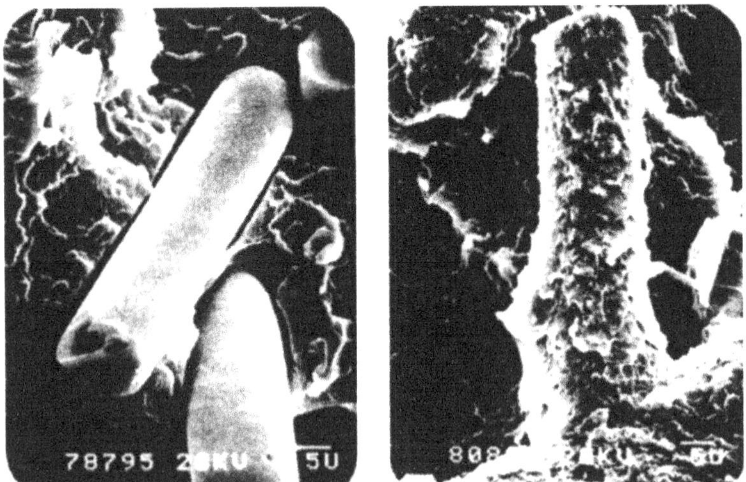

Impact fractured surface 35% filled
Scanning Electron Micrograph
Filled material, not reinforced

(a)

Impact fractured surface 35% filled
Scanning Electron Micrograph
Glass-fiber reinforced

(b)

FIGURE 3.3. Scanning electron micrograph of impact fractured surface (a) 35% filled material, not reinforced; (b) 35% filled; glass-fiber-reinforced (Ref. [1]).

up in the flow and fill direction. This causes more shrinkage in the transverse versus the flow direction. Mineral- and fiber-filled materials will experience less of the flow versus transverse dimensional flow problem, which will result in less dimensional variance in the part with more uniform overall part shrinkage.

The molder will also need to know whether regrind is allowed in the part. Typically, 25 percent or less will not appreciably affect the performance and processing of the product. But, some resins can be degraded by successive melt histories through the molding machine that decrease the materials' viscosity and impact resistance by breaking down the molecular chains in the resin. Regrind, rejected parts, runner, and sprue, ground up and fed back into the hopper with virgin resin, is used to keep material and part cost down. But, with increasing successive regrind cycles, heat histories reduce the physical and dimensional properties of the base resin. If a capability study, Cp, is being run, observe the data and determine whether a noticeable change is found as the regrind is continually fed back into the system. If a change is noted, then stop the use of existing regrind and purge it from the system. Then, begin collecting parts for new regrind as before, and when enough is available, begin mixing it into the virgin resin as before. Also, if allowed, be sure the regrind is kept dry as hot polymers have an affinity for moisture pickup. Regrind should be used as soon as possible and fed back into the hopper dryer system in the correct proportions of 25/75, regrind to virgin resin. Therefore, the use of regrind must be discussed before a pricing and specification decision is made for the product.

Custom injection molders are very accommodating in trying to meet their customers' "reasonable" part tolerances. They are often aware of part quality standards regarding part tolerances. Lower requirement part types, such consumable products, dunnage items, throw away after one use parts, meter closure tags, spacers, and covers, can provide a valuable service and savings to their customers. For these parts, tolerances are often said to be "open" meaning not critical. What may be critical is that no flash occurs on the parting lines, color is controled, that the snap and press fits the work, the information on the part is readable, and that part properties meet end-use requirements, such as cable ties, clips, and so on.

What customers may find more important are the following items: the product diameter, round not elliptical; no voids in the thick section of the part; clarity is achieved; no scratches on the part; smooth surface; and no weld lines or warpage is visible. These are specific and critical items for plastic products.

FORM, FIT, AND FUNCTION (FFF)

Some parts may have only form, fit, and function requirements as those just described. These parts have the lowest quality requirements, and the method of acceptability must be decided between the designer and part supplier, which uses the injection molder, before signing a contract to furnish the product. The main consideration is whether the part can be manufactured as designed and in the specified material with the part tolerances and requirements presented to meet FFF and at the price estimated to be profitable.

If any of these part tolerance and specification questions cannot be answered, then the program should be reevaluated. The greater the number of mold cavities in a mold, the more latitude must be allowed in the part's tolerance. Therefore, there are really only two types of plastic part tolerance requirements, as listed on the drawing specifications and FFF.

The important item to remember when discussing the manufacture of a plastic product is whether the tolerances are realistic, attainable, and capable of being produced repeatedly from the tooling (mold) and material as specified for the product? These items are negotiable, and an injection molder should not accept metal-like tolerances on a part drawing. The variance of plastic should be discussed and a compromise reached on exactly what is required for the part dimensions and end-use function.

It is difficult for some injection molders to discuss tolerance, as they often feel this reflects on their ability of manufacture. Molders can use the Society of the Plastics Industry, Inc. (SPI) molding tolerance specification for different plastic resins, as shown for acrylonitrile butadiene styrene (ABS) in Figure 1.3 as a part tolerance capability molding guide. Then when the part is molded in a multicavity mold, the dimensions will be in agreement with the standard and the part will be assembled and can function as required.

PRODUCT REQUIREMENTS

Many parts must meet agency, government, military, automotive, electrical, medical, food, and plumbing standard requirements for products in specific consumer and business areas. These agency publications state what standard the part must meet beyond even the drawing specifications. Plastic materials are used in parts that go into almost all of today's products.

These standards are very specific allowing only specific company-approved and certified materials to be used in an application, medical, electrical, and plumbing that were formulated for a specific standard specification. As a result, the supplier has a responsibility to inform the designer if they are not aware of the standard requirements for their particular application. In like terms, only specific materials are listed as approved materials for like applications, especially many automotive parts.

EXISTING MOLD CONSIDERATIONS

When a customer wants the injection molder to take over an existing mold from either their operation or another supplier, several areas must be explored.

The first is, why is the mold being moved? There can be several reasons why the mold is being moved, such as follows:

- Machine size not available for running mold
- Obtained a better part price with a requote
- No time or machine to run mold internally
- In-house or outside molder was not able to make good parts
- Molder wanted mold removed from their plant
- Mold was poorly designed and manufactured, last resort trial

It is important to know why an existing mold is being moved or requoted. Was there a problem with the mold? Were good parts ever made from this mold? A whole list of questions can be asked to obtain the information on why the mold is being requoted. It is important to evaluate an existing mold for its capability to produce acceptable products.

Custom molders are often asked to quote existing molds. Never accept a new job with an existing mold without first evaluating it or, at the least, talking with the last person to run the mold, if possible. The mold should be evaluated with a molding trial to produce an acceptable part for verification of the mold and materials quality and moldability. Unfortunately, not every mold built can make acceptable parts. The mold should be evaluated for operation, temperature control, balanced cavity layout, material flow/gate size, freeze off time, and capability of maintaining uniform part weight, cavity to cavity.

I have seen a brand new mold built with the cooling channels 4 inches from the cavity surface. The acceptable steel thickness to the inside channel surface would have been 0.375 inches for this single-cavity mold. This result is totally unacceptable as supplied by the lowest bidder! See the mold section for the correct spacing and layout of cooling or heating channels for a mold.

A mold trial will also determine the capability of the mold to produce parts on a uniform cycle and will establish the molding cycle for quote purposes. If a trial is refused, you did not want the program at all, because it possibly has too many problems.

Atypical intercompany flowchart for developing the requirements for a new mold to produce an acceptable part is shown in Figure 3.4. Development begins after the order is received and the part is designed. With the material selected, the following operations occur with sizing the mold cavity for material shrinkage plus determining the requirements, which include gate size, material flow in the mold cavity, number of mold cavities, cooling for dimensional control, and other mold and part design considerations before moving on to processing. See the checklists in Appendix C, specifically Mold Design Checklists, number 15 and 16, for the questions needing answered for the building of the mold. Once the mold is completed, the process control variables are established by trying out the mold. Once completed, any minor mold modifications can be made in preparation for production.

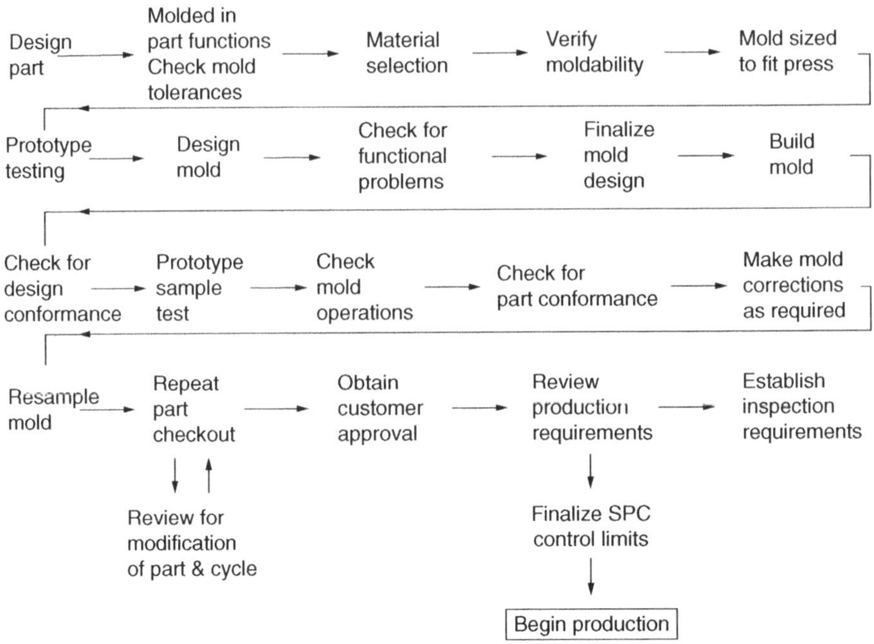

FIGURE 3.4. *The preproduction process. (from Ref. [2].)*

ESTABLISHMENT OF RESPONSIBILITY

Producing plastic products by injection molding is the responsibility of the entire organization. Referring back to Figure 3.1, the flow of responsibility for the product and its quality passes through the entire organization, from sales to shipping and back to sales, for follow-up and maintaining customer satisfaction.

Each department and company manager has their specific input for the quality and process control of the product. Each manager must perform their tasks as required for the product to traverse through the organization to achieve product realization. Their actions and responsibilities are shown in Figure 3.5 for a typical company departmental organization and responsibilities for operations. To ensure all operations are completed, checklists are recommended.

Checklists should be used to ensure all the information is available for their department's operations. The checklist should list all the duties that are to be performed in the department, even though all may not be done each time an order is received. It is easier to bypass a requirement, if not needed, than to try and remember it each time an order is received. Therefore, a list of the major items each department may perform is listed as a guide for implementing the checklists.

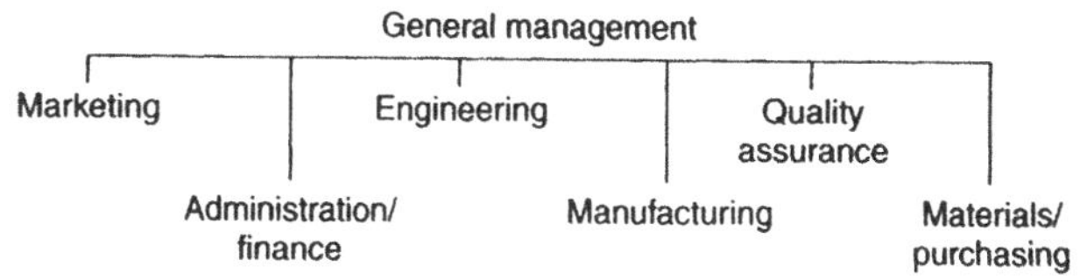

Department responsible for quality:

GM = General management
MK = Marketing
AF = Administration/finance
EN = Engineering
MF = Manufacturing
QA = Quality assurance
MP = Materials/purchasing

Responsibility code:

R = Responsible
M = Must contribute
O = May contribute
I = Informed

Area of responsibility:	GM	MK	AF	EN	MF	QA	MP
1. Establish business quality objectives	R	M		M	M	M	
2. Implement quality process control	I	I		M	R	M	
3. Plan quality systems	R	M		M	M	R	M
4. Determine customer requirements		R					
5. Establish product design specifications		M		R	O	O	
6. Establish manufacturing process control				M	R	M	M
7. Determine process capabilities				M	R	M	O
8. Quality vendors			M	I	I	M	R
9. Plan inspection and test procedures		O		M	M	R	M
10. Design inspection and test equipment				M	M		
11. Tooling reviews and purchases	O	I	I	R	M	M	M
12. Schedule production		I		R	M		
13. Material evaluation in-house						R	M
14. Produce products to specifications	I	I		M	R	R	I
15. Quality feedback information	I			M	R	R	I
16. In-process quality audits	I				M	R	
17. Gather quality cost datas	I	M	H	M	M	M	M
18. Analyze quality cost information			M		M	R	M
19. Final product quality	I	I		M	R	R	
20. Gather customer complaint data	I	R		M	I	M	
21. Analyze customer complaint data		M		M	M	R	M
22. Corrective action		M		O	R	R	O
23. Customer satisfaction	I	M		M	R	R	O

FIGURE 3.5. Responsibility and task guide (Ref. [3]).

1. Sales: contacts customer, gather information and needs, gets order
2. Contracts: obtains order, negotiates, for product with price established
3. Development: establishes and finalizes part requirements

4. Design: designs product to meet end-use requirements of customer
5. Specifications: customer and design input to establish part tolerances
6. Plan for Manufacture: determine method of manufacture and operations
7. Purchasing: order material, support items, and select vendors
8. Tooling: design, build, and tolerance mold to meet specifications
9. Production: select injection molding machine and auxiliary support equipment
10. Process Control: establish manufacture control points and tolerances
11. Inspection: verify control of process and parts meet specifications
12. Test: perform end-use tests on parts to ensure requirements are met
13. Assembly: may occur before test, ensure parts are assembled correctly
14. Decorate: added value to product if required, information on part
15. Ship: pack product for shipment as required for customer
16. Sales and Service: follow up with customer maintenance and service

This is essentially the process and flow of departmental major actions needed to proceed through the organization for a new order.

DEPARTMENT TQPC RESPONSIBILITY

Based on the task guide presented, it is important that each department participates in the responsibility of the product's development. Each must do their share of the work for the program to be successful. This implies the use of checklists, procedures, and a repetitive and/or specified method for doing their job to ensure no item is left undone or forgotten.

In a typical custom injection molder, there may be only one employee to cover several departments and operations, which is a stronger case for the use of checklists. The benefit from this is that the employees are more knowledgeable in more areas than an employee in a tightly controlled department in a larger company. Know more, do more, and forget/omit less is the key to this operation.

Often, the supplier is invited to participate in the development of the customer's product. This gives the supplier a strong area in which their knowledge and experience can affect the performance and quality of the product. By being proactive in working with the customer, they can influence the product's design, and type of mold, while adding value with the molded-in-part functions to give them advantages over their competition.

Too often in large companies when work is completed by one department and transferred to another, the objective of the product and what was done earlier to facilitate quality or savings may get lost in the transfer. This is why at the start of a program, the product's end-use objectives are documented, and all additions to the program are also documented. The manager travels with the documentation to ensure everyone is knowledgeable and the product will meet these objectives.

This product requirement list travels with the documentation, is reviewed by each department involved, and is considered for improvements at each step of the development process. As each item is met, it can be crossed off as completed. Should new ideas influence the product's development, which includes manufacture, tooling, injection molding, assembly, and decoration, the item is documented on the product manufacturing documentation. Affected personnel are notified, and the revised item is reviewed for incorporation into the program. If a material change is considered for the product, now is the time to do it, not after the mold cavity is sized. Also, under consideration is the end-use environment the part must endure. All of these and more items must be explored. The Design and Development Checklist number 3 in Appendix C is mandatory to avoid overlooking an important item during the initial design phase.

Program Development

Program development begins with the order or the company entering into negotiations with their customer involving the product. At this time, the use of the checklist, Program Development number 1 in Appendix C, is appropriate. The development checklist will assist the company in gathering the information needed to win the order by meeting the customer's requirements and needs. As discussions proceed, use the checklist with the customer to gather information by asking the questions on the checklist. Depending on the customer, they may be knowledgeable in plastic design or will rely on your expertise in providing them with information on how to best design and lay out their product. In some situations, your customer may be talking with a material supplier who has offered assistance, in the hope their material will be specified for the product. Working with a material supplier can be helpful as long as each of you are in agreement.

During the design phase, consider adding value to the part by molding in secondary functions. These end-use functions can be clips, snaps, threads, flexing and/or open and shutting panels, and so on. Also, consider assembly methods as using screws, press/snap fits, thermal welding, and so on, and decoration as color, molded in instructions on the part, use of decals, metalizing, and other methods. There are separate checklists for these items. The only consideration is to not weaken the part by incorporating these add-on benefits. Be aware that some color systems can lower the physical properties of some materials, plus sharp corners cause high stress concentrations, whereas the use of ribbing and section cutouts can reduce part section thickness and material usage, which conserves material and cost. Listen to the best options of your suppliers and designers and select the best for your program. The next step is to estimate a piece part price for the product.

ESTIMATED PIECE PART PRICE

Determining the parts estimated piece part cost is usually the next step, with material and molding cycle estimates made to price out the part. This usually occurs after the parts section thickness is determined and ribbing is considered to improve stiffness and strength with the use of molded-in ribs and other part design considerations. An example of a multifunctional symmetrical part is shown in Figure 3.6 with molded-in ribs, cams, bearing, gear teeth, shaft drive slot, and springs.

To finalize part design, a material is selected so the strength of material calculations can be completed. Materials in the amorphous family of plastics have lower physical strength properties than the engineering plastics. They are also less expensive but require thicker sections to carry the same amount of load as the engineering materials. But, by adding ribs on a part, they can be property and price competitive. Therefore, once the design is fairly well along, material selection begins and price estimating can be used to evaluate design, material, and pricing for different materials for the product.

Typically, if the part is straightforward, not complicated by additional part ribbing requirements, press/snap fits, under cuts, ratchets, etc., the material part volume is calculated and you proceed to estimating the finished price of

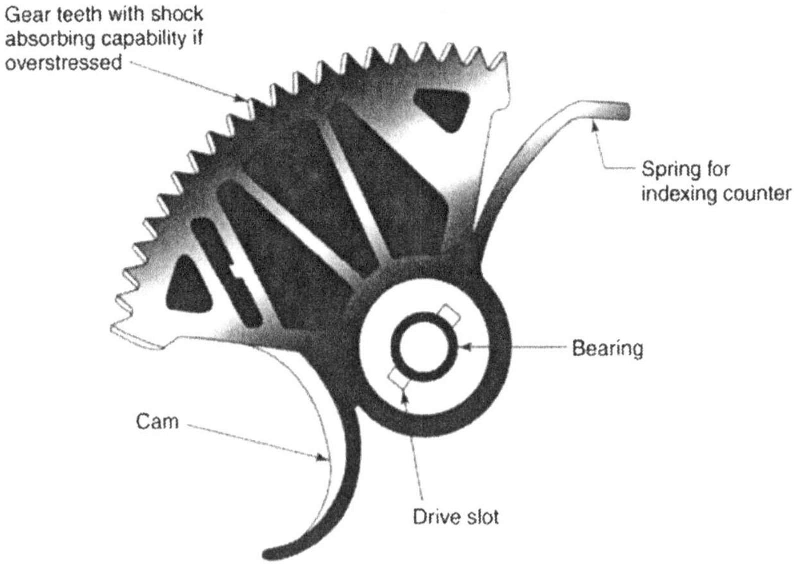

FIGURE 3.6. Multipart functions in a molded part.

the part. If not, then the two different part volumes are determined, molding cycles are estimated on section thickness and known molding variable differences, and pricing continues.

The Piece Part Price Estimating form number 8 is located in Appendix C. You will have to contact the material supplier to obtain its values for section thickness and material setup times, as well as any other information required. The estimating form is discussed in a detailed example for the part, material, and processing variables in Chapter 6.

In analysis, a thinner and physically stronger engineering material (nylon, acetal, etc.) with a faster setup time, even with a more expensive material price per pound, may be more economical. This will be determined during the design pricing study. Material selection may be determined by both physical and/or processing properties. The manufacturing cycle may be the deciding factor by being able to produce more parts using a faster cycle time, which results in a lower part price. This is one consideration the injection molder has to make when quoting a program.

A guide for determining the minimum cycle time while obtaining the necessary part dimensions and tolerances is by molding to the maximum part weight. This is achieved by lengthening the ram forward time on the mold runner system until the part weight is maximized. Once the part weight stabilizes, the hold time for maximum part weight is achieved. This method of establishing the minimum cycle time ensures the part cavity gate is always frozen off before the screw is retracted and builds up material for the next shot. At this minimum ram forward time, no more material can get in the cavity, and it will not depressurize on release of packing pressure and cause a dimension problem.

Also, when determining part cost, the number of mold cavities in the mold is an important factor. The greater the number of mold cavities, the lower the part cost, but the less control of part dimensions results. Therefore, there is a trade-off between part cost and quality requirements when the mold is designed and the cycle times are determined. The piece part cost estimate can be run every time a change is made in the mold design analysis. Once the mold and cycle time are optimized, any assembly operations are considered along with decoration, color, and/or information on the part.

Each plastic material expands during heating and on cooling, and then it returns to its original amorphous or crystalline molecular structure. This requires the mold builder to estimate the amount of mold shrinkage the plastic material will exhibit based on the parts molding conditions, such as melt temperature, cycle times, gate freeze-off time, mold cavity cooling, and part thickness. The thicker sections retain more heat, which causes longer material setup time, and with the engineering materials, greater material shrinkage. Amorphous materials require more heat extraction before they become solid enough to be ejected from the mold cavity so they do not distort. Other mold design considerations will be considered and discussed in greater detail in the mold section of the text.

MULTIFUNCTIONALITY

Plastics have the material capability to perform multifunctional tasks. By selecting the right plastic material, many operations may be accomplished with one part, as illustrated in Figure 3.6. Always consider the plastic part as being multifunctional and evaluate it and the material selected to determine whether it can perform additional functions.

This could cause the individual piece part price to be higher, but it may eliminate other parts, thus reducing the total products cost. This could involve the change of an amorphous material like ABS versus a nylon or acetal engineering material that has physical properties and capabilities exceeding other plastic material. These items are mentioned on the development and design checklists for consideration.

ASSEMBLY AND DECORATING

Plastic parts can be molded with self-locking snap and press fits for assembly with other plastic parts of materials. Permanent assembly is performed with sonic, spin, vibration, and heat welding applications. Assembly with screws, press and snap fits, and clamp methods can ensure repair is possible. Dissimilar materials, with close melting points, within a few degrees, can also be permanently assembled by heat welding methods.

Plastic parts can be colored, painted, printed on, foil and metal coated, dyed, and decorated in a host of many possible ways. Plastic parts are colored for safety reasons, such as to match company products colors, for identification purposes, and any other reason one can think of to color a product. Clear materials are colored as tail lamps and parking lights using acrylic and polycarbonate plastics and others. Basically, it is left up to the part designer to find new and different applications for plastic materials.

MANUFACTURING CAPABILITY

When using TQPC methods for manufacture, control of the program is not left in the hands of a few, but it is the responsibility of many. As described, many employee operations and processes are necessary to get the most value out of a pound of plastic material. The manufacturing department must now have the best possible machines and controls to produce the product within the time schedule and calculated price. Production is responsible for controlling the manufacturing process and for gathering real-time process quality data to ensure the manufacturing process is and remains in control during the entire production operation. The production team must also ensure their equipment is in good repair, it is clean, it has regular maintenance, the air filters are cleaned or replaced, the controls are calibrated, the machine wear is within limits, and all other items are taken care of to ensure quality manufacture of the product occurs.

The use of checklists and equipment startup instructions should be used to guarantee no items are forgotten and available during this stage of production. It is very important that plant systems and auxiliary equipment can supply their services as needed. Preplanning production startup is important so all the necessary equipment and systems are available for the production run.

The molding machines log book or molding data record sheet (Figure 3.7) is used for recording the startup conditions, ongoing process changes, and final production run settings. This includes all changes to the system before production equilibrium is reached for steady-state operation. Any changes made after this point should be recorded in the molding data record sheet for the run and at scheduled intervals on the system. The exact information should always be recorded; do not use dittos.

Then, as production proceeds, the operator will monitor and document the control settings as necessary at established time intervals while ensuring the process control checkpoints keep the system in control. Should there not be available closed-loop, continuous feedback support, the operator may have to collect data on the process and record the results on a real time run chart. The operator should be trained by quality assurance to perform this monitoring correctly.

COMPUTER-INTEGRATED MANUFACTURE (CIM)

Computer-integrated manufacture is used extensively in companies involved in TQPC. It uses the configuration management system (CMS) as its data storage system for the company's manufacturing operations. CIM is a real-time operating/control and information system for controlling and monitoring the business and manufacturing operations of the company in real time. Most CIM systems today record data in real time at manufacturing and monitoring stations, and the data are continually updated and available to management. It can inform specific personnel when event-based "triggers" occur and need attention. This will give any department within the company the actual results of its operations and ongoing order progress. CIM systems track and control orders through the system and out the shipping door to the customer's dock. In today's management environment, it is often called "real-time performance management" and can be coupled to "continuous improvement."

The CIM system can provide the following types of information and services:

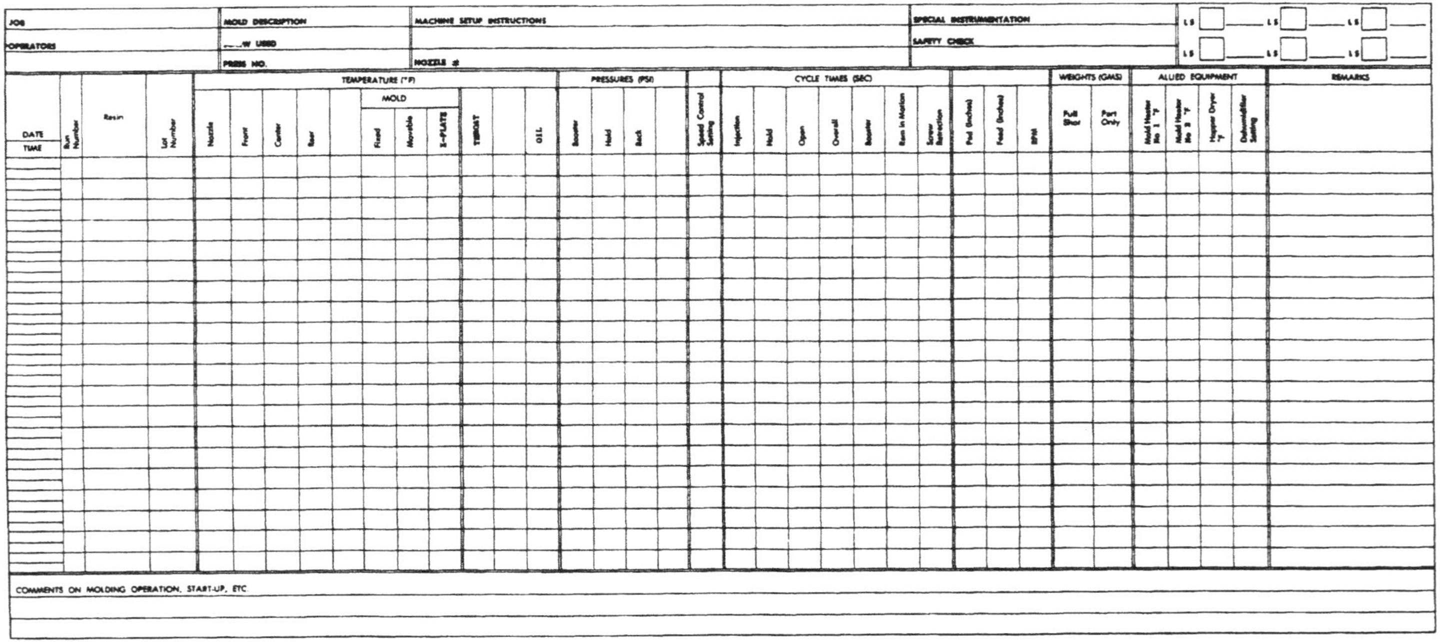

FIGURE 3.7. Molding data record sheet.

1. Centralized document and record control, protection, and retrieval
2. Control of product and mold design, computer-assisted design (CAD), mold cool, mold flow
3. On-time purchasing and material control for customer part numbers
4. Receiving documentation, inspection, and recording/storage
5. Inventory control of material and equipment usage
6. Maintenance control of all equipments and systems
7. Scheduling of equipment and calibration control
8. Production control and data retrieval and documentation
9. Auxiliary equipment control for production
10. Material control
11. Mold design
12. Finishing and assembly control of products
13. Finished part lot control and storage
14. Packing and shipping control and billing

Plus, there are other software suites that handle other business and molding areas of responsibility. These applications are listed for reference as follows:

- Estimating, pricing, and cycle time calculation
- Tracks production in cycles to handle multicavity molds
- Order processing/invoicing
- Integrated electronic data interchange (EDI)
- Inventory/lot tracking/bin location
- Purchasing [order and bill of material (BM) control]
- Bar code/radio frequency identification (RFID) material and labor control and reporting time
- Purchased material requirements planning finite/infinite scheduling
- Forward/backward—what-if—concurrent
- Schedules machines and molds
- Equipment/machine/mold maintenance
- Program and part pricing
- IS0 9001/TS 16949 quality control standards
- Scrap and regrind tracking
- Assembly and decoration actions
- Work orders and production plans
- CAD
- Calculate mold costs
- Accounting
- Payroll and human resources

TRACKING MANUFACTURE

Bar coding is used to identify products, equipment, material, tools, and so on in a plant and even personnel who perform operations and to monitor equipment usage. Tracking systems can serve multiple purposes, such as follows: recording received items, recording items going through manufacture, locating and indentifying items placed into storage, identifying tools, recording equipment, and tracking personnel performing operations in specific locations. Essentially, any place or item that must be identified, during or after an operation is performed, can be entered into this tracking system as long as a bar code number and label is attached to the item or person's badge. A bar code label scanner will perform the operation. Inventory control of molding cell equipment usage is essential for scheduling work. Knowing what and where equipment is used and when it will be free is essential information for scheduling production and keeping customer orders flowing and on time.

RFID

A technology introduced in the late 1990s and said to soon replace bar coding is RFID technology. The advantage of an RFID tag on an item is that the target and reader do not have to have an unobstructed line of sight. Because radio waves do not travel in straight lines but reflect off surfaces, they can be bounced around and read, but not necessarily viewed by the reader or person operating the reader used to identify the tag.

We are fairly well versed in the technology because of our garage door openers and car starters from inside the office. RFID is an identification device, not a finding item. It is used to determine "where is my device."

The RFID system is composed of two basic items, a reader and a transponder, which could range in size from a grain of rice to a hockey puck. The reader sends out a signal that frequency wise is compatible with the transponder, and when queried, it sends back a return signal. The price of the transponder is still a costly issue, but with more large retailers going RFID, such as Wal-Mart, their price will decrease. The real benefit is that an area can be queried and that a return will identify all the tagged items in the area searched. With bar codes, you have to find the item and then scan the bar code to record the item.

RFID technology can identify a series of different products, containers, personnel, machinery, tasks, and so on and can allow data to be collected by employees more accurately, efficiently, and reliably than by any paper-based system. Several RFID standards and technologies are available. Many are proprietary, but a growing number are not.

EDI

Electronic data interchange is a set of standards for structuring information that is electronically exchanged between and within businesses and other groups using an independent, third-party [value added network (VAN) or e-mail direct using protocols such as file transfer protocol (FTP) or AS2] to receive and then relay the information to the addressee. The standards describe structures that emulate documents, for example, purchase orders to automate purchasing. The tern "EDI" is also used to refer to the implementation and operation of systems and processes for creating, transmitting, and receiving EDI documents. Despite being relatively unheralded, in this era of technologies such as the Internet, EDI is still the data format used by most electronic commerce transactions in the world.

Just-In-Time

Just in time (JIT) is the manufacturing methods used by many custom and in-house molding organizations. JIT reduces inventory of product, produces parts for orders with sufficient lead time to buy the material, molds the parts, and ships them to the customer in lot sizes to meet the customer requirements. The main requirement is that all items, materials, molds, machines, and personnel are available and ready to produce the product as required. JIT is a precursor to the lean style of manufacture to be discussed later.

The use of these technologies, CIM, JIT, RFID, EDI, and bar coding will uncomplicate, speed up, track, transmit, and locate information and materials before, during, and after operations have been completed. Accuracy will be enhanced along with creating records and documentation of the operations. This will allow more time to be spent in building the business and improving product quality.

CONTROL OF OPERATIONS

Operations can be monitored and controlled as described by the five methods discussed: CIM, JIT, RFID, EDI and bar coding. Each has its place in the TQPC system, Tracking orders through design, manufacture, and shipping and keeping a tight schedule for making JIT shipments is a difficult task if the right tools and trained personnel are not in place and performing as required.

Savings of inventory costs have been as great as 50 percent with production improvements of 20 to 30 percent realized through better planning and use of existing equipment and personnel. Just reducing the daily stress in an organization can yield many benefits as the work place is easier to manage. Data are real time information that can be acted on as soon as it is generated. The reduction of errors by just being able to find and know what is available is a major positive change for many companies. The elimination of problems and being proactive in seeking out and performing preventive actions is a major hurdle to overcome.

The correct use of the operating system coupled with a quality system that is proactive and kept current with documentation and records is required for TQPC to perform the functions developed for it. ISO 9001 and its automotive counterpart ISO/TS 16949 will assist by providing the information and requirements that are necessary for a good operating quality system.

PROCESS CONTROL

Process control is often categorized as just monitoring a selected set of manufacturing variables. It is more than this; it is the control of the entire product system. It begins with product design, prototyping, molding, assembly, and all other operations and processes for manufacture by injection molding.

Process control begins with identifying all the product's variables to ensure that they are identified, considered in their effect on the process, and controlled to produce the product. Variables must be controlled for the entire operation, which include the machine; mold design; material selection; plant, auxiliary, and secondary equipments; mold setup and operating conditions; operator training; and personnel knowledge in injection molding. The latter is often not considered until a key person leaves the company and the design and/or manufacturing program begins to suffer a series of problems related to the prior care and knowledge of the person or personnel who left and took the information with them. This implies that nothing was written down as a part of the daily operation of the manufacturing department and followed for accuracy.

The control and use of documentation and records is vital for all businesses. How these elements are used to understand and interpret information and business and manufacturing knowledge are critical for everyday business operations. Process control is based on using the existing quality control methods in a selected manner to realize the greatest benefit from their application in the business. The recognized quality methods available are shown in Table 3.1. These methods are used for analyzing, monitoring, and improving any type of business and/or manufacturing system. Most of these are recognized as having had their time "in the spotlight" and as having lost the allure, not the confidence of the quality engineer to provide them with the "instant success" they recognize as a reward when using one of the newer quality methods, such as Six Sigma, Lean, and now Lean Six Sigma.

CONTROL CHARTING

Data monitoring started with Walter A. Shewhart, the "father of quality" who developed control charts and demonstrated that common cause and special cause variation exists in every system. His use of control charts illustrated how stable or unstable a system was with simple charts and graphs. Many managers today, if asked whether their company's system is stable, will not have a clue what you are talking about. In fact, most would not have an idea how to use data to demonstrate that stability. At this time, only the methods used to

TABLE 3.1. Quality Improvement Methods.
Quality Methodology Understood:

Program Name	Worker Involvement	Specialist Oriented	Group	Individual	Procedure	Work Methods	Quality	Prod Design	Moral Enhancement	Motivation
Quality Circles	X		X		X	X	X		X	X
Zero Defect	X			X			X			
Employe Suggest	X			X		X	X	X	X	X
Work Simplify	X		X		X	X				X
Qual of Work life	X		X			X	X	X	X	X
Scanion Plan	X		X			X				X
VE/VA		X	X		X			X		
IE Work Study		X		X	X	X				
QA/QC		X		X	X		X			
Org Developmt	X	X	X	X	X				X	X
Fish Bone	X		X	X	X	X	X	X	X	X
SPC	X			X	X		X			X
DOE	X		X	X		X	X	X	X	X
CP/CpK		X		X	X		X		X	X
FMEA		X		X	X	X	X	X	X	
PAP	X		X		X	X	X	X		X
PPAP	X		X		X	X	X	X		X
QFD		X		X	X	X	X		X	X

TABLE 3.1. (*Continued*)

Program Name	Worker Involvement	Specialist Oriented	Group	Individual	Procedure	Work Methods	Quality	Prod Design	Moral Enhancement	Motivation
GMP	X		X		X	X	X		X	X
Kaizen	X	X	X		X		X		X	X
ISO 9000	X	X		X	X	X	X		X	X
TS16949	X	X		X	X	X	X		X	X
CEA	X		X		X	X	X		X	X
8-D	X		X		X	X	X		X	X
Poka-yoke	X		X		X	X	X	X	X	X
VSM (value Stream mapping)	X		X	X	X	X	X		X	X
CTQ	X			X	X		X	X	X	X
VOC	X		X		X		X		X	X
TPS (Toyota)	X	X	X	X	X	X	X	X	X	X
FEA		X		X	X	X	X	X		
TQM	X	X	X		X	X	X	X	X	X
Lean	X		X	X	X	X	X		X	X
JIT	X		X	X	X	X			X	X
5S	X		X	X	X	X	X			X
C&A	X		X	X	X	X	X		X	X
Triz		X	X	X				X		X

provide process control data are involved in our analysis, the specific type of chart comes later after determining what is to be monitored and by what means. A good source of information located on the Internet is the following website: http://www.isixsigma.com/offsite.asp?A=Fr&Url=http://www.skymark.com/resources/tools/cause.htm, which reviews the different types of quality control charts and quality methods.

Personnel who use and understand control charts often use a sample size of five for data analysis on typical X-bar and R charts. The reason for this is that during the Second World War, the U.S. Department of Defense had to teach untrained personnel to measure the quality of the products they made. The sample size answer was five, because if you take any group of five numbers, add them up, double the sum, and then move the decimal point one place to the left, you will have the average. Shewhart preferred a sample size of four. The same type of reason was used for the time interval for collecting data, every hour, because teaching personnel to calculate a true sampling plan would have distracted them from their output, which was more important.

INTERNATIONAL ORGANIZATION FOR STANDARDIZATION (ISO) ACCREDITATION

Selecting or being required to have your company implement a specific quality-control method (e.g., ISO 9000) or one of the other industry-specific accreditations, to do business with a specific customer, is often required. Certification from specific quality accreditation agencies or meeting a company's supplier requirement, such as auto company suppliers, who must be accredited to ISO/TS 16949:2002 or higher when revised, is often essential. This implies to the requester that certification will improve or ensure a consistent product quality. In most cases this is correct, but if accreditation is only acquired to become more competitive or a supplier to a company, without management's follow through for actually meeting and maintaining the requirements, then it is worthless. Unfortunately, this is what happened initially with ISO 9001:1984. Quality was not improved, and in some instances, it was even worse and these companies were still supplying products. This has changed, and the standards are now being adhered to with quality improved as intended.

PROGRAM MONITORING—COMMUNICATION

If you cannot communicate, you cannot successfully operate a business. One of the most important quality functions is learning to listen to your customer. Without quality in communication, a company can lose business share. Poor listeners create a loss of customer confidence by not knowing what their customers want, which results in their company's inability to provide the products and services needed for future growth. Therefore, customer and supplier communication are quality active programs to ensure they keep the paths of information exchange open and healthy.

Quality educator W. Edward Deming said, "The customer's definition of quality is the only one that matters." How the supplier reacts to his customer's feedback is in direct proportion to its success in business. What investigation has shown is that we need better communication between the seller and the buyer. Communication can be taught, but it must be practiced to be successful with questions prepared before the interview even begins. This is a major consideration for ISO 9001:2000, with Customer Satisfaction having 11 sections in the requirements, for just communication. ISO 9000 will continue to be improved to meet to days & the futures requirement for producing consistant quality products.

COMMUNICATING QUALITY IN BUSINESS

Communication in business is a two-way street. Information is received from the customer on their anticipated needs for a product. The supplier then provides feedback, as a quote, on what it can provide based on its capability that may include prototyping, design assistance, mold and product manufacture, and so on.

Often at this initial contact you may not know exactly what the customer wants the product to do. This requires finding out what the customer wants and, if able, assisting them in determining whether the plastic product will meet their requirements. In some case, the customer's design is still in the initial design and trial stage, and the customer needs to know whether the part design can use the preliminary material selection and manufacturing methods to make the product and have it perform its intended end-use functions successfully.

Therefore, the communication link between the two is extremely important if the supplier is to assist and possibly assume some liability for the product. It is very important that this information gets documented and fully understood by the customer and the supplier to ensure the business relationship is a success for both.

COMMUNICATIONS

Communication is the key to a successful business relationship. Just talking between sales and purchasing is not enough in today's markets. Once a channel of communication is established, methods must be developed to monitor the success of the program using communication and to ensure the "words spoken" reach the ears of the personnel who need to know the information exchange communications among them, the customer, and the supplier. The evaluation of the quality of communication is very important and must extend to the highest levels of each company. Management needs to know what is being discussed between the customer and the supplier, and what service and support is discussed and to be supplied. Therefore, a reliable and tested method must be used. How much effort is inputted into it is in direct relationship to the output realized.

Questions and information shared with the customer should be well documented and decided on before the communication takes place, so that each can share, learn, and gain confidence in the communication exchange. Try to keep the information response in real-time feedback to be of any real value in the guidance of your business relationship and interaction with the customer. Also, meeting and business conversations should be kept a part of the program and filed in the customer's program section of the CMS.

Next, be proactive and act on the information gathered to prove you understand the customer's needs and you know what the customer wants. This will assist in determining what you can learn about how the customer feels about your company. Next, share this information promptly with your key personnel and management to obtain the maximum shared results. Do not wait for reports to communicate important information learned; instead, e-mail it right away. Also, try your best to verify key information that is critical to your company and that you know will be questioned by senior management. More than one source is harder to refute than a single source, and use their names if they were your source.

One area to remember is when information is given to you in confidence, treat it as such, and only share it with personnel who can be trusted and know how to handle this type of information.

Communication methods to consider are listed in Table 3.2 to assist your communication input and customer output.

face-to-face interviews with direct, specific questions are recommended. The data obtained need to give the requester the ability to act on the information when received and provide an internal quality input that will better the working relationship with the customer.

Also, show appreciation when the information is used successfully to help each company; only be sure the source gives you permission to do this as he or she may not have had permission from management in providing you with the specific information, even though it was successful. In these case, the source may want to remain anonymous until the company waters are tested, before revealing he or she was the source of the positive outcome.

SURVEYS

Surveys should not be used for obtaining information as it is tedious, requires valuable time, and meets your needs, not theirs. Surveys are too long and

TABLE 3.2. Communication Methods.*

Method	Output
1. Examine existing	Use existing actions, in person, e-mail, mail interactions; fax, phone, etc.
2. Interaction to collect	Contact neutral, routine, and candid feedback
Neutral	Contact not related to problem, neutral query, order placement situation
Routine	Contacts on regular, routine basis, information will be fresher and more meaningful
Candid	Contacts trust each other, communication is open with freeexchange
3. Tools to match	
Customer interaction	
Telephone	Short direct questions at end of a routine business call
In person	Routine feedback of information after service; flash feedback based on current actions
E-mail	Live link in message taking customer to a fast loading review of its recent experience
4. Focus on open-ended questions	
Likes and dislikes	Keep questions open-ended and easy a) Do you have problems with product you have not told us about? b) Is there anything you think we do very well? c) What can we do to make your job easier?
5. Act on these opportunities in real time and let the customer know the results of your actions and show appreciation in your discussions.	

*Cochran, C., Georgia Tech. Enterprise Innovation Institute, *Quality Digest* September 2006, Available at: www.innovate.gatech.edu/quality

boring to elicit real-time information needed from a customer. If the surveys mailed to your customer are too long, ambiguous, require response ranking, and have too many questions, they will never be completed, much less, returned.

Directing your survey to the wrong personnel to obtain a response to a question, such as the technical capability of your sales and service engineers, can prove difficult, as there may be no contact between these individuals, and worthless information will be produced. Finally, should you respond to a survey, be sure what you communicate is approved by your management and is accurate. Ask whether the information is proprietary and how it should it be treated.

Remember, communication in business goes two ways; you present your company to the customer and in return learn about them. Sales makes the first contact by identifying a customer need for your product or service. Sales representatives contact the customer and present the product for its consideration. Discussions then begin that will lead to more discussions and possible negotiations with product information; trials and pricing will be presented.

QUALITY FUNCTION DEPLOYMENT (QFD)

QFD was developed to ensure an open, planned, and effective method of exchanging and gathering information needed to manage and grow a business. ISO 9001:2000 has a specific requirement for customer information exchange. The section in chapter 7 called "customer communication" directly relates and requires a company to develop effective arrangements for communication with customers in relation to product information, enquiries, contracts, or order handling, which includes amendments, customer feedback, and customer complaints.

QFD is a proven quality method for ensuring that good communication is established between the customer and the supplier, as well as internally between their company departments. This is often an unrecognized intercompany problem until a serious problem occurs that internal communication should have eliminated. Companies that are successful in communication skills use QFD communication techniques. Repeated success with existing and new customers is the standard for their use of the QFD method. Training is available for individual QFD users in the Green and Black Belt expertise areas offered through websites on the Internet.

It is necessary to know what the direct influence on your business relationship thinks of your company. This information will give you insight as to what actions are needed to maintain your company as their prime supplier. What your customer's actual needs and wants are involved in your business relationship. Contacts within all levels of your customer are necessary for a successful business relationship, and how you get reliable information is very important.

QFD is implemented immediately after customer and supplier contact occurs. When the program has commenced past the initial negotiation stage, the customer may provide a competitive product sample or design idea for the supplier to review for manufacture. Pricing, prototyping assistance, and specifications are discussed with a pending order and delivery date established. Then, depending on subsequent discussions, a contract or purchase order is written. The final details are worked out, and a contract may be signed before business begins for the program. QFD is used to both establish and maintain good communications with the customer and to keep it strong and continually improving for continued business growth.

QFD is a reliable communication system used to establish good communication and is a recognized quality method that assists in building and opening up the customer communication bridge. This information collection method is shown in Figure 3.8. As designed, it replicates a house, thus its name, the house of quality (HOQ). The QFD method of communication between customer and supplier is called the "voice of the customer" in matrix form.

QFD was designed to develop the information needed from any level of company management by just varying the type of questions asked, primary and secondary, and the type of information needed, planning, procedures, and

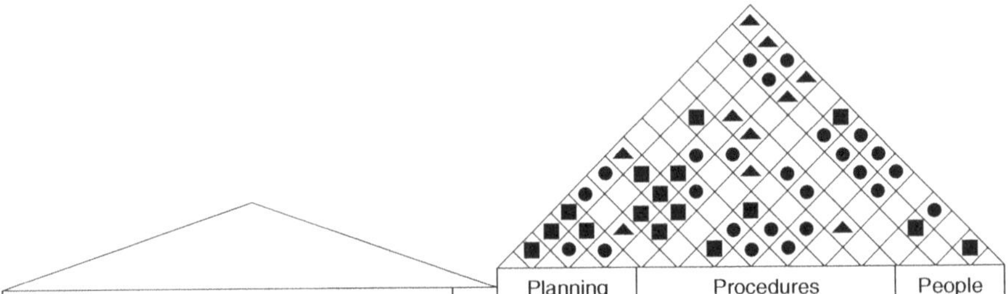

■ Strong ● Medium ▲ Weak

What? Customer quality criteria (Primary)	(Secondary)	Relative	Planning: Layout	Planning: Resources equipment	Planning: Resources(personnel)	Planning: System capacity	Procedures: Housekeeping	Procedures: Customer handling	Procedures: Documents handling	Procedures: Information handling	Procedures: Nonroutine situations	Procedures: Inventory	Procedures: Job/people schedule	People: Selection	People: Skills/training	People: Attitudes/morale
Reliability	Accuracy	5			■				■	■						
	Dependability	5		■	■					■	■	●	■		●	
Responsiveness	Willingness to help	4			■			■		●						■
	Prompt service	4	▲		■	■		■	■	■			●			■
Assurance	Knowledge and courtesy of employees	3			■			●		▲	■				●	●
	Ability to convey trust and assurance	3			●			●								●
Empathy	Caring of and attention to customers	2			●	■	▲	■		▲	■				■	■
Tangibles	Facility; equipment, people, and materials	1	●	■								▲	●			

(Service facility facets — How?)

FIGURE 3.8. *House of quality. (Adapted from Ref. [4].)*

people. When the roof of the house of quality is completed, additional information is gained and is identified as the performance measure between the relationship items of planning, procedures, and people for this particular HOQ. As can be observed, a great potential of information is possible when completed. The HOQ can consider a host of variables and have them compared with others in a developing matrix of information that can then be used to build on a business relationship for success.

QFD IN OPERATION

Customer satisfaction depends on the supplier determining what the customer really wants and needs to be successful as well as on how the supplier will meet the quality and product specifications and requirements. To do this, the supplier uses QFD to develop the action items needed for meeting the customer's needs and wants through acceptable responses to questioning.

Customers often are not aware of the QFD function occurring and are generally very open in their response and assistance to their potential or current supplier. To obtain this information, the supplier must decide what is thought to be the customer's most important needs, until the supplier is told otherwise. Also, the higher up the decision chain, the more valuable the output to the questions asked.

Considerable planning is required for a successful questioning session. QFD is designed as a house (matrix) of many specific questions with primary and secondary needs determined through analysis of the initial customer solicitation. From this involvement, questions are selected as an identified requirement on the customer request for information side, and at the top (roof), the supplier's measure of performance as perceived needs and concerns, when the service is provided.

What is established is a customer and supplier "relationship grid." The grid is used to request information and then, when answered, to record the response of the customer. The matrix is used to quantify the importance of customer perceived service and quality items or any other items the matrix is set up to analyze. All industries and businesses can use the matrix by simply adjusting the relative importance factors to suit their business. The matrix can be narrowed down to specific analysis areas of the business, and when modified with information requests, it can be used in other areas for customer input to your personnel.

Knowing what you think your customer needs and wants before developing your business relationship is a plus for any customer/supplier arrangement. QFD can focus on the analysis of your customer business relationship as to its ability to meet the customer's needs. A flow of customer and supplier communication plan is shown for the design and product realization process in Figure 3.9. The information developed in the initial house of quality does not stop but continues with new and more specific questions supplying the information to improve and encourage a discourse of information between each party.

This is the first step in establishing both business and quality relationships with your customers, knowing what they want, need, and expect from you, their supplier. Table 3.3 shows response scoring to the questions in the relative column. Scoring can be by symbol or numerical value where a score can be generated and used for evaluation of your customer response system improvements over time.

A major reason why QFD and specifically customer contact is so important is an example of a major supplier/customer experience.

CUSTOMER FEEDBACK

Customer feedback is composed of five stages of action as shown in Figure 3.10. To accomplish the goal of transforming loyal customers into dedicated

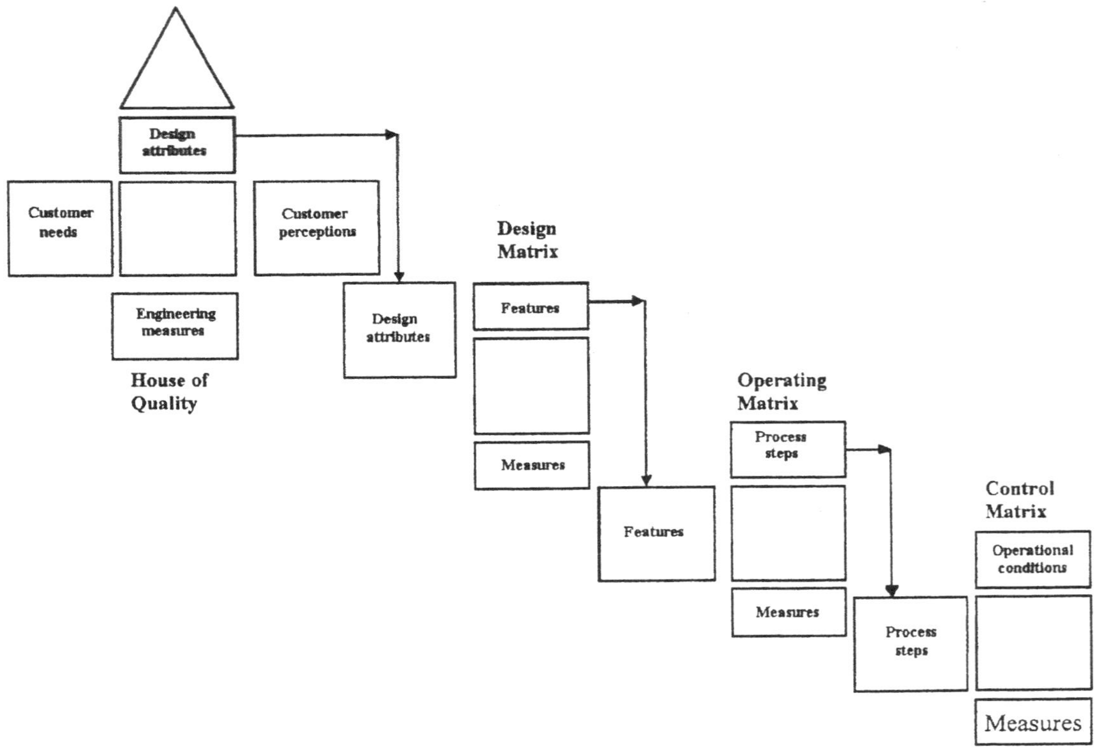

FIGURE 3.9. The flow of communications in translating customer needs into operations using QFD interaction matrices. (Adapted from Ref. [5].)

TABLE 3.3. Example of QFD Customer Scoring.

Score	Critical Response Need
5	Required to sell and hold as customer
4	Major benefit to hold customer
3	Good to have to maintain customer
2	Could offer if needed to get business
1	Not recognized as a benefit to sell

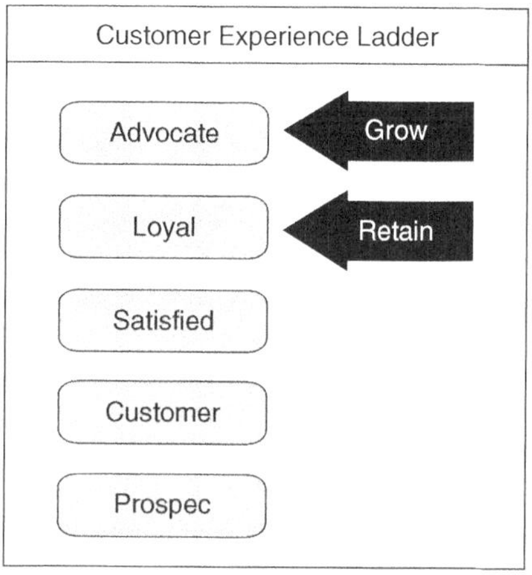

FIGURE 3.10. *Customer experience ladder. (Adapted from Ref. [6].)*

customers requires the company management team to turn customers into loyal customers first by delivering what it promised. This begins with satisfied customers who are pleased with the service provided and become loyal customers. To move to the next level, you must strive to deliver positive experiences continually and create a strong relationship that developed the loyal customer who turns into the dedicated customer.

The dedicated customer is an advocate of your services and will recommend your company to others and will even base its reputation on your services. The goal to achieve the dedicated customer is shown in the customer experience ladder, with the highest being the customer advocate (slated for growth), next the loyal customer (to retain), satisfied (meeting expectations), the customer (initial sales begin) and the opening level, the prospect (convince them to be your customer).

The goal is to have the customer elevated in its recognition of your supplier skills by moving up the ladder of customer satisfaction. This is accomplished by your supplied support to create a dedicated customer for life. You accomplish this, along with product and quality improvements, by using communication and information feedback to the customer on what you are doing for them.

This begins when the customer is still a prospect. The use of your dedicated customers in advertising your business successes by their endorsement and use of personal reference is one step closer to the prospect trying your product. Once they try it, you must complete the transaction by supporting and servicing the product, supplying additional information if required, and developing a fast, lasting, and final solution that is conveyed to the customer as a preventative fix.

The business goal is to obtain good solid information on which to build a business relationship and create a friendly working atmosphere for conducting business. Most businesses do this already but may never have realized that it had such an important, defined purpose, goal, and agenda.

CRITICAL TO QUALITY (CTQ)

Critical to quality is another quality communication method to improve the understanding of your customer's needs and wants. Determining what the customer really needs and wants is a step above most companies' involvement in obtaining information and using it to increase their business growth.

CTQ represents the key measurable characteristics of a product or process whose performance standards or specification limits must be met to satisfy the customer's (internal or external) wants. They combine design and manufacturing improvement efforts with customer requirements. They may include the product's attributes or specification limits plus any other factors related to the product. A CTQ is often interpreted from a customer qualitative statement to an action and a quantitative business need or specification. CTQs align improvement or design efforts with customer requirements.

CTQ products are what the customer expects of a product through its communicated needs with a supplier. The customer should list and express its needs in plain language, and it is up to the supplier to convert them to measurable terms using quality methods such as design failure mode and effects analysis (DFMEA).

The CTQ methods use a lean enterprise force to make the conscious decisions of which customer strategies to champion and provide process and product improvement with 100 percent buy-in.

CTQ information can be obtained using the QFD method by asking the questions you believe are critical to the customer. By asking each customer contact in different areas of interest, you can determine the specific area of criticalness that individual is concerned about.

Examples include a product specification that may have the following questions asked for it as:

1. Listed
2. Realistic
3. Required

4. Needed
5. Customer desired
6. Cost inhibited

Requirements are other specific questions that must be answered to determine the true needs of the product for the market.

In fact, when talking with several people in the customer's company, there will be as many different CTQ answers as people interviewed, and each will have their own idea of what is critical for the product, such as price, cost, function, method of manufacture, and so on. It is now up to the supplier to get the customer to reduce its want list into a need list to determine how the new product will meet the many specified requirements of its customer.

BUILDING ON TQPC, PRODUCT MANUFACTURE

Numerous items or "variables" must be considered when establishing a new or updating an existing quality program. Once the business and design of the products is finalized and approved, the method of manufacture variables must be considered, because there are many options and all are important to the success of the program.

The manufacturing program is identified for all the components and parts for injection molding separated into their specific requirements. This will include selection of the molding machine, mold, support equipment, secondary operations, if any, and quality requirements and equipment to ensure the product meets requirements. This includes listing the information and variables determined for the manufacturing program.

Based on the known or to be completed information, the remaining operations can be completed. These are the mold design and build, manufacturing equipment selection, secondary operations, decoration, and assembly requirements. The final inspection as well as packing and shipping methods must be considered. Schedules with timelines are established, and personnel are given their responsibilities to ensure the program remains on track and all components come together at the right time to have a successful program. To assist in this area of quality, checklists will cover most of these areas by asking the necessary questions to ensure all is planned and selected when the company is ready for manufacture.

CHECKLISTS

Checklists are a quality tool that assists employees to obtain and maintain accurate results. Checklists save time, increase accuracy, ensure the correct questions are asked, and require the information to be collected in a timely manner. The only negative is the impression some personnel have for the usefulness of checklists. One negative often assumed with using a list is, "It implies the employee does not always perform the operation repeatably." This is often true, and they and I forget things!

Checklists are an aid in "not forgetting" the important steps for performing an operation. When discussing the use of checklists, always be positive and project their use will make each person's job easier and will prevent future problems in their work.

Checklists are used for each launch of our space shuttle, and the checklist used has more than one million items that must be completed, monitored, or signed off before each successful space launch. The problem is that management has not always correctly responded to a situation or possible safety occurrence.

To assist in eliminating problems, a series of specific checklists, for each operation, were developed. These are both business and manufacturing checklists specific to injection molding and all are provided in Appendix C.

1. Product development
2. Sales and contracts
3. Product design
4. Material
5. Purchasing
6. Quality
7. Design and development schedule
8. Price estimating
9. Program scheduling for manufacture
10. Manufacture
11. Assembly
12. Decorating
13. Pack and ship
14. Warranty problems
15. Mold (for injection molding)
16. Mold specification (for injection molding)

Checklists should be included in the framework of all quality operation instructions, including ISO and other quality methods. The task is to ensure personnel actually follow these instructions. This is accomplished by training personnel to use the checklists during their training program and then to monitor their use in performing their job.

Major equipment and material suppliers provide troubleshooting guides for solving equipment, material, and processing problems for their most common problems. This type of checklist provides a corrective reaction to try when an unwanted action occurs. These guides are very helpful in quick problem solving with an experienced machine operator to get the manufacturing process back in control.

QUALITY CIRCLES

The earliest information exchanges occurred on the factory floor and were called quality circles. Quality circles were developed to encourage workers to share their knowledge in operations to assist in solving problems and in general to recommend improvements in working conditions and productivity. When a proactive leader leads a quality circle team of employees, he or she can accomplish goals in a minimum of time with excellent results.

In 1962, Kaoru Ishikawa from Japan developed the quality circle concept. The quality circle was used to tap the creative potential of workers. A quality circle is a small group of employees, usually 6 to 12, from the same work area who voluntarily, or are directed to meet at regular intervals to identify, analyze, and resolve work-related problems. Quality circles have improved the performance of many organizations in both business and manufacturing, and they have also aided in motivation and enrichment of the daily work life of employees.

In fact, quality circles are alive and well at NACOM, Griffin, GA, a Division of Yazaki, where a team realized $95,000.00 in savings in 2005 in their department within six months. So if someone tells you the older quality methods do not work any more, ask a Six Sigma Black Belt what quality methods he or she uses to make quality improvements. Also, one can conduct an Internet search for more positive examples of the older methods, doing what they did then, even better now.

FISHBONE ANALYSIS

A quality method used for assisting in determining all of the variables of a system and the actual "root cause" of a problem is the Ishikawa "fishbone" diagram. A typical fishbone diagram is shown in Figure 3.11. The fishbone analysis lists the basic variables or components (procedure, polices people, etc.) that interact with the main effect.

Other variables, which are shown as branches and multibranches coming off the ribs forming sub-branches on the diagram, are added until no more variables are possible for analysis or the necessary information is determined. The size and branches off the main trunk will be as many as necessary to reach the last controlling variable in an analysis. Some fishbone diagrams have had more than 300 variables for a specific effect.

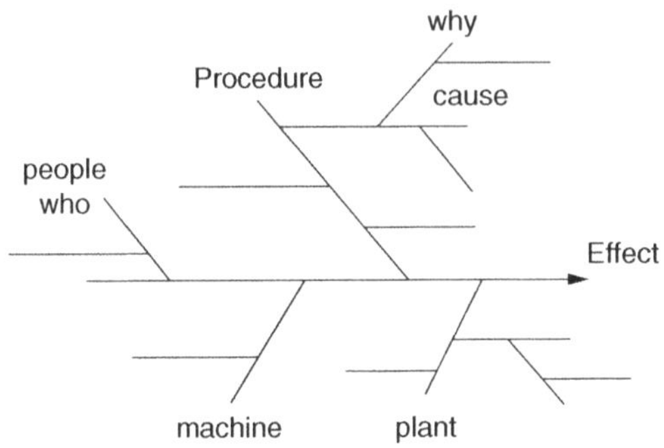

FIGURE 3.11. *Fishbone analysis.*

FAILURE MODE AND EFFECTS ANALYSIS

Failure modes and effects analysis is a proactive evaluation technique used for identifying potential product and/or processing problems. Potential problems can be identified and then traced to their root cause and eliminated. FMEAs are used by manufacturing personnel to analyze their production plan by identifying any potential problems, at each stage of an operation, that may occur when the operation is performed during each step (e.g., in the manufacturing process). Using a form designed for this analysis, information is recorded for each operation performed.

Once the entire process is documented, an analysis is conducted at each operation point for what might cause a problem at this step in the process. Any potential problem areas are identified, and when the analysis is completed, each problem area is investigated for the cause of the identified potential problem, if it could occur, and its effects on the process.

This method is also used when a Parato chart shows that problems are occurring and by using the FMEA determine the cause of each problem. The FMEA is run on the entire process, as the root cause of the problem may have occurred prior to where the problem was discovered. The obvious indicator is not always the root cause of the problem. The lean manufacturing system uses the FMEA as a guide for streamlining a new or existing process into a lean system.

The FMEA provides greater confidence in the manufacturing operation performing error free, resulting in higher yield and product quality. The Ishikawa fishbone diagram in Figure 3.11 is used for identifying all variables in each step of the operation.

The fishbone diagram, combined with the FMEA analysis, will aid in determining where potential problems may occur and will allow the analyst to evaluate all variables. This method of analysis provides information for determining whether a potential problem area exists and what is the cause, and then it aids in developing a lasting solution.

Besides identifying potential manufacturing problems, the FMEA can be used to evaluate different variables acting on the product that were never anticipated. This analysis will assist in identifying potential end-use actions that could cause a failure of the product. A crucial step is anticipating what might go wrong with a product once it is in the consumer's hands or workplace. The early use of an FMEA in the design process can allow the engineer to design out the potential and unplanned use of the product that may result in failures to produce a more reliable and safe product.

TYPES OF FMEAs

Several types of FMEAs are used for analysis. An FMEA should always be run after the product's design and/or manufacturing instructions are written. This will allow the design, manufacturing, and quality engineers to evaluate the effectiveness of the product and production line for problem prevention. The types of FMEAs are as follows:

- System—focuses on global system functions
- Design—focuses on components and subsystems
- Process—focuses on manufacturing and assembly processes
- Service—focuses on service functions
- Software—focuses on software functions

The FMEA is a very analytical, informative, and supportive quality tool. It can be replicated in many forms to suit all facets of a company's business, manufacturing, and service operations.

It is a metric that examines all operations in a process in a detailed and sequential order. This permits operations and process variables to be evaluated that act on a specific operation or process at a specific point in the product's manufacture for the occurrence of a potential, up to this point, undiscovered problem.

Historically, engineers have done a good job of evaluating the functions of products and processes in the design and manufacturing phase. They have not always done so well at designing in reliability and quality. Often the engineer uses safety factors as a way of making sure that the design will work and protect the user against product failure. As described in recent article:

> A large safety factor does not necessarily translate into a reliable product. Instead, it often leads to an over designed product with reliability problems.
>
> "Failure Analysis Beats Murphy's Law,"
> *Mechanical Engineering*, September 1993.

Because an FMEA helps the engineer identify potential product or process failures, he or she can use it to assist in product design.

Design Analysis of: ____________ Failure Mode Effects Analysis Prepared by: ____________
Item Description: ____________ FMEA Date: ____________
FMEA Number: ________ **Design FMEA** Page ___ of ___

Design Function Require-ments	Potential Failure Mode	Potential Effect(s) of Failure	Severity	Class	Potential Cause(s)/ Mechanism of Failure	Occurrence	Current Design Control	Detection	R.P.N.*	Recom-mended Design Action(s)	Person responsible Target completion Date	Actions Taken Results	Severity	Occurrence	Detection	R.P.N.*

FIGURE 3.12. Design FMEA. (Adapted from Ref. [7].)

Development of a design FMEA (see Figure 3.12):

- Develop product or process requirements that minimize the likelihood of product failures.
- Evaluate the requirements obtained from the customer or other participants in the design process to ensure that those requirements do not introduce potential failures.
- Identify design characteristics that contribute to failures and design them out of the system or at least minimize the resulting effects.
- Develop methods and procedures to develop and test the product/process to ensure that the failures have been successfully eliminated.
- Track and manage potential risks in the design. Tracking the risks contributes to the development of corporate memory and the success of future products as well.
- Ensure that any failures that could occur will not injure or seriously impact the customer of the product/process.

- Investigate manufacturing and material variables that could cause a problem during the manufacturing phase.
- Analyze the assembly and shipping of product to customers to ensure all problems are eliminated.
- Evaluate business practices to ensure order taking, information sharing, and monitoring are documented and recorded for accuracy and customer agreement in invoices and business relationships.

To develop a useful FMEA requires a thorough understanding of the product's intended operations, as well as how the product is to be manufactured and each operation will occur in the process. It is important that a team is used to map out the specific design FMEA flow plan to ensure no potential problem or operation point is overlooked.

Data to be recorded include all possible functions and operations the product may be subjected to, and the manufacturer must list what variables are injected at each point.

Plus, an analysis is needed of what potential problem types and their effects, severity, cause, occurrences, methods in place to control the operation, actions to take when they occur, as well as who is responsible and results of actions taken to eliminate the problem from occurring.

FMEA is designed to assist the engineer in improving the quality and reliability of the product's design and manufacture, as well as in detecting potential problems. Properly used, the FMEA provides the engineer several benefits. Among others, these benefits include the following:

- Improving product/process reliability and quality
- Increasing customer satisfaction
- Identifying and eliminating potential product/process failure modes early in the process
- Prioritizing product/process deficiencies
- Capturing engineering/organization knowledge
- Emphasizing problem prevention
- Documenting risk and the actions taken to reduce risk
- Providing a focus for improved testing and development
- Minimizing late changes and associated cost
- Serving as a team and offering idea exchange between functions

FMEA TIMING

The FMEA is considered a living document because throughout the product development cycle, change and updates are made to the product and process. These changes can introduce new failure modes if they are not fully investigated. It is therefore important to review and/or update the FMEA when the following events occur:

- A new product or process is being initiated (at the beginning of the cycle).
- Changes are made to the operating conditions in which the product or process is expected to function.
- A change is made to either the product or the process design. The product and process are inter-related. When the product design is changed, the process is impacted and vice versa.
- New regulations are instituted.
- Customer feedback indicates problems in the product or process.

This is illustrated in Figure 3.12.

IMPLEMENTING AN FMEA

Manufacturing engineers write the product's manufacturing procedures for their department. When completed; it should be trialed to ensure all operations are performed correctly, work instructions are accurate, and anticipated results are obtained. Also, the trial should ensure that the product meets the necessary specifications for this point in its manufacture. At this point, an FMEA is conducted to ensure the quality department has not missed any unknown problems; the quality department is assisted by the department's engineers who are knowledgeable of the product and operation. Knowledgeable personnel must be used here to ensure no operation or item is left unanswered in the FMEA.

The FMEA is the responsibility of the department where the operation is performed with a team composed of personnel who have a stake in the process. The form for a maintenance FMEA is shown in Figure 3.13, and a Process FMEA form is shown in Figure 3.14.

FMEA DEVELOPMENT

The process for conducting any type of FMEA is straightforward. The basic steps are outlined below. What varies are the headings under which information is described and later collected and evaluated as to the outcome of certain variables that act on the product or process.

The FMEA is filled out as explained in the following description.

A. Describe the operation to be performed and fill out the top section of the FMEA form, items 1 through 9. Modify the headings to suit the type of FMEA.

Potential Equipment Failure Mode Effects Analysis Process FMEA

Line description: Semi-Automated Assembly Line
Machine description: Transfer Unit
FMEA Number: 64-5791

Prepared by: QM (Quality manager)
FMEA Date (Org.): 12-Nov. 2001 (Rev.) A
Page 1 of 1

Core team: MM (Maintenance Manager), MB (Maintenance buyer), LHM (Lean hand mechanic), PS (Production supervisor), LO (Line operator), QM

Equipment Process function	Potential Failure Mode	Potential Effect(s) of Failure	Severity	Potential Cause(s) Mechanism of Failure	Occurrence	Current controls predictive methods	Detection	R.P.N.*	Recommended actions	Person responsible/ target completion date	Action Results Action taken	Severity	Occurrence	Detection	R.P.N.*
Transfer unit-main drive Drives the printer mechanism	• Chain breaks. P/N X 03-31-34	• Transfer unit down • Post-repair set-up problems	5	• Misses limit switch • Broken limit switch • Stretching due to normal wear	8	None	10	400	• Replaces limit switches with heavy duty type. • Install tension detector • Install cycle counter. • Monitor cycles. • Evaluate changes to improve accessibility. • Spares: chain & blocks mounting to chain. • Write repair procedure: describe tasks, change sprockets with chain, periodic lube and preventive replacement.	MM MM PS and MM MM and MB MB MM, QM, and PS					
Transfer unit lift Lifts the fixture support	• Chain breaks • Chain stretches. P/N Y03-31 Y03-30 2 each	• Unit down	2	• Wear • Misalignment • Lube	1	None	8	16	• Monitor cycles • Write and train repair procedure	PS and MM MM, MB, and QM					
Stroke control Controls the length of the stroke	• Chain breaks • Chain stretches P/N X-04-43	• Unit down Post-repair set-up problems	1	• Wear • Misalignment • Lube		None	8	8	• Monitor cycles • Repair and set-up procedure • Change chain with backup rolls	MM and PS MM, MB and LHM					
Turn-around drive Turns the fixture around and back to original position	• Sprocket wears • Sprocket breaks P/N 31-56	Sprocket becomes loose causing misalignment Causes chain to break	1	Set-screw loosens due to shock and vibration	8	None	10	80							
Frame Supports the fixture when it holds the part	Frame twists Frame bends P/N X-03-88 P/N Y-03-77 Two each P/N 03-03-99	Gradual decline in assembly quality Gradual decline in productivity	1	A failure which induces unusual shocks - chain breakage Improper storage	10	None	10	100	Check after a chain breakage Spares P/N 19 & 15						
Carriage Moves the fixture back and forth	Wear	Machine down	1	Friction	9	None	10	90	Periodic inspection and replacement as needed Spares	MM and LHM					

FIGURE 3.13. Maintenance FMEA. (Adapted from Ref. [7].)

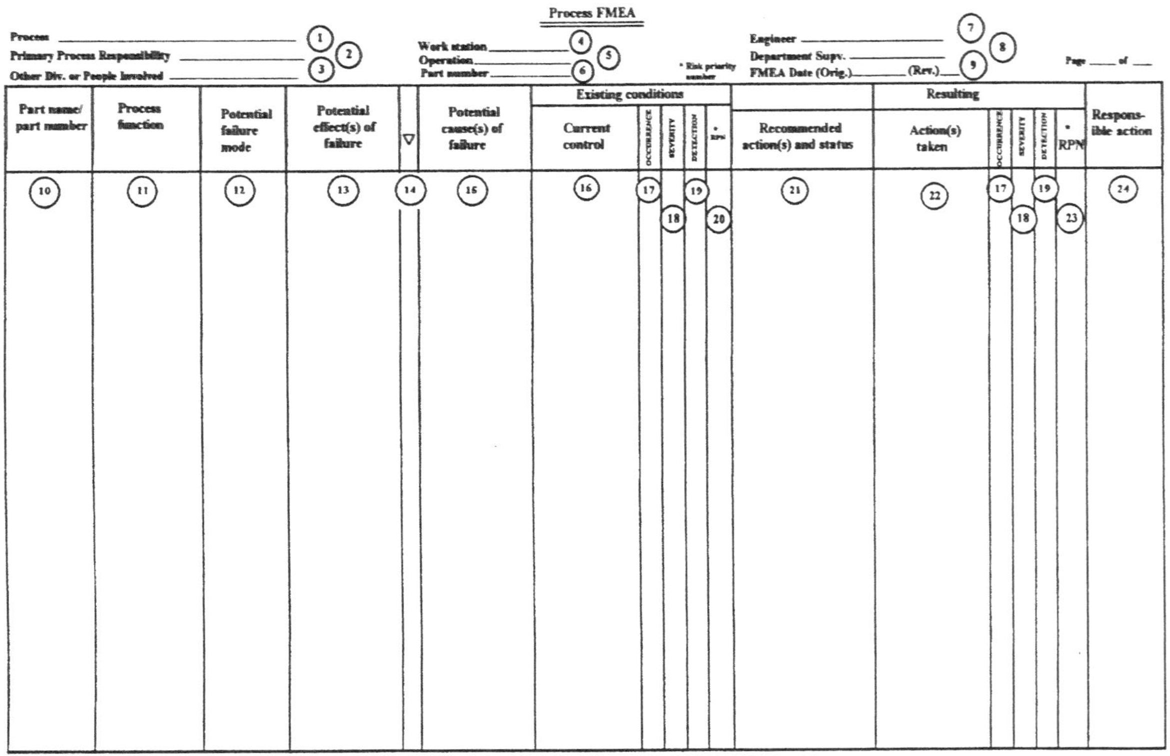

FIGURE 3.14. Process FMEA. (Adapted from Ref. [7].)

B. Create a block flow diagram of the process, or use the manufacturing procedure as a guide, to develop for the operation being analyzed and identify the operations. Connect together by lines those operations that are sequential and others that may be side items or operations that are fed into the process stream, as they affect the product's manufacture. Indicate how the items or steps are related. The diagram will begin to show the logical relationships that establish a structure around which the FMEA can be developed. Establish a system to identify the elements.

C. Beginning with items 10 through 15, fill in sequentially all operations performed and any potential problem areas that the team can identify in descending order. The intent is to identify all potential problem areas associated with the product manufacture and evaluate each as a possible problem. Use the "what if" method of assuming whether a problem will or will not occur. From this list, you should consider all the effects and whether they could cause a problem. Continue filling out the form for the required information.

   Depending on the complexity of the product, it may take several FMEA forms to analyze the process for potential problem areas. Commercial software is now available to simplify this operation.

D. Identify each line item's particular failure mode as if it could occur. A failure mode is defined as the manner in which a component, subsystem, system, function, and so on could potentially fail to meet the design intent. Examples of potential failure modes may include the following:

   - Abrasion
   - Temperature change
   - Fluid temperature variability
   - Poor cooling transfer
   - Deformation
   - Cracking
   - Torque settings of tools
   - Contamination

F. A failure in one operation may not necessarily cause a problem right away. It could cause a failure to occur at the current location, in the next operation, or even at another operation further into the manufacturing process. It is all problem-type sensitive to the operation. Should this occur and the root cause is not immediately found, the fishbone diagram can assist in determining exactly where the problem originated. At this point, the failure type should be identified and should indicate whether the failure is likely to occur. Looking at similar operations and estimating whether the failures could occur is an estimation of the risk probability.

G. (16) Identify current control used to prevent the cause of the failure from happening and determine whether it can detect the failure before it occurs.

H. Enter the probability factors (17 and 18). A numerical weight should be assigned to each cause that indicates how likely that cause is to occur. A common industry standard scale uses 1 to represent not likely and 10 to indicate inevitable.

I. Determine the probability of defect detection (19). Detection is an assessment of the likelihood that the current controls for the item will detect the cause of the Failure mode, thus eliminating it. Based on the current controls, consider the likelihood of detection using the following calculated value for guidance.

N. Review risk priority numbers (RPNs) (20). The risk priority number is a mathematical product of the numerical severity, probability, and detection ratings estimated by the engineer for the product:

$$\text{RPN} = (\text{Severity}) \times (\text{Probability}) \times (\text{Detection})$$

$$\text{RPN} = (1 \text{ to } 10) \times (1 \text{ to } 10) \times (0 \text{ to } 100\%)$$

The RPN will identify and assist in prioritizing potential failure items that will require additional quality planning or action to eliminate a problem occurring with the process.

O. Determine recommended action(s) (21) to address potential failures that have a high RPN. These actions may include specific inspection, testing, or quality procedures in effect, making the process as failure proof as possible with known or estimated information.

P. Assign "responsibility" (24) and a target completion date for these actions. This makes responsibility clear cut and facilitates tracking. (17 to 23)

Q. Indicate actions taken. (22) After these actions have been taken, reassess the severity, probability, and detection and review the revised RPN.

R. Update the FMEA as the design or process changes, the assessment changes, or new information becomes known.

The use of FMEAs can continue throughout the business as just seen starting with the design FMEA and continuing through manufacture, maintenance, shipping, etc.

4

Customer Satisfaction

Quality is defined as customer satisfaction. If the customer is not satisfied, then business ceases. The goal of any good supplier, in-house captive molding operation, or outside custom molder is to satisfy their customers. Senior management must always make a quality product for their customers. But, sometimes management is not capable or competitive. Therefore, many organizations subcontract outside molding companies.

Manufacturing organizations must provide the means and capability to perform this duty. They should have an input as to their capability to manufacture the product designed by the engineering team. Manufacturing departments must input their requirements for the product, so the product can be manufactured and can perform its end-use function. Plastics are now performing more stringent functions than before, such as under the hood, engine parts, and space applications. The use of reinforcing mediums of glass fibers, mineral, and others have made plastics a universal material, and a list of physical and chemical property modifiers is provided in Table 4.1.

Composites, glass and carbon fiber, and other materials impregnated with different resin bases are used for multitype applications, airplane bodies, and wing and control surfaces, such as Boeing's new 787 aircraft and other structural parts. Composites today are stronger and lighter than aluminum. These products are not discussed in this text, but the Internet is an excellent source of information.

Total Quality Process Control for Injection Molding, by M. Joseph Gordon

TABLE 4.1. Typical Fillers, Reinforcing Fibers, and Modifiers.

Fillers	Reinforcing Fibers	Modifiers
Glass spheres	Glass fibers	UV stabilizers—Processing aids
Carbon black	Carbon fibers	Plasticizers—Preservatives
Metal powders	Aramid fibers	Lubricants—Smoke suppressants
Silica sand	Jute fibers	Colorants—Impact modifiers
Wood flour	Nylon fibers	Flame retardants
Ceramic powders	Polyester fibers	Antioxidants—Foaming agents
Mica flakes		Antistatics—Fungicides
Molybdenum disulfide		Viscosity modifiers

MANUFACTURING AND SUPPLIER INPUT

When is the last time you purchased or used a product made from plastic and it broke or failed? When the failure point was examined, you realized someone with little knowledge must have designed the product, and the mold assisted in creating the failure site. Did anyone review the part for a potential failure? In most case, the answer is "no." Time was not spent on examining the design, mold, material, assembly, decoration, and so on. Obviously, if it failed from normal use, then no one checked the design. This is the primary reason all departments provide input into the design and manufacture of a new or existing product to ensure it can to perform its end-use function as required. Sharp corners and a lack of adequate radii in a mold are the prime reasons for most plastic part failures.

How this is done is not always easy, depending on the structure of the customer's company and departmental interaction. In this analysis, I shall assume they are involving each department and their input is welcomed. If the company is an outside supplier, it is hoped it will exercise its influence to assist in initial design analysis. Or during contract negotiations, the supplier should influence the company to review the product for manufacture, and the customer should accept the knowledge the supplier can offer from its experience in the molding business. This process will involve the extensive use of checklists and interaction of employees from both companies to ensure the product produced will be as good as it can possibly be and will still stay within the expected pricing guidelines.

VENDOR SELECTION

Before moving too far along in the production process, the selection of your product supplier has to be decided. Look for experience, not price, when selecting a molder. The same goes for in-house molding. If your production department is not knowledgeable in the molding of a new material and/or part design, get a second opinion. It could be very expensive to learn on your time if the product introduction schedule is too tight.

It may serve your best interests to use an experienced custom molder and have them break in the mold and establish the cycle for producing good parts for market entry. But, before you do this, let the molder know your intent is to later bring the mold in house. Be upfront with your suppliers and give them a higher part price for their experience and capability. You may end up letting them continue supplying the part, at a requoted price, when they bring the mold online sooner than anticipated and meet all your requirements.

VENDOR SURVEY

Selecting a qualified custom molder is very important for the success of a new program. The molder can bring experience to assist your engineering department to create and make the best plastic product possible within cost guidelines.

All outside vendors should be surveyed for their knowledge, capability, and experience in manufacturing injection molded products in the material selected. Selecting a molder not experienced in the materials needed could cause a problem, such as degrading the material and being unaware of this occurring. Experience is critical for structural products as are parts with visible show areas. A high-gloss surface or surface to be decorated must be protected and not contaminated during the manufacturing process. Contamination could occur if the operator transferred hand and body oils on the part surface, if the product was poorly handled, if fans blew dust and fumes around, and mold release was used, even in the next molding cell, as air movements could carry the spray over to your machine. This list includes just a few items to be considered before the vendor is selected.

Get references, talk with their customers, and perform an audit of the molder's facility to learn how it performs quality and molding operations. Also, examine whether the molder can assist you in part and/or product design reviews and mold design. Selecting a custom molder is like adopting a partner into your business plan, as it will be an integral partner in your business.

Evaluation begins by using the Supplier Evaluation Survey for becoming an approved supplier (see Appendix D). The audit should provide sufficient information on the supplier, along with references, that will allow you to select the best custom molder for the quoted price. One of your best sources of information will be one-on-one talks with your audit team and the supplier's personnel. These talks will provide a good understanding of how it will perform as your custom molder and how it will share knowledge to assist in product manufacturing.

It is important to request quotes from several custom molders, compare the information from the other interviews, and audit the results. Be sure you get similar information from each molder, and if the information is different, ask why the variation. The audit results from well-recommended custom molders should be almost identical when discussing the product and mold design; slight variances may occur based on the molder's experience and knowledge, which is to be expected.

When a contract is awarded, there are legal agreements that include product price, delivery, procedures on how disagreements are to be settled, penalties, and so on. It is best if these elements are reviewed and a determination is made whether they should be included in the contract. Discuss this with your team and determine what elements are necessary.

CUSTOMER AND SUPPLIER AGREEMENTS

The customer discussion should include the following tasks:

1. Ensure the supplier is aware of all quality regulations, tolerances, and specification.
2. Schedule specific discussion meetings during the course of development.
3. Provide engineering and technical experience to the supplier as needed.
4. Agree on the mold design, material, and source of tooling.
5. Place orders within the supplier's capability to supply.
6. Agree to quality requirements and document this information.
7. Agree to the method for handling nonconformance parts and solutions.
8. Determine packing, shipping, and payment methods.
9. Acknowledge and respond with solutions if mistakes are made.
10. Determine how nonagreements can and will be arbitrated.
11. Maintain a professional business relationship at all times.

The supplier discussion should include the following tasks:

1. Provide design and mold assistance at an agreed-on rate.
2. Meet customers needs and requirements for the quality system.
3. Exhibit a quick reaction time for problem solutions and inventory reserve.
4. Provide 30 to 60 days notice of price increases for material and items.
5. Produce the product using equipment similar to startup samples provided.

6. Keep efficiency and quality improvement as customer benefits.
7. Expand accordingly to meet customer product requirements.
8. Provide technical assistance for current and new programs.
9. Maintain strong communication links between companies.
10. Promote the business relationship as a lasting partnership.

These are extra items that can be considered in or as an addendum to the contract or just discussed during negotiations of the business contract.

VENDOR CLINICS

Major companies now hold vendor clinics and invite their key, current, and possible new suppliers to attend a 1-day affair of gathering and interchanging information. It is an excellent time to discuss with their suppliers what the current and future product demands are and what their plans are for new items. Plant tours and scheduled presentations with time for meetings of existing vendors with their counterparts are recommended to exchange information.

These visits often result in cost savings and product improvements. Usually, this communication exchange improves the business climate but also leads the supplier to understand what, where, and how its parts are used in the customer's products and marketplace. The suppliers may be able to offer cost savings with part and function combination to reduce product costs.

The tour can show the reasons for the quality requirements and possible ways the quality can be improved to provide a greater benefit to the customer. The customer and supplier are a team in producing a product at a reasonable cost so that each can profit and grow its business. Each must meet the other's criteria to do business together, and the common binder will be the production of a quality product to meet the requirements of the marketplace.

PRODUCT REQUIREMENTS

The requirements of the product are determined by its end-use application. The design and development checklist has questions that will determine the product's capability and whether structural analysis is required based on the loading, frequency of use, environmental considerations, and other requirements it might be exposed to during its service life.

Also be sure there is not an agency or industry requirement the product must meet that has an influence on the material selection. If in doubt, contact the agency and discuss the application to obtain its requirements before proceeding. The intent is to eliminate supplier liability and any chance the product will not meet its intended requirements when put into service. Some of us can relate to a situation where a little more investigation could have avoided a recall or costly lawsuit.

Be sure the part requirements are realistic and can be achieved. These expectations may be in the area of cost, delivery, tooling tolerances, material, code and agency requirements, or part performance. The plastic part supplier should help find answers to its questions before part design and production are initiated. The supplier is also responsible for any product design and material recommendations it provides that could result in legal action. Product liability is the responsibility of all involved parties, but this should not preclude supplier assistance in end-use testing and market trials to prove the design, product quality, and performance. Obtain a waiver of responsibility if this is a concern.

Many customers are relying more on their custom molders who are exposed to more and varied products and can provide design and suitable material recommendations. Likewise, companies with in-house molding operations have come to rely on their material suppliers for knowledge in areas of tool/mold design, mold cavity, material tolerance capability, cycle times for estimating part price, and assembly methods and decoration of the product. Customers are forming a partnership of expertise and knowledge with their suppliers.

Supplier and department interaction and support are needed for a successful program. Many customers are leaving the final fine tuning of the part with their suppliers who have the software, mold fill, and mold cool that can assist in determining the parts section thicknesses for flow and avoidance of weld lines, which are potential weak points.

PRODUCT PREPRODUCTION REVIEW

When the product design review is held, either in-house or with a supplier, the following checklist information should be completed. Before the meeting, all preliminary information should be shared with the review team to obtain the maximum benefit during the review. Be sure a copy of the contract is available to ensure nothing is forgotten. Any other discussion item can be added to the review meeting. The intent is to be sure all information is presented and each party, supplier, and customer is in agreement.

Contract Checklist

The checklist is a very useful tool for the prebid and after contract review for ensuring all business, contractual, statement of work, and manufacturing information is considered, provided, and reviewed as an important part of the premanufacturing agreement. This review process should be used when preparing a bid for a customer, when responding to a request for quote, and after an award is made to ensure all areas have been discussed, questions have been answered, information has been made available, and the business agreement, price, delivery, and quality are all understood and attainable within the terms of the contract and/or purchase order. Any other discussion item can be added to the review meeting. The intent is to be sure all information is presented and that each party, supplier, and customer is in agreement.

5

Organization Responsibilities

Senior management is in charge of the total quality process control (TQPC) system. They provide the direction and assets that drive the quality program. The communication of recognizable benefits is key to the program. Management must ensure all personnel adopt the program ideals and follow its procedures and values.

The plan for establishing a TQPC program begins with understanding the business and manufacturing responsibilities. Personnel duties are determined, and a progression of steps are followed to ensure the correct information is available and used by all involved departments. How well information flows to the responsible personnel in an organization is a statement of the quality health of the company.

All companies should have a quality manual that outlines the responsibilities and requirements to be performed and documented in the business and manufacturing process. Procedures are then developed with ISO 9001:2000, and the following six procedures are required:

1. Control of documents
2. Control of records
3. Planning, conducting, and reporting internal audits
4. Control of nonconforming product
5. Corrective action
6. Preventative action

Total Quality Process Control for Injection Molding, by M. Joseph Gordon

Two additional procedures that may be considered to ensure additional quality operations may include the following:

1. Management review
2. Supplier selection

Instructions are then developed if any operations require subsequent in-depth information to ensure they are performed correctly. These instructions could include special operations performed as an assembly operation, job-specific instructions, or a filing directory for the configuration management systems documents and/or records. An example of such are the following:

1. List of quality records and what should be recorded during operations
2. Storage of quality records instruction and ease of finding the information

Organization and flow diagrams will be helpful to explain 1) how the flow of information moves through the company and 2) the responsibilities to perform specific operations.

It is important that the quality manager reports directly to senior management and the president/CEO, not the plant or production manager. This structure allows the quality manager to demonstrate an unbiased responsibility for product approval, to give guidance and support for improving quality operations, and to ensure no nonconformance product is shipped to a customer. Figure 5.1 shows the functional relationships for business operations in the flow and control of quality operations.

Based on these operations performed by the designated personnel, a total quality process control flow diagram of operations through a company is presented in Figure 5.2.

The typical business operation begins with the request for quote (RFQ) and concludes with the shipping and billing of the product. In some situations, as with an original equipment manufacturer (OEM), service support may be required. Service support depends on the product and customer, and if a problem develops, the supplier may be obligated to provide support to solve the problem.

It is important that all operations required for the program are identified and mapped out, which include the requirements of equipment, labor, procedures/instructions for the manufacturing operation, training if necessary, and methods to verify that the required results are obtained for each operation. Records must be kept with the correct forms, and the means to generate and record their input into the system are determined for each operation. All sizes of in-house and custom molders, even with only one or two machines, must document the manufacturing data for control of the process.

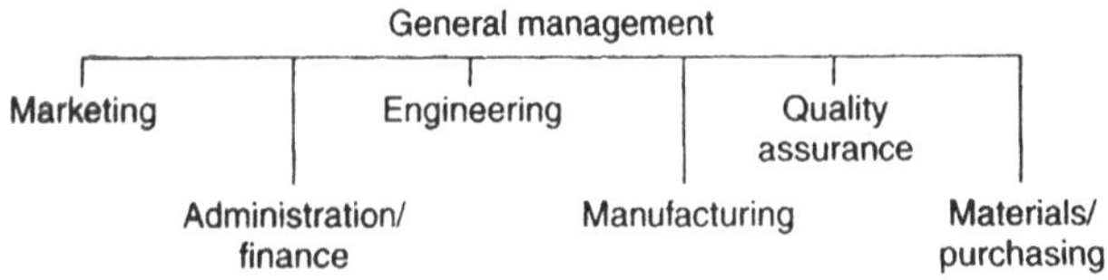

Department responsible for quality:

GM = General management
MK = Marketing
AF = Administration/finance
EN = Engineering
MF = Manufacturing
QA = Quality assurance
MP = Materials/purchasing

Responsibility code:

R = Responsible
M = Must contribute
O = May contribute
I = Informed

Area of responsibility:	GM	MK	AF	EN	MF	QA	MP
1. Establish business quality objectives	R	M		M	M	M	
2. Implement quality process control	I	I		M	R	M	
3. Plan quality systems	R	M		M	M	R	M
4. Determine customer requirements		R					
5. Establish product design specifications		M		R	O	O	
6. Establish manufacturing process control				M	R	M	M
7. Determine process capabilities				M	R	M	O
8. Quality vendors			M	I	I	M	R
9. Plan inspection and test procedures		O		M	M	R	M
10. Design inspection and test equipment				M	M		
11. Tooling reviews and purchases	O	I	I	R	M	M	M
12. Schedule production		I		R	M		
13. Material evaluation in-house						R	M
14. Produce products to specifications	I	I		M	R	R	I
15. Quality feedback information	I			M	R	R	I
16. In-process quality audits	I				M	R	
17. Gather quality cost datas	I	M	R	M	M	M	M
18. Analyze quality cost information			M		M	R	M
19. Final product quality	I	I		M	R	R	
20. Gather customer complaint data	I	R		M	I	M	
21. Analyze customer complaint data		M		M	M	R	M
22. Corrective action		M		O	R	R	O
23. Customer satisfaction	I	M		M	R	R	O

FIGURE 5.1. *Company departmental organization chart and responsibilities. (Adapted from Ref. [1].)*

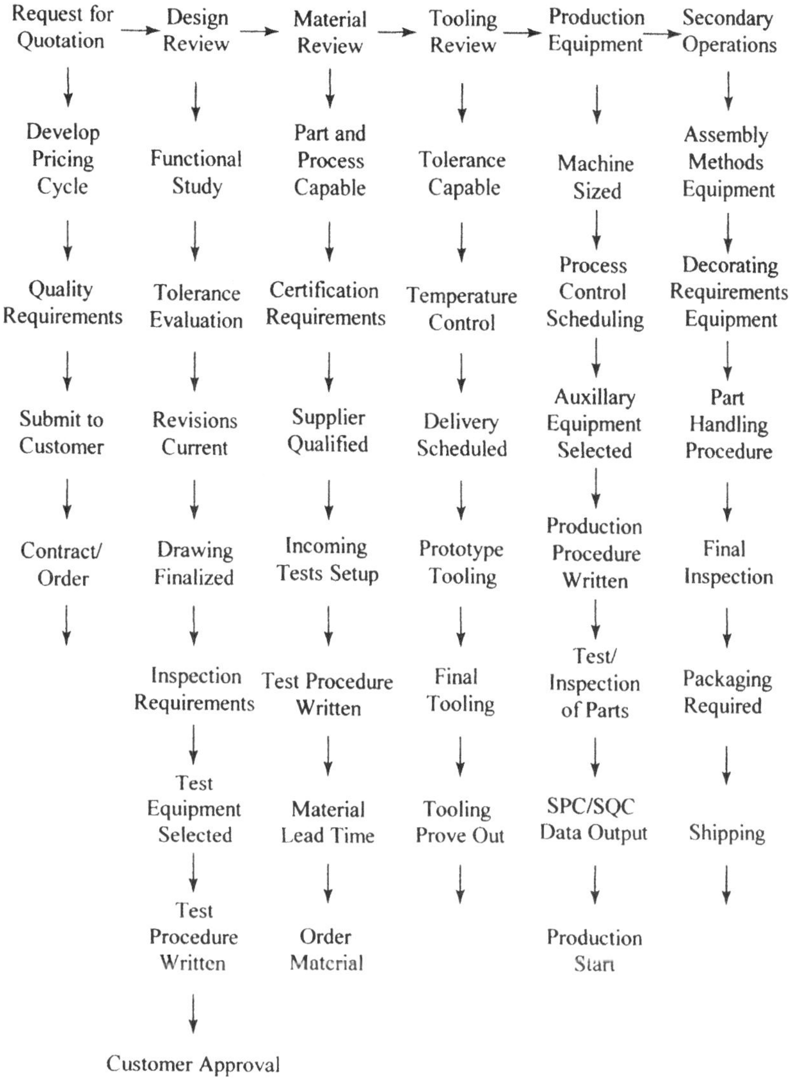

FIGURE 5.2. *Total quality process control flow diagram.*

QUALITY OPERATIONS

Quality is involved in all business and manufacturing operations to the degree required to ensure only a quality product is produced. Quality managers will audit the configuration management system (CMS) section for each department to ensure their employees are following procedures, and the documents and records of the operations are completed and filed in the system.

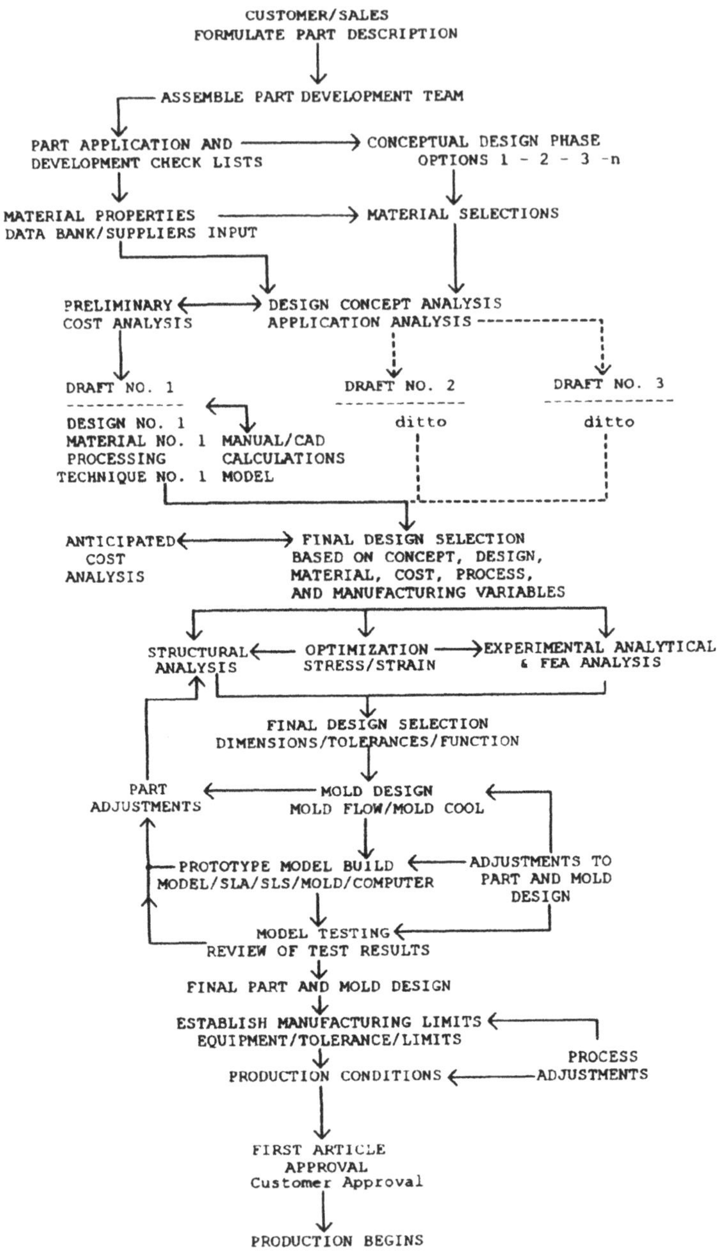

FIGURE 5.3. *Application development flow chart.*

Documentation and records must be uploaded to the company CMS for permanent filing, control, maintenance, and protection. The length of time these documents and records are retained is established by senior management and documented in its respective quality system procedure.

QUALITY UNIFORMITY

Once the company's quality operation is established, it is assumed their quality of manufacture will never vary, only the requirements of individual products. Some customers may require more stringent requirements or inspections than another customers. No lessening of company quality standards is allowed for any product. Some products may only require an aesthetic inspection (e.g., for color, no flash, etc.), whereas other parts may require two to three measurements per part. The quality system is established to give each customer the same degree of manufacturing quality. Too often, when a part has fewer specified quality requirements, a form of apathy develops and the lower part quality requirement means that anything is acceptable, when it is not. Never forget that customer satisfaction is the goal for quality and product acceptance. The quality system is established to ensure all customers receive the quality necessary for the product to meet their product requirements. Quality provided can be more, but never less!

Figure 5.3 illustrates a successful way to bring a program to market. It is very important to review all programs for the quality requirements that are specified. To fail to recognize an important quality inspection point could cause the product to exhibit poor performance. Quality is an everyday operation.

COMPLIANCE AUDITS

Mention the word "audit," and most employees think the worst. Audits perform a necessary function, telling employees whether they are performing operations right and whether improvement is needed. Audits are fact-finding missions to assist personnel in doing their job correctly. A good definition of aduit is, "determining conformance to an instruction or standard."

Audits should be considered as learning events. Finding fault is not the intent of an audit. Audits are used to verify if you are actually doing what you say you will do. Checks and balances are used to evaluate an operation and to verify it is proceeding as developed and the final results are positive. A good audit will tell you what you are doing right and where extra effort is needed to meet requirements. Train your personnel to follow the procedures, instructions, and guidelines for performing operations. Also, be sure the records of this work are kept for reference and audit verification.

SIX SIGMA INTRODUCTION

Bill Smith of Motorola Corporation developed Six Sigma in the mid-1980s to improve processes systematically by eliminating defects by using the existing quality methodologies. This resulted in producing major cost savings within the operating systems of the company. Cost reductions and process improvements with recognizable savings of at least $175,000 were identified. Although a time of learning was involved and not all programs were immediately successful, these programs eventually yielded the required results of savings and increased productivity. Therefore, Motorola focused considerable time, personnel, and resources into this money-saving program.

Six Sigma programs were to reduce process variation and reduce defects to 3.4 defects per one million parts produced. It did not address customer wants and needs. In any good quality program, the focus must be on both the customer and how well the supplier, who furnishes the products and the customer who uses them, interact with each other to obtain positive mutual results.

Walter A. Shewhart of Bell Labs developed the continuous improvement program called Plan, Do, Check, and Act (PDCA), which was later changed by Dr. Edward Deming to Plan, Do, Study, and Act (PDSA) (see Figure 5.4). This program followed his own method of quality analysis and is an excellent guide to follow for continuous improvement. Because a circle has no end, the PDSA cycle is repeated again and again for continuous improvement. Planning always precedes doing.

The PDSA cycle should occur during the following times:

- When beginning a new improvement program
- When developing an improved or new design for a process, product, or service
- When analyzing and defining a repetitive process

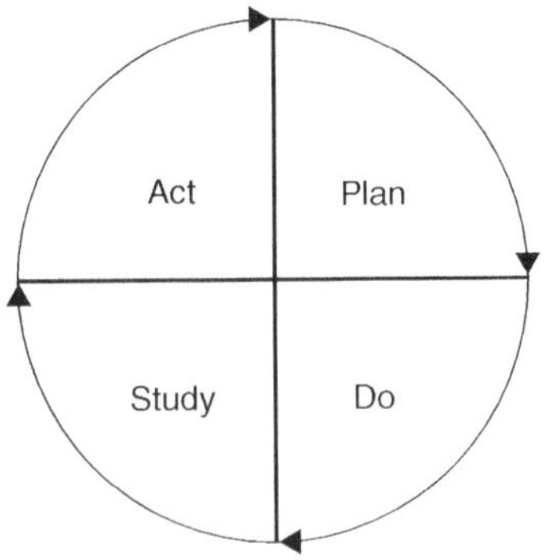

FIGURE 5.4. *Use for a continuous business and/or manufacturing improvement model.*

- When planning data collection and analysis to verify and prioritize problems or root causes of problems
- When implementing all types of change

PROCEDURE

1. Plan. Recognize an opportunity and plan a change.
2. Do. Test the change. Carry out a small-scale study operation.
3. Study. Review the test results, analyze, classify, and identify what you have learned.
4. Act. Take action based on results from the study step. If the change did not work, repeat the cycle with a different plan. If successful, incorporate what you learned from the test into wider changes. Use what you learned to plan additional improvements by starting the cycle over again.

Continuous improvements:

Another acronym of Six Sigma is define, measure, analyze, improve, and control (DMAIC). This procedure is best used for explaining the process for establishing a process control procedure and/or instruction.

The procedure for establishing TQPC are defined as the DMAIC operations, which are explained as follows:

1. Define what is required or desired from the process.
2. Measure the results of the process variables.
3. Analyze the results and determine whether the controls used will produce desired results.
4. Improve if the process needs tighter control or other items required.
5. Control the system and reevaluate the process for conformance.

It is a good idea to reevaluate an existing process to ensure the instructions are correct in describing all the operations necessary for completing the process satisfactorily. In some cases the process is okay, but the parts are not in specification for the operation under investigation.

Measuring the results of the process includes evaluating the output and testing the parts. When physical measurements are taken, the operator is trained in the correct manner to perform the measuring. Often the training step is forgotten, and it is assumed the operator knows how to perform a measurement but actually is doing it incorrectly. In this case, the results may not be in agreement with the customer's measurements.

The data are then analyzed for compliance, and the operator notes any trend that may show the process not remaining in control. An analysis may result in needing additional data, making a required change, and retraining the operator.

If improvements are necessary, they are analyzed and are implemented if found to be adequate. Just be sure before a change is made the analysis truly requires the change and is not a normal variance in the process.

The last stage is to establish control of the process to the new requirements and monitor the results. During the evaluation, be sure the process has reached equilibrium before new data are taken. Any process with temperature as a variable must be allowed to attain equilibrium before new data are taken for evaluation. When possible, be sure that temperature is the last variable to be investigated, as more time is required for the system to reach equilibrium after a temperature change is made.

Personnel new to quality and even experienced quality personnel and their counterparts in design and manufacturing departments need a little guidance and training before they begin using new and even older quality methods. Training should always precede the use of new methods.

Just as rebooting your computer often cures 95 percent of software problems, reviewing your procedures and instructions for your operations is an excellent way to begin your quality analysis. Even when a process is deemed good, it could probably be improved. A simple idea of preheating a mold prior to a cold startup can save valuable molding time. But be sure to insulate the mold from the machine's platens so the benefit is not quickly lost because of heat transfer to the colder platen.

QUALITY PROBLEMS

Essentially five types of quality problems can be improved or eliminated by the use of preventative action. These problems occur with unsatisfactory performance and when developing new products and processes. They are shown in Table 5.1.

Often, these problems are difficult to categorize. Solutions are sought, but the results may not be as anticipated. Problem identification is just as important as problem solution. In today's fast-paced supply and demand, an incorrectly diagnosed problem is as serious as the problem. Time wasted in researching out a solution is time lost. Therefore, be sure the problem is understood and correctly analyzed before a solution is sought. Not all problems may fall into these categories, but most will and these solution techniques will make preventative problem actions easier and able to develop a lasting solution.

Too often, the older quality methods are forgotten, and when needed for solutions, these methods have to be relearned. These methods are not all bad because they will now be added into the inventory of problem-solving tools that more personnel have used to their advantage. Keep the list of quality tools handy as presented earlier in Table 1.1.

TABLE 5.1. Types of Quality Problems.

Problem type	Defining characteristics	Key problem solving tasks	Strategies and techniques
Conformance problems	Unsatisfactory performance by a well-specified system; users not happy with solutions	Diagnosis; determining why the system is not performing as intended	Use statistical process control to identify problems cause and effect diagram to diagnose causes
Unstructured performance problems	Unsatisfactory performance by a poorly specified system	Setting performance goals; diagnosis; generating viable solution alternatives	Diagnostic methods; use incentives to inspire improvement; develop expertise; add structure appropriately
Efficiency problems	Unsatisfactory performance from the stand-system owners and operators	Setting performance goals; localizing inefficiencies; devising cost-effective solution alternatives	Use employees to identify problems; eliminate point of unnecessary activities; reduce input costs, errors and variety
Product design problems	Devising new products that satisfy user needs	Determining user requirements; generating new product concepts and developing them into viable products	Quality function deployment translates user needs to product characteristics. Value analysis and "design for" methods support design activity
Process design Problems	Devising new processes or substantially revising existing processes	Problem definition, including requirements determination; generating and developing new process alternatives	Use flowcharts to represent processes, process analysis to improve existing processes, new processes, reengineering to devise new processes and processes from others; use benchmarking to adapt

TQPC MANAGEMENT OPERATIONS

Managers of a company know how the company operates and what is expected from their departments in the manufacturing operation. If the operations produce a problem, it is carried into the next operation and subsequently will cause the product to fail or not pass final inspection. Therefore, methods have been developed to eliminate these occurrences. In batch types of manufacture, this can be a serious problem as many parts may be judged as nonconforming. This can stop an assembly line and cause serious problems throughout the organization. Therefore, this type of problem must be eliminated, and contingency plans must be developed if it occurs. All planning at this step of the operation is valuable, as all personnel are involved and a solution is developed to prevent it from happening.

The use of checklists and pre-kickoff project meetings with involved personnel discussing their participation in the program and sharing information is required to ensure that a key item in the program does not get overlooked or not planned. Management should have a similar review meeting prior to submitting their bid to the prospective customer.

The space shuttle program has a one-million-item checklist that must be completed before each launch. Even then, an item could be overlooked or not investigated sufficiently, and a problem could result in a disaster. Avoid disasters and use the program and quality tools available to ensure the program will be a success.

The following 17 operations for TQPC will assist in the evaluation of process and quality programs:

1. Organization and management policies structure
   Provide an organizational chart for program workflow through the company, department signoffs, reviews of specific contract requirements, and general required operations. These are the standard operations that begin the program, which indicate who does what and where responsibility resides until an item or operation is completed. Personnel need to know who to consult if an item is not completed and where a problem could occur or be originated.

   If special forms or operations are necessary to convey information, make sure they are identified with an example showing the work needed and routing to the next operation. This could be you "Manufacturing General Workflow" instruction for each job if different from normal workflow in the company. This is the workflow that all operations follow and referenced to your job manufacturing operation sheets (MOSs). The MOSs are specific operator workflow instructions for how each job is performed.
2. Specification review and design assurance
   A preproposal meeting should be held prior to making a bid. This meeting will discuss the program requirements, specifications, and

quality, plus all aspects of the bid so no information is overlooked. If any questions need answering, now is the time to do it. Make sure the bid covers all possible contingencies, and if a change is inserted, each party must approve it.

Any change after quote submission should be negotiable if a major increase or decrease in the bid quote should result. Schedule reviews for the type of tooling, who provides the funding, and other items that may or may not be included in the bid package but are deemed necessary to complete the program. Now is the time to take any exceptions, and each party should agree to how any changes to delivery and purchase orders will be handled and now disputes will be negotiated.

3. Design assistance
   When product design assistance is requested, there should be an agreement, contract, or written development plan outlining the total participation of the supplier as requested for the product's design and possible testing. This may include a full-size model or a molded prototype. The supplier's liability issue must also be considered when design assistance is contracted.

4. Manufacturing planning and controls
   Effective control and scheduling of a company's assets, machines, auxiliary equipment, and personnel is necessary to plan adequately for manufacture. Accepting an order and not having the resources to produce the product is prohibited. The injection molding machine and system should be evaluated for their capability OR Cp index, which is a process capability index of the ratio of the tolerance (specification or permitted amount of variation) to the process variation. A value of 1 indicates that the process variation exactly equals the tolerance for maintaining continuous process control output over the run time of a program. Machines should be evaluated at scheduled periods with different molds, to ensure they are as capable as possible for the existing job and future work.

   Each program should have individual setup instructions the technicians can follow to ensure all operations have been completed prior to startup. The instructions should require the following: the plastic resin is dried to a specified moisture level, mold at required operating temperature, the resin in the barrel is operating at a specified temperature, and all required items (pyrometer, scale, operator gloves, etc.) are available at the molding cell.

   The test and inspection equipment, as well as all instructions, must also be established, and the operator must be trained in how to use the equipment and perform the inspections. The operator must have the forms or access codes to the computer system to record the data and any other information required for the program.

5. Process control established
   Process control during the manufacturing operation is necessary to ensure the manufacturing variables in the process will stay in control to produce an acceptable product. Depending on the equipment, this may be an automatic process with the molding machine's computer using a closed-loop continuous feedback control system or one of the operators controling the process and recording the machine and process data during the production run. Automatic documentation of data is preferred, as there is less chance of an operator error and it will be recorded as programmed.

   The goal is to use real-time data to control the manufacturing process. These data are used in the "closed-loop continuous feedback" fashion for the injection molding machine control system. Limits are entered for the variables, and as the machine operates, any out-of-tolerance controls are corrected and the system is brought back into control. If the control system or a variable goes over the limit and is not correctable, the system can shut down or signal an alarm for operator assistance to correct the problem. In these cases, there is a more serious problem occurring, and an operator is needed to make adjustments and find a solution. A good process control system will allow the process to have its natural variation during the molding process. Too tight a control will only result in the process never being in control and in adjustments continually being made that are not necessary. This is shown in Figure 5.5, in which the closed-loop control system and/or operator holds the process control in the control limit range and no further than the process limit range for the process.
6. Measurement and test equipment
   All inspection and test equipment must be calibrated to a known standard at least once a year. Some equipment must be calibrated more frequently as required by the manufacturer. A list of equipment calibra-

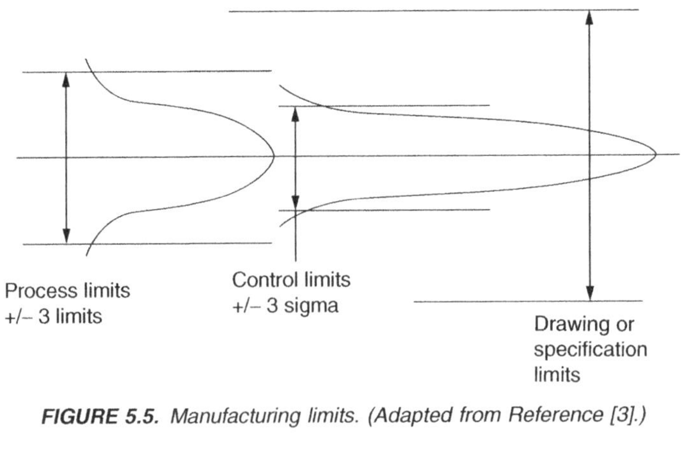

FIGURE 5.5. *Manufacturing limits. (Adapted from Reference [3].)*

tion requirements and the due dates must be maintained to ensure only correct data are generated when the equipment is used. Any time an item is suspected of not giving correct readings, it should be evaluated against a known item in calibration. If out of calibration, it should immediately be sent out for calibration at a certifiable calibration laboratory. Also, any product inspected with this item must be quarantined, and material must be separated until it can be certified as correct using an in-calibration piece of test equipment.

It is important that anyone using the inspection equipment, especially mechanical measuring equipment, is certified or at the least trained in the correct way to take a measurement.

The results of all measurements taken during the production run must be documented as a record of the operation. These records must be kept to show the process and product control during manufacture. It is very easy to train the machine operator to take measurements during manufacture, especially part weight, as this is a good control of the molding process repeatability should closed-loop continuous feedback control not be available. The data should also be taken continuously or at specified intervals to prove control and product variables are within specifications. If in a multicavity mold, a single-cavity part weight is recorded. An alternating cavity number should be selected & recorded to ensure there is even gate freeze off time between cavities to show the process is in control.

7. Maintenance of equipment

Maintaining quality during manufacture is dependent on control of many variables, such as equipment, plant systems, and material. A dirty machine barrel and screw, clogged filters, out-of-calibration controls, wet material from desiccant contamination, and so on will cause processing problems.

Using a checklist, the operator can verify whether all items are ready and operations are completed with variables within tolerance, the filters are clean, and the molding machine can provide the services required for production. Once the process variables are verified within specifications, production can begin, and essentially only good parts should be produced. Some systems have replacement parts, filters, relays, instruments, and so on that wear out over time. The upkeep and replacement of these items is the responsibility of the maintenance department and is not a result of a process or material problem. Keep these items separate and corrected accordingly. Keep a maintenance log for each machine that lists when each item of equipment was replaced or when the last maintenance was performed in the equipment's log book at the machine, or in your CMS for each respective machine. Then, when a job is to be run, the equipment is selected and a review can easily be conducted to ensure the machine is available and maintenance is completed as required.

8. First article inspection
   First article inspection typically involves an approval from the customer who reviews the product and manufacturing data to verify the product was made to specifications and the follow-on product will also conform to the same requirements.
   When required, all related documents, measurements, retained samples, and data are presented for comparison and checking of part production. Sufficient notification also should be provided to allow this review either at the manufacturer's or the customer's plant. Above all, first article inspection should establish the production standards to be maintained to meet customer quality requirements. Often, the first acceptable part is saved to compare with and the last part at the end of production as a validation TQPC control was maintained during manufacture.
9. Consigned material
   The supplier is responsible for all parts and/or materials furnished by the customer. This includes ensuring the items are correct and not damaged on receipt. They are then placed into storage and inventory control. The supplier absorbs any mishandling of consigned material because of negligence. The responsibility for the quality of incoming consigned material must be negotiated at the time of contract.
10. Supplier purchased material control
   A record system is established and implemented to ensure that purchased parts for the product, used in manufacture or assembly, will meet the purchase order, drawing, and quality specifications. This should be documented and verified through audit inspections, certification, and/or quality in records supplied with purchased items. Incoming material tests, as deemed necessary to control quality of product, should be performed. Some plastic materials will experience melt flow property variances caused by lot changes of molecular weight. This can cause both a manufacture and end-use property problem if not tightly controlled. If this is the primary material of manufacture, then this value should be determined as a quality material specification and only resins within a specified molecular weight range should be provided and accepted from the material supplier. Impact testing can be used for some materials to show whether melt viscosity has been lowered to avoid overheating during manufacture. The Sharpy or falling dart drop test are possible tests that can be conducted.
11. Handling, preserving, packaging, and shipping
   To ensure quality protection of product after manufacture, the supplier should use specific instructions, materials, and preservation methods to guarantee the finished product arrives at the customer's final destination within product requirements and contract delivery schedules. Show surfaces must be kept clean and free of scratches. Negotiations

with the customer for this degree of protection are necessary as it may require an operator or robot at the press to collect the parts from the mold and place them in a protected area to cool prior to final packing for shipment. Critical parts are often shipped on special dunnage racks that hold the part securely and protect it during shipment. Once emptied, the racks are returned and reused to save packing costs and materials. This can also eliminate the cardboard dust problem if parts are to be surface decorated. A small speck of dust on a metalized automobile headlamp reflector shows up like a large grain of sand after metalizing.

Some parts may require postmolding conditioning, such as a nylon zipper that must be moisture conditioned, to ensure it will meet a later assembly requirement and not fail as a dry as molded part could.

12. Employee training, motivation, education, and certification
Senior management is responsible for ensuring that personnel are trained and can control the equipment in their area of operation. They must be aware of the maintenance requirements and able to perform these tasks as required. Any certification required for special operations must be maintained, and a program must be provided to ensure that employees have the motivation and education to continue adherence for improving the quality of the product. When possible, in-house training programs, specifically the quality area, should be implemented to assist newer employees to advance in knowledge, education, and responsibility. All training must be documented.

13. Control of nonconformance
Any nonconformance is not allowed to continue and must be corrected as soon as possible as a corrective action. Any nonconformance is immediately segregated and held for future determination or committed to regrind in the molding operation, if regrind is allowed in the part. Regrind is only allowed if the customer agrees and the amount will not jeopardize the part's physical properties, dimensions, and appearance. Regrind use should be determined, and if allowed in the part, the allowable percentage of regrind to be used should be indicated.

Should a noncompliance be reported by a customer, a corrective action response is usually requested to determine the source of the problem and action taken to eliminate it in the future. When corrective action is required, always do the best job to locate the root cause of the problem. If not in your area of control, go to the source and have them make the necessary corrections. Verify the action taken has actually solved the problem.

14. Documents and records
A CMS can be established that will be the control and storage center for all documentation and records. Separate files will be established for each customer, with subsections for each program. The company will

also use the CMS as a repository for its quality system documentation. An index will be established that is representative of the business operating to file, store, preserve, protect, and locate and retrieve documents and records. The quality manager has the prime responsibility for the CMS and shall select a document controller for the day-to-day operation of the system.

The CMS will be a real-time system with computer access at the machine side for the recording of information and molding data. Select personnel will have access to the system with a full manufacturing, inventory, and shipping status of all programs.

15. Price of quality
A price is paid for your quality system, typically as part of overhead, in the product's piece part price. Quality is not a liability but an asset that will grow in value as the company grows. Senior management will designate specific quality and business objectives that will be met on an annual basis. Quality personnel will monitor these objectives for the areas they are involved and issue an annual report.

16. Corrective action
When a nonconformance occurs, a corrective action response is necessary. The source of the nonconformance must be determined so it will not happen again. Each corrective action requires a separate form and response. Each nonconformance is documented and assigned to an employee with the capability to determine the cause and source of the problem, as well as to assist in recommending a permanent solution through the use of corrective action.

An example of a corrective action might be one of the following: material contamination; color not correct, if a salt-and-pepper blend; or any problem even a dirty screw and/or barrel causing contamination of the product or material missidentified; or the material handler used wrong lot number. These problems should be listed, investigated and corrected, and a memorandum must be issued to the plant personnel as to how it happened.

A standardized customer corrective action form should be used to ensure all the available information is collected. If the action involved a part, be sure representative samples are returned, hopefully with a lot number and date of manufacture for subsequent problem analysis. Required information is as follows on the corrective action checklist:

Customer Corrective Action Information List

1. Customer/location/contact/phone/e-mail address
2. Product number/identification/description
   a. Cavity number, on part
   b. Purchase order for part/date/lot number

   c. Time part in customer inventory
   d. Any customer secondary operations performed on part
   e. Part of an assembly
   f. Method of assembly/forces/tool settings/etc.
3. Lot number/inspection number
4. Type of defect/failure/complaint
5. Seriousness
6. Application/use conditions/temperature/humidity/chemicals
7. Environmental history and installation date
8. Unusual circumstances
9. Percent defective
10. Whether this is first time or repeat problem
11. Possible root cause
12. Appropriate internal and external contacts
13. Obtain/send both failed and current inventory samples, minimum of 3
14. Investigator/date

PREVENTIVE ACTION

Preventative action is used to solve suspect and potential problems in operations. Preventative actions may result from an audit or an employee suggesting the review of a process with the potential to cause a future problem. A failure mode effects analysis (FMEA) is a preventative action diagnostic that explores an operation for potential problems. Once identified, the fishbone analysis can be used to determine the root cause, and then a preventive action response is initiated to eliminate the potential problem.

Preventive actions should be documented and kept as a record for subsequent investigation, should it occur again. Being proactive in searching out problem areas is an important function of the plant quality and maintenance personnel in all areas of the business. Informing customers that this is one of the key objectives for improving product quality will be an asset.

17. Quality audits of systems
    Management must be kept informed of the status of their business and manufacturing operations on a scheduled time frame. An audit will show whether the staff and operations personnel are maintaining compliance. Audits should be performed on all operations. Business practices often become lax, and the same occurs on the manufacturing floor. Some operations, which are deemed as not necessary, are forgotten. Maintenance is often the first item to go when schedules are tight and

the operations are in danger of getting behind. To get back into the flow, there may have to be more innovative scheduling and planning performed. Unfortunately, when audits reveal that not all operations are being performed, or are done incorrectly, it has already cost the company valuable production time.

Audits are a valuable tool for finding areas that are not getting the required attention. Charting problem area defects without searching out their root cause is a waste of valuable time and manpower. Audits performed with well-planned and thought out checklists will prove their worth in developing information and in assisting in elimination of problem areas. They will also show how well departments and manufacturing areas are performing.

Audits are a valuable tool of the TQPC system, and careful attention to the details and correctness of conducting an audit is required to obtain the maximum benefit. An excellent source of information on auditing and how to perform audits is available in a new publication available at www.theauditoronline.com.

6

Establishing the Limits for Quality Control

Everything manufactured has a set of specifications. Specifications are used to establish control, and control is used to produce a product repeatably, each and every time. We know that to make identical items, operations and manufacturing must stay within specified control and processing limits.

Therefore, if we stay within the limits established, the product should replicate its predecessor. Prior to the industrial revolution, craftsmen produced items individually, and there was little control other than form, fit, and function being attained. Therefore, as the demand for like products grew, control was necessary to produce the product the same way each time. Limits or specifications were established for mass-producing like products. To do this, it was necessary to determine what the limits must be and to then how to stay within the limits, which is the goal of total quality process control (TQPC). These three sets of limits are shown in Figure 6.1.

The tighter the processing limits, based on the specifications, the more costly the item, but the closer each item will be to the others. The product specification limits are established on what the product must do for the customer. Therefore, the customer establishes the product specification limits by defining its intended use.

The "control" limits are established by the capability of the process. The process limits are determined by the capability of the manufacturing equipment and personnel to replicate their operations each cycle of the manufacturing process. To define the control limits, we need the following:

Total Quality Process Control for Injection Molding, by M. Joseph Gordon

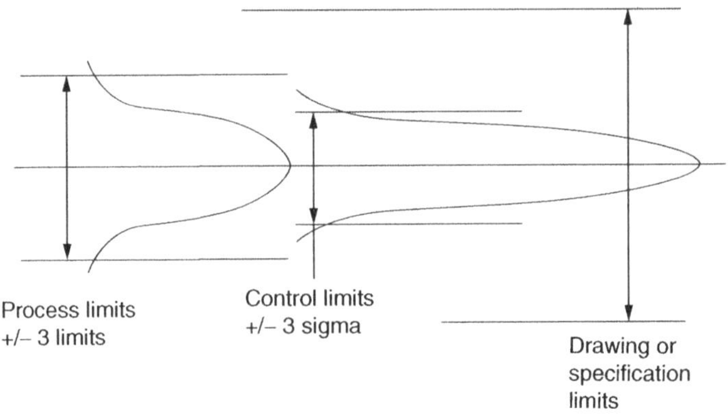

FIGURE 6.1. *Manufacturing limits. (Adapted from Ref. [1].)*

1. An ample history of the process to define the level of common cause variation
2. A basis for determining how to set the control limits to remain within the process limits of the system

To perform the product's end-use requirements, a specification is developed. It must be established so that as the variables that act on the manufacturing process change over time and conditions, the manufacturing process can produce the product within the specification established by the customer. This requires the manufacturer to back into the process and control limits based on their manufacturing system's capability. The variables that the supplier must consider during manufacture of the product are as follows:

1. Material variability, lot-to-lot
2. Mold cooling, cooling tower water, and chiller settings
3. Injection molding machine internal variables
   - Hydraulic oil temperature and viscosity, injection speed
   - Barrel, type of screw, and ring wear
   - Pump seals
   - Environment at molding machine
   - Controls, temperature, and timers
   - Heater bands, aging, insulation
   - Heat/cooling loss through platens for the mold
   - Melt generation capacity and capability
   - Other variables of manufacture (mold, feeders, regrind, etc.)
4. Moisture content of resin
5. Plant operations and secondary operations, etc.

These are just a few of the many manufacturing variables; the molder must consider more than 33 variables for the molding machines operation alone. When time permits, perform a fishbone analysis of one of your manufacturing operations, from start to finish. Plot how the work begins and then moves through the operation, who is involved, what they do, where they get their information, what information is needed, where is it documented, what happens next, and so on and on, until the product ships and billing occurs. When you have finished, you will have mapped the entire company's organizations, for just one job.

The manufacturing process begins with the mold manufacturer making some calculations, judgments, and knowledgeable decisions to ensure the product from the mold will be repeatable. There are specific areas of dimensional control that will assist in attaining the product requirements on each cycle.

Based on product requirements and established specifications, the supplier and mold designer begin by knowing what the finished drawings dimensions must be and establish a tolerance for the mold cavity(ies). The mold builder determines the mold cavity dimensions based on material flow and fill assumptions, knowledge, and material shrinkage. They also should use a mold checklist and software packages for estimating the best and average dimensions possible from the mold, material, and processing. Software packages for estimating requirements for mold fill (pressure and fill pattern) and mold cooling (temperature) are very helpful.

Each material has its basic material shrinkage rates as reported by the material supplier and the Society of the Plastics Industry, Inc. (SPI) in its "Standards and Practices of Plastic Molders" for generic plastic materials, for reinforced, filled, and nonreinforced. The cavity gate size must be calculated to remain open until all the mold cavities are packed out (to maximum part weight) and then must be freeze-off. This is necessary to eliminate cavity depressurization when the packing pressure is released and the screw retracts for the next cycle. This is a key variable for estimating minimum cycle time. When the gate freezes off, the screw retracks to build up the material in the barrel ahead of the screw for the next shot (cycle). Then, once the part has cooled sufficiently, it can be ejected from the mold cavity without distorting or warping.

This is just one of the reasons that only a few critical part dimensions should be held to the tightest molding tolerances and to the critical dimension; the others should be allowed to float within a specified tolerance range. Reinforced and even unreinforced materials have a differential shrinkage in the flow versus the transverse direction when the mold cavity is being filled. Trying to hold more than three part dimensions to a very tight tolerance is very difficult, one and two part dimensions at the most can be held to tight tolerances. The *Modern Plastics Encyclopedia*, which is published yearly, is a valuable reference book.

PREPRODUCTION PRODUCT ANALYSIS

Before a product is released for production, the production team should sign off on the product's design, the mold, the material, the supplier and all other items that will impact the quality and success of the parts program. Determining the number of mold cavities, gate size, runner layout, tolerances, cooling, flow in the mold cavity, product production limits, and material capability, and so on, all go together for a successful program. Gaining an agreement for establishing the manufacturing limits is necessary so all know what can be expected after production begins. To correct an in-process production problem on finalized tooling is very costly, as much as 25 to 30 percent of the product's development budget. The production team's responsibility does not end until the program is completed.

Then if a problem develops and a multiple of variables need to be evaluated in a short period of time, a simple method is available for evaluating specific variables in a planned orderly method for the establishment of a logical solution. This is Dr. Genichi Taguchi's "Design of Experiments" (DOE) method; an example is given in Appendix B. If an agreement cannot be reached on specific items and several options need to be evaluated, the DOE method should be employed.

A DOE can save countless hours of debate, many hours of labor, and considerable amounts of money. The DOE uses an established method for evaluating the key variables for the solution of the problem. By using Taguchi's techniques, these major variables are identified and evaluated against each other in an ordered experiment to determine the main controlling variables and their effect on the current problem. The Taguchi experiments are orthogonal matrix layouts that evaluate variable's high- and low-effect factors against the other variables in a very short period of time.

TAGUCHI METHODS

When multiple variables are involved, a series of trials could take days. Using DOE, the same number of variables may only take hours using a planned and ordered series of tests. The contributing factors are isolated, and the cause and effects are analyzed. The Taguchi method concentrates on tighter control to make a product that meets customer requirements. This is illustrated in Figure 6.2, which uses the premise that the design and manufacturing process should be insensitive to factors beyond and not under the direct control of the supplier. The Taguchi philosophy tries to make all products as close to target specifications as possible by identifying the major factors that cause product variation.

Through simple, noncomplex mathematics, these factors can be evaluated and their influence on the design, material, tooling, or the manufacturing process can be identified. To understand this even better, a course in Taguchi

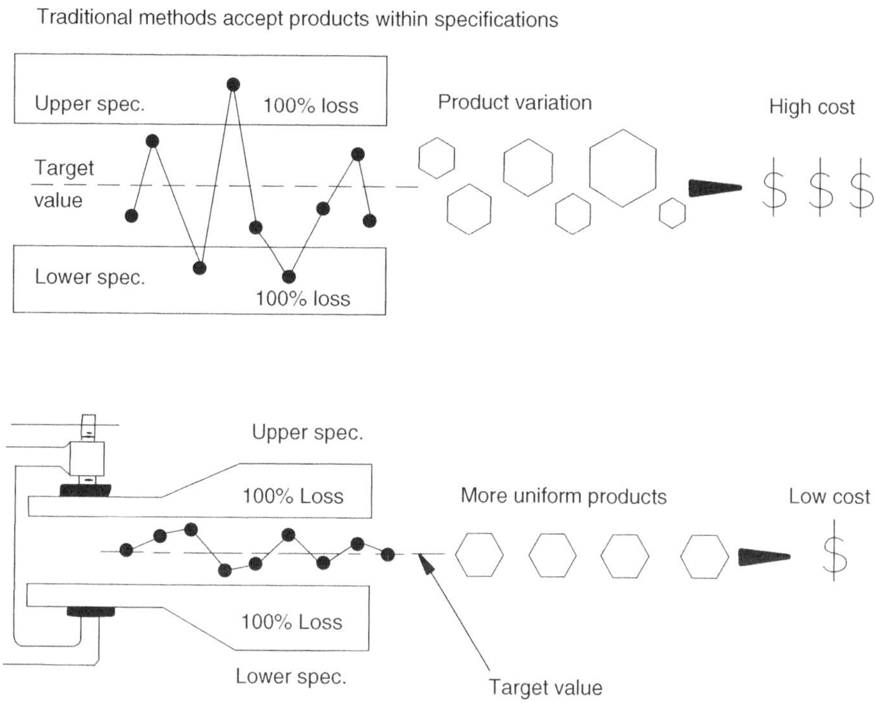

FIGURE 6.2. The Taguchi methods use tighter control to reduce variations between products. (Adapted from Ref. [2].)

DOE methods is recommended and additional information can be obtained from the American Supplier Institute, Incorporated, Dearborn, Michigan.

PROTOTYPING

Once the product is designed, the next step is usually to develop a model or prototype for testing and end-use evaluation, if possible. Most plastic products are prototyped either by being machined from bar stock, three-dimensional (3-D) modeling, or prototyped or preproduction manufacturing in a single mold cavity cut in the production mold base. If the program is big enough and a material has not been selected, then a rough-cut aluminum mold prototype cavity is often made and used to qualify several different materials for the product. Depending on the size of the part, a prototype mold base is used with a machined or cast aluminum or Kirksite mold cavity.

There are methods used to produce a 3-D plastic part in different materials, such as acrylonitrile butadiene styrene (ABS) and nylon using fusion deposition modeling. Although these parts are not as strong as an injection molded part, they can be used in various ways to evaluate the future product and how it may behave in service. Various types of prototyping methods are as follows:

Stereolithography
Fusion deposition modeling
Selective laser sintering
Composite injection molding
Kirksite injection molding
Vacuum cast tooling

Some of these prototype parts may be able to be tested or used for form, fit, and function analysis depending on the method and material. The only drawback from a machined part is that the flow of material in the machined part will not replicate a molded part that will be, typically, much stronger and tougher. Voids will also be a problem; even though the part seems solid, there is microporosity in the extruded bar stock material.

Prototype molded parts are very close to the finished article. There may be some minor variations, but for the most part, they can be tested as a finished part. The advantage is the evaluation of various materials in the same mold, even though the dimensions will not be exact because of differences in material and in-mold shrinkage, usually, no mold temperature control is in the prototype mold. Therefore, the decision to build a prototype molded part is valid, as it can be used to speed the development of parts and save money when bringing a new product to the market. Often, a semifinished mold cavity is used as a prototype cavity and put into a prototype mold base for evaluation. Prototyping is not inexpensive, but it has proven repeatedly that it is the best way to ensure the product will accomplish its end-use function. The areas prototyping can assist in are as follows:

1. Testing parts to prove the structural capability of a design, form, fit, and end-use function
2. Selecting material to meet the product requirements
3. Identifying potential molding problems of a design
4. Verifying shrinkage, obtainable tolerances, weld line problems, number and location of gates, and anticipated cycle tunes
5. Determining part wall thickness, warpage problems, part fill, and mold temperature requirements
6. Providing samples for market evaluation

Another prototype mold advantage is being able to mold the part in a clear material, such as acrylic, styrene, or polycarbonate. This will show where the high-stress points are in the part, when put under a load and viewed, using polarized glasses. All molded-in stress points will be visible, especially at the gate area and at any sharp corners. Material flow stress patterns will be visible, and the gate location can be evaluated for location and stress inducement. This is often never considered during the part's design design and only is considered if the part has a problem in these high-stressed areas. Weld line joining is also very visible and easily evaluated.

There are only a few disadvantages with using a prototype mold. Typically, there is very limited cavity temperature control. Its useful tool life is limited if made of a soft material, such as Kirksite or a soft aluminum. Usually no moving mold functions are included, so if an undercut must be pulled, a core has to be removed after being molded. Cycle times can be estimated, even without cooling, but parts may not be as well packed out as with the production tool, and dimensional considerations must be taken as very coarse if shrinkage values are trying to be developed. But, when all is considered, this is the least expensive part that will replicate the end-use item.

MOLD LIMITS

The mold used to make the part must be well designed and built to achieve the specifications established for the part. The mold should have a good temperature control system and hardened replaceable gate blocks if an abrasive material is used or a long run is anticipated. The draft must be adequate so the part releases easily on ejection. It must have a balanced cavity layout with a runner system correctly designed for the material and flow in the filling operation. A poorly designed and built mold may never achieve the desired product requirements. This is where the tooling money should be spent to achieve the desired results for the program.

The process involves selecting the type of mold and tool base, steel type, cavity layout, runner and gate sizing, and related mold components. The mold base size is selected for the number and layout of the part cavities and cooling/heating arrangement. The mold must be able to fit a selected injection molding press size or range of machine platen sizes. The mold must fit within the area of the machine's tie bars for maximum clamping during manufacture.

The mold assembly and rough machining are begun on all cavities; early tool tryout and material tests are conducted to find any problem areas, such as weld lines and part dimension tolerance capability. Final cavity tolerances are then cut with draft selected for the part's depth and material selection. Mold cavity finish is important for a texturized surface with sufficient draft to release correctly from the mold and any final polishing, always in the direction of part release. The venting should be sufficient so the fill and flow in the mold cavities is not compromised for the evacuation of cavity air during the injection cycle. The mold is then checked for operation and released for production tryout.

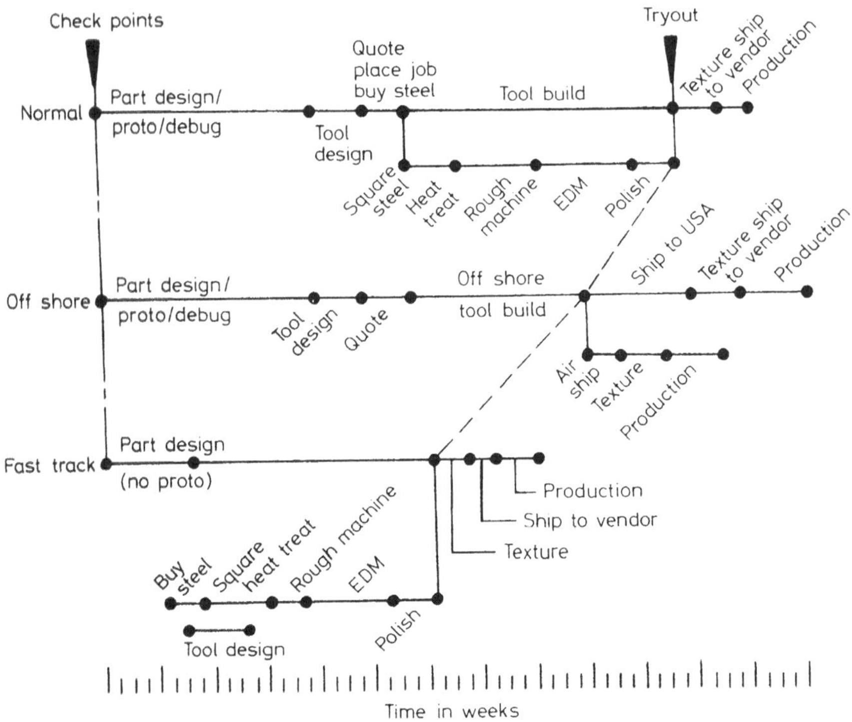

FIGURE 6.3. *Mold design and construction paths. (Adapted from Ref. [3].)*

How all of these operations are accomplished is involved with the tooling flow paths as illustrated in Figure 6.3 for normal, off-shore, and fast-track tooling delivery schedules. Each method is good as long as the following principles are kept in mind.

The typical domestic tooling time has been estimated at 16 weeks, but this will vary with the economy and complexity of the tool. When going offshore for tooling, the quotes are dramatically lower in cost, with quality and capability of rework often questionable. Just be sure your source comes well recommended and has the capability needed to produce the mold to your specifications. Also, make sure the type of steel, cooling and part layout, and runner system are sized correctly. When filled and reinforced resins are specified, take into effect the wear associated with these materials and harden the runner system and gating or provide gate replacement capability as applicable. This is where a good checklist of items to consider is very important. Your tooling source should also have the ability to trial the mold and supply you with molded samples before final buyoff on the mold.

1. Design—resist change. Do all of the part development upfront. Do not make engineering revisions to the tool unless a benefit in quality is made.

2. Make all "key" decisions at the start of the program. Know the product's end-use objectives, required tolerances, material requirements, and limitations. Design cooling circuits for uniform cavity temperature control.
3. Use conservative design principles. Combine functions, use material assembly features, and keep it simple. Use the Mold Design Checklist, number 15 and 16 in Appendix C.
4. Draw on the advice of suppliers early in the program and follow their advice.

The flow of typical custom and in-house injection molding startup operations is shown in Figure 6.4. The evaluation and selection of mold criteria is a joint department operation. The design team wants to ensure the product's end-use intent and operation is not jeopardized by the manufacturing operation based on material, processing, and molded-in product functions. This implies that any bearing surfaces have the correct size and finish, the functional items will perform as required and are sized to work correctly, and so on. The details for good mold design will be discussed in the mold design chapter.

A question often overlooked is whether mold release is allowed should a sticking problem occur. When this occurs, a mold problem must be addressed as it will not go away by itself. If a part sticks in the mold cavity, then ensure

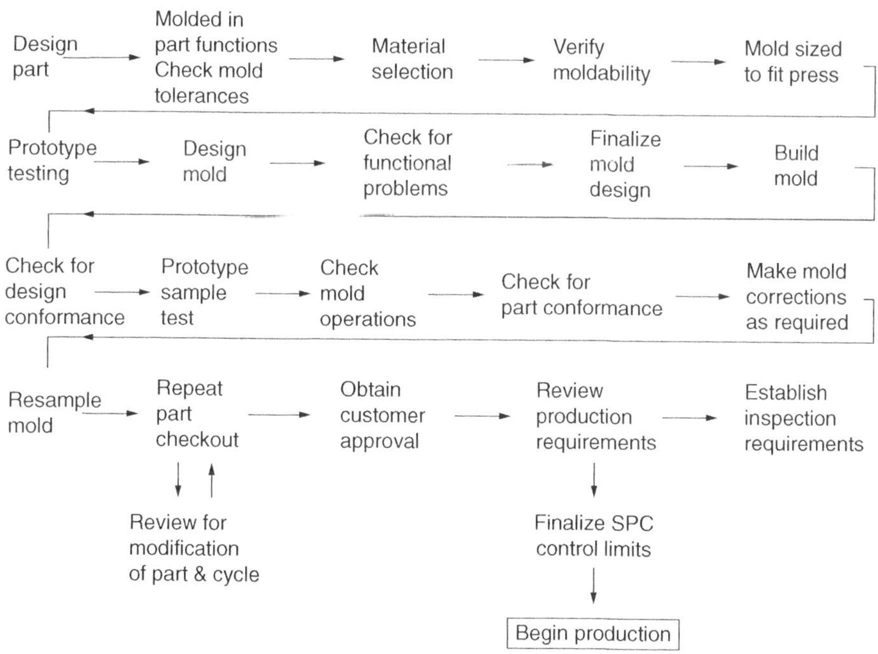

FIGURE 6.4. *Molding startup operations.*

the final polish is in the direction of the core pull and no undercuts or flashing is holding the part in the mold cavity. These problems should have been corrected before production was begun and must be corrected as soon as possible as it affects the cycle time.

But, as molds age, they often develop a condition where a part may not eject correctly. This problem must be corrected as discussed above. Flash and undercuts can occur as the mold wears and can affect the cycle and part quality.

MATERIAL SELECTION

Designers are often prejudiced with preconceived ideas regarding plastic materials. They select materials they are familiar with based on past experience for similar products. With the growth of materials for plastic products, the majority of material selections remain with the classic standbys such as ABS, polycarbonate, polystyrene, acrylic, polyethylene, polypropylene, polyethylene terephthalate (PET), nylon, and acetal. These are basically the main families of the amorous and engineering-grade plastics used today in most injection-molded products. Within these materials, there are thousands of material variations, such as reinforced, filled, flame retarded, impact resistant, and so on.

With today's software design programs, the part designer has the option of evaluating several materials in their designs. They can specify ribs, thick and thin sections, molded-in assembly methods, snap-and-press fits, screws, sonic assembly, and so on, as well as spring and other design operations that are not available in all materials. Often, a more versatile material is an engineering plastic with a higher price per cubic inch but with greater design possibilities.

A simple calculation to evaluate this is material cost equals part weight multiplied by resin price. To calculate part weight, use the factor 0.0361 multiplied by the resin's specific gravity (SG) multiplied by the cubic inches of the part. To estimate the reduction in the commodity resin's part weight resulting from reduced section thickness using an engineered resin, use a ratio of the resin's physical properties. Use tensile strength, modulus of elasticity, or a similar physical property ratio multiplied by the engineered resin's estimated weight. This will reveal a savings is possible.

Example

A medium-impact grade of ABS is proposed for a bolt, but acetal homopolymer is being considered as a possible cost savings.

	Price per lb		SG		Material in part				Material cost in part
ABS cost	$\$.90/\text{lb}$	$\times$	1.04	$\times$	$2\ \text{in}^3$	$\times$	0.0361	=	$0.0675\ \$/\text{lb}$
Acetal cost	$\$1.25/\text{lb}$	$\times$	1.04	$\times$	$2\ \text{in}^3$	$\times$	0.0361	=	$0.1260\ \$/\text{lb}$

To evaluate section thickness reduction, use the following:

$$\text{Tensile strength ratio of } \frac{\text{ABS} = 5000 \text{ PSI}}{\text{Acetal} = 9700 \text{ PSI}} = .51^{1}$$

Evaluating a thinner part section with equivalent ABS part strength yields:

$$\text{Acetal price \$/lb} : .126 \times .51 = .064 \text{ \$/lb}$$

ABS yields 0.0675 $/lb. Therefore, this analysis shows acetal at an equal or slightly lower material cost than ABS. More savings may be possible if the design of the dead bolt's shape is reevaluated to make the part stiffer and stronger. Production part rates will also be faster, because a thinner part section and a faster material setup in the mold will reduce the injection molding cycle time for acetal. These calculations will assist in estimating the cost of each material used for a part; each application can be evaluated on its own end-use product requirements.

Limits should not be placed on material selection during part design, as end-use requirements must take priority over price. As in the example, a heavier (SG), more expensive material (an engineering plastic) may sometimes offer a cost savings. The possibility of combining part functions should also be considered. These may include assembly and piece-part reducing features that use press-and-snap fits, bearing surfaces, integral springs, and so on, in the finished part. These may reduce the number of parts in the assembly process. The company should also consider the ease of repair of the finished article if something in the assembly fails.

CALCULATION OF PLASTIC PART COST

Plastics in a cost basis of cents per cubic inch are very economical compared with metal. Plastics cannot do everything that metal can, but the newer plastic alloys are fast, narrowing the gap in all markets. The design team is responsible for ensuring the finished piece-part price is competitive. The part's material is based on end-use requirements as well as on the designer's knowledge and prior history of what has worked in similar parts. Often, more than one plastic material is considered with the designer evaluating different material properties to reduce the part's section thickness. The part's thickness is related to the part's material properties as to tensile and flexural strength and impact properties. The part's thickness will affect the manufacturing (molding) cycle time. Material properties are considered before the final

[1]Reduction in section thickness of ACETAL part is possible to obtain equivalent part strength in ABS.

material selection is made to keep the part's cost low for both material and manufacturing.

An example showing the comparison of two different materials for the same part illustrates the possibility of using plastic materials in a more efficient manner.

The product's final cost will be determined when production begins, but the initial cost calculations will show whether the program is within anticipated cost goals. The case study for piece-part cost estimation is developed, which should be within 2 to 5 percent of the product's actual final price.

CASE STUDY OF PRODUCT COST ANALYSIS

An electric hand power tool housing (Figure 6.4) has been designed with internal motor supports, ribbing, and assembly screw bosses. An ABS impact-modified plastic resin will be evaluated for part cost.

ABS (impact modified)

Specific gravity = 1.04
Modulus Elasticity = 450×10^3
Thickness = 0.125

Density = 0.0437 lb/in^3
Resin price \$/lb = 0.95
Part weight = 0.35 lbs
Part volume is 250,000/year

The part weight in ABS was calculated at 0.35 lbs from a similar size drill housing.

ESTIMATING PART CYCLE TIME

The part's manufacturing cycle is controlled by two main variables. First, the cycle time begins on closing the mold, injecting the plastic and holding packing pressure until the gate is sealed, and retracting the screw to build up melt for the next cycle while the part cools and solidifies for ejection from the mold; then, the cycle repeats.

Cycle Time:

Cycle Time (CT) = 8 + T(200)
8 = mold open factor
T = wall thickness in inches
200 = cooling factor in seconds

Cycle time selection uses the graphs in Figure 6.5 for estimating amorphous, crystalline, and engineering resins cycle time. Select the value closest to the part conditions, material, and section thickness. The molding cycle time for the parts material is calculated:

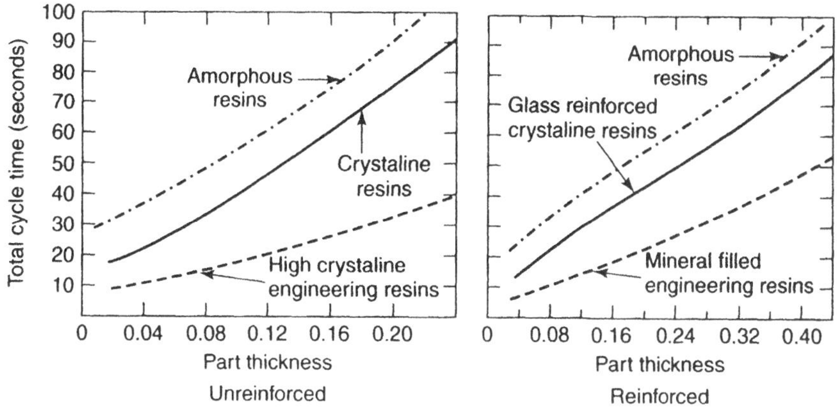

FIGURE 6.5. *Average estimation of total cycle time.*

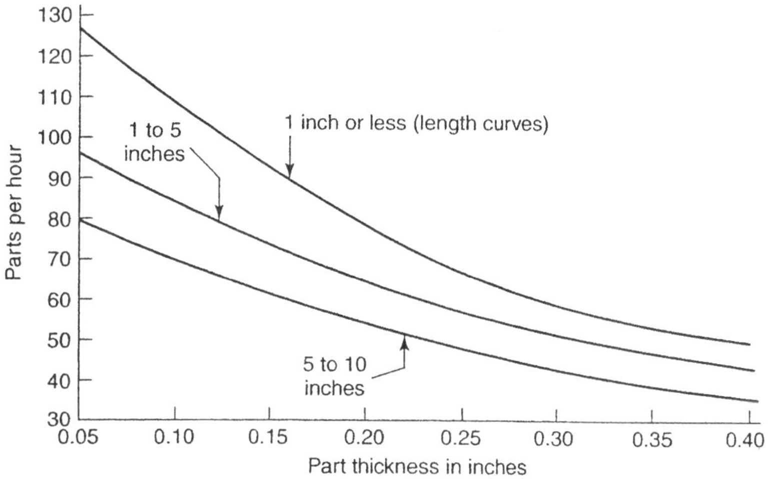

FIGURE 6.6. *Estimated production per hour: length versus thickness.*

CT Comparison:

$$CT = 8 + T(200)$$

$$CT = 8 + (0.125)(200) = 33 \text{ sec}$$

Because all parts are not equal, the flow length and section thickness must be considered in determining estimated cycle time. Figure 6.6 relates part thickness to part length and is used for estimating whether the cycle time is compatible for the part's geometry. Therefore, the part's section thickness of a 0.125-inch and a 10-inch long flow path in the mold yields a parts per hour cycle time adjusted (CTa) to:

$$CTa = (\text{seconds/hour})/(\text{Parts/hour}) = \text{adjusted for flow length cycle time}$$

$$CTa = 3600/66 = 54 \text{ sec (actual cycle time estimated for part)}$$

The longer cycle time is used based on material flow length and section thickness, as the calculated cycle time of 33 seconds did not consider part flow length. The runner system in the mold is kept as short as possible and balanced for uniform pressure drop at each cavity.

MOLD PART CAVITY ESTIMATION

Determining the number of part cavities in the mold involves the annual estimated volume, monthly ship requirements, part type, and size.

$$\text{Number of cavities (NC)} = (\text{annual number parts})\left(\text{CT cycle in seconds} \times 10^{-7}\right)^{2}$$

The tool housing example uses a two-cavity mold, based on part configuration and layout for a balance mold runner configuration, producing one complete tool housing per cycle with one right and one left half. The monthly sales volume is estimated based on a projected annual sales of 250,000 drills. The monthly manufacturing volume is (250,000 units/year) / (12 months/year) equals 20,834 units/month.

Based on the example, the number of mold cavities must be determined.

$$\text{Number of cavities (NC)} = (\text{annual number parts})\left(\text{CT cycle in seconds} \times 10^{-7}\right)$$

$$NC = (250{,}000)\left(54 \text{ sec} \times 10^{-7}\right) = 1.35 \text{ or 1 unit, 2 halves}$$

The NC is based on required monthly product volume. The mold can be either (two) or (four) cavities, with separate cavities for each half of the housing, producing either one or two complete housings per cycle. The number of cavities will depend on the number of shifts the company plans on running the part.

Always round off the NC to the next closest even number. This number must provide the required volume of parts per the schedule, with the sequence of mold cavities being 2, 4, 8, 16, 32, etc., for a balanced layout and pressure drop for all cavities.

If the part size is very large or requires core pull in the plane of the cavity, a single or two-cavity mold is built. The parts tolerances have a definite effect on the number of mold cavities. If part tolerances are critical, the number of cavities selected must be able to produce the required tolerances from each cavity.

[2] Assumed: three shifts/day, 6 days/week, 95 percent yield, and 80 percent utility.

Part production per hour (PH):

Using one as the number of cavities (actually two with one left and one right half of the housing) producing one complete housing per cycle

$$PH = NC/CTa \times 3600$$

$$PH = 1/54 \times 3600 = 66.666 \text{ use 66 complete housing per hour}$$

$$\text{Units/month} = 20{,}834/66 \text{ units/hour}/120 \text{ hours/week} = 2.6 \text{ weeks}$$

Production time requires a little more than two and one half weeks based on a three-shift operation of 5 days per week plus an initial setup time. Adjustments may have to be made to increase the number of cavities to four, with two complete housings per cycle. This analysis can be run later if needed. (The example continues based on the single unit, two-cavity mold.)

MOLD SIZE CONSIDERATIONS

The mold is sized to fit between the tie bars of the molding machine and not extend beyond the machine platens. The mold width and height are a minimum of ½ inch wider per side than the part cavity dimensions and typically 1 to 2 inches for structural support strength. The molds tack height (depth of mold closed) is 2.5 times the depth of the part cavity plus 4 inches. This allows for the part depth plus 2 inches of steel safety stock for the support of the base of the cavity, plus a minimum of a 1-inch plate on the core, an ejector stroke equal to part depth, and a 1-inch thick ejector plate. This also provides sufficient steel for the cooling line routing around the cavity for the temperature control of part dimensions.

INJECTION MOLDING MACHINE SELECTION

The machine selected is usually based on what molding machine will be available in the schedule, plus two evaluation methods involving the mold clamping pressure and the machine's melt and shot capacity.

The first method is based on machine clamp pressure determined by adding the exposed part surface, sprue, and runner area, times the resin's recommended clamp pressure, in tons per square inch, as shown in Table 6.1.

Mold clamping force is determined as follows:

Machine Clamp Force (MCF) = PA (projected part, runner, and sprue area)^2 × Mtl CF (material clamp force) tons/in^2

MCF = [PA] in^2 × [Mtl CF] tons/in^2

TABLE 6.1. Material Mold Clamp Requirements [Material Clamp Force (MHCF)].

Material	tons/inch^2 [part and runner surface area]
PE/PP	1 ½
ABS/styrene	2 to 3
PC/nylon	2 to 5
Polysulfone	3 to 5

PC, polycarbonate; PE, polyethylene; PP, polypropylene.

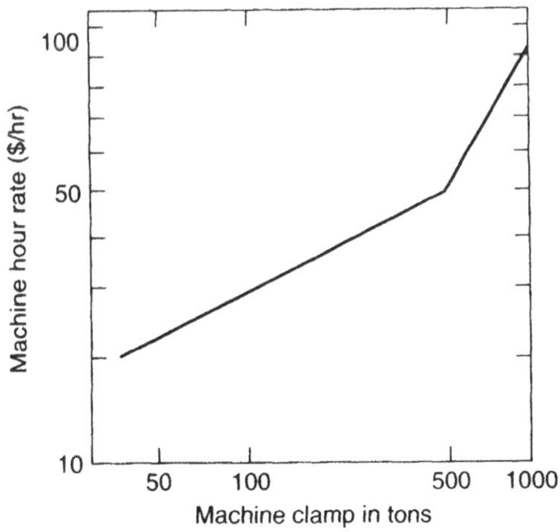

FIGURE 6.7. *Machine hourly rate per ton of clamp. (Adapted from Ref. 2.)*

Surface area of the two mold halves and runner and sprue = 84 in^2

MCF = [84 in^2 × [2.5 tons/in^2] = 210 tons of clamp force required.

The hourly rate of manufacture is obtained using Figure 6.7. It is approximately $35.00 per hour. The graph must be updated yearly to adjust for costs.

When estimating a machine's clamping requirements, the clamping tonnage should be 20 percent greater than the MCF calculated to hold the mold halves closed. For this example, the machine must have a minimum of 250 tons of clamp force, and 300 tons, if available, of clamp force is preferred.

Depending on injection speed and material viscosity, the clamp pressure may have to be increased to prevent the mold breathing, opening slightly during injection, which can flash the mold. Also, thin wall section parts, 0.050 inch or less and very fluid resins, as nylon, may require higher clamp pressure to compensate for the higher fill pressure necessary to fill the thinner cavity sections and prevent flashing.

MELT GENERATION

The second method of sizing the molding machine is determining the melt capacity required for the parts each cycle and the capability of the machine's screw and barrel combination to create and have two to four shot weights of melt in the barrel each cycle. For example, a 30-ounce melt capacity machine would be selected to mold parts with a shot weight of 8 ounces. This leaves two to three shot weights of melt in varying molten stages in the barrel being prepared for the next cycle.

Shot size is the amount of resin required to fill the mold cavity, runner, and sprue, and it should be between 25 to 85 percent of the machine's rated melt capacity. The lower melt percent (25 percent) is recommended so that the resin in the barrel is not overheated while the machine is generating melt for the next cycle. Also, if the machine's melt capacity is too large, the resin is subjected to a longer residence time in the barrel. If residence time exceeds 5 minutes, for heat-sensitive resins, it may cause degradation in physical properties and color quality.

The estimation of shot weight is shown in Figure 6.8 with the volume of the runner and sprue added to the part volume to estimate shot weight correctly. Determining the machine size for shot weight and melt generation capacity requires comparing the shot weight volume for the mold and screw cushion with the molding machine's melt generation capability.

The tool housing part weight in ABS is 0.35 lbs, or 5.6 ounces plus 1.5 ounces added for sprue and runner, bringing the total shot weight to 7.1 ounces of material. Using the three times factor for resin in the barrel requires a machine with a melt capacity of at least 25 ounces. Continuing the molding machine evaluation, we next consider the melt generation capability of the machine.

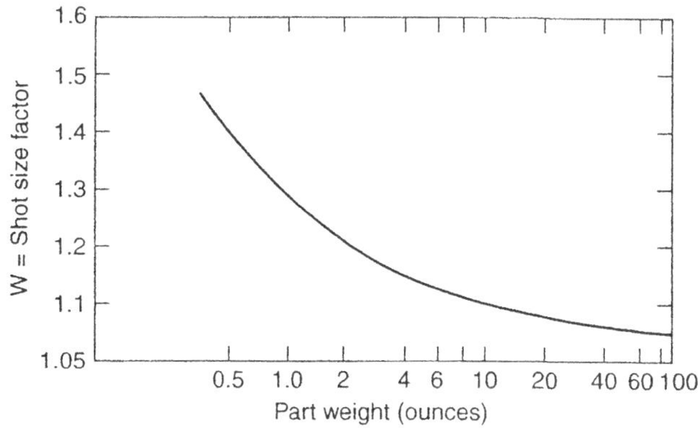

FIGURE 6.8. Shot weight factor.

A machine in the 300-ton range of mold clamp capability has an average shot weight capacity of 50 ounces of polystyrene. Polystyrene is used as the melt gauge capacity for all injection molding machines when melt capacity is estimated. The melt capacity of other resins must be adjusted to polystyrene to arrive at the injection molding machine's melt capacity for a particular resin.

Impact-modified ABS compared with polystyrene is only 80 percent of the machine's rated melt-generating capacity or 40 ounces. The shot volume weight in ounces falls within the 25 to 85 percent shot size range or 62 percent of the barrel and screws capability to produce the required melt per cycle. This range of melt capacity will allow the machine to prepare the plastic resin correctly, without degrading it, for every cycle.

MOLDING MACHINE SCREW-TYPE CONSIDERATIONS

An area often overlooked is the type of screw in the molding machine.

The type of screws used is shown in Figure 6.9 with general purpose; high compression, nylon screw, and the length-to-diameter ratio (L/D) and diameter of the screw are often misunderstood by manufacturing personnel. The type of screw design used is critical for some heat-sensitive, high-temperature melting, and crystalline and amorphous, reinforced, and general-purpose resins. The screw generates the shear heat in the barrel as the material is compressed and melts on the barrel walls as the screw conveys the material down the heated barrel. Always check with your material supplier for the screw type and the compression ratio recommended for their resins.

The type of molding machine clamp system (electric, hydraulic, or toggle) should be considered for cycle time estimation. Toggle machines (Figure 6.10) with an over center clamping design have a slightly longer opening and closing time for both the hydraulic and the electric clamping operation.

MACHINE HOURLY RATE

Each size of the molding machine has an established hourly rate of operation based on its size and is based on the company's overhead and manufacturing costs. Machine cost, like inflation, increases annually and is usually very competitive among custom molders. The average injection molding machine hourly costs (MC) for 2006 are shown in Table 6.2 and are developed by the following equation:

$$\text{Machine Cost (MC) in \$/hr} = 0.112\,(\text{MCF}) + 12$$

In our example:

MCF = (part area inch^2) × (tons of clamp force per in^2) A 3 ton/in^2 pressure was selected for impact modified ABS.

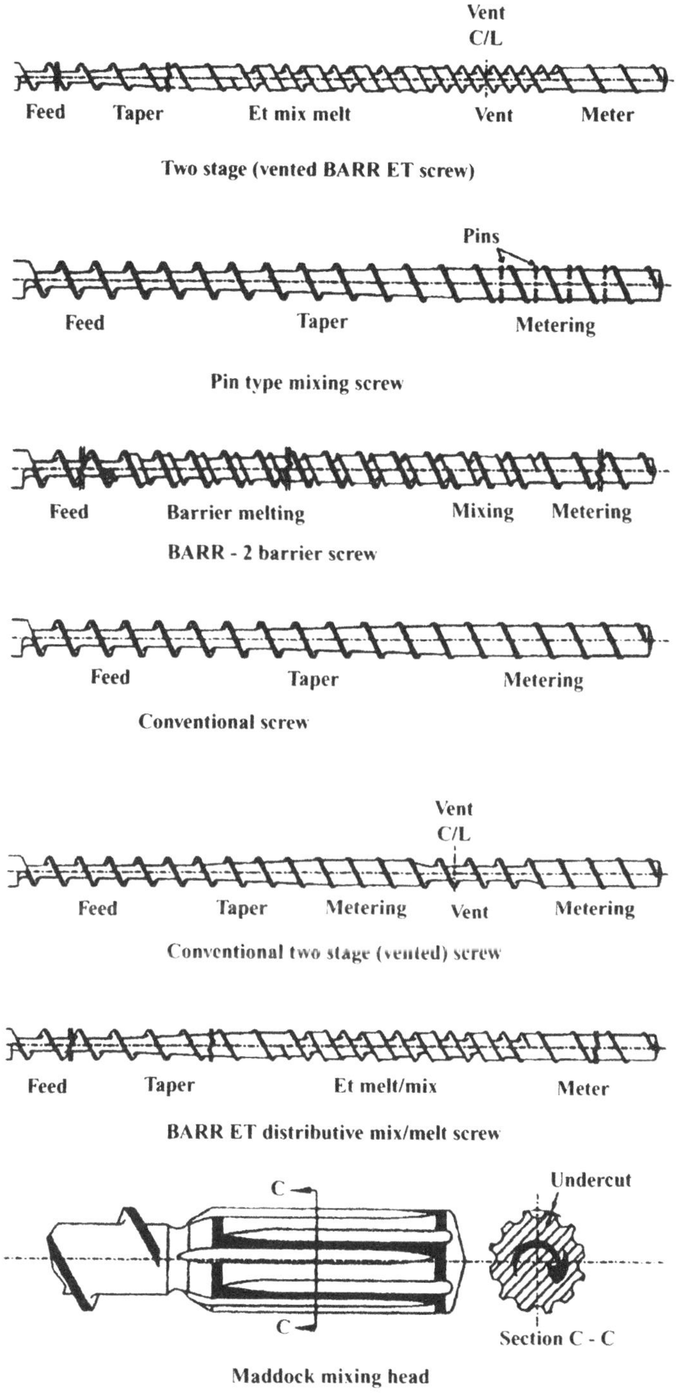

FIGURE 6.9. *Screw design (Courtesy of Robert Barr, Inc., Virginia Beach, VA.).*

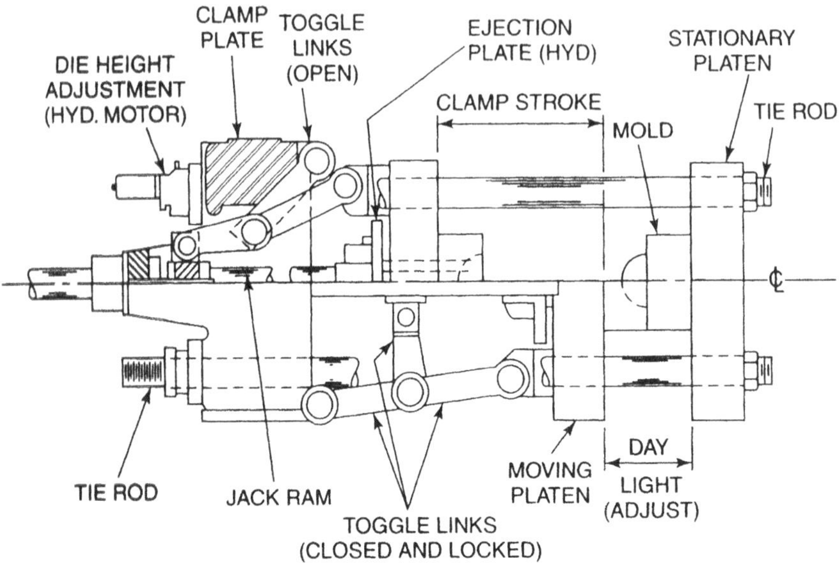

FIGURE 6.10. *Toggle molding machine (Adapted from Ref. 4.).*

$$\text{MCF} = (72)(3) = 216 \text{ tons of clamp force}$$

For a safety factor, use 1.2 times the CF calculated equals 260 tons of clamp force. The closest machine in this range is a 250-ton machine that will be adequate.

With the above established, the machine cost comparison continues

$$\text{MC} = 0.112(\text{CF}) + 12$$
$$\text{MC} = 0.112(250) + 12 = 40.00 \text{ \$/hr}$$

This number is more in line with today's machine hour cost, which includes the cost estimate using the hourly rate versus the machine's tons of clamp obtained from Figure 6.7.

Compared with the national average hourly operating rates in Table 6.2, a figure of $45.00/hour is suggested. The MC equation gives a good estimation of the average machine hour costs, and it must be adjusted annually for inflation and operating expenses, power, and maintenance.

MACHINE SETUP CHARGES

Setup (SU) charges (mold setting, material preparation, process establishing, etc.) are determined by the press size and by the time required for installing

TABLE 6.2. Average Machine Hourly Operation Rates for Custom-Injection Molders with Operator and Profit Dollars per hour (average) Reported for Second Quarter 2001.

Machine Tonnage Range	<50	50–99	100–299	300–499	500–749	750–999	1000–1499	1500–1999
Northeast	35.11	37.23	41.76	48.31	55.86	71.88	87.50	90.00
Southeast	30.27	37.33	41.05	48.80	70.31	132.00	—	—
North central	32.78	34.59	40.55	48.65	65.98	79.18	93.81	159.34
South central	24.20	28.80	36.43	44.51	62.92	75.96	98.91	132.51
West	35.57	38.98	43.59	48.03	61.23	80.93	102.00	207.00
National average[a]	32.73	35.71	41.01	48.05	62.74	82.65	94.18	145.95

Correction factors for Table 6.2

Press Tonnage and Operator	Without Profit (%)	Without Operator (%)	Without Either (%)
<100	12.2	16.6	30.3
100–299	8.2	12.0	24.1
300–499	7.5	10.1	20.5
500–749	5.4	9.7	17.3
750–999	5.9	11.9	21.3
1000+	2.1	11.3	15.5

[a]Data weighted geographically according to Plastic Technology's Manufacturing Census.

Deduct these amounts (cumulative national averages over several surveys) from the values in Table 6.2 for rates without profit, operator, or both.

Source: http://www.plasticstechnology.com/articlews/hrates.html; "Your Business Hourly Rate Survey—October 2001" (online article).

TABLE 6.3. Setup Charge.

Press Size, Tons	Charge $
25 to 125	150.00
125 to 500	250.00
600 and up	300.00 to 500.00

the mold and equipment with trained personnel necessary to have a successful production run. Essentially two methods are used to establish SU charges. The conventional and the quick mold change (QMC) method. QMC may take only 15 to 30 minutes using QMC methods. Or it may require several hours for conventional change over the mold.

Reducing the setup time is essential and dependent on the complexity of mold operations, auxiliary equipment support availability, and experience of technicians. Also, consider whether the molding machine needs to be purged and the barrel and screw pulled and cleaned for the new material. The SU charge for different size molding machines is shown in Table 6.3.

Completing the analysis, a setup charge based on a press size of 250-ton clamp capacity yields a setup charge of $250.00 per production run.

CALCULATING PRODUCT MANUFACTURING COST

The initial part quote for a program includes many factors, such as the machine, resin, volume of parts, mold cost, operating overhead, profit, and the method, to amortize these items over the life of the product. Using the information developed, the product's manufacturing costs can now be calculated.

The final product manufacturing costs are calculated using the Piece Part Estimating form in Figure 6.11, and the Total Cost is calculated in Figure 6.12.

Using the part cost estimating form can also show the cost differences between other materials and can show whether savings are possible. The mold cost is typically amortized over the first year or two of product production to obtain the true part–cost relationship. An important item to remember is that the lowest quote is not always the most economical. If the molder is not familiar with running the material selected for the part or able to obtain the part tolerances consistently, then a more experienced molder with higher costs may be the better choice. Many original equipment manufacturers (OEMs) qualify their subproduct suppliers, custom molders, and in-house molding capability for all their molds before selecting their molders to quote on their product program. Many custom molders are now ISO 9001 or TS16949 certified if they want to participate in the OEM's product supplier programs. One always ensure must the molding operation, as well as the in-house or custom molders, are qualified and approved manufacturers for the company.

No. 8 PRICE ESTIMATION CHECKLIST

CUSTOMER: DATE:
ADDRESS:
CONTACT: PHONE: FAX: E-MAIL:

PART NAME: JOB NUMBER:
DRAWING NO.:

PIECE PART COST ESTIMATING PER 1000 PARTS

A. MATERIAL	:	:	:	:
B. RESIN COST ($/LB)	:	:	:	:
C. SPECIFIC GRAVITY (Sg)	:	:	:	:
D. PART WEIGHT (lbs)	:	:	:	:
E. PART WEIGHT (D × 1000)	:	:	:	:
F. MATERIAL COST (B × E)/0.95	:	:	:	:
G. CYCLE TIME (CT)	:	:	:	:
H. NUMBER OF CAVITIES (NC)[a]	:	:	:	:
I. PARTS/HOUR (H/G × 3600)	:	:	:	:
J. CAVITY AREA (PROJECTED)[b]	:	:	:	:
K. CLAMP FORCE[c] (CF) TONS × (J × MATERIAL FACTOR)	:	:	:	:
L. SHOT WEIGHT (oz) (D × H × W[d] × 16 oz/lb)	:	:	:	:
M. MACHINE HOUR COST OUT (MC[e])	:	:	:	:
N. PROCESSING COST ($/1000 PARTS) M/I × 1000	:	:	:	:
O. ADJUSTED PROCESSING COSTS[f] [N/(0.95)(0.80)]	:	:	:	:
TOTAL COST (PROCESSING PER 1000 PARTS)	:	:	:	:

[a]Assumed three shifts/day, 6 days/week ([f]), one years production produced.
[b]Projected cavity area and runner/sprue, mold cavity in square inches × number of cavities, plus runner and sprue area of mold surface in square inches.
[c]80% to 20% maximum shot weight of resin, use material clamp factor to estimate tons of clamp required.
[d]Use reference chart for shot weight.

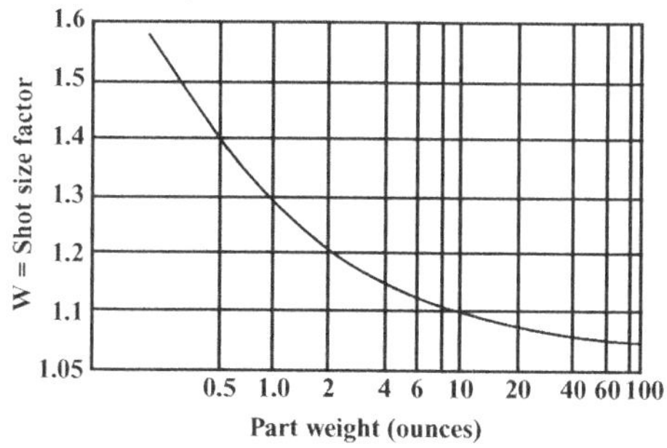

[e]Adjust for current machine rates.
[f]Assumes 95% yield and 80% utility of molding process.

FIGURE 6.11. *Price estimation.*

No. 8 PRICE ESTIMATING CHECKLIST

CUSTOMER: DATE:
ADDRESS:
CONTACT: PHONE: FAX: E-MAIL:

PART NAME: JOB NUMBER:
DRAWING NO.:

PIECE PART COST ESTIMATING PER 1000 PARTS

A. MATERIAL:	Impact ABS
B. RESIN COST ($/LB):	$0.95
C. SPECIFIC GRAVITY (Sg):	1.04
D. PART WEIGHT (lbs):	0.35
E. PART WEIGHT (D × 1000):	350
F. MATERIAL COST (B × E)/0.95:	35:
G. CYCLE TIME (CT):	54
H. NUMBER OF CAVITIES (NC)[a]:	1
I. PARTS/HOUR (H/G × 3600):	66 complete housings
J. CAVITY AREA (PROJECTED)[b]:	84
K. CLAMP FORCE[c] (CF) TONS × (J × MATERIAL FACTOR):	250
L. SHOT WEIGHT (oz) (D × H × W[d] × 16 oz/lb):	7.1
M. MACHINE HOUR COST (RATE × (MC[e]) (Machine Clamp):	40.00
N. PROCESSING COST ($/1000 PARTS) M/I × 1000:	606
O. ADJUSTED PROCESSING COSTS[f] [N/(0.95)(0.80)]:	797
TOTAL COST (PROCESSING PER 1000 PARTS:	$797.00

[a]Assumed three shifts/day, 6 days/week ([f]), 1 year's production produced.
[b]Projected cavity area and runner/sprue, mold cavity in square inches × number of cavities, plus runner and sprue area of mold surface in square inches.
[c]80% to 20% maximum shot weight of resin, use material clamp factor to estimate tons of clamp required.
[d]Use reference chart for shot weight.

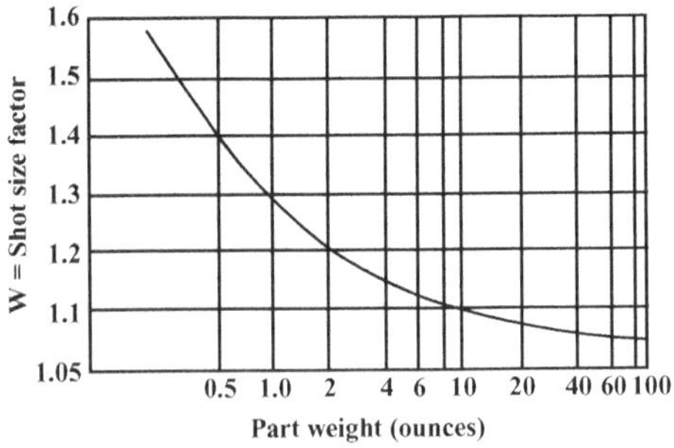

[e]Adjust for current machine rates.
[f]Assumes 95% yield and 80% utility of molding process.

FIGURE 6.12. Price estimation.

MATERIAL SUPPLIER LIMITS

Material suppliers are evaluated on their ability to supply consistent quality products. Their plastic resin material source, distributor, or major resin manufacturer are the major suppliers of a consumable product for the injection molder. The injection molder relies on a consistent product supply that will continuously meet its processing and customer's product requirements.

Variation in melt flow, viscosity, and other material variances must be kept to a minimum with good heat resistance properties and processing characteristics. The molder wants to receive a consistent product with minimal variability, a dry product, and a product without any contamination.

But, all materials will vary lot to lot and even within the same coded lot. Plastic resins are made in either a continuous or a batch process. Each supplier has its own formulation for its products. Each lot will vary to a certain degree and is tested at periodic points in the manufacturing process with data collected for the typical physical, chemical, electrical, and other properties as listed in their specification sheets. Each of these production runs is certifiable to the results obtained, and many have to meet specific company specifications to be sold into an area requiring these properties. This is where the supplier will certify that the material of a lot meets the properties required by the final customer.

Whenever a new lot of material is being introduced into the production cycle, it must be recorded on the manufacturing molding sheets. Then, if a problem occurs, there is a record of the material used for the product under investigation. When the lot of material is changed, it may even require a change in process variables to compensate for a variation in the product's processing properties. Always stay alert when a material change is in process, for these slight variations can require a process adjustment.

This also applies to other items purchased, such as assembly items plus equipment support items as heater bands, screw and check ring assemblies, cooling and heating systems, grinders, feeders, dryers, conveyors, part separators, robots, and any other item used and/or consumed in the manufacturing process. Many of these items control temperature, and as wear occurs, these items should be checked periodically for variance as they are not the same as when first installed. The cleanness of the mold cavity, oil in the mold, and other items used (e.g., as molded in inserts) can cause a problem. A simple item such as inserts with cutting oil remaining on them can cause contamination and stress cracking in some materials, such as acrylic, polycarbonate, and styrene. Before the use with a product, it must be checked for contamination. In some materials, it may not matter, but in these, it is a site for a catastrophic part failure.

ESTABLISHING MANUFACTURING LIMITS

The process of actually producing the product is filled with a multitude of additional variables. Once the design is set, the mold is built, and the material is selected, the actual process of manufacturing the product begins. The limits for the product's manufacture have yet to be established. Knowing what the product must do and how the mold was designed and built can aid in establishing the manufacturing process.

Often, the manufacturing variables for the product are never considered until a problem happens. Then the responsibility of these items creates a greater interest with management as the process or operation does not have the necessary control or ability to produce good repeatable products.

The variables of manufacture need to consider everything that has happened from the time of the contract signing forward, such as the product's final price, shape, material, function, method of manufacture, tooling, specifications, operations to be performed, operator training, tooling, assembly, and decoration; then, the process must be considered through any conditioning and shipping to the customer's point of sale. The manufacturer must also consider other items such as purchased parts used in the product's manufacture. Each operation involved in the part's manufacture should have its own checklist of requirements that may have to be performed before the product moves to the next step in its manufacture.

Just selecting the correct size of the injection molding press to mold the product is often not considered really critical, but it should be as it has an effect on the product's quality of manufacture. Melt generation capability, ounces of melt generated per cycle, screw type, and check ring assembly are all variables to ensure the correct consistency and temperature of melt is used to fill and pack out the mold's cavities. The freeze off of all cavity gates within a second of each other is necessary so that a mold cavity does not decompress when the screw retracts for the next cycle and packing pressure is released on the cavity gate area.

The molding machine size and clamp capacity must be sized for the mold and melt generation required to produce the product. This is not often considered but is very important to the quality of the product. Each machine is sized for melt generation based on polystyrene. If the melt generation capability or screw size are too big, the resin in the barrel could degrade because of excessive holdup time. Not enough melt generation capability and the mold is starved and may not pack out each cavity during the cycle, as it has to retract and build up melt for the next cycle. This is discussed in detail in the molding chapter.

Sufficient clamping capacity is also required so the mold will not breathe, open slightly under injection pressure, and cause a decompression of the mold cavities. The mold must be contained within the platens and not stick out beyond the clamping surface. Failure to do this will create unequal clamping pressure and may spring the mold.

Too often, press size is not considered and the molding machine is too small or generates too much melt and problems occur. Even a fan blowing on the molding machine's heater bands can change the heating profile of the barrel because of heat loss. An important reason to have heater band covers, which act as a thermal blanket, is to keep heat in and save energy. This is one area that is usually forgotten in most shops.

Problems such as material feed and contamination, mold temperature variation, heater bands wearing out, and dusty or cardboard trays holding parts that must be kept dust free are just the tip of the iceberg for consideration of variables and their effect on the process and product. It is very important that the molding manager or assistant review what is required for the part and then ensure that it is provided and working.

AUXILIARY EQUIPMENT

The auxiliary support equipment to control the product's manufacture is just as important. One must examine the equipment to determine whether the material feeders are clean from their last use, the water temperature is consistent, the tower water is within specifications, and the assembly equipment is within torque ranges as required and verified.

With the use of more electric injection molding machines, it is important the plant is checked for adequate electrical power capability. Some plants have had to increase their power sources to ensure their feed-line voltage was constant during all stages of daily manufacture, especially at peak times. Changes in line voltage can cause fluctuations in power supplied to equipment, which cause changes in its operation and output. Your staff electrician should be trained to take readings of power consumption and demands during peak operations.

IN-PROCESS INSPECTION

When process control is used during production, parts must be periodically inspected to be sure they are in tolerance. The gauges and fixtures used to check parts for conformance should be specified in the job instructions and employees must be trained to take these measurements. Inspection measurements are recorded and are traceable to the lot of material, processing time period, mold, operator, machine, and inspector. The method of part testing must document the location on the part and the results according to the customer's procedures. The tests should duplicate any customer spot checking at their incoming inspection. It is always best to over inspect than to inspect not enough, just perform the inspection in real time with any necessary corrections made if parts are still cooling down after molding. If this is done, predetermine the allowable dimension, which on cool down and par stabilization, the correct dimensions are achieved.

This is especially true when colored parts are made that must match other parts molded or decorated separately. Color matching of parts from the same lot of material is easier than matching from different lots. Materials that absorb moisture and materials that are molded several weeks prior to another batch can exhibit shade variations and may not match when the pieces are assembled together. Always use a color-matching checklist to achieve the best color matches possible.

Colored resins are typically made in batches with lots of material. They are either salt-and-pepper blends, compounded colors from the supplier, or batch mixed by the personnel at the plant, and there will always be a slight variance from batch to batch and/or from lot to lot. When color matching is critical, always specify the type of lighting to be used to judge the color match of the items. Different light sources have different wavelengths that will create part color variance as will varying section thickness and surface finish to the observer.

ESTABLISHING TOTAL QUALITY PROCESS CONTROL

TQPC is composed of two parts: The first establishes limits that meet the supplier's capability and the customer's requirements. The second part is applying methods that address specific items in the daily business and manufacturing processes and operations. TQPC must become a method for all personnel working together within an organization for the customer. All the typical buzz words apply, such as, excellence, zero defects, statistical process control (SPC), Six Sigma defect objectives, lean manufacturing, and total quality management. These concepts are at the heart of the TQPC quality program. The use of ISO 9001:2000 accreditation to strengthen a quality system is key to continual improvement. All the above concepts are necessary and are even more important depending on each company's situation and customer mix. To be the best manufacturer in the industry is a worthy goal. To win the highly prized Malcolm Baldrige National Quality Award is even more exciting in the quality area.

TQPC is like ISO, in which a manufacturing manager says what he/she will do and then does what he/she says, in a repeatable operation. No changing, no cutting corners, no limiting what will be done for one customer versus another. Keep your internal "in-house quality" as good as possible and the customer's quality will follow. Changing the company's quality methods from customer to customer never works, as in time operations will revert to the lowest denominator of service and suffer the consequences. Equality for all is the key method for ensuring the best daily output.

Keep operations in real-time process control, monitor output, make corrections, calculated corrections when necessary, and otherwise let the process float within the limits established. Too many machine adjustments will cause an out-of-control situation and can take hours to get back into control. Train operators to monitor and observe the process and let the process vary as the conditions change based on normal conditions. On long-running jobs, observe the daily variance as it is typical for a process to drift within the established

TABLE 6.4. Master Sampling Table.

Lot Size	Sample Size	0.25		0.5		0.75		1		1.5		2		3		4		5		6		7		8		9		10		12	
		A	R	A	R	A	R	A	R	A	R	A	R	A	R	A	R	A	R	A	R	A	R	A	R	A	R	A	R	A	R
499 or less	40	↓		↓				0	2	0	2	0	3	1	4	1	4	1	6	2	6	2	7	3	7	3	8	4	9	4	9
	50							0	2	0	3	1	3	1	4	2	5	2	6	3	7	3	9	4	9	4	9	5	10	5	11
	60							0	3	1	3	1	3	2	5	2	6	3	7	4	8	4	9	5	10	5	11	7	12	7	13
	70							1	3	1	3	1	4	2	5	3	6	4	8	5	9	5	9	6	10	7	11	8	13	8	12
	80							2	3	3	4	3	4	4	5	5	6	7	8	8	9	8	9	9	10	10	11	12	13	12	13
500 to 799	40	↓		↓		*	1	*	2	0	3	0	3	0	4	1	5	1	5	1	6	1	7	2	8	2	8	2	9	4	10
	60					*	1	0	2	0	3	1	4	1	5	2	6	2	7	3	8	3	9	4	10	5	11	5	11	6	12
	80					0	2	1	3	1	4	1	5	1	6	3	7	3	8	5	10	5	11	6	12	7	13	8	14	9	15
	100					0	2	1	3	2	4	2	5	2	6	4	8	5	9	6	11	7	13	8	14	9	16	10	17	12	13
	120					1	2	2	3	3	4	4	5	5	6	7	8	8	9	10	11	12	13	13	14	15	16	16	17	13	19
800 to 1,299	40	↓		*	1	*	1	*	2	*	3	0	3	0	4	0	5	0	6	1	6	1	7	1	8	2	8	2	9	2	10
	60			*	1	0	2	0	2	0	3	0	4	1	5	1	6	2	7	2	8	3	9	3	10	4	11	4	12	5	12
	80			0	2	0	2	0	3	1	4	1	5	2	6	2	7	3	8	4	10	5	11	5	12	6	13	8	15	8	15
	100			0	2	0	2	0	3	1	4	1	5	2	6	3	8	5	10	5	11	7	13	7	14	9	15	10	17	10	18
	120			0	2	1	3	1	3	2	5	2	6	3	7	5	9	6	11	7	13	8	14	9	16	11	18	12	19	13	21
	160			1	2	2	3	3	4	4	5	5	6	7	8	9	10	10	11	13	14	15	16	16	17	18	19	19	20	22	23
1,300 to 3,199	50	*	1	*	1	*	2	*	3	*	3	0	4	0	4	0	5	0	6	1	7	1	8	2	9	2	10	3	10	3	11
	75	*	1	*	1	0	2	0	3	0	4	0	5	1	5	2	7	2	8	3	9	4	10	4	12	5	12	6	14	6	15
	100	*	1	0	2	0	2	1	4	1	4	1	5	2	6	3	8	4	9	5	11	6	12	6	14	8	15	9	17	10	18
	125	0	2	0	2	1	3	1	4	2	5	2	6	3	7	4	9	5	11	7	13	8	15	9	16	11	18	12	20	13	21
	150	0	2	0	2	1	3	2	5	2	5	3	7	4	8	6	10	7	13	9	15	10	17	11	19	14	21	15	23	16	25
	200	1	2	1	2	2	3	4	5	5	6	6	7	8	9	10	11	13	14	17	18	17	18	20	21	22	23	25	26	27	28

(Column groups 0.25–12 are under the heading "Acceptable Quality Level".)

A—Acceptance number; R—Rejection number; *No acceptance at this sample size.
Arrows: When there is an arrow under a given AQL use the first sampling data below arrow. (Form larger lots if possible.)
Adapted from Ref. [1].

limits. Let the system reach equilibrium, monitor the variance, and then leave it alone and only adjust it if it goes out of process control limits.

Whenever possible, set the molding operation to automatic. This will assist in ensuring that the cycle is uniform. If this is not possible and an operator must open and close the mold to remove the part, then a systematic timed operation can be established. Even a change of a few seconds in cycle time affects the molding cycle. Over a period of time changes can occur that cause cycle shifts to cause and out of control situation. Keep the molding cycle as consistent as possible. Some molders have established a flashing timing light for cycle consistency. The operator opens and closes the mold, removes the part based on the timing of a flashing light, and establishes and ensures a uniform cycle, no matter which operator is running the machine.

ACCEPTABLE QUALITY LIMITS

Acceptable quality limits (AQLS) should only be used for determining product acceptance or rejection for mass-produced lots of noncritical items. Hot runner 64-cavity molds and larger are an example. Using AQL inspection and acceptance methods, an AQL master sampling (Table 6.4) is used, and a sample size is selected that represents the lot size chosen to be inspected. The sample size is selected from the lot and inspected. If a predetermined number of defects, which is a key number for defects chosen for the sample lot size, are not found in the inspected sample size, then the entire lot is accepted. If the key number of defects is found for the sample size, then the inspection stops and the entire lot is rejected. Then based on the criticalness of the product, a decision is made by management to inspect the lot at 100 percent or scrap it. The use of the AQL method is the customer's decision. The customer must be made aware that lot acceptance using the AQL inspection method is knowingly accepting a lot of material with a statistical estimated amount of defects in the lot. This method of quality control was implemented during war time when mass-produced items were needed and not enough time was available for 100 percent inspection.

Always discuss with the customer the quality inspection method to be used for accepting the product. Determine what dimensions or other condition(s) are necessary to judge the product acceptable and have the customer agree that this method will be used for the inspection. Develop an instruction and train your quality inspectors in the correct way for accepting the product. Be sure all inspection information and records are documented and become a part of the customer's program.

7

Material Selection and Handling

One of the most important items necessary for maintaining total process quality control in your plants is to ensure that the materials received for the manufacture of parts are consistent and of high quality. If material variability occurs, you will have to adjust process control parameters continuously to compensate. To consistently make good parts, the materials received must meet the customer's specifications and be within your equipment's processing window. Some variations are acceptable. For example, all resins of the same grade or product type will vary slightly from lot to lot, even when produced by the same supplier in a continuous polymer manufacturing operation. Many plastics, which are alloys of different polymer families, are still made in batch lots. These are blended together later in the production cycle to even out any inconsistencies that could affect the product's ability to meet final specification limits and properties. These variations are the result of manufacturing process variables; blending reduces normal lot-to-lot variances. These variances may be more noticeable when evaluating equivalent grades from different suppliers, as the feed stocks and process used will vary. Each, however, should meet the final material specifications.

The major concern of most plastic part manufacturers is a lack of detailed knowledge about the more than thirty generic families of plastic resins. Within these generic families, there are hundreds of different product types. They

Total Quality Process Control for Injection Molding, by M. Joseph Gordon

range from pure polymers (homopolymers) to copolymers (blends of different polymers) to blends and alloys of different polymers with different additives (e.g., glass, minerals, carbon, elastomers, nucleators, lubricants, processing aids, and flame retardants) and color pigments. Any change in the level of any of these additives will also result in a new product designation. These changes are less noticeable in the base polymers, because changes in molecular weight are reflected in viscosity differences that affect the ease with which they are processed. If you modify the additives used in a base polymer to more meet part end-use requirements effectively, you automatically create additional quality control problems and processing limits.

Plastics are organic materials constructed from large molecules that form continuous chains based on specific building block molecules and their atomic attraction for each other. During polymerization, under pressure and elevated temperature, starter molecules for a specific generic polymer react and, as a result of valence bonding of the carbon atoms, form chains of the new polymer. Initiators are often added to begin and/or continue this reaction, until the desired polymer molecular weight is reached. Once the molecular weight is obtained, based on molecular chain length, the polymerization reaction is terminated. The molten polymer is then extruded into pellets; additives then enhance or modify the product for its end-use function. The terms polymer and resin are the same in context and are used interchangeably to describe the basic plastic material. There are basically two types of plastic materials, thermoset and thermoplastic. They are so named because of the effect of temperature on their behavior during processing and their resulting end-use properties.

THERMOSETS

Thermosets are polymers that when first heated, usually under high pressure, will melt, flow, and then fuse into an insoluble form. After this reaction takes place, they cannot be melted again or softened by heating. Thermosets, which are the most rigid plastic, are so stable that they are inert to almost all chemicals. In physical properties, thermosets are related to ceramics. They are usually filled with other materials (e.g., talc, glass, minerals, and rags) to yield the desired end-use properties. Common thermoset plastics are phenolics, melamine, urea, alkyds, and epoxies. Even at elevated temperatures, these materials will char but not burn. They have excellent creep resistance, are good insulators of heat and cold, and are nonconductors of electricity. They also have low coefficients of thermal expansion and little-to-no moisture sensitivity or pickup. For certain applications, they are unbeatable. Their biggest drawback is low elongation and an inability to absorb impact loads. They are brittle in nature.

THERMOPLASTICS

Thermoplastics are the main polymers used in the plastic industry. They are heat sensitive. Depending on the specific type—amorphous or crystalline—they will soften or melt when heated to specific temperatures. They can be melted repeatedly and cooled but will suffer a loss in physical properties after each melting as a result of heat degradation of the polymer. Thermoplastics are divided into two main classes, amorphous and crystalline.

Amorphous Plastics

Amorphous plastics do not have a sharp melting point. When subjected to increasing temperature, they will soften and begin to puddle. They have lower physical properties, which rapidly fall off as temperature increases. They are very resistant to impact loads because of their molecular structure versus thermosets. Many are transparent in their base resin form, with low solvent and chemical resistance. Because they are easily alloyed with other polymers and additive systems, to enhance their properties, and are less expensive to produce, they are the mainstay of most consumer products used today.

Amorphous plastics are usually molded in cold molds (40 to 60°F). You have to first remove the heat from the molten plastic, so that it will solidify and become rigid enough to be ejected from a mold. Cold molds also keep the manufacturing cycle to a minimum period of time.

Crystalline Plastics

Crystalline polymers have a sharp melting temperature. They remain solid until that temperature is reached and then melt like an ice cube from the outside to the center. They behave like an amorphous polymer in the molten stage but usually have a higher flow rate. On cooling, they have much greater mold shrinkage because of the degree of crystallinity of the polymer. Crystallinity in any material creates a more ordered molecular packing structure in its solid form. The higher the degree of crystallinity of a polymer, the more dense it becomes on cooling. The result is higher mold shrinkage.

Crystalline polymers require mold temperatures in the range of 150 to 300°F, depending on the base polymer. An increased mold temperature will initiate crystallinity, resulting in a faster plastic resin setup time (solidification), higher polymer physical properties, and greater mold shrinkage. Higher mold temperatures also increase flow length and cavity fill rates, as the crystalline polymers are more fluid than the amorphous polymers. Crystalline polymers can be molded in cold molds, below 150°F, but will exhibit greater postmold shrinkage if subjected to temperatures above their molding temperature.

CLASSIFYING THE POLYMERS

To assist in classifying the different polymer families, Table 7.1 lists the basic polymer family information. When these basic polymers are altered with additives, their physical property data changes. Good references for these polymers are found in McGraw-Hill's *Modern Plastics Encyclopedia* and in reference data from individual material suppliers.

Resin suppliers must consistently produce the basic polymers, and the hundreds of variations of each, to meet a wide range of specific manufacturing specifications and tolerances. To achieve this goal, suppliers have instituted statistical process control of their manufacturing process, with tighter control of incoming raw materials and testing to verify lot-to-lot uniformity of the finished product.

Many resin suppliers are working closely with customers to reduce testing on incoming resins. For specific applications and larger volume users, resin suppliers are offering information, running specific material tests, and providing specific lot data for these customers on a need-to-know basis. The normal chemical information supplied on these polymers is moisture, specific gravity, melt flow, molecular weight distribution, and data on specific internal additives, such as percent of lubricants, filler content, heat stabilizers, antioxidants, plasticizers, and other key additives.

In addition to chemical tests performed during manufacture, the material supplier also conducts physical tests after the material is finished. The chemical tests performed during manufacture control polymer quality and are used to indicate how they will "process" in their equipment. Based on these data, the plastic part manufacturer can make any necessary equipment adjustments. Most material suppliers have assumed that because products are made into so many different parts, customers will adjust processing variables to suit manufacture of their parts. But plastic processors are becoming better educated about polymer variables that can and will affect part manufacturing and process control. Therefore, they are demanding, and in most cases obtaining, better chemical data on specific lots. This enables them to determine whether a lot of material will process properly on their equipment and meet customer requirements.

PRODUCT CERTIFICATION

The typical product certification data are obtained from the product's physical properties, that is, polymer/product properties that are essential for the end-use product. The OEM specifies the property data based on the physical property values required for the part.

How often these data are generated is based on the supplier's confidence in the manufacturer's data, along with the results of chemical tests obtained during manufacture, information in statistical process control (SPC) records,

TABLE 7.1. Basic Thermoplastics Property "DAM" Dry as Molded (Range).

Generic Polymer	Crystalline (T_m) Amorphous (T_g) Degrees (C)	S_g	Mold IN/IN Shrinkage	Elongation (%)	Modulus (M PSI)	Tensile (PSI)	Drying Required	H_2O (%) Range
ABS	A	1.01–1.08	0.004–0.009	1.5–125	300–1800	2500–16,000	Y	0.10–0.15
	110–125							
Acetal	C							
Copoly	160–175	1.41	0.020 avg	40–75	370–450		Y	0.10
Homo	175–181	1.42	0.180–0.025	25–75	380–430	8800–9700		
Acrylic	A							
	85–105	1.09–1.20	0.001–0.008	2–70	200–500	7000–11,000	Y	0.02–0.10
Cellulosic	C							
Acetate	230	1.22–1.34	0.0030–0.0010	6–70	1200–4000	1900–9000		0.40 (max)
Acet/Butyrate	140	1.15–1.22	0.003–0.009	40–88	90–300	2600–6900		
Polyamid	C							
Nylon 6	210–220	1.12–1.14	0.003–0.015	30–100	390–2000	6000–24,000	Y	0.05–0.20
Nylon 6/6	255–265	1.13–1.15	0.007–0.018	15–80	410–2000	13,700	Y	
Nylon 11	191–194	1.03–1.05	0.012	300	150	8000	Y	
Polycarbonate	A 140–150	1.2	0.005–0.007	110	340	7000–20,000	Y	0.02 (max)
Polyester	C 220–267	1.30–1.38	0.009–0.022	50–300	330–400	8200–27,500	Y	0.02–0.20
Polyethylene	C 122–124	0.918–0.94	0.020–0.022	100–965	40–105	1900–4000	Y	0.02 (max)
Polypropylene	C 160–175	0.90–0.91	0.010–0.025	100–600	175–250	4500–6000	Y	0.10
Polystyrene	A 74–105	1.04–1.05	0.004–0.007	1.2–2.5	380–490	5200–7500	Y	0.02
PVC	A 75–105	1.16–1.58	0.003–0.050	40–450	300–500	1500–7500	Y	
Polyimid	A 310–365	1.36–1.65	—	3–10	450–500	10,500	Y	0.02

ABS, acrylnitrile butadiene styrene; PSI, pounds per square inch; PVC, polyvcnyl chloride.

and historical manufacturing data. Data obtained when each lot is tested for all specified property values are known as specific lot data. Data based on periodic testing after a set number of lots are produced are called periodic lot data. The OEM and material supplier must determine what property data will be required on their certification sheets concerning the additive and property values of the specific resin required for their product. Normally, a minimum or a range for each physical property value is selected, with specific additives given a maximum or minimum value depending on how they affect the polymer's end-use properties. These material values are then used to qualify other vendors to become approved suppliers.

When the mold, part, and resin are first qualified for a new product, specific lot data are required. Once production begins, however, periodic lot data are normally all that is required for certification. There are two ways to assess this. First, the customer wants to be sure the right material is used and end-use testing is fully documented. Second, once production begins, they have total confidence that the material will meet specifications. Unfortunately, during these relatively short trials—a few hours at most—no regrind is fed back into the system. Normally if regrind is allowed and properly handled, it can be fed back and mixed thoroughly with virgin resin, so that there is very little effect on end-use properties. But it may affect the processing cycle and part quality. To determine this, the effects of regrind should be tested during tool qualification trials.

MATERIAL SPECIFICATION

In determining the quality of a product, understand that the final product will only be as good as the material used to make it. Therefore, the following eight program steps should be used, coupled with the supplier's statistical process control procedures, to ensure a product's consistency.

1. Establish functional specifications for purchased products.
   A. Engineering and manufacturing should establish material specifications that are no more or less rigid than necessary.
   B. Note values that are easily measured and typical during manufacture. Leave measuring of American Society for Testing and Materials (ASTM) physical property values to the material supplier. Concentrate on values that will affect the processing, melt flow, viscosity, and molecular weight.
   C. Limit those values that critically affect the product. If necessary, establish ranges on these values but obtain supplier agreement that these values are typical or can be met consistently during manufacture. Values that are too tight could result in shut downs when material falls outside of this range and the supplier cannot certify the product.

2. Collect and analyze supplier's data to determine incoming test capability and stability using tests and equipment similar to the supplier's. When the product is received, inspection should test each lot of material for the first five deliveries, maintaining control charts on the critical resin characteristics. These data should then be compared with the supplier's test and certification data for each lot, which must contain the actual test results for each lot shipped. Comparisons can then be made to analyze capability and lot-to-lot uniformity. Any discrepancies between test data should be resolved immediately to add credibility to the respective test methods and results. Once concurrence and stability of results are achieved, every other lot should be tested until a confidence level is reached, and then, test every fourth or fifth lot to show continued stability. During this trial period, the supplier will continue to provide specific and periodic lot data to establish that each is a true indicator of material consistency. At the same time, the parts manufacturing process control data should be examined for any variances noted in the lot data. This verifies that the right characteristics are being monitored and determines whether any resin swings are appreciably affecting processing conditions.

3. Determine critical resin characteristics and control methods. Based on experiences with similar customers, suppliers can assist purchasers in determining what resin characteristics to monitor initially. You may discover that other items also need monitoring. This is based on manufacturing methods and customer part requirements. Incoming testing should be minimized once the critical characteristics are determined and the supplier shows a consistent ability to control the resin's uniformity. Critical resin characteristics, defined on the specification documents, should be tracked as described in item two.

   The Taguchi methods are ideal for determining whether any critical resin processing characteristics require monitoring; these can be determined during processing and end-use testing. If such characteristics exist, the material supplier can recommend tests to verify that the material remains within the required control ranges.

4. Require identical specifications from multisource suppliers. Unless a material is specified, most processors have multiple sources for similar materials. All may meet the end-use requirements for the part, but each supplier's processing may be different. Generic materials from different suppliers will often process slightly differently, and you must be aware of the factors that can cause these variances when processing resins. Therefore, qualify each supplier and have them center their material characteristics to meet your processing requirements.

5. Review each supplier's quality program for cost effectiveness. Suppliers should be using SPC to control product variation and reduce cost. Avoid suppliers who do not use SPC, as materials and processings cost will never be under control.

6. Evaluate each supplier's control plans. As defined in the contract for critical characteristics, monitor each supplier's program to ensure compliance with your requirements. Make sure you both understand what is required and how it is to be obtained.
7. Supplier certification programs. The larger material users and original equipment manufacturers (OEMs) have regular supplier certification programs. Ford's "Q1," Chrysler's "SCORE" and General Motors' "Targets for Excellence" are prime examples of programs that evaluate material suppliers for SPC compliance and other factors. These quality programs were positive in reducing cost and supplier quality as is Ford Motor Company's ongoing quality program. But, as with all programs, times change and Chrysler's "SCORE" (Supplier Cost Reduction Effort) program went away when they were acquired by Daimler-Benz. It was abandoned in favor of a more simplistic and combative approach of demanding concessions with price and cost from its suppliers.

   As with all programs, ISO/QS-16949 remains the major quality program in the manufacturing arena, and it seems that it will remain there no matter how many new quality improvement programs are developed.

   Many plastic processors require similar compliance from their suppliers. The intent is to reduce, and eventually eliminate, their own incoming testing and inspection. Once a supplier proves continued product reliability, you should be able to rely on incoming lot data. Only occasional audits and incoming material tests need be conducted to prove compliance. The suppliers can either keep their SPC data on file or submit periodic reports on product stability.
8. Successful suppliers should be recognized. Many companies are now awarding certificates, plaques, and letters of commendation to suppliers who successfully meet their goals. These are based on in-house formal rating systems that evaluate product quality, pricing, delivery, technical support, and overall cooperation in meeting customer goals and objectives. In these awards, individuals and departments should be credited with helping achieve this success.

The material processor is no longer at the mercy of suppliers. Increased emphasis on SPC and compliance to make a more uniform product has reduced the number of complaints. Processors now know or are becoming aware of the material property values they need to control. Their manufacturing equipment responds more successfully to off-specification materials and signals the operator if changes occur. More processors are now requesting material certification to ensure the material will meet part end-use and processing requirements. Quality is now a basic material requirement.

Even with approved material suppliers, it is still the molder's responsibility to pick up material differences during incoming inspection. This allows you to adjust machine process control variables to produce good parts. Many material suppliers now send copies of process control charts, developed during the manufacture of the resin, to show that their process was in control. This is good, as long as the data were generated for the material you are now testing and are not periodic process lot data that may not cover the material you received. This is expensive for the supplier but most are complying with this new request, which increases customer confidence that the material will process successfully.

Often, however, the process control records do not pick up subtle differences in resin variability, and production personnel are tasked with adjusting the manufacturing process. This is even more important, in higher throughput, lean manufacture, zero defects, and just-in-time (JIT) product manufacture. With these, setup personnel have less time to adjust the process variables, which is harder to do as resin types proliferate. Therefore, before production begins, define the variable properties of the resin that can cause process problems and determine which incoming tests will uncover them. This is done most effectively when first setting product quality limits and determining the range/spread of resin variables that could affect processing and part quality.

This assumes the part design, mold, and manufacturing equipment can produce a product to meet customer requirements. Therefore, even with the establishment of material limits and supplier certification, your incoming inspection department must be capable of testing the material prior to the start of production.

PRODUCT VARIABLE SPECIFICATION

Each resin has specific product variables that can be analyzed to classify a material's processing capability. Your material supplier can be of great assistance in identifying these resin processing variables. This is not the time to evaluate physical property testing variables, such as tensile strength or modulus; these should have already been established. You now need to identify any product properties, such as viscosity and molecular weight variations, which could cause variations during production. With limits established for these variables, tests on incoming resins will note any product variations. If necessary, minor process control changes can then be made. The data should then be compared with prior lots to evaluate how consistently the material meets the established limits. The production staff should never maintain manufacturing rates and part quality by continuously adjusting the machine variables because of lot-to-lot variability in the resin. If problems like this continue, a talk with the material supplier is justly warranted.

INCOMING MATERIAL TESTING

For incoming material test analysis, thermal analysis (TA) and gel chromatography (GC) are used commonly. They increase the customer's confidence that the resins received will process correctly. These methods pick up differences in material characteristics from lot to lot and determine how successfully a material will perform in your manufacturing equipment. You can also use this data to verify the quality of the incoming resin and lot-to-lot uniformity.

MATERIAL TESTING EQUIPMENT

The TA and GC systems analyze different aspects of a material. They can show variations between lots and can be used to accept or reject material at incoming inspection with a minimum amount of test time. The method you select depends on your own manufacturing requirements. Each instrument analyzes different characteristics of a resin and provides specific information on polymer properties. Therefore, discuss your needs with the material and equipment suppliers in order to select the system best suited for your application. Remember to select the equipment only for resin inspection, not analysis, unless you are concerned with the actual composition and percentage of a supplier's constituents.

Types of Tests

When the instrument is selected, define the type and degree of material analysis required. Three types of analysis are possible. Although they are often interchangeable and have some application overlap, the procedures satisfy different requirements. The material analysis can consist of resin inspection, resin analysis, and property testing.

Resin Inspection. This test compares the resin received with prior lots of resin that were successfully processed. It shows the number of constituents in a compound and their general type and percentages but not their specific identity. This is a go/no go type of analysis that compares the two measurement graphs or profiles to determine whether they match closely enough to accept the incoming lot for production.

Resin Analysis. Resin analysis identifies the actual polymers, additives, and ingredients, as well as the percentage of each item, in the compound. This type of analysis is usually reserved for research and development (R&D) work to identify specific modifiers and additives in a compound, but it can also be used for incoming material inspection.

Physical Property Testing. This test is usually reserved for analyzing a material's design and application potentials. Tensile, flexural, impact, creep, heat, moisture, and electrical effects are tested. OEMs who require these resin tests on their parts should rely on the data generated by the material/resin supplier who systematically tests the physical properties of each lot of resin they manufacture. This is a poor, costly, and time consuming method to use for accepting or rejecting material at incoming inspection.

Analyzing the Tests

TA and GC both offer the processor a highly sophisticated analysis system that classifies the polymer properties and anticipates processing information. The inspection data is collected from the supplier's initial lot data and from data on enough successive lots to establish the acceptance limits. The number of lots required for this will depend on their variation and the processing results of each. From this data, a master graph or curve is generated. This is used as an overlay to check the high and low data generated by tests of later lots of resins to determine if the material should be accepted or rejected. These curves can also be used to show thermal decomposition temperature, degree of crystallinity, level of impact modifiers, and molecular weight distribution.

A typical test, which takes 30 to 60 minutes, uses computerized software packages to format the data into usable information and graphs for lot comparison. The cost of incoming inspection quality test equipment, which begins at $30,000, varies with individual instrument suppliers.

If these instruments are to be used in your incoming material receiving laboratory, you need to know which instrument is to be used for each type of analysis; what material parameter is to be measured; the procedures for testing, calibration, preventive maintenance, and operator training; and the number of samples required to obtain reliable data. It is also very important to know if the supplier uses these analysis tests to qualify material before shipping. If not, what testing is done and on what type of equipment? There must be agreement on the tests necessary to determine acceptance or rejection of the material. Otherwise, misunderstandings will occur, and agreement on the standards for resin acceptance or rejection will be hard to rectify if a dispute occurs. Therefore, encourage your material supplier to run tests identical to yours on resins.

If you run identical tests and your supplier agrees on test specifications, you will have a high degree of certainty that shipments of resin will process almost identically. Table 7.2 lists the information that each instrument will provide on the various tests for composition, processability, or physical property listed.

The customer should not participate in physical property measurement testing of materials. This is the responsibility of the material supplier, who must note these values in its material certification sheets. But should you be forced to do this testing, your material supplier should mold and test samples from the same lot of material as you do. Request test bars and samples to compare the supplier's test data with the results from your test lab. Because some test variation always exists, the supplier's data and molded test bars will give you a standard for comparison.

TABLE 7.2. Material Test Methods.

Information	Process	Data Obtainable
Composition		
Resins, polymer modifiers	DSC	Types present, plus ratio if copolymer, blend, or mixture.
	GC	Polymers, oligomers, and residual monomers; amounts present.
Additives	DSC	Deduce presence and amount from thermal effects (stabilizers, antioxidants, blowing agents, plasticizers, etc.).
	GC	Detect any organic additive by MW.
Reinforcements, fillers	TG	Amount (from weight of ash).
Moisture, volatiles	TG	Amounts (from weight loss).
	DSC	Amounts (if sufficient heat absorbed).
Regrind level	DSC	Deduce amount from melting point shift.
	GC	Deduce amount from shift in MW distribution.
Processability		
Melting behavior	DSC	Melt point and range (each resin if blend); melt-energy required.
Flow characteristics	GC	Deduce from balance of high and low MW polymer in MW distribution.
Physical, Mechanical		
Glass transition temperature	DSC	Shown by step-up in energy absorption as resin heats (marginal for some semi-crystalline resins).
	TM	Detected by sample's expansion.
Cystallinity	DSC	Calculate percentage from heat of fusion during melting, also can find time-temperature conditions for desired percent crystallinity.
Tensile, flexibility impact	GC	Deduce from balance of high- and low-MW polymer in MW distribution.
	DSC	Deduce from overall composition.

Source: Adapted from Ref. [3].

It is the responsibility of incoming inspections to know what percentages of what additives are required to make a good product. Thermal analysis instruments, which can perform this type of incoming inspection, are the differential scanning calorimeter (DSC) and the thermogravimetric analyzer (TG).

Differential Scanning Calorimeter

The DSC (differential scanning calorimeter) test looks at and measures heat absorbed and released by a polymer sample as it is heated beyond its melting point and then cooled under controlled temperature conditions. All changes in a polymer's state, from softening to melting, recrystallization, or thermal degradation, are signaled by a heat flow change. These heat flow measurements are represented in a graph. As different constituents react, a "fingerprint" of the compound is produced. The typical equipment for DSC analysis is shown in Figure 7.1.

Based on experience, acceptable and unacceptable generated curves have been established. Therefore, using these graphs, a decision can then be made to accept or reject the lot. DSC for thermoplastics shows the melting range and melting point for each component in a resin and will indicate the percentage of each component present. For semicrystalline resins, it will also indicate glass transition temperature, percentage of crystallinity, and the optimum cooling conditions to yield maximum crystallinity for maximum part strength.

DSC can also monitor for proper additive content and confirm the presence of additives, plasticizers, blowing agents, heat stabilizers, and antioxidants that may affect end-use or processing properties. An example is given in Figure 7.2, which shows a typical DSC curve for a polyethylene (PE)/poly-propylene (PP) copolymer. The first of the two dips (endotherms) in the heat flow curve confirm the lower melting component PE, whereas the second dip confirms the higher melting component polypropylene. The percentage of crystallinity and FE/PP ratio can also be determined by computer analysis if required. The usefulness of DSC is illustrated in Figure 7.3. An overlay of an acceptable polymer curve compared with the test lot shows the latter lacked a sufficient quantity of ingredient (PE). This might cause a problem in end-use properties.

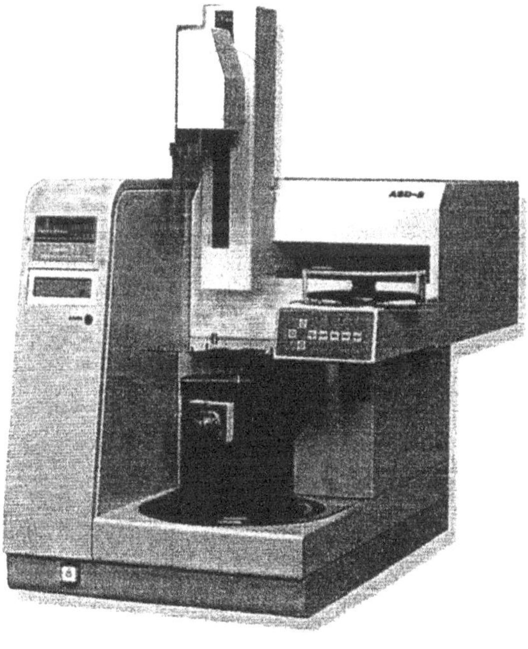

FIGURE 7.1. *Differential scanning calorimeter (DSC). (Courtesy of Perkin-Elmer Corporation, Norwalk, CT.)*

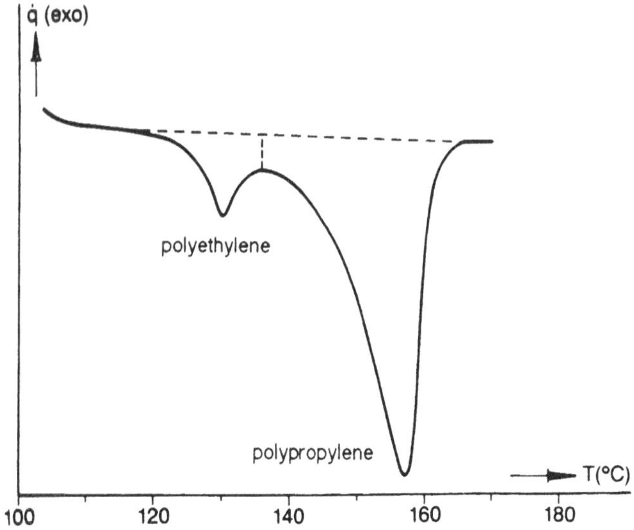

FIGURE 7.2. *Polyethylene/polypropylene copolymer curve. Polyethylene melts at a lower temperature than polypropylene, and consequently the relative peak areas can be used to determine the percentage of polyethylene in the blend. All blends of thermoplastics with different melting points can be analyzed this way. It is a very attractive feature of the DSC technique. (Courtesy of Mettler Instrument Corp., Hightstown, NJ.)*

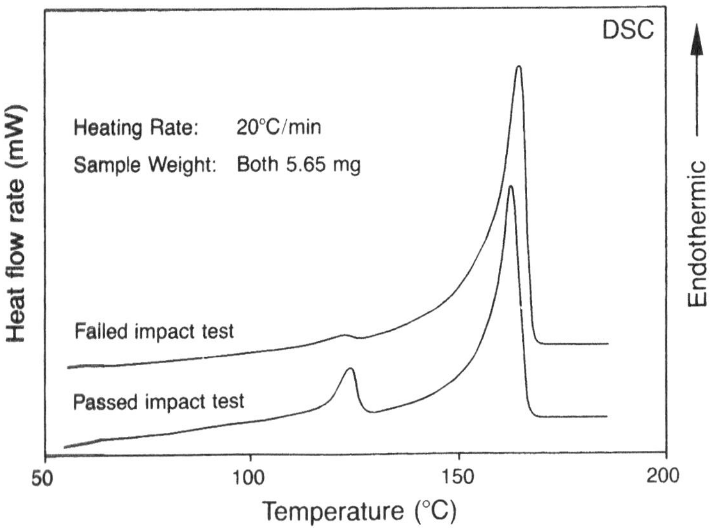

FIGURE 7.3. *DSC analysis of polyethylene/polypropylene copolymer. (Courtesy of Perkin-Elmer Corp., Norwalk, CT.)*

With proper inspection, this material lot should be caught when incoming and rejected.

Thermogravimetric

Thermogravimetric (TG) testing uses weight loss to analyze the compound additives that volatilize at specific temperatures. The equipment used is shown in Figure 7.4. The tests are done in an air or inert gas atmosphere up to temperatures exceeding 1000°F. With this method, inorganic fillers are analyzed by measuring the weight of the sample and comparing it with the resulting weight of ash of the remaining material from the volitalized organic polymer. Knowing the percentage of filler content is important not only for end-use physical properties but also for material verification. Filler content also affects the processing of amorphous and crystalline resins. Furthermore, the process ability and flow of amorphous resins is affected by filler content and processing temperatures. Therefore, increased heat or higher pressure settings are required to push the polymer more easily into the mold cavity.

For crystalline polymers, with their sharp melting points, this also applies to melt flow, which is a function of the base resin's molecular weight and its resulting viscosity. Crystalline polymers usually have good flow, even when highly loaded with fillers and additives. In fact, these additives aid in reducing molding cycle time, as they act as nucleation sites for the start of crystallization when the resin is cooling in the mold cavity. Therefore, many fast-molding

FIGURE 7.4. 1020 Series TGA7 Thermal Analysis System. (Courtesy of Perkin-Elmer Corp., Norwalk, CT.)

crystalline polymers have talc or other ingredients added to minimize the setup time.

As with crystalline polymers, amorphous polymers must also cool sufficiently in the mold before ejection. Similarly, filler in these resins aids in stiffening the part enough to allow earlier ejection without distortion. Therefore, the molding cycle for reinforced and filled amorphous resins can be shortened if the appropriate percentage of filler is present. When a filler is used, highly filled amorphous polymers require higher injection pressures. This is caused by their increased viscosities. Therefore, knowing the correct percentage of filler is important for adjusting processing conditions.

Gel Chromatography

Gel chromatography (GC) is one of a family of liquid chromatography (LC) techniques used to analyze the composition of a polymer compound by separating its components on the basis of molecular size or molecular weight. Each component in a polymer is unique in its individual molecular weight. Items such as polymers, oligomers, and monomers, or additives, such as plasticizers, impact modifiers, processing aids, and antistats, can be separated by type and size. The process requires that the sample be in liquid form; most solid organic materials can meet this form and be identified. This includes virtually all thermoplastics and crosslinkable compounds, such as crosslinked PE.

The analysis begins by pumping the liquified sample at a controlled rate through a filter medium made with a precise range of pore sizes. The larger molecules exit the filter first, with the smaller exiting later, in order of molecular weight. This occurs because the larger the molecule the less deeply it penetrates into the pores of the filter.

From the separated stream of molecular compounds, the instrument develops a strip chart. Figure 7.5 shows a peak for each different material in the

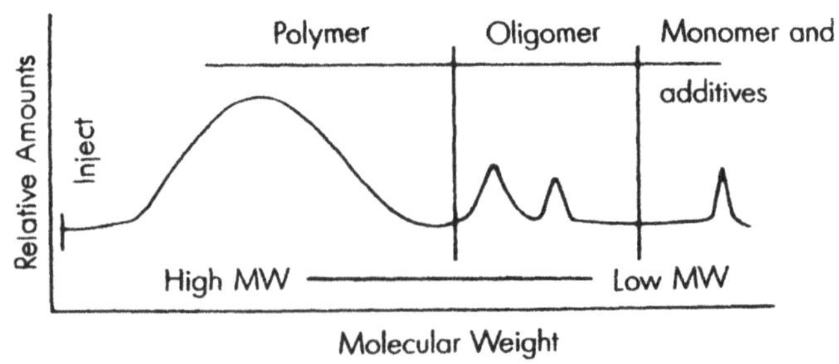

FIGURE 7.5. An idealized chromatogram shows the elution order of various components. The largest molecules elute first, followed by the oligomers, and finally by the unreacted monomer and additives. (Courtesy of Millipore Corp. All rights reserved.)

sample. The height and area under each peak correlates to the concentration/percentage of each item in the resin. The strip chart produces a "fingerprint," similar to the DSC plot, which can be used to accept or reject the material. If identification of components is required, reference curves of known molecular compounds can be compared with their molecular-weight peaks.

GC offers valuable insight into the processability and performance of a resin that cannot be obtained by ordinary resin tests. The analysis of molecular-weight distribution (MWD) of a polymer shows how the molecular weights are distributed around the average MW value and identifies the high and low limits of the MW range. Molecular-weight distribution is a good predictor of how a resin will process and of its end-use properties. An MW skewed to the high side indicates higher viscosity (harder flow), tensile strength, flexural strength, and melting point. A lower skewed value would reverse these values as shown in Figure 7.6A.

A good rule is to set a specification for MW values comprising the upper and lower 20 percent of the distribution curve. Figure 7.6B shows an example of molecular weight shifts that affected molded part quality. This could have been detected at incoming inspection, allowing the molding conditions to be altered appropriately. Resins B & C were skewed in their MW percentages from the high and the low side of the required resin curve A. This means that an alteration of the processing parameters is required to meet part quality requirements.

Besides these tests, which require more expensive and analytical equipment, other tests for incoming materials can help insure resin uniformity. These are not as specific as the analytical tests in determining how the materials will process but can identify the materials and verify that they are within the processing window. Some resins look very similar, and you do not want to mix dissimilar materials. These tests can identify and classify resin regrind quality before the mixture is blended with virgin resin. Most plastic regrind can be blended back at a 20 to 25 percent level, depending on finished part requirements, as long as it is dry and not contaminated.

In more critical parts allowed regrind, usually after the fouth complete regrind pass through the material happen, it is not saved. This ensures any regrind remaining from the initial first pass is used up and will not affect the physical properties of the molded part.

This method of regrind controls the parts with sprue. And runners saved, ground up, and fed back into the machine's hopper/dryer in metered proportions to keep the virgin resin and regrind in the correct ratio to produce acceptable parts.

Regrind should be used as soon as possible and protected from excessive moisture pickup of polymers have a high affinity for moisture pickup if not kept in moisture barriers logs and/or containers.

The small custom molder, who handles many generic plastics, has a need for less expensive and more easily operated equipment to run

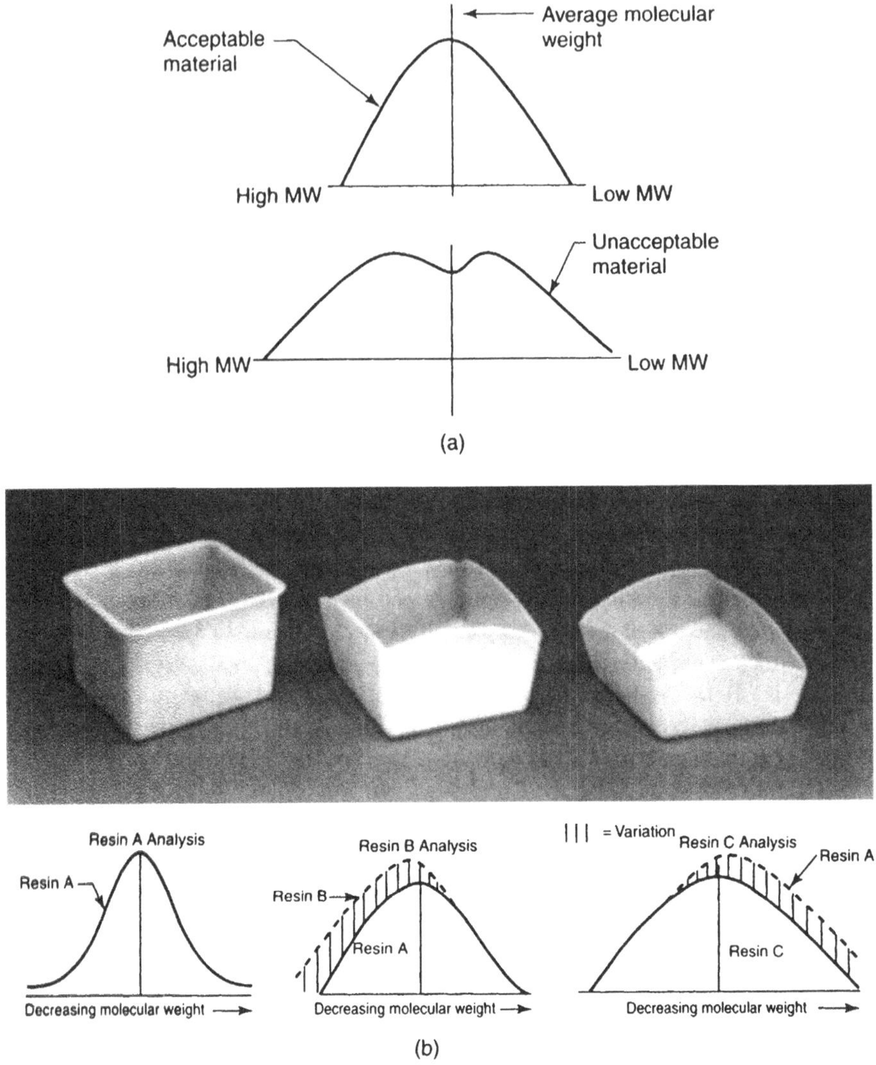

FIGURE 7.6. *The effects of molecular shifts on part quality: (A) Molecular-weight distribution; (B) GC comparison of high-density polyethylene resin variances after meeting incoming melt index specifications. Slight shifts can cause part quality problems. (Courtesy of Millipore Corp. All rights reserved.)*

specific tests that will qualify and identify incoming materials. The tests used by these processors should easily pick up resin variability. If resin differences are noted on testing, they can then reorder the material or contact their material suppliers to determine how they can adjust their process to obtain good parts.

Test Methods

These tests are:

	Automatic Analysis By:
Moisture analysis	TG/DSC
Melt index (viscosity)	TG
Spiral flow	GC/TG
Glass/mineral/filler content	TG
Specific gravity	GC
Melt and softening point	TG
Resin colors	—
Resin contamination	—

Moisture Analysis. Plastics that are moisture sensitive or hygroscopic must be within a specified moisture level for processing. As a result, many molders automatically dry all their resins. But, if the incoming moisture percentage is too high, it may not be practicable or possible with their equipment.

Excess moisture in the polymer can cause aesthetic surface damage, internal problems, processing problems, and may also reduce the physical properties of the molded part. Excessive moisture in resins during processing can cause the polymer chains to fracture, thereby reducing melt viscosity and toughness. Some polymers, even if wet, will process normally, but the parts will become weak and brittle. Examples are polybutylene terephthalate (PBT) and polyethylene terephthalate (PET) resins. The only way to be sure of moisture content is to test all incoming materials. This is especially true if the material is to be stored in large silos or bulk systems, as you do not want to contaminate the resin previously accepted with wet resin.

All resins are shipped from their manufacturers at or below their recommended processing moisture levels. The moisture level of each resin will vary with its generic composition. When modified with fillers, reinforcements, plasticizers, and lubricants, the moisture level will have to be even lower for satisfactory processing. For example, begin with a generic unreinforced resin with a processing moisture level of 0.2 percent. If filled, it must have a lower moisture level. How much lower depends on the percent of filler, because filler lessens the amount of resin available to absorb moisture. What is an acceptable moisture level for one material may not apply to another. This applies to even other resins in the same generic family. Always consult your material supplier as to the maximum and minimum allowable moisture levels a resin can have before processing. Also, remember some resins require a minimum moisture level and, if overdried, can cause processing problems.

The important information to determine is the acceptable moisture level for processing. If the level is too high when a resin is received, the resin can be dried. Most molders dry their resins to flash off surface moisture to avoid processing problems. It is important during drying that the correct temperature and time are used so that helpful additives are not driven out of the resin. Overdrying can also affect the color of heat-sensitive resins and pigmented grades, as they may oxidize and turn brown. Resin received at the correct moisture level should only require hopper dryers at the machine to maintain the resin at the correct moisture level while waiting to be processed. Care must also be exercised with organic pigmented resins, which have a lower heat sensitivity and can fade or burn out if exposed to extreme heat.

The easiest and most common incoming test for moisture analysis is to weigh out a specified sample into a dry container, dry it at a specific time and temperature using dry air, and reweigh the sample using a digital scale. Then calculate the difference in weight loss as a percentage of moisture in the sample. The correct time and temperature profile must be used, so that any friendly volatiles in the resin are not driven off and interpreted as moisture loss. This is a common mistake made by technicians if they do not have the correct temperature and time values for conducting this test.

All resins are shipped by suppliers to customers in moisture-proof bag and box containers (gaylords), including bulk tankers and railcars. These packages are proven moisture barriers, but there is no guarantee that proper sealing of package was obtained and that the moisture barrier of the package was not damaged during transit. Therefore, before processing, moisture level checks should be conducted. In bulk container shipments (tankers/railcars), it is even more important, as condensation may occur within the container, depending on the season, and water can collect at the bottom. In some cases, if the wet material can be diverted or segregated, the remainder of the material in the container may be dry and acceptable. Only testing will prove whether this is true.

Melt Index. Melt index: ASTM D-1238 (Viscosity), or the measurement of the melt flow of a material, measures lot-to-lot resin uniformity precisely. This test, which provides a good indication of how the material will flow in the tool under typical molding conditions, is a good test for determining acceptance or rejection of materials. It is important that resins are dry before testing, as moisture can cause variations in flow rate. Therefore, a moisture test should always be run prior to a melt test. If the resins are dried prior to processing, this should also be done before performing this test. This will correctly classify the material's flow for identification and aid the production people in understanding how the resin will process.

The test requires an extrusion plastometer (funnel-shaped device), which is a heated container with a die opening of specified diameter and length that has a piston to fit the barrel. To perform the test, resin is loaded into the funnel. Under specified conditions of temperature, piston weight, and position in the barrel, an amount of polymer is extruded during a specified time period. The sample is then weighed, and the results compared with standards and previous lot melt index data for acceptance or rejection. Figure 7.7 shows the

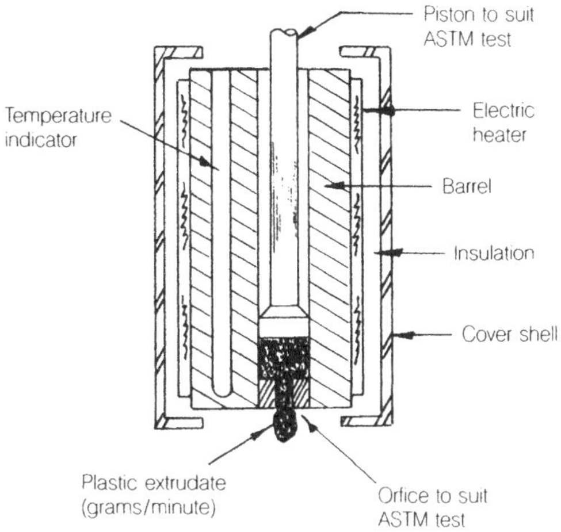

FIGURE 7.7. Melt index testing procedures. (Adapted from Ref. [1].)

TABLE 7.3. Melt Index Conditions of Thermoplastics.

Material	Temp (°F)	Load (gm)
Acrylics	446	1200
	446	3800
ABS	392	5000
Polystyrene	392	5000
	446	1200
	446	3800
Nylon 6	455	1000
Nylon 6/6	527	325
Polyethylene	257	325
	257	2160
	374	325
	374	2160
	374	21600

Source: Adapted from Ref. [1].

basic test process and Table 7.3 lists some generic resins along with test conditions indicating temperature and load requirements. The actual equipment to perform this test is shown in Figure 7.8.

In performing these or any other tests, at least three tests should be run to determine the slope of the data curve. Usually, five samples are run from each incoming lot to determine the results and make sure the data are reliable. Before and after each test, check that the equipment is clean and in calibration. The failure to do this can result in erroneous flow data.

These data are useful in many ways. A minor shift up or down may not always be cause for rejection, but it will enable the production setup team to

FIGURE 7.8. *Melt index testing equipment.*

adjust processing conditions, barrel heats, screw speed, back pressure, and mold temperatures to produce the product required.

Another test, similar to melt index, is called the capillary rheometer test (ASTM D-3835). It measures flow (viscosity), plus shear rate and shear stress on the resin. This rest is used for more viscous resins used to produce sheet, film, tubing, and profile shapes, where lot-to-lot variability could cause more and different problems for injection molding.

Spiral Flow. The spiral flow test is a form of viscosity measurement used to measure the flow of resins in a mold. Using a special mold, as shown in Figure 7.9 with a specified section thickness and circular flow path length, materials are evaluated under typical injection molding processing conditions. The length of flow in the mold is a function of a resin's viscosity versus injection pressure, fill rate, melt temperature, and setup. All these factors affect the resin's flow length. The lower the viscosity, the longer the flow. This test is less sophisticated than GC, but under standard molding parameters of pressure and temperature, the resin's flow length can be measured. Variations in flow (viscosity) will give an indication of the material's flow properties, crystallinity and freeze-off time for each lot of material. In hard-to-fill tools, it can indicate whether the resin has the flow length necessary to fill the tool under standard molding conditions. Any major flow variances in lot-to-lot testing would justify the use of more sophisticated viscosity measurements and a discussion with the material supplier.

As an example, a standard grade of ABS was evaluated in a spiral flow mold with varying melt temperatures and fill rates or injection speeds. Another variable was mold temperature, which also has an effect on flow length but is not as significant for ABS. The flow length changes are given below.

FIGURE 7.9. Spiral flow test mold. (Courtesy of DSM Engineering Plastics North America.)

Melt Temperature (°F)	Fill Rate (in/sec)	Flow Length in Inches for Mold Temperatures —°F		Change in Percentage
		85	175	
425	0.25	12.1	13.7	13.0
425	1.75	20.5	21.6	5.4
525	0.25	22.0	25.4	15.5
525	1.75	37.3	39.6	6.2

An example of generic resin flow qualities is the viscosity of flow length of nylon 6 versus 6/6.

Cable ties vary in the length and the number of individual cavities in a mold. Nylon 6/6 is the preferred material based on flow to fill qualities. Nylon 6 has higher elongation but lower flow qualities versus 6/6. Each can perform equally well but 6/6 is the prime material based on flow and economics.

One must also remember that the mold cavity gate size has a definite effect on resin melt temperature (shear heat). This affects flow as well as the physical properties of the ABS and other shear sensitive polymers, such as PVC. If a resin picks up too much shear heat when pushed through a small gate, the physical properties of the resin and molded part will be lowered dramatically.

Glass/Mineral/Filler Content. Resins with inorganic fillers, such as glass and minerals, are tested for filler content by heating a weighed sample of material at an elevated temperature to yield only the filler or ash. This ash is then compared with the sample weight, and a weight average is calculated to determine the percentage of filler.

In discussing a resin's filler content, most processors think only of glass, mineral, talc, or other nonpolymer additives. They often forget the ingredients added to impart toughness as well as impact and tear strength to polymers. It is the responsibility of the material suppliers to ensure that these copolymers and reinforced or filled resins have the correct amount of the right additives when shipped. Verifying this data is often done to avoid processing and end-use part problems.

Specific Gravity. This test is normally not run during a processor's incoming inspection. This property is controlled by the material supplier during production. The GC, DSC, and TG tests will accurately measure those polymer properties necessary to verify specific gravity. It is important to know the material's exact specific gravity, because it affects your material costs and final piece-part price. The factor 0.0361 lb/inch^3 is multiplied by the polymer's specific gravity to calculate part weight in cubic inches.

Melt and Softening Point. The standard used in the plastics industry to measure a polymer's melting or softening point is the Fisher–Johns method. It requires a heated stage and a hot plate with a direct reading thermometer attached. The sample is wet with a drop or two of silicon oil and sandwiched between two glass slides, with the oil forming a miniscus on the glass, and placed on the heating surface. The temperature is then slowly increased and viewed under a magnifying glass. As soon as the miniscus moves, when the pellet melts or softens, the melting point of the plastic is reached. This ASTM D-789 test is suitable for all thermoplastic resins, both the amorphous and semicrystalline, and is shown in Figure 7.10.

The Vicat test shown in Figure 7.11 measures the softening point of both amorphous and crystalline thermoplastics. This test is run in a temperature-

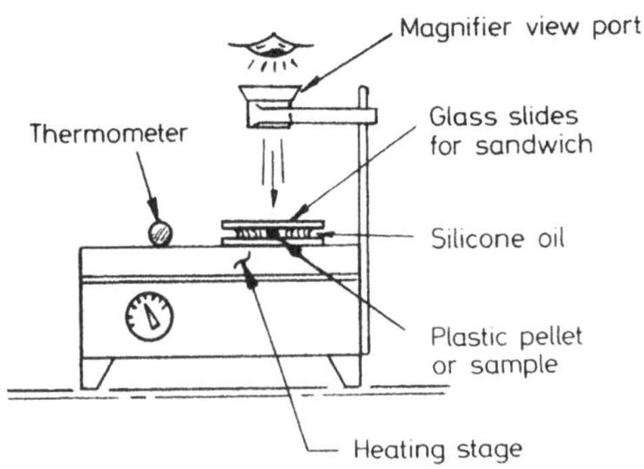

FIGURE 7.10. Fisher–Johns method: Measuring the melting or softening point of thermoplastics. (Adapted from Ref. [1].)

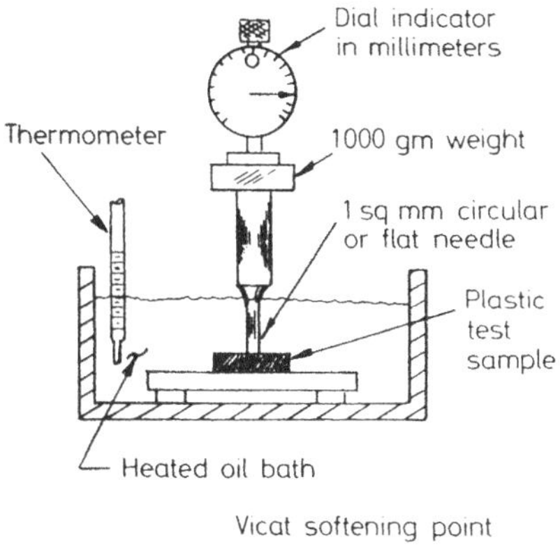

Plastic	Softening Point (°F)
ABS	190–257
Acetal	323–343
Nylon 6	420
Nylon 6/6	490
Acrylic	185–248
PE (low density)	179–210
PE (medium density)	221–237
PE (high density)	255–260
PP	291–302
PS (general purpose)	213–233

FIGURE 7.11. *Vicat softening test. (Adapted from Ref. [1].)*

controlled oil bath using a flat-ended needle (either circular or square cross sectioned) of one square millimeter. The test sample is placed in the oil bath under the needle, which has a load of 1000 grams on its shaft. The dial gauge reading is recorded. As the temperature is slowly raised, 50°C per hour, the needle will penetrate the sample when the softening temperature is reached. When it has moved exactly one millimeter, the temperature is recorded as the Vicat softening temperature for that material.

The Fisher–Johns test method can also be used to estimate the dryness of a resin or regrind prior to molding. To check this, the sample is put between two glass slides coated with silicone oil. The slides are preferably preheated on the heating surface to reduce test time. After complete melting of the sample, the glass slides are compressed together to form a layer of resin on the slides. If bubbles form in the resin puddle, it is wet. This indicates that drying is required, but will not indicate the percentage of moisture present in the resin. After drying the material, the test should be run again to verify dryness. This test takes only a few minutes and is a fast reliable check on the dryness of the resin prior to molding. The Fisher–Johns test is not reliable for super moisture sensitive resins such as PBT, PET, and PPS.

Resin Colors. Each generic plastic has its own distinctive color in its virgin state—pellet or powder. The type of additives and its content, however, affects this characteristic color. Some are crystal clear, such as acrylics, polycarbonates, and polystyrenes; some are semiclear, such as polyethylenes, polypropylenes, and nylons; some opaque, such as acetals, ABS, and filled or reinforced polymers of the above.

Each generic polymer may also differ in color from different suppliers, because some materials, such as nylon, may tend to surface oxidize and turn color during manufacture. Suppliers grade these resins as to their yellow index or degree of surface oxidation. Nylon tends to oxidize on the surface if not cooled properly; other resins exhibit similar pellet color problems if not properly manufactured. Therefore, depending on the final use of the part and whether it requires an appearance finish, resin pellet color is important.

When molding a compounded color or using color concentrates blended with virgin resin, the molded part's color becomes very important. If a part's color is important, then a color sample must be used as the standard for grading the color of the molded part. These color standards are available from your material supplier and should be obtained at regular intervals from standard colored resin production. If you do the blending, then a colored part, acceptable to the customer, must be retained as the color sample for current and future production evaluations. When not in use, the color standard should be stored in dry containers in a temperature-regulated room, away from direct lighting. The color standard is then compared with a color plaque received with an incoming lot under a known light source by a specific quality control (QC) inspector, who is trained to notice resin color variations. Color standards should be replaced at least every 6 months, as they will fade with time and exposure to light and humidity.

The same QC person should also inspect incoming color-compounded resin samples. Because working with colored resins is more difficult, some resin suppliers now furnish color chips molded from the subject lot plus color analysis data and chromatography readings of color values from the sample. The same is true for salt-and-pepper (S&P) color blends (pellets of concentrated color blended with virgin resin) to verify that the correct blending ratio was used with the virgin base resin to meet the color standard. Because cadmium and heavy metal pigments are being replaced for health and environmental reasons, processing conditions are even more critical when using organic pigments. These pigments are more "heat sensitive" and any prolonged delay in the molding machine (usually over 10 to 15 minutes depending on pigment type) can cause a color change (burn out of pigments) in the resin.

Care must also be taken when selecting pigments, so that the base resin additives, heat stabilizers, antioxidants, and plasticizers do not cause a color change during processing. When doing a color match, always confirm that the match is being conducted with the base resin that will actually be used to make the parts.

When purchasing these pigments in concentrate form, such as powders, liquids, or color concentrates, have the supplier certify the pigment loading levels. This is to be sure the correct let-down blend ratio will be used when blending the pigment with the virgin resin to produce the correct product color.

Both the color concentrate supplier and compounded resin supplier should understand that any change in the type of pigments used or their additive levels must be submitted to you and the customer for testing and approval. Even a slight change in pigment types and percentages can sometimes lower the material's physical properties.

Some pigments will have an adverse affect on a resin's properties even in percentages of less than one. Consult with your material supplier if you are not sure of a pigment's effects on a resin before using it for a critical part.

Resin Contamination. Checking incoming resins for contamination usually consists of a visual check on a 1 to 2 lb sample drawn from an incoming lot. The sample is spread on a large well-lighted surface, which will adequately expose the resin for inspection. If possible, the samples should be obtained from the center of a box or at frequent points within the size of the lot of material received.

The common forms of contamination are as follows:

1. Burned resin or off-color pellets
2. Die drips—carbonized resin drool from the die face
3. Marriages—pellets poorly cut and sticking together
4. Foreign material in the resin
5. Excessive fines or flakes.

Because all forms of contamination are bad, it is a judgment call whether to accept or reject the lot of resin. Slightly off-color resin may not affect the physical properties of a part, but if appearance is critical, any carborized or off-color resin may cause rejection.

If any of the above problems are found, these lots or packages should be questioned. Draw additional samples to determine whether it is just a local problem or whether it exists throughout the lot. Samples of the contamination should be sent to the supplier and noted in your records by material description, lot number, package number(s), date received, and customer contact person. If necessary, your supplier should replace the lot.

All too frequently, supplier error or poor housekeeping results in contamination in either virgin resin or regrind. A speedy analysis is required to avoid further quality problems. Table 7.4 lists identification tests that can be performed by burning pellets, parts, or regrind, and then by observing the color of the flame, smoke characteristics, and odor of the gases once the flame is extinguished.

The test is easily performed by safely holding the test specimen and igniting it. Let it burn for a few seconds, observing the flame and any smoke or particles given off. When extinguished, use your hand to fan the fumes carefully to your nose. Then, based on your observations, you can draw a conclusion as to the generic base of the test specimen. This simple test can save time and hopefully solve or eliminate a future quality problem because of cross contamination of different materials. Note that this test, especially the breathing of the fumes, should not be performed by anyone with a high degree of sensitivity to any plastic resin or compound. Other more sophisticated tests will be required for positively classifying the specific material type and grade.

TABLE 7.4. Identification Tests for Thermoplastic Materials.

Resin	Burning	Odor	Melt (°F)
Acetal	Blue flame, no smoke, drip may burn	Formaldehyde	323–347
Acrylic	Blue flame, yellow top	Fruitlike	374
Acrylic rubber mod.	Yellow flame—spurts	Use control	279
ABS	Yellow flame, drips black smoke	Use control	—
Acrylonitrile styrene (As)	Yellow flame, black smoke, clumps of carbon in air	Illum. gas	268
Cellulose acetate	Yellow flame, sparks, drip may burn	Acetic acid	446
C.A. butyrate	Blue flame, yellow tip sparks, drip may burn	Rancid butter	365
Cellulose nitrate	White rapid flame	Sharp Decom	—
Cellulose propionate	Blue flame, yellow tip	Fragrant	456
Cellulose triacetate	Yellow flame, drips	Acetic acid	572
Ethyl cellulose	Yellow flame, blue top, drip may burn	Burned sugar	—
Ethylene	Blue flame, yellow top, drip may burn	Paraffin	221L 246H
Methylstyrene	Yellow flame, black smoke, carbon in air, softens	Illum. gas	349
Polyester film	Yellow flame, smokes and drips	Use control	482
Propylene	Blue flame, yellow top swells and drips	Sweet	334
Styrene	Yellow flame, dense smoke, carbon in air	Illum. gas	374
Styrene	Yellow flame, smoke	Use control	351
Vinyl acetate	Yellow flame, smoke	Acetic acid	140/190
Vinyl alcohol	Yellow flame, smoke	U.C. soapy	446 DECOM
Vinyl butyral	Blue flame, yellow top melts and drips, may burn	Rancid butter	345
PVC–Polyvinyl acetate (PVACET)	Yellow flame, with green	Hydrochloric acid (HCL)	261
Carbonate	Decomposes	—	430
Chlorinated ether	Sputters, bottom green, top yellow, black smoke, carbon in the air		358
Nylon	Blue flame, yellow top, melts and drips	Burned wool	
6/6			490
6/10			415
6			420
11/12			351
Vinyl chloride polymers	Yellow flame, green at edges, softens and chars	Hydrochloric acid (HCL)	302
Vinylidene chloride polymers	Yellow flame, ignites hard, green spurts	Hydrochloric acid (HCL)	313
Fluorocarbons	Deforms	—	
Fluoroethylkene polymer (FEP)			554
Tetra fluoroethylene (TFE)			621
Vinylidene fluoride polymers	—	—	340
Fluorochlorocarbons	Deforms, slight melting, drips	Weak acetic acid	383

MATERIAL SAFETY DATA SHEETS

The Occupational Safety and Health Administration (OSHA) now requires that Material Safety Data Sheets (MSDS) for all resins and other materials used in a plant be on file. These MSDS are provided by the material suppliers and state whether precautions and special handling must be taken during processing of these resins. They provide information on base ingredients that is consistent with the OSHA directives that must be followed when the materials are used in your plant. Because some people are sensitive to these materials, OSHA guidelines aid in protecting the health of your employees. They are usually filed in a binder with a duplicate copy available for review on the production floor. Purchasing is responsible for obtaining MSDS for all materials processed in the plant. These sheets should always be updated annually in case a question or problem arises in the plant. Some resins also have warning labels printed on their shipping packages to insure proper handling of the materials. Thus, if a problem or question occurs, the MSDS can answer the concerns of the workers.

RECORD ACCURACY

Accurate recordkeeping during incoming material inspection is very important. It is necessary to document and verify that the material meets requirements. It is also compared with other lot data to guarantee that the material received is within specification. In addition, if a problem should occur, you can compare the data on earlier lots with the problem lot samples. Often, when problems occur during or after manufacture, the supplier and part's manufacturer do not have complete information to document the lot of resin used for the part's manufacture. This complicates problem solving and assigning responsibility for the solution. Resin documentation must always begin at receiving, and the material identity should be traceable through production to the shipped article. The customer may also require the part's supplier to furnish resin lot traceability for the materials used in their parts for the following agencies: Underwriters Laboratories (UL), National Sanitation Foundation (NSF), and the U.S. Food and Drug Administration (FDA).

Setting up your records for material testing and identification of material flow through your plant's manufacturing system need not be difficult if procedures are properly established, personnel trained, and directions followed. This procedure will also assist you in keeping control of your regrind, which may be recycled during production or reused later by blending it with virgin resin for other less critical parts.

The material record sheet should follow the production of the part and become part of the job order documents. At the end of production, you should be able to trace each material's path through the plant, including its lot number, package type and number, day and time processed, and box number of parts when shipped. This is only complicated if procedures are not set up and followed diligently.

BAR CODING: AN AID IN TOTAL QUALITY PROCESS CONTROL

Bar-coded labels placed on the material packages aid in accurately tracing their flow through the manufacturing system. You may be able to use the supplier's bar-coded labels, as most will include the product code identification and package numbers. Check with your suppliers to determine whether their labels are adaptable to your bar-code reader system.

The bar-coding system can also be used to track and identify tools, processing equipment, personnel, and workflow within your production system. This helps verify the accuracy of manufacture and production output. A basic material incoming inspection receiving and material flow sheet is illustrated in Figure 7.12. It can be revised as required for your operation.

Once the material is received, inspected, and found acceptable, it should be stored in an area reserved for virgin materials and properly identified for the job and its location in the warehouse. If for any reason, it is suspect and

Job Description ________ Date ________
Part Number ________
Quantity to be produced ____ Tool No. ____ Machine No. ____
Dryer No. ____
Regrind allowed __ % ____ Time/temp __/__
Blend back how __ When __ Feed system ____
Container No. ____
Material ____________ Lot No. ____ Package type ____
Package No. __ to __
Accept/Reject ____ If No, reason ______ or ________
Stored ______ Disposition ____________
Incoming Inspection Certification received ____ From ____

Test Data			Tested By	Supplier Lot Test Data	Customer Requirement data
Test Type	Results	Standard		Specific/Periodic (List data or attach cert sheet)	
Moisture	____	____	____		
Viscosity	____	____	____		
Filler (%)	____	____	____		
____			____		
____			____		

PRODUCTION START

Date/Time	Lot No.	Package No.	Operator	Part No. Box	Assy w/Part No. Box	Ship Box No.

FIGURE 7.12. Inspection and material flow sheet.

must be returned to the supplier for investigation, it must be segregated and labeled in a visible location so it is not used in error. Avoid writing on the package or defacing the label, as it may later be found satisfactory. Such writing may also cause problems for the supplier, when material is received back at its warehouse. Usually a red warning label, with information written on it to indicate the reason for rejection or reexamination, will suffice. This does not deface the package.

REGRIND CONTROL

Regrind should never be stored with virgin resin, as it is often stored in used packing containers that still have the original material identification on the container. A different "colored label" should be placed over the original package label to avoid any misidentification of the material. Regrind should also be stored in moisture-proof containers and properly covered and protected to avoid contamination.

MATERIAL HANDLING AND STORAGE

The handling of resins from receiving through production is critical for a TQPC system to function correctly. All material must be properly identified and stored in a known location, with the intended program identification information affixed to the container. This will help ensure that the correct material is always used for the intended job. These can be either specific part number labels that the supplier affixes to the material packages before shipping or labels the plant puts on after receiving and inspection are completed. To prevent contamination, care must be taken that the package is not damaged. The material must also be stored away from heat to avoid degradation during storage. If stored in a nontemperature-controlled warehouse, the material should be brought into a warm area of the plant during the winter at least 24 hours prior to use, so it can warm up. This avoids condensation on the resin prior to manufacture. Plants with bulk silo storage facilities and closed feed systems can pump material directly to the machine hopper if it is kept dry. There is often an in-plant storage system to hold, dry, and condition (warm up) the material before feeding it to the molding machines. When resin is delivered to the molding machine in bags, drums, or gaylords, the packages must never be opened and left exposed to the plant environment. The resin must be protected. Only the number of bags required to supply the hopper should be opened at any one time. Gaylord box liners should only be opened to accept siphon tubes and then tightly sealed around the tube to keep out moisture and contamination. There are drum and gaylord covers available to provide this extra degree of protection for the resin. Do this even when hopper dryers are used, as moisture sensitive resins will pick up moisture very rapidly in high-humidity level plant environments.

Material is conveyed to the machine hopper in several different ways, and care must be exercised in preventing contamination during transfer. Depending on the size of the hopper and material usage in pounds per hour, no more than 1 hour's worth of resin should be loaded into the hopper. This is done to minimize moisture pickup if a hopper dryer is not used for hygroscopic and nonhygroscopic resins.

When using the hopper dryer to dry the resin, the volume of resin in the hopper must be adjusted for throughput rate and the time and temperature required to dry the resin adequately. Too high a drying temperature for too long a time can degrade the resin. If the dryer is only used to maintain the dryness of the resin and avoid subsequent moisture pickup, then the temperature of the dryer should be reduced accordingly. In all cases, the hopper must be covered to prevent contamination.

Before loading the hopper with resin, make sure the material specified on the job description matches the material package exactly. Many resins look alike, and positive verification of the correct material is essential by the operator and the job setup person. Subsequent lots brought to the machine must also be verified by the operator and quality assurance inspectors. The use of bar-coded labels can aid in preventing use of the wrong resin.

When blending colors at the hopper with virgin resin, the correct pigments or color concentrates and blend ratios must be verified. If mixed elsewhere, the correct weight ratio must be established and verified and mixed properly. For two reasons, any S&P blends of resin, should be reblended prior to loading into the hopper. First, the resin and concentrate are often of different pellet sizes and density. Second, during shipping these pellets are shaken in their packages. As a result, the denser pellets will tend to settle on the bottom of the container, which is usually the color concentrate. Therefore, if not properly mixed prior to loading, color variations could occur. Although there may be a cost savings when using S&P color blends, in some cases they are only made by the supplier and the processor must be sure that the resin and color concentrate blend is homogeneous.

Operators should be trained to recognize the correct color of parts produced during molding. Often an approved colored part is kept at the machine for this purpose. But with some pigment systems, the true color may not become stabilized until the part has cooled down. In most cases, incoming inspection has verified the correct color of the blend. There is also the supplier's certification that the color match is correct. But if a doubt exists, use a TA test to verify the type of pigment and its percentage in the molded parts. To avoid destroying a good part, a runner sample can be used for this test.

REGRIND USAGE

Regrind from out-of-specification parts and runners can be used in most parts without drastically affecting physical properties as long as it is not contaminated and kept dry. In some cases, however, it may not be permitted because of critical part tolerances and possible effects on the processing parameters and end-use properties of the part.

Typical regrind levels used with virgin resins are 20 to 25 percent. Studies have shown that with good regrind feed control, the effects of feeding first pass regrind back into the system are, after the fifth pass, negligible.

However, with glass reinforced resins, each pass of regrind continues to break down the glass fiber lengths. This causes a lowering of the resin's physical properties. The glass fiber's length has a definite effect on the retention of physical properties of a resin. With these polymers, some end-users restrict the percentage of regrind to a lower level. Testing will show the allowable level that can be used based on end-use requirements.

Measurements and tests can be performed on the parts to be sure the processing, dimensional, and end-use properties are not affected by using regrind. What usually occurs when using regrind is that the polymer chains and bonds are broken during subsequent reheating cycles. This lowers polymer viscosity and thereby affects the strength and impact qualities of the resin. Figure 7.13, which shows a 100 percent regrind study of a homopolymer resin, illustrates these effects. With other resins, the effects may differ depending on processing conditions and each resin's thermal stability.

With colored parts, regrind will dilute the pigment system. Color concentrate must therefore be added to keep the correct color. Most molders do not use regrind for colored parts. If they do, they use such a low percentage that the cost may not be justified with a critical color. In addition, if using organic pigments, which burn out faster with each pass through the machine, additional concentrate may have to be used to control color. As a result, most

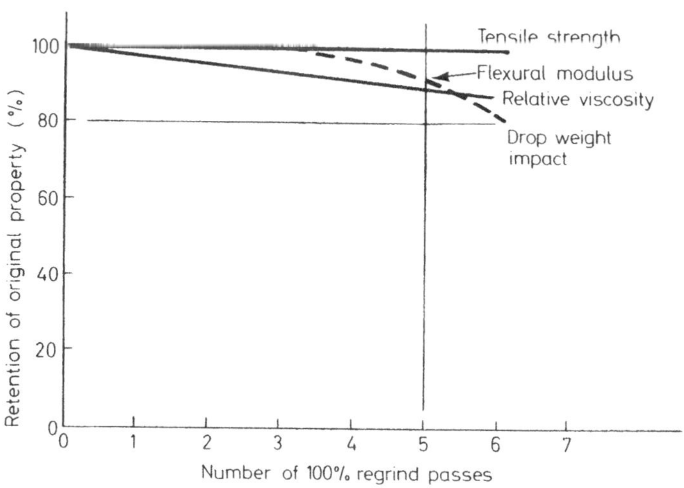

FIGURE 7.13. *Homopolymer resin regrind study.*

molders of colored parts refuse to use regrind. It is reserved for black parts and noncritical parts, where its effects on properties are of minor consequence.

Regrind and that does not feed back into the system should be well identified and kept dry and free of contamination for a later reuse. Identification of regrind should consist of product description, type of regrind-sprues and runners, rejected parts, and whether it is first, second, or third-pass regrind.

Parts and runners with defects of contaminated resin should be thrown away. Never mix this regrind with virgin resin. Any resin dust and fines should also be thrown away, as it can cause feed problems. Basically, regrind can be reused if it is properly handled and protected, as long as part quality is maintained and processing is not handicapped by having to consistently adjust the machine parameters to keep the part in control.

PROCESSING AIDS

All resins are formulated with processing aids—some for mold release, some to prevent polymer degradation in the machine, and others to aid material flow. The processor should not be concerned initially with what internal lubricants and resin stabilizers are used. They are formulated by the supplier to yield a good product during processing and the levels of these additives are controlled within a specified range of values. But if processing problems occur, processors need to know what they may add to assist in making a good product. Your supplier can provide this information. Addition of these additives may be required when using regrind blended back into virgin resin. TA analysis will tell you whether this is necessary. You can specify limits, if necessary, on the job setup sheet.

If you must use additives, try to obtain a pellet concentrate. Powders, oils, and waxes can be very difficult to feed and handle, and they can contaminate a large area in your plant if improperly handled. You should also try to avoid using mold spray releases, as they contaminate part surfaces, are only temporary fixes, and can later create finishing problems. If excessively used, they can also contaminate other parts being molded nearby. The use of mold releases indicates there is a material or mold problem that should be fixed.

A properly designed tool, plus material from a quality supplier, is the best route to follow. Selecting a quality resin supplier and keeping open the lines of communication is one of the best methods of controlling the quality of resin. When you buy material, you also buy the supplier's technical expertise. A wise plastic part manufacture will call on and use such technical and processing experience.

In essence, always deal with qualified and capable material suppliers, who certify their materials, and work with your plant personnel to solve any problems. You and your suppliers are a team working to meeting customer part requirements for total quality process control.

8

The Mold

The mold, which is more commonly called the "tool" in the injection molding field, is another key link in the chain of total quality process control. Without a quality built tool, the best part design and material and manufacturing processes are handicapped. The design and construction of the tool must be of the highest quality to produce the plastic parts to meet the customer's requirements. The key to this task is early communication with all parties responsible for the part's finished specifications. This includes sales, design, purchasing, tooling, and manufacturing, in varying degrees. Part quality and its manufacture should never be placed in jeopardy by negotiating for a bargain tool price. The tool is more than 60 percent responsible for the quality of the finished part.

The part designer must assume some responsibility for the tool's design and construction. This involves selection of part section thicknesses, dimensional tolerances, the shape and complexity of the part, and how the material for its manufacture will enter, fill, and flow in the mold cavity. The part designer should not leave these decisions to the tool designer, who will not know all of the requirements for the part's end-use function. To obtain a quality finished part, these two departments, part and tooling design, must work together.

Therefore, during the tool's design stage, the part designer must ensure that all necessary information is available to the tooling designer. This early involvement will save time, money, and future manufacturing and part problems. It will also result in a successful and profitable program.

Total Quality Process Control for Injection Molding, by M. Joseph Gordon

COMPUTER-INTEGRATED MANUFACTURE

High-technology computer and software programs are aiding the designer in the initial design of a part, in addition to handling layout features of adjacent parts of complex shapes. These programs, which are part of the computer-integrated manufacture (CIM) process, can also furnish computer-controlled machine-cutting software for the toolmaker to use in cutting the part cavity in the mold. These software programs also have solid modeling capabilities that allow the part to be viewed in two and three dimensions as well as sectioned at critical points. These programs also have directories for part stress analysis, material property selection, mold fill and cool, and the selection of mold and tool components. Other systems will produce a three-dimensional solid prototype for subsequent examination and analysis, known as stereolithography (SLA).

These new programs are specifically designed to assist in the following areas:

- Determining material flow in a part (two-dimensional analysis only and then limited).
- Determining material flow in a runner system, plus sizing.
- Selecting the number of gates, including location and type.
- Analyzing the capability of a part to be molded under specified conditions (i.e., mold and melt temperature, fill times, and wall thicknesses).
- Balancing a runner system for the same part or a family mold.
- Analyzing cold and hot runner systems, the latter to minimize regrind and residence time of the material.
- Analyzing tendency of part to warp based on pressure gradients in the part, fill pattern of the material, polymer temperature as the mold fills, residual shear stress, and shear rate of polymer.
- Balancing material flow so all sections of the part fill at the same rate to prevent overpacking and warpage.
- Eliminating or reducing the number of weld lines.
- Identifying surface aesthetic problems related to processing (overheating of resin due to polymer shear and weld lines).
- Analyzing mold cavity temperature control to aid filling and reduce molding in material stresses.

These computer-aided engineering (CAE) systems will also assist the part and tool designer when coupling computer-aided design (CAD) with computer-aided manufacture (CAM), now combined into CIM. The CIM system can greatly reduce the lead time for mold building from months to weeks. It uses the design information database to begin the tool design and anticipate manufacturing process prior to the finalizing of the actual part design. This

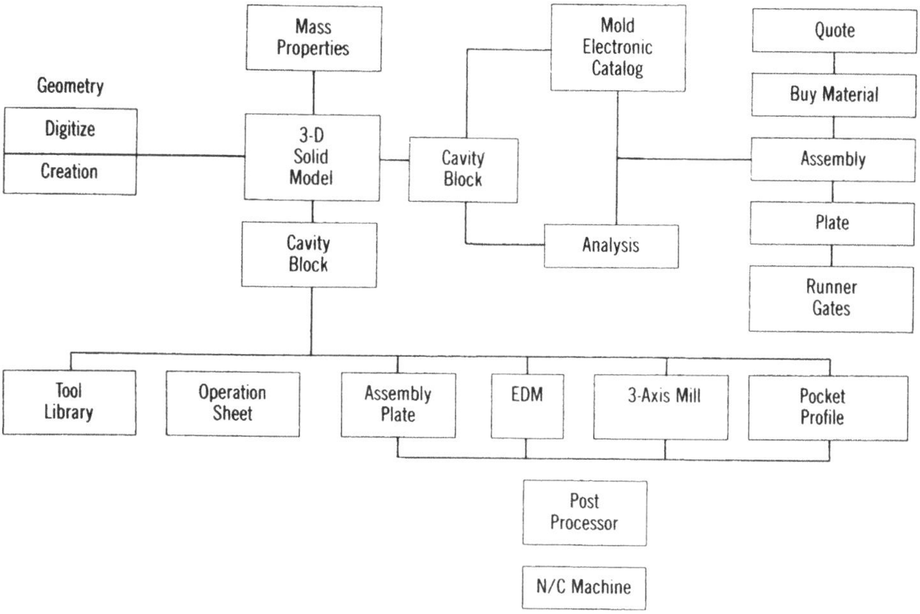

FIGURE 8.1. Computer-integrated manufacturing flow program. (Adapted from Ref. [1].)

allows standard mold components to be purchased; the analysis of flow, fill, cool, and so on, to be run; and corrections to be made. A typical CIM flow program is illustrated in Figure 8.1. These systems will reduce tool lead time, evaluate the part's design parameters, plus consider the material, tool, and manufacturing variables associated with plastics part and tool design.

A good tool design requires more than just a good part design. Even with computer technology, the designer and toolmaker must consider certain basics if the tool is to produce the required part. Making a good tool is an art as well as a science.

The designer and the toolmaker make a major portion of these decisions at the start of a program. The proposed manufacturer's (outside molder or in-house production) personnel should also be consulted, as they have to consider the processing parameters required by the molding machine that the tool will be run in. They also have extensive knowledge and experience on how similar tooling behaves in their equipment and can offer helpful suggestions to make the program a success.

The following pre-mold design checklist should be used to evaluate areas of PART and tool design and manufacture. To use this list, the part design should be far enough along that its form, function, and end-use requirements are clear. This is the time for tooling and manufacturing personnel to add their expertise. The best way is to consider the following items, each of which will, in its own way, affect part quality and the ability of the tool, material, and manufacturing process to meet customer requirements.

Two, more comprehensive, mold design and checklists and injection mold specification sheets are in Appendix C (See Figures 8.6 and 8.7).

PRE-MOLD DESIGN CHECKLIST

1. Part design. (Can it be molded, is there a better form of manufacture, is volume justified?)
2. Resin selection. (Is it processable to meet part requirements?)
3. Shrinkage (of resin used).
4. Molding machine compatibility. (Will tool fit and is melt capacity adequate?)
5. Strength of materials for mold.
6. Fluid flow in mold (resin and cooling).
7. Venting the mold.
8. Heat transfer. (What tool materials are necessary to control dimensions and cycle time?)
9. Thermal conductivity (tool materials).
10. Thermal expansion of the mold.
11. Coefficients of friction (mold steels).
12. Abrasion resistance (resin and tool wear).
13. Corrosion resistance (resin effects).
14. Ejector system (part and runner knockout).
15. Draft and shutoff.
16. Part drawings and dimensional stackup.
17. Mold setup (number of cavities).
18. Secondary operations (cams, inserts, etc.).
19. Maintenance/repair/operation.
20. Methods of construction (type of tool).

The questions that need to be addressed before actual tool design begins are discussed below.

Part Design

Is the design finalized to the point where all aspects of the part's function and operation have been considered? Can the part be molded as a single unit or must it be broken down into separate parts and assembled in a secondary operation? If prototypes were made, do they perform as required? Are there still unanswered questions that could change the part's design, shape, or function? Is the manufacturing process and anticipated tooling capable for making the part in the primary resin? Have all secondary operations required for the part been considered for assembly, fusion welding, fasteners, snap-or-press fits, plating, painting, or other types of decorating? Are the part's tolerances in line with the method of manufacture and the tooling to produce the volume of parts required within anticipated cost guidelines?

Each individual part must be considered independently to determine whether individual molds are required or whether a family mold (similar parts of the same material in the same mold base) can be used. These are all questions that need to be addressed based on part design, tolerances required, and the material selected to make the part.

Material Selection

The resin or plastic selected will have a definite bearing on tooling requirements. Each material requires careful consideration in tool design. Some are corrosive, others are abrasive, and each has its own unique thermal and flow properties. The tooling layout, materials of construction, and design must conform to the resin's requirements and limitations during processing. Often, the original resin of choice has been changed after the tool is completed and production was started. This can require a total rework of the tool to meet part requirements.

Shrinkage

Each material has its own shrinkage behavior. Crystalline polymers shrink more than amorphous polymers. When these are enhanced with fillers, reinforcements, and modifiers, they exhibit different shrinkage rates. Mold temperature and processing parameters will also cause different shrinkages to occur in the same material. The number of cavities, runner system, and gating will also affect this area in varying degrees. In addition, the flow of the resin in the part cavity has a definite effect on part shrinkage in some reinforced and filled materials. Differential shrinkage can cause part warpage and must be considered in all parts—both big and small, thick and thin.

In some cases, before cutting the cavity's finished dimensional tolerances, the tool is sampled under anticipated production conditions. Sample parts are molded with regrind, if allowed, to determine the actual material shrinkage values on the molded part. This permits the final part and tool dimensions to be determined and adjusted to compensate for material flow and shrinkage variations. This also allows production personnel to recommend any tooling changes they feel will assist in optimizing molding process conditions to make the part to tolerance.

Molding Machine Compatibility

The tool must be compatible with the injection molding machine that will make the part. The tool must fit between the machine's tie bars, be fully supported by the machine's platens, and not extend beyond the edges of the machine's platens. The machine must also allow the tool to open sufficiently (known as daylight space) to allow easy removal of the part. If robots are used, space must be allowed for their mounting and operation on the machine's surfaces. In essence, any auxiliary equipment must fit in or operate within the machine's space without hindering the machine's or mold's operation. Other items to be considered are as follows:

1. The knockout pattern of the tool should match that of the machine.
2. The shot size and melt capacity of the machine should match the tool's part and runner system in order to supply an adequate melt of resin. A minimum of two shots of resin in the barrel is necessary to ensure a good melt supply.
3. The clamp force necessary to keep the tool locked during injection must be indicated. The force required is based on the projected surface area of the mold face of the part and runner system multiplied by the material supplier's recommended clamp pressure for resin in tons per square inch of projected surface area. This can vary from 1 to 5 tons per square inch of surface area. There should always be a 20 percent safety factor of machine clamp force to avoid having the tool open under injection pressure.
4. The mold's sprue bushing must match the machine's nozzle and its opening size.
5. Some resins may require a special screw design for processing and ensuring a good quality of melt to the tool. Check with your material supplier to determine whether this is required.
6. The machine must have process control capability, with continuous feedback process adjustment control of the molding cycle.

Strength of Materials for the Mold

The tool must be built to withstand injection and clamp pressures, so it will not deflect or crush during production. The steel must adequately support the cavities and cooling lines in the tool. There should be a minimum of 1½ inches of steel from the furthest cavity point to the edge of the tool. Materials for the tool's construction are selected for the number of parts to be produced, tool life, thermal conductivity and cavity temperature control, and the resin used for the parts.

Fluid Flow in Mold

Two types of fluids will flow in the tool. One is the polymer that uses the sprue and runner system to feed the cavity. The other is the cooling or heating medium that flows around the cavity. It is used to control the cycle and assist in controlling the size of the part. Both factors are critical in their sizing and placement in the tool. A runner system sized too large results in a pressure drop. As a result, increased injection pressure and resin is required to fill the cavity, and more regrind is generated. Too small a runner can result in premature freeze off and poor part packout pressures. These result in undersized parts that are full of voids or are porous.

The cooling system is critical to ensure a uniform temperature gradient around the part cavity for dimensional control and cycle uniformity. This will be discussed in greater detail throughout this chapter.

Venting the Mold

The proper venting of molds is typically thought of as a tool designer's function; it is independent of part design. Good venting is important when critical tolerances, surface aesthetics, and part function are considered. Air must be vented quickly from the tool during the injection, or burns, non-fill, poor weldline strength, and surface problems may occur. How the part is laid out in the tool determines its parting line location. Any areas with blind pockets must be vented through the ejector system or by another effective means. All of these areas must be considered during the mold design phase to ensure that the polymer flows uniformly into the mold cavity with minimum resistance.

There are now systems to evacuate air from the tool before injection. In critical parts, this has improved fill and weld line strength, reduced the cycle time, and yielded better and stronger parts. However, this mold evacuation system is costly. In many cases, it may not be required, particularly if the tool is designed correctly with adequate venting. This area will also be covered in more detail later on in this chapter.

Heat Transfer

Heat transfer is basically the mechanism used to remove heat from the molten plastic. This enables the plastic to become a solid that can be ejected from the mold. Through the conductive properties of the mold steels, the heat from the polymer is transmitted from the mold cavity into the cooling water. Each steel has its own heat transfer rate, which will also be discussed later on in this chapter.

To control dimensions properly, the part's cavity must be maintained at a uniform temperature, be it high or low, depending on the resin and dimensions required for the part. Unbalanced cavity temperatures can produce parts with varying dimensions. The parts may possibly be warped because of differential shrinkage or distorted because one section is softer and bends on ejection from the tool.

Cooling lines, which are usually the last item considered for new tooling, must be routed around the ejector system. Often, the cooling system is undersized for the volume of cooling required, inadequately designed to maintain a uniform temperature gradient across the part, made of the wrong materials to maintain the desired temperature control, and plagued by problems resulting from poor decisions made in designing the system to suit the resin and part tolerance requirements.

The same factors apply to hot runner molds, where heat control for the system is critical for controlling the melt flow into the cavity.

Thermal Conductivity

The correct selection of tool steels for heat transfer retention or elimination is very important. If heat must be removed from a critical area, a material must be selected that can extract it and then transfer it to the cooling medium while maintaining the temperature balance in the mold cavity. The part designer must make sure that the tool designer is aware of critical sections, so that the correct tool materials and heat transfer conditions exist to meet the part's dimensions.

Thermal Expansion of the Mold

Because all tools will be subjected to high temperatures during production, thermal expansion must be considered. It is essential that cavity and mating tool areas will shut off once the tool heats up. The tool is essentially two independently temperature-controlled masses of steel that must mate and seal off during each cycle. For parts with nonuniform sections, some areas of the tool will run hotter if the cooling layout cannot control the temperature gradient across the cavity surface. This may require the use of materials of different thermal conductivity. Suppliers should be contacted to determine the desired cavity temperature to be used to control final part dimensions with each resin.

Coefficients of Friction

All molds have moving or sliding parts, including ejector pins, slides, and lifters and cores (both rotating or collapsing), which can develop high-frictional forces. These forces are a major cause of wear and galling of mating parts. The use of "dissimilar" materials in these areas is best, because typical lubricants cannot be used in areas that come into direct contact with the part. The same steels can be used if different hardnesses, finishes, or specialized coatings are specified. Excessive sliding friction causes wear that results in part tolerance and surface problems, as well as premature failure of the tool.

Abrasion Resistance

All plastics will be abrasive to some degree when heated to their melting point and injected under high pressure into a mold. This is even more true of filled and reinforced materials, as the fillers are often glass and minerals that wear away the surface of the tool, particularly where initially injected and along the flow (fill path) within the mold and cavity. For tools using these resins, the use of abrasion-resistant, heat-treated, hardened, and special coated materials are required. Replacement of tool or cavity sections in these high wear areas should also be considered where the melt impinges or flows. This would be primarily the gate area, which wears more rapidly than other sections on the tool, or a section of the cavity wall or core pin if the melt is impinged on it during injection. As gate wear progresses, the freeze-off time increases. This affects part size and tolerances and increases the cycle time necessary to keep parts within specifications.

Corrosion Resistance

All plastic resins are corrosive to some degree because of their chemical makeup. The correct selection of tool steels for the cavity and runner system are crucial for extending tool life. If the right tool steels and protective coatings are not used, then pitting and surface erosion will occur. Material suppliers can assist in this area by defining the corrosive products in their resins and recommending materials that will provide protection.

Ejector System

How the part is ejected from the tool is determined by the geometry of the part, the resin, and the placement of the part in the tool. Part surface appearance, as well as parting-line and mating-part shutoffs, are also important. The part must be uniformly ejected from the cavity. Cantilevered sections should have ejector pins behind them, so that they follow the main part section out of the tool at the same time and do not bend. A poor ejector system can cause part warpage or surface and dimension problems. In addition, the ejector system can place limits on the routing of the cooling system. The tool design, resin, and cooling system have a definite affect on the type and number of ejector pins required to insure that the part is ejected uniformly from the mold cavity.

Draft and Shut-off

For proper release from the mold cavity, all molded parts require draft (the angle of surface slope from the vertical in degrees for a core, ribbed, or walled section) from ¼ to 7 or more degrees. Without draft, the part may stick in the mold and lock the tool. Draft is required to ensure easy release of the part from a perpendicular cavity surface. These draft sections must not have tool marks; they must be smooth and final polished in the direction of tool/part movement to avoid part hang-up during ejection. Draft is therefore critical for good part release from both the core and cavity and at shutoff areas where mating tool steels meet. This is shown in Figure 8.2 in metric (A) and English (B) values.

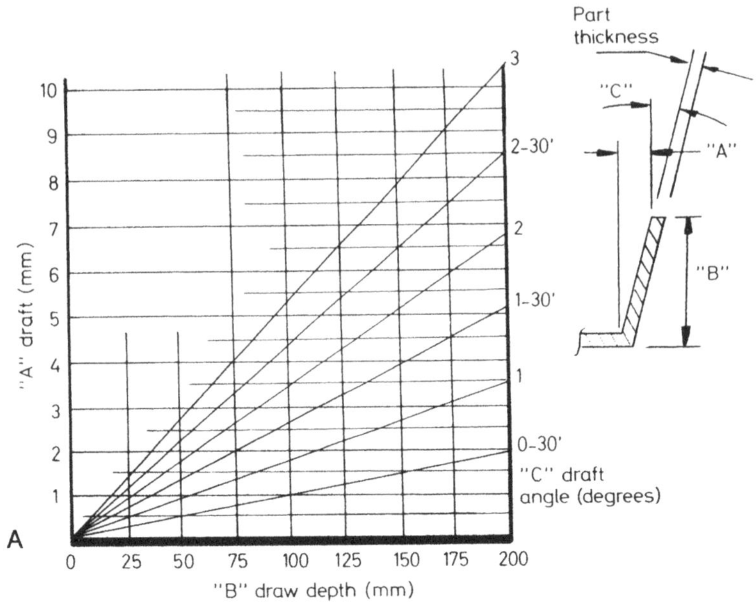

"C" draft angle (degrees)

"B" draw depth (in)	1/8	1/4	1/2	1	2	3	4	5
1	0.002	0.004	0.009	0.017	0.035	0.052	0.070	0.080
2	0.004	0.009	0.017	0.035	0.070	0.105	0.140	0.175
3	0.007	0.013	0.026	0.053	0.105	0.157	0.210	0.263
4	0.009	0.018	0.035	0.070	0.140	0.210	0.280	0.350
5	0.011	0.022	0.044	0.088	0.175	0.262	0.350	0.437
6	0.013	0.026	0.052	0.105	0.209	0.314	0.420	0.525
7	0.015	0.031	0.061	0.123	0.244	0.367	0.490	0.612
8	0.018	0.035	0.070	0.140	0.279	0.419	0.559	0.700
9	0.020	0.040	0.078	0.158	0.314	0.472	0.629	0.787
10	0.022	0.044	0.087	0.175	0.349	0.524	0.699	0.875

"A" draft (in)

FIGURE 8.2. *Draft angle graph: (A) Metric, and (B) English. (Adapted from Ref. [8].)*

Tight shutoff is required to avoid a witness line at mating part surfaces. On mating tool steel sections, a draft of 5 degrees is preferred to minimize the wear of mating surfaces. If this cannot be achieved, then hardened steels are required. The size of the shutoff is also important, as linear expansion of the tool steels in these local shutoff areas must be considered. Because witness lines that occur at these junctions may affect part aesthetics, the tool and part designers should consider selecting a surface texture to disguise these lines. At witness lines, the designer can also add a surface effect to disguise any mating surfaces created by a side action.

In cases where minimal or no draft is required (not recommended for deep sections or high-shrink resins), consider a light (glass-bead) shot peening of the area or a very slight texturing of the surface. This can provide just enough surface discontinuity on ejection to break the suction effect of the resin on the cavity surface.

Part Drawings and Dimensional Stackup

The part design drawings must be complete in all details, including tolerance selection of dimensions based on actual requirements of the part's function and fit. Depending on a part's function, only one or two very tight tolerances should be specified. As the number of cavities in a tool increases, the tolerances must open up. The design and part fit must allow for this spread. Do not hold the tool designer to a general overall tolerance on all dimensions on the drawing. Where tolerances are critical—as between hole spacings—specify realistic, required tolerances. Where tolerances can be more open, so specify. Do not use the typical metal working tolerances for a molded part. These are usually much tighter than required for a plastic part and will only increase the piece part cost.

Note the critical tolerances actually required. Then, give the toolmaker as much leeway as possible with the other dimensions. Reference dimensions should be given from only one side or surface of the part to avoid misfits. This alerts the toolmaker to use this same surface for tool dimensioning, thereby avoiding a stackup of dimensions that can result in mismatches or poor fits in mating parts. Not working from the same starting surface often happens when using CAD/CAM mag tapes, which have no dimensions. When the tool designer determines part dimensions by starting from a surface other than the one the designer used, stackup tolerance errors can occur. Therefore, the designer must provide dimensioned drawings, with critical surfaces and to-fit dimensions specified. With more complex three-dimensional shapes, it is necessary to have a three-dimensional CAD database, with corresponding drawings. There enable the toolmaker to make sound decisions and allow quality control personnel to verify critical dimensions.

Design and tooling drawings should also specify all radii, including internal corners. Never assume that the tool designer will add them in a general note on the drawing. The radii specified should be sized to suit the part function and end-use requirement. Well-defined part drawings are a necessary preliminary for making valid tooling decisions when building the mold. Meetings should be held with the part designer whenever question occur that are not answered by the drawing.

Mold Setup

Molds need to be designed for fast installation and hookup to auxiliary mold support equipment. Connections on the mold for temperature control and secondary tool operations should be standardized to suit the typical fittings found in a molding plant.

Quick disconnect and standard electrical connectors are required and designed with safety in mind. They should be easily accessible and fit within the machine tiebars and safety gates. The same is true for mold clamp slots; they should be easily accessible and not obstructed. This ensures fast mold changes, easy hookup in any molding plant, less downtime, and fewer hours searching for special connectors. The manufacturing operation thus becomes more efficient and profitable. A Kaizen is often performed to evaluate the capability of "quick mold change," which leads toward lean manufacture.

Secondary Operations

Secondary operations required to make special features on a part should be reviewed by the part and tool design engineer. These may include core pulls, side action, inserts, or unscrewing operations.

Based on experience, the tool designer may be able to offer suggestions and recommendations to simplify the tool design and still keep the same part function. As shown in Figure 8.3, these simplifications can reduce mold cost, eliminate expensive in-mold functioning operations and out-of-the-tool add-on components, and help reduce mold maintenance and repair operations. In addition, by selecting another material or altering the design slightly, an in-mold operation, as stripping an undercut, can be performed without the necessity of a mechanical operation before part removal is possible. Checking with the material supplier may save adding these costly secondary operations to a tool.

With some complex parts, a secondary out-of-mold operation may prove better and less costly than if attempted in the tool. The designer and tool engineer need to discuss these areas to identify problems and possible simplifications of the tool.

Maintenance/Repair/Operation

A good mold is designed for ease of assembly, maintenance, operation, and repair. Each mold should come with well-defined mold drawings and a part's list. Materials used in the mold construction and cavity finishes should be specified. Parts that experience wear should be designed for ease of replace-

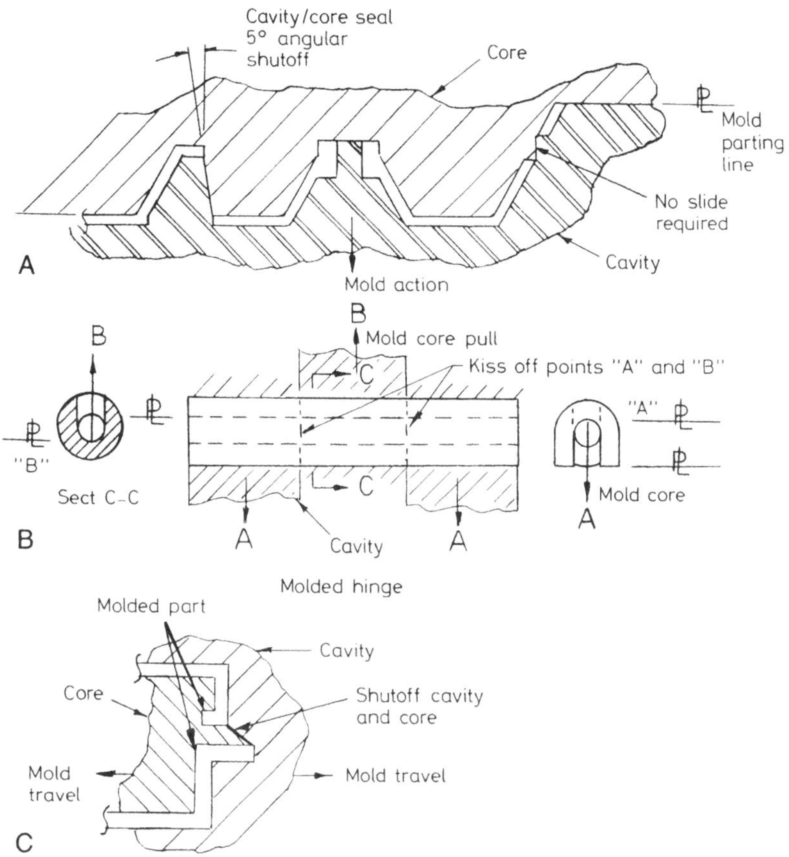

FIGURE 8.3. Tool design simplification: (A) cavity and core shutoffs, slides not required; (B) two-plate hinge design; and (C) complex cavity core shutoff.

ment. The tool should also have detailed assembly/disassembly and maintenance instructions.

Tool modifications are sometimes required. Therefore, tool steels with weldability are desired. Whenever a tool is removed from production, it must be torn down, checked for wear and dimensional accuracy, cleaned and lubricated, and preserved.

Methods of Construction

Once all of the mold design questions have been answered, construction can begin. The use of standard mold base steels and components is preferred. They not only assist in building the tool, but also allow necessary spare or replacement components for repairs to be obtained easily and fitted to the tool.

Conventional machining techniques are used to shape the cavity and the runner and gating system, including the cooling system. On multicavity molds, the exact duplication of each cavity is critical; each should be identified with a cavity number. This is necessary for fine tuning or repair because of dimensional variance. If any modifications are made to the tool, then the tool drawings should be revised to illustrate "as-built" construction and dimensions.

All seals in the tool required for cooling lines and pneumatic operations, such as O-rings, should be machined to the contour of the seal. This provides additional support to the seal and allows them to function as designed without becoming unseated and causing leaks that would require that the tool be pulled from production and repaired.

TOOLING

Processing

By calling on the experience of your plant personnel or outside molder, you can build a tool that functions and produces parts in the resin at the part volume required, with a minimum of processing problems. Many tool engineers have had experience with setting molds, processing, and problem solving. Their experience is invaluable for understanding and considering all the variables in building a quality mold.

It is imperative that a top-quality mold be designed and built. With a poorly designed tool, you will never achieve the cycles and cost structures required to meet the customer's specifications. Your parts will only be as good as the design and quality of the tool. In a high-quality tool, with a correctly sized runner and gating system and the ability to control cavity temperature, the processing parameters can be determined and consistently controlled.

Reviewing Existing Tooling

Because plastic processors do not always have the luxury of assisting, in the design of new tools, but rather bid on existing tooling, they should review the tooling and part design very carefully. Poorly designed tools that cannot meet customer requirements have a habit of continually moving from molder to molder in the customer's attempt to produce "good" parts. Even molders with process control systems may not be able to make acceptable parts with these tools. This is often not known until the tool is run, and the results prove no better than previous efforts. Many original equipment manufacturers (OEMs) often send their harder to produce parts to custom molders, if they do not have the in-house capability to produce the part.

Therefore, a tool review, using the mold questionnaire, is imperative. The customer should be questioned about the quality of the tool's construction and the part quality required. If problems are found, then part and/or tool redesign may be required before accepting the job. If redesign is not possible, the customer must lower part requirements to match the tool's capability or the molder should refuse the job.

A mold design questionnaire and checklist for customer response is shown in Appendix B, along with an injection mold specification sheet. Use of these questionnaires will supply the answers to most part and tooling questions. For an existing tooling problem—with available parts and mold drawings—these problem may be more easily identified and solved. The molder needs to know the precise conditions of the job on which he is bidding to be confident of being able to manufacture parts to meet customer requirements.

Part Cost and Cavity Optimization

Manufacturing piece-part cost is the basis for calculating the finished part cost. The customer must determine the estimated number of parts required to create the tooling necessary for production. There are general guidelines that establish the tool design and the number of cavities needed to meet part cost estimates. The number of cavities selected for a tool is based on part size and the dimensional tolerances obtainable with the tool manufactured in the selected material. This is further controlled by the cycle time estimated for production and the volume of parts required for the marketplace.

If the number of cavities selected falls short of volume requirements, then additional tools may be required to meet the customer's volume requirements. If so, build one mold first and prove it out before building the second tool.

Prototype Tooling

Many plastic parts are first made in low-cost prototypes to prove out design and part function. This technique allows analysis of material flow as well as gate and runner location. Molded prototypes are more representative of how the part will function than are parts machined from a plastic stock shape. Furthermore, it is often difficult, if not impossible, to obtain a resin stock shape in the size required. As a result, models or patterns are often made and a tool cavity cast either from Kirksite or machined from soft aluminum or steel to produce sufficient parts for testing. These cavities are put into a standard mold base and parts are then molded. Often, hand-loaded inserts are used to replicate complicated part sections or functions that would operate automatically in a production tool. In prototypes, the gate location should be the same as in the production tool. Runner sizes are usually dictated by the mold base selected. Because cavity temperature control may be restricted, depending on the type of prototype tool used, temperature control of the cavity may be variable at best.

The objective of prototype tooling is basically to determine whether the part and material selected can perform the end-use function. However, additional information is also gathered on such topics as the shrinkage effect on tolerances, flow-and-fill patterns, weld-line strength, warpage, and cavity cooling line requirements.

Most prototypes are made from one of three groupings of materials: castable alloys, such as Kirksite; zinc, brass, or aluminum; or the new mass cast epoxies that are filled with metal materials, such as aluminum, or machined from soft aluminum alloys. The last presumes an understanding that only a limited number of parts can be made from the resulting tools. With some of the above tools, ejection systems are installed to ease part removal, but the ejector system may be limited by the mold base used to support the cavity. The life of the tool is based on the material used, the amount of handling required to remove the part, and the base resin and injection pressures used to make the part. Some tools can produce several thousand parts, if treated carefully; others can produce only a few. The decision to use prototype tooling is predicated on not having enough information or confidence to go directly to a hard tool. It is considered a somewhat inexpensive way to prove out a part's design, material, and molding capability before committing to the expense of a production tool.

Production Tooling

Once a decision is made to proceed to production, the type of tool needed to produce the part must be selected. Four basic types of tool design are currently used. They are two-plate, three-plate, hot-runner, and hot-manifold tools. Recently, the hot-manifold tool has lost ground to the hot-runner tool, in which resin melt temperature is better controlled. Typical mold configurations are shown in Figures 8.4A to I.

Two-Plate Mold. The two-plate mold (Figure 8.4A) is most widely used. It has a conventional runner system at the mold parting line, with a single drop from the sprue to the runner. It is used for simple parts requiring limited cam actions. The parts are usually ejected while attached to the runner and are gate design dependent and limited to edge gating. This is less costly than three-plate and hot-runner molds and has fewer restrictions for cooling channel layouts. It is also easy to maintain.

Three-Plate Mold. The three-plate mold (Figure 8.4B) is becoming more widely used, with gate placement not restricted to the edge of parts. It uses separate plates for the runner system and cavity/core. The part and runner are automatically degated when the mold opens; the runner systems can be downsized, thereby obtaining shorter flow paths. Such central gating of parts can reduce weld lines and mold-in stresses, as well as provide shorter flow paths for the material filling the mold/part cavity. It uses conventional ejection techniques, with minor limitations on cooling channel placement. Although more expensive than two-plate molds, it can run automatically, with no operator required for part/runner separation.

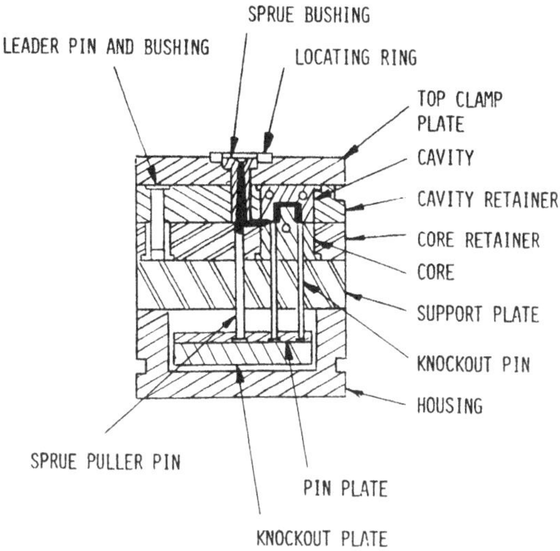

A TWO-PLATE MOLD

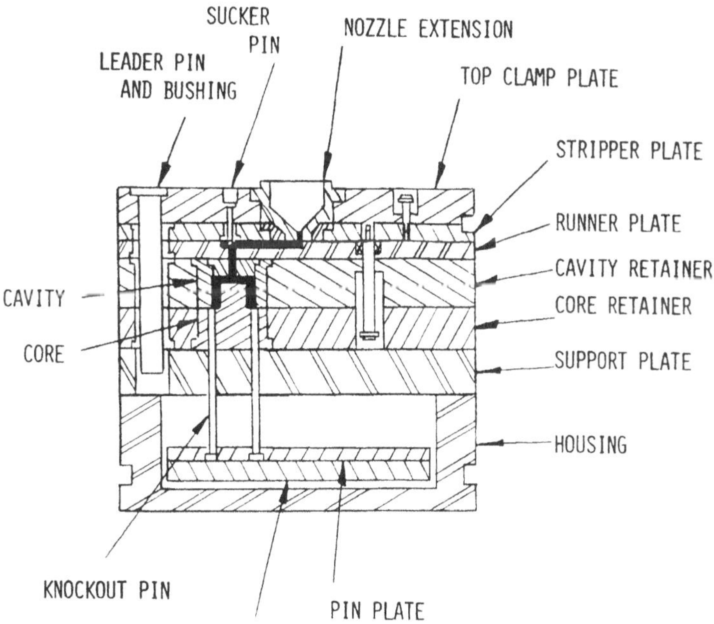

B KNOCKOUT PLATE

FIGURE 8.4. *(A) Customer questionnaire for plastics moldings. (B) Internal supplementary questionnaive on plastics molding. (C) Hot-runner mold. (D) Hot-manifold mold. (E) Inside center-gated modified hot-runner stripper ejection mold. (F) Edge-gated conventional ejection mold. (G) Stripper mold. (H) Submarine-gated stripper ejection mold. (I) Outside center-gated three-plate stripper ejection mold. (Adapted from Ref. [24].)*

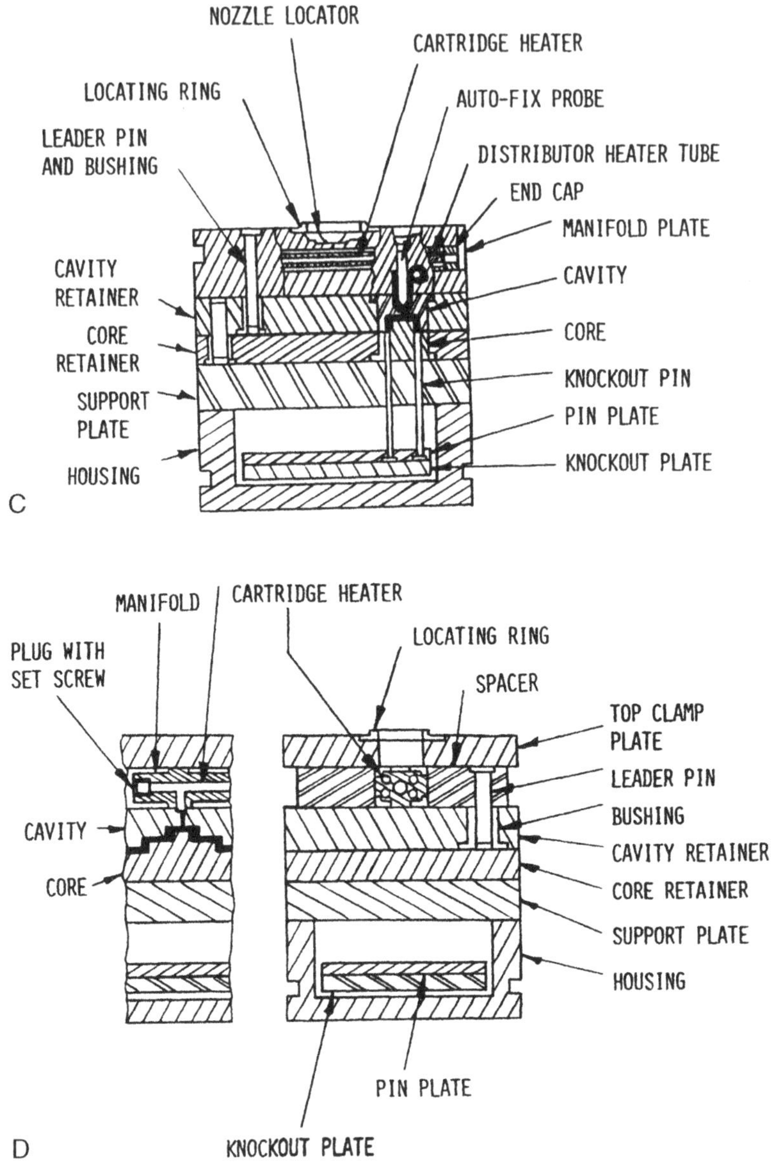

FIGURE 8.4. *(Continued)*

Hot-Runner Mold. The hot-runner mold (Figure 8.4C) is used for long-running programs. Many advantages offset its higher cost and tighter control of operating conditions. The lack of a runner system, in conventional terms, results in efficient material use, fast cycling, automatic degating, and clean gate aesthetics. It is designed for multicavity molds and runs automatically. It requires more technical equipment to operate, control, and monitor opera-

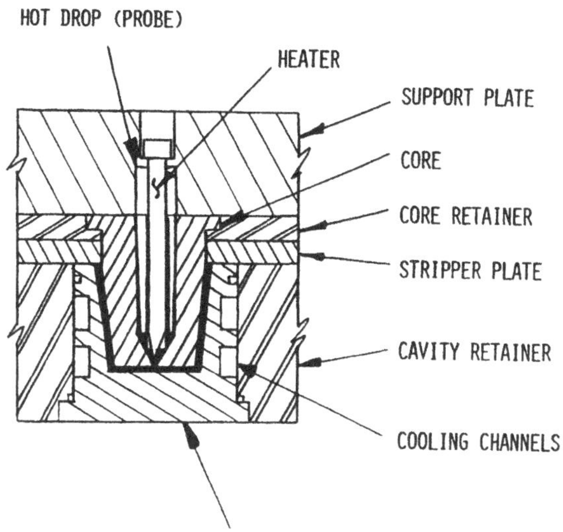

E

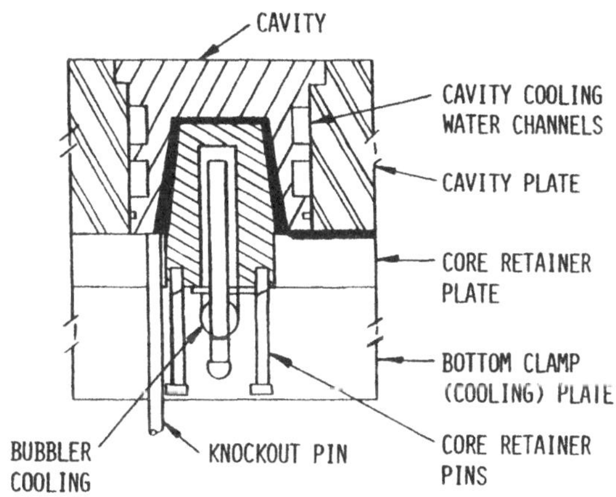

F

FIGURE 8.4. *(Continued)*

tions. Although split-plate manifolds can be obtained for frequent material/color changes, solid-plate manifolds are recommended. Cost savings can be achieved, along with increased part quality.

Hot-Manifold Mold. The hot-manifold mold (Figure 8.4D) uses cartridge heaters to keep the material fluid in the runner system, but has less control at the cavity gate area than does a hot-runner mold. The elimination of the heavy center section runner permits the use of smaller runners that are broken down into quadrants. Its extra manifold and tighter temperature control does

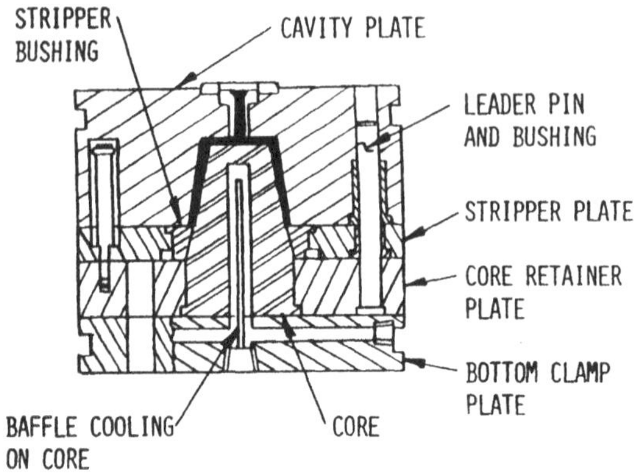

G

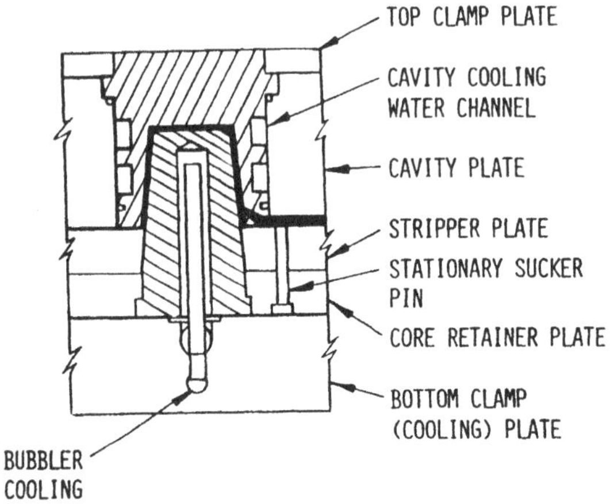

H

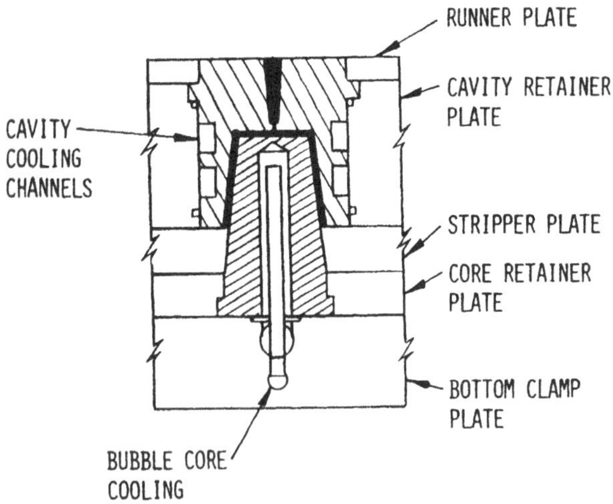

I OUTSIDE CENTER-GATED THREE-PLATE STRIPPER EJECTION MOLD

FIGURE 8.4. *(Continued)*

increase costs because of the need for more sophisticated equipment. Maintenance costs are higher for these molds.

Inside Center-Gated Modified Hot-Runner Stripper Ejection Mold. In this mold (Figure 8.4E), the exact gate placement inside the parts is attained when aesthetics so demand. The core and cavity are reversed, with hot drop going through the core. This can cause part distortion or sink marks, because the core cannot be cooled. Uniform part sections are required, with no thick sections at the end of the flow path to fill the cavity.

Edge-Gated Conventional Ejection Mold. In this mold (Figure 8.4F), the runner system is located on the parting line, with material flow to the edge of the part. It is used with two-plate molds; construction is simple, and the mold is easy to maintain. The cooling system must be adapted for knockout-pin placement. Because the size of the runner can control cycle time, the smallest possible runner size is preferred.

Stripper Mold. The stripper mold (Figure 8.4G) is a three-plate mold design with stripper bushing that pushes parts off the core after the mold has opened. It is generally used for round parts, because it provides uniform part ejection around the circumference. It is simple, easy to maintain, and relatively wear free, but more expensive than pin ejection. It uses bubbler style cooling for cores.

Submarine-Gated Stripper Ejection Mold. In this mold (Figure 8.4H), the part and runner are automatically degated on the mold opening, with the runner and part ejected on stripper plate action. The runner is held in the core section by the sucker pin until the stripper plate is actuated. Because the runner and parts fall in the same area, they must be separated.

Outside Center-Gated Three-Plate Stripper Ejection Mold. In this mold (Figure 8.4I), the runner is independent of the part cavity. When the mold opens, it degates from the part. As the stripper plate operates, the part and runner are ejected from the mold into separate collection areas underneath. It can also use a double-ejection pin or sucker-pin design, with a second stripper plate to eject the runner from the mold as shown in the figure.

Summary. Because each part is unique, the type of mold selected must be based on part geometry, resin, number of cavities, part tolerance requirements, and anticipated tooling costs and maintenance. The quality of the tool built will be reflected in the quality of the finished part.

The most economical tool is a two-plate mold that may require an operator to degate the parts and separate sprues and runners. The three-plate mold, although initially more expensive, can be run automatically. A hot-runner tool produces only the part, without generating regrind, and has more temperature controls built into the tool. The latter requires closer startup and tighter process control.

When selecting the type of tool to be used, talk with tool design and manufacturing firms, their customers, your molders, and the material supplier before making the final selection. They can use their experience and knowledge to determine which type of tooling would best suit your part and production requirements.

Pricing the Tool

The cost of a tool depends on the part complexity, part size, number of cavities, tolerances, tool type, steel used, caming actions in the tool, anticipated tool life, resin selected for the part, and part aesthetics. Although these costs are amortized over the number of parts to be produced, the real cost also includes the molding cycle and manufacturing operations needed to produce the part.

Thus, although a three-plate or hot-runner tool will cost more initially, they will usually not require an operator or a part and runner separator to degate and trim parts as would a two-plate tool. Therefore, piece part price may be lower in return for a slightly higher tool cost.

The total responsibility for buying the tool should not be placed solely on your purchasing department. There are too many factors it might not know. The tool's requirements should be determined and specified in writing before purchasing the tool. The goal is the lowest cost for the right tool to produce the parts to specification by a competent molder. The checklist for mold design and injection mold specification sheet (Appendix B) for the mold's design and specification show the areas of interest that need to be addressed when buying a mold.

The design and sales departments, with input from the other departments, should consider mold cost, production cost, and the final piece-part price during the initial product development. This will help management determine whether you are going in the right direction.

Estimating the cost of a mold is not difficult; all parties involved must be consulted. When preparing a request for a tooling quotation, the preliminary design, final drawings, and, if possible, a model of the part, should be completed. This information, which should be obtainable from the part drawings and specification sheet, includes the following data:

1. Number of dimensions on the print to determine cavity complexity and part tolerances.
2. Number of different surface finishes required.
3. Tight tolerance requirements and the number required.
4. Length and width of part in square inches of part surface area.
5. Any in-mold operating (caming, unscrewing) functions.
6. Balanced tool requirements, based on tolerances.
7. Resin for the part.
8. Number of cavities, based on tolerances and part volume required.
9. Tyne of tool, for example, two-plate, three-plate, or hot-runner.

Once the above questions are answered, determine the amount and type of materials and standard components that will be used and calculate costs from supplier catalogs.

Next, prepare a list of the manufacturing and assembly procedures that includes cavities required and whether CAD/CAM computer tape information is available for cutting the cavity(ies). Cooling requirements and the placement of cooling lines around the cavity for temperature control is then considered, along with the ejector system and any in-tool operational functions required.

Calculate the estimated time required for each procedure and the hourly wage rate for each. Add to this the external costs of heat treatment, finishing, polishing, surface texturing, and so on, to develop a feel for all aspects of the work. By collecting this information when buying the first few molds, the purchasing department can more accurately and efficiently analyze all factors in a specific job, thereby developing a realistic price range for budget and tool selection decisions.

For first-time mold buyers, especially those not knowledgeable in mold design and construction, this information will take time to develop. Often, the wage rates and time needed to complete operations may be only an estimate. The final costs must wait until solid information is obtained by talking to suppliers and analyzing the answers to your request for tool quotations. The gathering of specific information can be more useful if the buyer specifies how the tooling source should prepare its quotation. This includes a breakdown of components, functions, and schedule.

The time required for the mold design will add additional cost, even before a mold builder is selected. These costs can vary from $3000 for a straightforward design, with few or no changes, up to more than $8000 for a constantly revised design. Mold design costs are approximately 5 to 10 percent of the total manufacturing price for a simple job. An average mold requires between 800 and 1500 hours to design, machine, build, finish, and prove out. If possible, select a tooling source that has injection-molding capability to sample the tool for prove out. This avoids shipping it back and forth if revisions are required before final acceptance.

Experience is the best teacher in mold estimating, and all departments are responsible for assisting in preparing the tooling estimate. One of the best ways to learn is to visit mold builders who specialize in molds of similar size and complexity to the one needed. Review the quality of work produced by each. In addition, examine their equipment, processes used in construction, and range of services. Request quotations from the tool builders you visited as well as from molders who have their own mold building sources. At this stage, you must be able to describe part requirements in detail. Solicit recommendations that may allow you to add or change features on the part that will lower the cost or eliminate anticipated secondary functions.

The experience and knowledge one can gain from this procedure can save added expense and result in an even lower piece-part price. These may initially increase the cost of the tool, but the end result—based on piece-part price—is the bottom line to profitability.

Tool Scheduling

Once a contract is signed for the mold and the design has been finalized, the question of tool delivery becomes important. Because a quality part requires a top-quality tool, the delivery time is a major factor in keeping product introduction on schedule. If the preceding tool design program is followed, then all questions should be answered expeditiously. But changes can occur, and their impact on delivery depends on when they happen and their extent.

A good tool builder will present a schedule of events that details the different operations to be performed weekly as a percentage of the entire job. A good way to follow progress is with a spreadsheet program, such as Microsoft on Excel Lotus 1-2-3, which indicates progress made and the impact of any changes on the total program. Figure 8.5 shows a typical spreadsheet that outlines all pertinent information on a time-to-completion scale.

Good toolmakers can predict the time required for certain operations, such as machining, electric discharge machining (EDM), and polishing, etc., and can use this information in estimating tool cost and completion date. But since this is proprietary information, it is normally not shared with the customer.

For the customer's benefit in following the progress of building a tool, an abbreviated progress spreadsheet may be used to explain any delays or overtime required. As tool completion nears, the customer can also use the spreadsheet information to plan production schedules and product introduction.

The spreadsheet is an excellent tool for mold builders to use in developing and fine tuning their quotation system. It can be used to rank different tooling programs, log the hours spent on each operation, and determine how to remain competitive with other tool builders. They can also use this information, along with the mold questionnaire, photographs, drawings, and parts, to show potential customers their knowledge and experience in building tools to tolerance and specifications. This information, along with customer testimonials, is a very strong sales tool.

Tool Steel Selection

The selection of the components and tool steels are very important. Often, the toolmaker selects materials based on experience, but with not enough forethought about the new part's requirements based on tool life, specified resin, and tolerances.

The use of filled or reinforced resins that cause tool wear necessitates the use of harder tool steels. Over time, these can change dimensions and cause tolerance problems. Wear at the gate area increases the resin freeze-off time, and if the screw forward time is not increased, it can result in cavity depressurization and out-of-tolerance parts. High-carbon steels, with their increased

PROJECT NAME: GEAR BRACKET
PROJECT #: FAST TRACK #1
RES. ENGR: GORDON
TOOL NUMBER: A-78496
ESTIMATED COMPLETION WKS: 16
START WEEK: 3
COMPLETION WEEK: 16
REPORT WEEK: 16

COMPANY ABC
MATERIAL NYLON 6/6
ENGR. CONTACT B. BOWES / PHONE NO. 469-8000

Calendar in Numeric Weeks Description of Operations		Percent Completion/Calendar of Weeks 3	4	5	6	7	8	9	10	11	12	13	14	15	16
CAVITY HALF	PERCENT	CURRENT DATA													
BASE PREPARATION															
square/build-up	2	100	—	—	—	—	—	—	—	—	—	—	—	—	—
lift holes	1	100	—	—	—	—	—	—	—	—	—	—	—	—	—
waterlines	3						50	50	55	55	55	55	55	55	100
rough machine	2	10	15	20	30	40	80	100	—	—	—	—	—	—	—
CAVITY INSERTS:															
square plates	1	50	85	100	—	—	—	—	—	—	—	—	—	—	—
rough machine	2					0	80	100	—	—	—	—	—	—	—
finish machine	3				10	20	20	20	35	35	35	40	70	85	100
waterlines	5						50	50	55	55	55	55	55	55	100
EDM:															
electrode preparation	9			10	25	50	75	88	90	99	100	—	—	—	—
burn	7											30	60	100	—
stone	5											40	60	80	100
EJECTOR HALF:															
BASE-PREP:															
square/build-up	5	100	—	—	—	—	—	—	—	—	—	—	—	—	—
lift holes	1	100	—	—	—	—	—	—	—	—	—	—	—	—	—
waterlines	3							80	90	95	95	95	95	95	100
insert pockets	2	20	30	40	50	60	60	100	—	—	—	—	—	—	—
rough machine	2			30	50	80	80	90	100	—	—		—	—	—
CORE INSERT:															
square plates	1	100	—	—	—	—	—	—	—	—	—	—	—	—	—
rough machine	3	50	60	75	80	100	—	—	—	—	—	—	—	—	—
finish machine	5					10	40	50	55	60	70	80	85	90	100
EDM:															
electrode preparation	6							50	90	90	100	—	—	—	—
burn	2												100	—	—
stone	2												95	95	100
heat treatment	1												100	—	—
EJECTOR SYSTEM:															
build-up and fit	7	70	70	80	80	85	85	100	—	—	—	—	—	—	—
FINAL ASSEMBLY:															
runner and gates	1												100	—	—
fit and shutoff	5														
assembly	2												100	—	—
vents	1														
TRYOUTS:															
tryout #1	0														
fit and adjust	7														
tryout #2	0														
fit and adjust	0														
full inspection	0														
texture	0														
Tryout	0														
first article	0														
TOTAL MOLD	100														
PERCENT COMPLETE (MOLD)															

FIGURE 8.5. *Mold build spreadsheet. (Adapted from Ref. [20].)*

hardness, when alloyed with a higher content of chromium yield better wear characteristics in these areas. Air-hardened steels give improved dimensional stability over oil-hardened steels for cavities during heat treatment. Table 8.1 rates nine typical tool steels for manufacture, operation, and maintenance of

TABLE 8.1. Mold Material Selection Chart.

Poor ○ ◔ ◑ ◕ ● Best

	Machinability	Hardness	Toughness	Corrosion resistance	Abrasion resistance	Heat transfer	Thermal fatigue	Polishability	Heat treatment stability	Weldability	Comments
ALUM	●	○	○	◑	○	●	○	○	N/A	●	Prototype, short run and structural foam molding
P20	◔	R C 30–34	◔	◔	◔	◑	◑	◕	N/A	◕	Large cavities, cores, eliminates heat treatment process and associated warpage and cracking
H13	◑	48–52	◔	◕	◔	◑	●	◕	◑	◕	High thermal fatigue resistance and good polishability; generally chosen for zinc and aluminum die casting
420SS	◑	50–52	◔	●	◔	○	◔	◔	◑	◕	High corrosion resistance; poor thermal conductivity
P6	◕	56–58	◕	◔	◔	◑	◑	◕	○	◕	Easily machined, welds and repairs well; low-carbon steel, not dimensionally stable in heat treatment
0–1	◑	56–58	◔	◔	◔	◑	◔	◑	◔	◑	Oil hardened; pins, small inserts, etc.
S7	◑	57–59	●	◑	◑	◑	◔	◕	◕	◑	Shock-resistant steel; long cores where subjected to mechanical loads (slides and lifters)
A2	◑	57–59	◕	◕	◕	◑	◔	◕	◕	○	Good abrasion resistance and polishability; air hardening and heat treatment stable
D2	○	57–59	◕	◕	●	◑	○	◕	◕	○	Extreme abrasion resistance; used as gate inserts, etc., in filled resin applications

Source: Adapted from Ref. [16].

a tool. Usually combinations of these steels are used in the construction of a quality tool.

SELECTING MATERIALS FOR THE MOLD

The question often raised is which is better—a machined or a cast mold. The advantages of each are listed here for your consideration.

Advantages of Machined versus Cast Molds

Machined	Cast
• Ideal for single or multiple mold production. • Guaranteed density and properties. • More easily machined than tool steel or stainless steel molds. • Either wrought or cast plate, bar, or rod can be used.	• Ideal for producing large quantities of molds. • Rapid turnaround possible on large orders. • Good density if approved casting procedures followed. • Fine detail can be cast-in reducing machining time and cost. • Additional alloys are available with higher beryllium contents and hardnesses.

The method of selection must be based on tool life, tolerances, and cost, plus conversations with customers who use each type to arrive at the construction method.

Aluminum molds are also being built with new alloys that are harder and tougher than the standard 7075 aluminum alloy. In the past, aluminum molds were typically considered only for prototype and short-run production jobs. This was because of problems experienced in milling and finishing this alloy, its less than desired tool life, and tolerance problems. Tool builders are finding the new high-strength alloys, such as Alphastock-79, Alumec-89, and QC-7 from Alcoa, Inc., Pittsburgh, DA, more suitable for ease of manufacture. Depending on the resin used, they also result in long tool life, similar to that produced by steel. These newer alloys are more dense, have no soft spots, and are heat treatable. They can be polished for good part aesthetics and have found acceptance in low-to-medium pressure moldings. Other major factors to be considered when selecting the materials for a mold are as follows.

Corrosion and Abrasion Resistance

Each resin may form corrosive by-products during processing. Some resins often form strong acids when moisture is present. Others, such as flame-retarded grades, may form other corrosive elements that, if left to accumulate on the mold surface, can cause deterioration. Polyvinyl chloride (PVC) resins produce hydrochloric acid as a by-product; it can severely attack some tool

TABLE 8.2. Abrasion Resistance of Mold Materials Using Taber Met-Abrader Test (105 rpm).

Material	Rockwell Hardness	Weight Loss mg/1000 Cycles
D2 tool steel	C61	0.5
Chrome plate	—	1.7
Beryllium copper Alloy 20C	C43	3.0
Type 440C	C57	3.8
Silicon bronze	B95	5.6
H13 tool steel	C53	22
Type 420 stainless steel	C46	170
AISI 4130 steel	C52	260

Source: Adapted from Ref. [6].

steels, causing pitting of the cavity surface. This includes steel and aluminum alloy materials. Beryllium copper is a material that better withstands acid attack if properly cleaned and maintained prior to storage and periods of inactivity. But it too will, over time, be attacked. These copper alloys also exhibit fairly good abrasion resistance as shown in Table 8.2.

Beryllium copper, which is a high-heat conductivity material, can also be plated with nickel or chrome for additional corrosion resistance. This is similar to typical tool steels and aluminum alloys. Other advantages in using copper alloys are that no rust forms and cooling channels never become clogged with corrosive products.

In many cases, the use of stainless steels (SS) has proven a cure for rust and corrosion problems. Plating required for standard tool steels, such as chrome, which is susceptible to corrosive gas penetration generated by some resins, has been eliminated by using stainless steel. It has good machinability and tool wear with most resins.

The 420 SS has good stability in heat treating, can take a high polish, and exhibits good physical properties. Its main drawback is poor heat transfer when cooling line routing and temperature control across the part cavity is more critical. But this drawback can be compensated for in other ways and will be discussed in greater detail in the section on mold cooling.

Thermal Conductivity

To control part tolerances, thermal conductivity of the tool's material must be considered. The materials selected will affect production cycle times and part tolerances. In hard-to-cool sections, where water line routing is difficult, thermal hot spots requiring different materials with higher heat conductivity may have to be substituted to control dimensions and surface aesthetics. In these situations, the higher heat conductive material may be joined to or touching a cooling line. This will rapidly draw off heat to control the problem area.

The thermal conductivity of the tool's steels has a major impact on part quality. Cavity temperature control is critical for part tolerances, because an even temperature must be maintained across the tool's cavity surface. Otherwise, uneven shrinkage will occur as a result of trapped material stresses, nonuniform material packout, and variable material crystallinity. This will result in tolerance variations, possible warpage, and variable cycle times. Table 8.3 lists typical mold materials and their thermal conductivity. Figure 8.6 shows the rate of heat removal with different resin melt and part ejection temperatures.

TABLE 8.3. Material Heat Conductivity.

Material Type	Tensile Strength, PSI	Brinnell (300 UG)	Thermal Conductivity BTU/FT/HR, °F
S-7 (annealed)	93,000	214	21
S-7 (hardened)	275,000	526	21
P-20	93,000	263–344	19–21
Aluminum 6000, 7000	65–70	50–100 (500 kg)	80–90
420 S.S.	250,000	512	10–12
H-13	170,000–283,000	352–530	16.4
Ampcoloy 940	95,000	210	120
Ampcoloy 18	90,000	180	40
Beryllium/copper (2.0% BE)	115,000–125,000	336–380	62
Ampcoloy 22	95,000	311	27
A-10	303,000–355,000	560–680	18

Source: Adapted from Ref. [13].

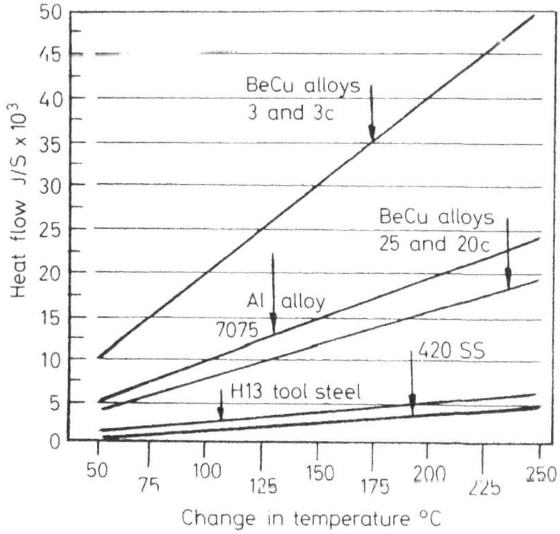

FIGURE 8.6. *Material heat removal. (Adapted from Ref. [13]).*

Amorphous plastics require a higher heat removal rate from the cavity steel to the mold's cooling system to obtain faster cycle times for part setup. This is necessary so that the part can he ejected from the cavity without distortion. This is not as critical with crystalline polymers, which have a sharp melting point. They can be ejected from the cavity at much hotter temperatures. The higher temperatures aid in the crystallization of the polymer, which results in faster part solidification.

To create a more uniform temperature profile across the tool, the use of nonferrous inserts in hard-to-cool areas is recommended. These are required if the main cooling lines cannot get close enough to the hot areas in the cavity or if the tool steel is unable to drain off the heat fast enough. Because these nonferrous inserts and core pins have less wear resistance, their best placement is away from major resin impact and high flow areas. By using these higher thermal conductivity materials, part quality is maintained and a better surface finish results. There is also less warp with better control of dimensions due to lower postmold shrinkage. Faster molding cycles can result. Often, these nonferrous inserts yield lower tool wear because of their natural lubricity. This aids in sliding and unscrewing areas of the tool. Cycle time reductions of 20 to 30 percent have been realized by the use of these nonferrous alloys in some molds.

Cavity Forming and Finishing

The degree of workmanship used in making the mold cavity is directly related to the final part's quality. Certain aspects of how the cavity is formed and finished are very important. This is the main responsibility of the toolmaker.

Most cavities are rough machined to form the basic shape of the part and are finished by fine machining to suit shrinkage and dimensions. After this is accomplished, heat treating, annealing, hardening, plating, texturizing, and/or polishing are performed. How well these operations are executed is directly related to a tool's strength, wear, and production life. Therefore, how the cavity is formed—how treatments to the tool materials are handled and finished during these operations—will have a direct bearing on the life and quality of the parts produced. Care must be taken to use the correct manufacturing technique for shaping the tool's steels to form the part cavity in the mold.

Microscopic cracks will occur during cavity formation from machining and EDM. The normal cavity formation goes from rough to fine machining. Surface defects can then be removed by subsequent rough-to-fine polishing. The polishing is always done after any surface treatments, annealing and hardening, are completed.

Polishing also aids the easy release of the molded part at the end of the molding cycle. It also reduces the formation of rust and corrosion on various parts of the mold surface. The Society of Plastics Industry (SPI) and The

Society of Plastics Engineers (SPE) have six numbered finish grades that mold builders and buyers can specify for an application. Usually, only the cavity side, show side, or wear surface require a special finish. For the core (nonvisible or nonfunctioning) side, a lower grade finish will suffice. For finishing extremely detailed molds, this will reduce total polishing costs to as little as 20 percent of the total mold making cost. Depending on the polymer selected, the mold, and the manufacturing process, different finishes can be specified to insure easy part release during processing.

Special considerations occur with cores that require minimal draft and resins that have a tendency to shrink down tight and have a low modulus or rigidity when hot. In these cases, shot peening the cores has resulted in easier part release and ejection. Shot peening the surface is often enough, requiring only minimal ejector pin pressure on the part, to break the suction between the part and the core. But in all cases, the core must not have any rough machine marked surfaces; these can impede part release. Normally a rough-to-medium polish, in the direction of the part release, will be adequate.

Electric Discharge Machining

Electric discharge machining, which is now preferred to finish or form cavities, is economical and very efficient. It can also be used in special forms for polishing and producing intricate shapes and undercuts.

Four main factors should guide the choice of EDM operating parameters on tool steels.

1. Stock removal rate
2. Resultant surface finish
3. Electrode wear
4. Effects on the tool steel

The first three variables must be selected properly, or the fourth, the working performance of the tool, might be jeopardized. During molding, the tool is subjected to thermal shock, polymer impact, and wear. This occurs during the injection stage and as a result of packing and clamp pressures. EDM subjects the surface of the cavity to very high temperatures that cause tool steel melting and/or vaporizing. During EDM machining, the surface microstructure is changed, hardness changed, surface stressed, and carbon content altered. As a result, surface brittleness can occur. As the temper of the steel is changed, cracks can form and the toughness of the cavity surface is sacrificed.

During the EDM process, microchanges at and just below the surface layers affect hardness and brittleness of the cavity steel. Generally, the outer steel layer is melted and resolidified. Below this is a rehardened layer, then a tempered layer, and finally the unaffected matrix of the material.

In the rehardened layer, the tool steel has changed composition because the temperature rose above the hardening point and formed martensite, which is a hard and brittle material structure. Below this, the temperature was only hot enough to temper the steel, causing the material to lose hardness. One problem that occurs during EDM machining is crack or fissure formation in the re-melted area. A crystal microstructure, which is formed in the metal during solidification, grows outward toward the surface, forming fissure sites or microcracks.

Different tool steels are not all candidates for EDM's rough-to-fine machining with graphite electrodes. Table 8.4 shows hardness versus machining variables tested in the hardened and tempered conditions. This information should assist in your selection of steels using EDM methods.

The thickness layers generated are largely independent of the steel grade and electrode used. In the annealed steels, the zone thickness is less, with fewer fissures and scarcely any hardened zone present. This is illustrated in Figure 8.7, which shows the surface and depth effects on an EDM machined surface.

Rough EDM machining shows greater thickness variation than fine EDM machining; the number of fissures in each zone of four steels is shown in Table 8.5.

Figure 8.8 shows the effects of noncontrol of the sparking process, which must be uniform and tightly controlled for uniform material removal. Poor control of sparking can cause local "arcing" between the electrode and workpiece. This causes craters to form in the surface of the steel. These are often confused with slag inclusions or porosity in the material. This is amplified in Figure 8.9, which shows the different effects of EDM machining on one side of A-2 steel bars of 57 Rockwell C hardness. After machining, relieving stress, or polishing, the bars were then bent with the machined surfaces on the outside. The results show that a fine EDM spark machining, followed by a fine polish, greatly improves the steel's strength. This information can be used to specify the finishing technique for cavity surfaces. For thin core sections, which

TABLE 8.4. Hardness versus Machining Variables.

Tool Steel, AISI	Austenitizing (°F)	Tempering (°F)	Hardness, Rockwell C Hardened	Hardness, Brinell Annealed
O1	1490	430	60	190
A2	1725	430	60	
D2	1870	480	60	220
L6	1545	480	54	
P20	1560	1075	30	
H13	1875	1040	50	180
M3:2	2010	1040	62	220–240

Source: Adapted from Refs. [3] and [5].

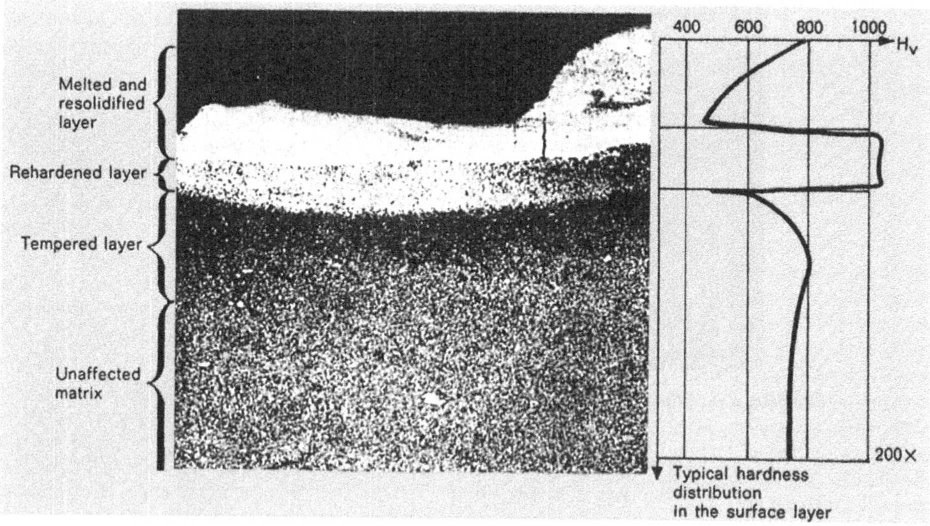

FIGURE 8.7. Effects of EDM machining. (Courtesy of Uddeholm Corp.)

TABLE 8.5. Fissure Occurrence.

	Number of Fissures		
Tool Steel, AISI	Melted Zone	Hardened Zone	Matrix
D2	20 to 50	2 to 10	0 to 5
H13	10 to 40	2 to 5	0 to 2
O1	10 to 30	0 to 5	0 to 2
P20	0 to 5	0 to 2	0

are subjected to bending under polymer injection pressure, the fine spark, machined, and polished surface may be required to ensure against premature failure of the tool.

The recommended procedures for EDM of hardened and tempered steels are as follows:

1. Conventional EDM machining. (Avoid arcing and high-stock removal rates.)
2. Hardening and tempering.
3. Fine sparking (low current, high-frequency rates).
4. Grinding and polishing EDM surface or tempering 30°F lower than the original tempering temperature.

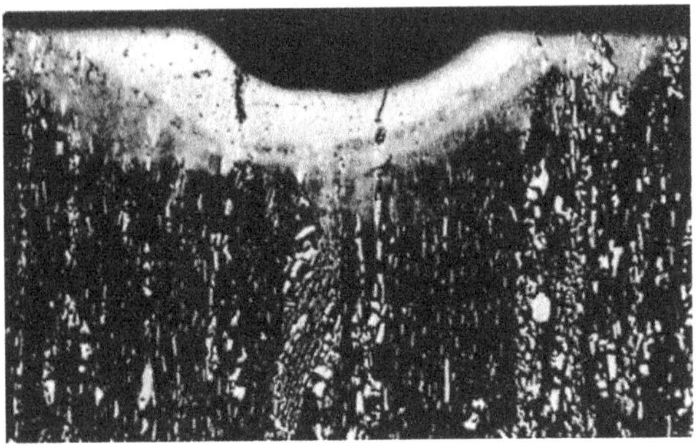

FIGURE 8.8. Effect of noncontrolled sparking. (Courtesy of Uddeholm Corp.)

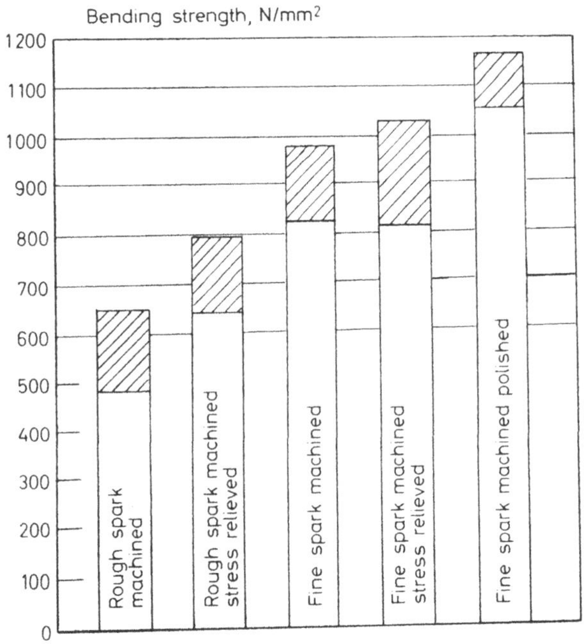

FIGURE 8.9. A2 steel bending strength at various machining points (lined area is spread). (Adapted from Ref. [3].)

EDM of annealed steels are as follows:

1. Conventional EDM machining.
2. Grinding and polishing EDM surface.
3. Slowly preheating, in stages, to hardening temperature.

Therefore, EDM rough machining, followed by fine machining, is recommended as the best method to produce a surface that is free from defects and cracks.

Polishing

Final polishing is required to restore the surface toughness and bending strength of the tool surface. Thinner sections will require extra polishing to reduce this problem.

After the cavity is to tolerance, the final polishing operation begins. It proceeds from a rough stoning to a fine stoning to a fine polishing. EDM machining, as opposed to conventional tool-bit cutter machining, usually requires only fine stoning and polishing.

The conventional polishing steps to finalize the cavity surface are as follows:

1. Coarse polishing paste with a hard polishing tool.
2. Coarse polishing paste and a medium-to-hard polishing tool.
3. Medium-to-coarse polishing paste, with a medium-to-hard polishing tool.
4. Medium-to-coarse polishing paste, with a soft polishing tool.
5. Fine polishing paste, with a soft polishing tool.

A word of caution is required. On vertical or raised surfaces on the core and cavity, always polish in the direction of the draw or mold opening. With some high shrink or tacky resins, even fine polish marks may retard ejection of the part from the mold.

Polishing is an art. Only experienced, well-trained, and competent people should be used to perform this final operation. Polishing can never be rushed; it takes many years for a skilled tool polisher to develop.

Texturing

Most plastic parts have some degree of texturing on their visible surfaces. Texturing is used to impart different surface finishes to a part. It also provides a gripping surface, provides contrast with mating parts of the same or different materials and colors, provides decoration to the surface, and hides such surface imperfections as sink marks from molded in-ribs and bosses on the underside on a visual surface. With parts that have complex shapes, the designer can use texturizing to aid release from the mold. Texturizing also eliminates the polishing required for show surfaces in a mold and reduces the cost and delivery time for tool. A few examples of the varying styles and patterns of texturizing available are shown in Figure 8.10.

Almost any surface effect can now be reproduced on a molded part, including wood and leather grains, matte finishes, geometrics, brick, splatter paint,

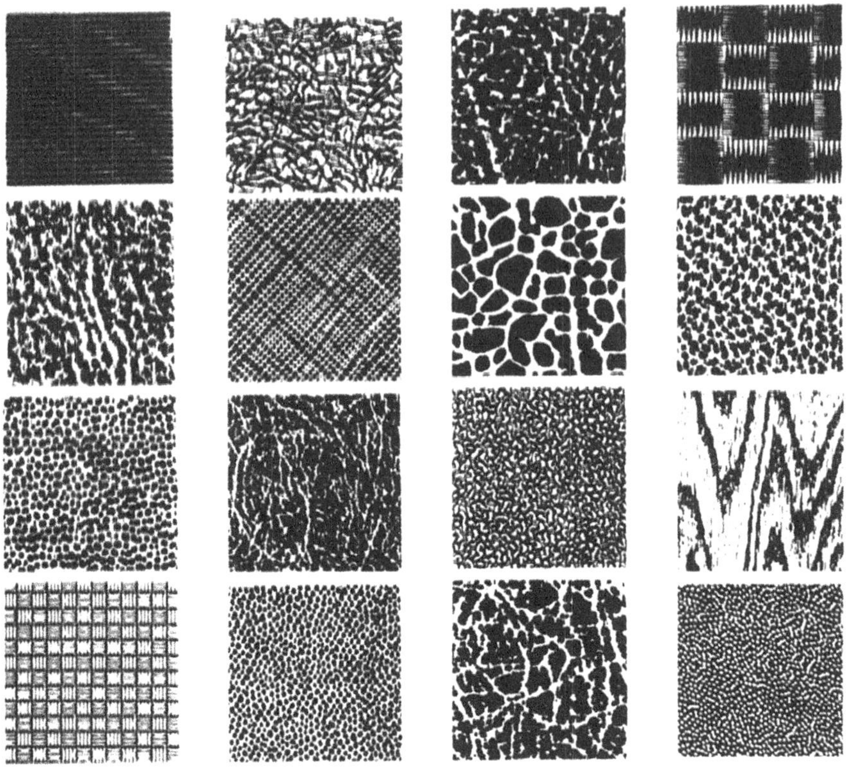

FIGURE 8.10. Texturizing pattern. (Courtesy of the Akron Metal Etching Co., Akron, OH.)

graphics, logos, and precisely controlled three-dimensional designs. The major mold texturizing sources have molded plaques showing the different patterns and texturized surfaces available—usually on P-20, a prehardened tool steel. The same patterns, when. applied to aluminum, stainless steel, and beryllium/copper, show a visible difference. This is because of differences in each material's grain structure and the etching reagents used to impart the pattern.

Most of the typical mold steels produce a very good pattern on the molded part. This is the result of tighter grain structures. Variations of surface finish. occur when high-chrome and nickel-content tool steels are used. To produce the best-textured surface, use only high-quality steel and alloys.

Texturizing a mold can yield benefits beyond a variety of finishes. It can hide flow lines, weld lines, shut off, and witness lines, all of which may be visible in some molds with different resins. The varying texturizing effect is often just enough to hide these blemishes, which would be visible on a polished mold surface.

The designer should decide early in the program whether any texture finishes will be used. This allows the designer and toolmaker to determine draft angles, contours, and other details prior to building the mold. There should not be any subsequent costly revisions to accommodate the texture requirements.

If the part must mate with another metal or plastic part, then the texture patterns must either match or be sufficiently opposite to be visibly acceptable. Because of differences in light reflectance, texturizing will also affect the color of mating parts, particularly if note act Therefore, the surface finish and depth of texture must be exact. For mating parts made in separate tools, it is recommended the same texturizing source do the work to ensure that the match will be identical for all parts.

Texturizing is done by a variety of methods such as EDM, chemical etching, grit blasting, etc. All these techniques employ the disintegration of surface metal to achieve the pattern. Therefore the surface finish of the tool prior to texturizing must be known before applying the textured surface finish. Normally, as long as cutter, file, and sanding disc marks are removed a finish of 240 to 320 emery finish is acceptable. All scale resulting from heat treating, hobbing, or the EDM operation must be removed prior to texturizing. On chrome surfaces, all plating must be removed prior to texturizing. Also because the chemical reagents may discolor polished areas on the mold, these areas should be polished after texturizing.

Normally, a 280-emery-grit or 320-grit stone surface finish is all that is required for the depth of texture to remove most of the tool surface marks. Surfaces with a finer texture may require a finish of 320-to-400 grit emery paper. The 400 grit is used for a very fine mechanical design. But if a tool is already highly polished, there are now methods available to engrave a design, lettering, or logo chemically without staining or harming the surface.

The depth of texturizing has a direct correlation to the draft angle required on the tool for areas in the line of tool opening or draw. This ensures that the textured area is not marred during part release. The standard rule of thumb for noncomplex patterns is 1½ degrees of draft for every 0.001 inch of texture depth. Most textures are 0.0025 to 0.0030 inches deep. The shrinkage of the resin used will determine the required draft on texturized cavity and core sections. As an example, a leather grain pattern that is 0.0035 inches deep in a cavity, molded with an amorphous resin, should have 4 to 5 degrees draft to avoid part scuffing on ejection. Whereas with a crystalline resin, with higher shrink, 3½ to 4 degrees draft may be sufficient.

Other items must be considered before adopting the 1½ degree rule of thumb. They are the resin's shrinkage, the texturized pattern's layout direction, and the depth of the texture. Once these are known, along with the pattern placement, the required draft angles can be determined.

The depth of the texture can be controlled and altered to meet specific requirements, such as blending a finish to meet a shallower depth or allowing certain areas to be shallower in depth when draft angles do not permit a pattern to be a full standard depth. However, as mentioned previously, this will affect the appearance of the textured finish and the color match between this and mating parts.

All injection molded thermoplastic resins will reproduce 100 percent of the textured pattern on the molded part. Foamed-in resins, with blowing agents, require more care. Patterns for these resins should favor medium to large type textures to aid in concealing the swirl patterns that often result from the foaming agent. Also, all undercuts in the pattern should be eliminated; use more generous draft if the texture is etched 0.003 to 0.004 inches deep to avoid the part locking in the mold. Painted parts may require a deeper etch to retain the pattern and required finish appearance. This is the result of paint buildup in the texture depressions.

A high-quality tool steel can produce a top-quality etch, with texture depth tolerances of within (+/–) 0.0003 inches. Texturized aluminum molds will have a greater variation. If etched to 0.015 to 0.020 inches deep, aluminum will have a mean etched depth of 0.0175 inch. Etching of aluminum is less controllable than steels.

With the advances in tool steels and the development of improved transfer and duplication processes, toolmakers can duplicate nature's most spectacular textures on to plastic parts.

CAVITY SELECTION

Part Layout

With the design of the part finished and the resin selected, a decision must be made as to the number of cavities, single or multiple, to be used to make the part. The items to be considered are as follows:

1. Number of parts required and delivery schedule
2. Quality control requirements (tolerances)
3. Piece-part price of part (to meet cost objectives)
4. Resin elected (runner and gate location(s) and size)
5. Part shape and size (parting line, tool size, internal mold operations, ejection)
6. Injection molding machine (clamp, shot, and melt capacity, as well as tiebar dimensions)

The part quality question must be answered first in cavity selection. Once the tool is built, it is too late to request quality levels that the tool cannot meet. This happens frequently with poorly designed and built molds that have too many cavities to properly control critical dimensions. Often, a decision to produce the latter is based on an unrealistic set of economic and part quality decisions.

Single-Cavity Molds. Tighter tolerances and uniform part-to-part reproducibility are the main advantages in the selection of a single-cavity mold. Part size, mold complexity, and molding machine size limitations may also be factors.

The advantages are as follows:

1. Simple and compact construction (less expensive and quicker delivery)
2. Tight tolerance control and part-to-part uniformity
3. Tighter process control (conditions need only be adjusted to suit one cavity)
4. Greater latitude in part layout and material selection
5. Selection of tool type, runner size, gating, and ejector and cooling system—which can be designed and achieved without compromising part quality.

Multicavity Molds. Multicavity tools have the advantage of producing more parts at a lower piece part price. They can produce complex parts, but at a sacrifice in tolerance control. Since all parts do not require tight control, multicavity tools may be satisfactory.

The key factors are based on the following items:

1. Part size and complexity.
2. Tolerance requirements (no two cavities will be exactly alike).
3. Material of selection (shrinkage factors).
4. Cooling requirements (may be limited because of part layout and ejector system; a longer cycle may result).
5. Injection molding machine that is sized for adequate melt supply, mold clamp, tiebar dimensions, and process controls.
6. Tool type (may consider automatic operation).
7. In-tool cam and slide operations may be limited.
8. Processing conditions must be averaged to obtain common tolerance limits for all parts.
9. Runner system sized to provide adequate melt flow to fill all parts within cycle times (regrind could equal part weight; a balanced system is preferred).
10. Cost of tool amortized over additional parts.

Family Molds. Family molds are often used for parts with low quantity requirements, very loose tolerances, and the same resin. In a family mold, part sizes can vary from very small, ¼ ounce or less, to greater parts of an ounce or more. For best results, all parts should be as close as possible to the same size to avoid overpacking and underpacking.

Family molds use individual cavity blocks inserted into a common mold base. To accommodate these inserts, the base is designed with precut runner systems. The cooling system is limited to the area around the cavity inserts and the ejector system to the cavity block area. In some cases, family molds have their cavities cut, like regular production molds, directly into the cavity steel without using the separate cavity inserted mold base concept. This is not recommended if part size varies a lot. Some molders use a wagon wheel cavity layout with a center sprue and spokes feeding round, square or rectangular cavity inserts. This minimizes and standardizes the runner flow length and provides a more balanced layout and resulting cavity fill pressure to each cavity at the mold's wheel rein end.

Parts with lower quality tolerance are molded in family molds with a fairly good degree of part uniformity—as long as the size differences are not too great. The gating for each part is cut into each cavity block insert and then sized to fill and pack-out each part adequately before gate freeze-off. Based on different gate sizes, there will be nonuniform freeze-off for each cavity, and packing pressures will vary from cavity to cavity. The overall cycle time is then dependent on the longest part setup time. Family tools should only be considered for low-volume, noncritical, toleranced parts. What can often happen with family tools is that one part may require tighter tolerance control. The molder is thus forced to block off the other cavities and use the tool as a single-cavity mold. This defeats the family mold process.

Cavity Selection Based on Molding Machine Size

If the mold is to be run in a captive operation on a specific injection molding machine, the number of cavities must conform to the machine's melt capacity, mold size capabilities, and clamp pressures. If the parts are to be molded outside, these items must still be considered along with the machine's hourly rate charge for producing the parts.

Whether a single or multicavity tool is selected., it mast be sized to fit a standard molding machine with specific material melting capacity and machine clamp to keep the mold closed under injection pressure. The higher the polymer's flow relative to the melt viscosity, the more clamp tonnage is required to keep the mold closed.

Melt Capacity. Each molding machine's shot capacity (in ounces) is based on the volume of polystyrene (PS) that can be generated in front of the screw when it has fully retracted during the melt generating stage. Because of the many different types of resins, never use more than 80 percent of the machine's rated shot capacity to calculate shot weight. No molding machine is ever 100 percent efficient, as will be discussed in the machinery section. Under optimum conditions, the number of cavities is based on the machine's melting and shot capacity and having in the barrel two to three equivalent shot weights of resin. Only in very rare eases, with very large parts, is the total machine's shot capacity used. It taxes the machine's melting ability and there is always the chance of unmelt being injected into the part.

The maximum number of mold cavities is therefore calculated based on machine shot capability as follows:

C = number of cavities.

S = 80% of barrel shot capacity of machine.

W = part cavity weight plus spree and runner.

Y = barrel melt capacity safety factor using 2 or 3 to suit the resin.

With a 10-ounce barrel melt capacity machine and a 2-ounce shot weight, the maximum number of cavities would be:

$$C = \frac{S}{WY} = \frac{8}{(2)(2)} = \frac{8}{4} = 2 \text{ cavities}$$

The number of cavities is also governed by the melting capability (P) of the machine, based on screw size and the use of polystyrene as the reference resin. If P equals 120 lb/hr, which is 32 oz/min, and X is 6, the number of cavities (C^*) is:

$$C^* = \frac{P}{XW} = \frac{32}{(6)(2)} = \frac{32}{12} = 2.6 \text{ or 2 cavities (rounding low for melt capacity)}$$

It would be best never to exceed the molding machine's limits, but if more cavities are required, a large machine with higher melt capacity should be used. This is where production can assist as it knows the equipment's capabilities. Production could, in some cases, put back pressure on the screw or even install a more efficient screw to increase melt capacity. However, with some resins, this could cause increased melt temperature through increased shear heat, and this could affect, the part's appearance or performance.

Clamp Capacity. Other factors affecting the number of cavities in a mold are the machine clamping force needed to keep the mold closed during the injection and packing stages. The mold clamp force is based on the number of square inches of projected part surface area, including the sprue and runner system. The clamp force can vary from 1 to 5 tons per square inch of projected area. Each machine is rated for its ability to apply maximum clamp force on its platens, and each resin has its recommended clamp force per square inch of surface area. A general rule of thumb is: amorphous resins, 1 to 3 tons, and crystal line resins, 3 to 5 tons. Also, the easier the resins flow and the faster the injection rate, the more clamp tonnage is required. Each resin supplier should be consulted as to its respective resins' required clamp force.

Another requirement that is often overlooked is whether the mold will fit between the machine's tiebars. If the tool is too large or has externally attached mold operations, a larger press may have to be selected. This may affect processing conditions and melt quality. The mold must also open up enough for easy access to parts. This is referred to as the daylight dimension of a machine's platens.

Mold Cavity Layout

In multicavity molds, consistent part-to-part tolerances require that the cavity layout be balanced in the tool. This means that the runner lengths have equal flow distances for the resin to reach each cavity at the same time and with the same injection pressure. There are many ways to achieve this, and they relate to part shape and positioning in the tool.

Most multicavity molds are designed for an even number of cavities. Only if tolerances are not critical can an unbalanced layout be considered. If an unbalanced layout is used, part tolerances must be relaxed. Examples of balanced and unbalanced cavity layouts are shown in Figure 8.11.

Layout 1 shows a six-cavity tool laid out symmetrically around the central sprue. Each cavity is fed by a separate runner. A better runner layout is shown in layout 2, where three runners feed two cavities each. This allows for downsizing of the runner system and less regrind, as well as a cold-slug well where the runner divides to trap the partially cooled polymer while maintaining

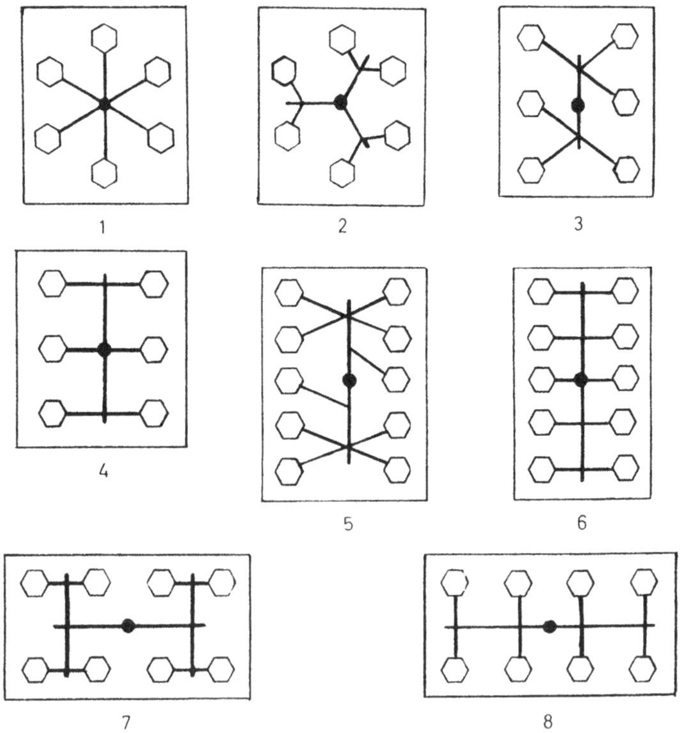

FIGURE 8.11. *(1, 2, 3, 7) Balanced and (4, 5, 6, 8) unbalanced multicavity molds.*

equal feed and packing pressure to each cavity. In cases where a circular layout is not possible, because of part size or side cores, cavities are arranged in two rows (layout 3). This is a very poor arrangement, as the two center cavities will fill first and then become overpacked while the other four fill. This layout does not permit uniform dimensions to be obtained from all cavities and, as a result, part quality will vary. By being overpacked, the center cavities may be oversized, while the end cavities are undersized, and could experience porosity by not being properly packed out. In layout 4, the runner system is balanced to feed and fill plus pack out each cavity more uniformly. Layouts 5 and 6 show examples of 10-cavity molds with similar designs.

Most molds used today are built with a balanced number of cavities, as shown in Figure 8.12. When the number of cavities varies from even numbers, it is difficult to obtain a balanced runner system. Such a system is required to obtain equal packing pressure for each mold cavity. If tight part tolerance control is not required, then an unbalanced runner system may be permitted (see Figure 8.13).

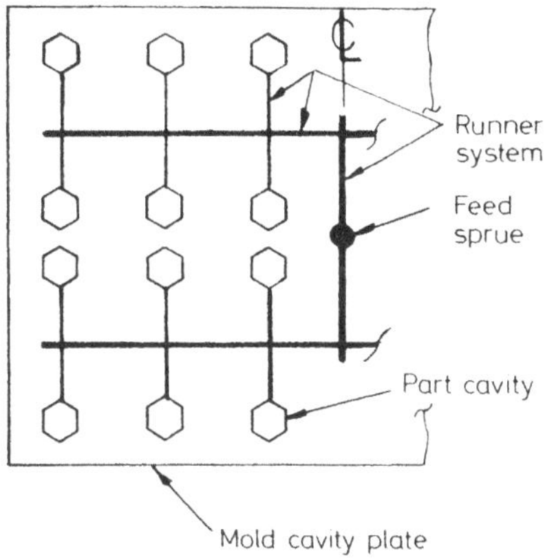

FIGURE 8.12. Balanced multicavity mold.

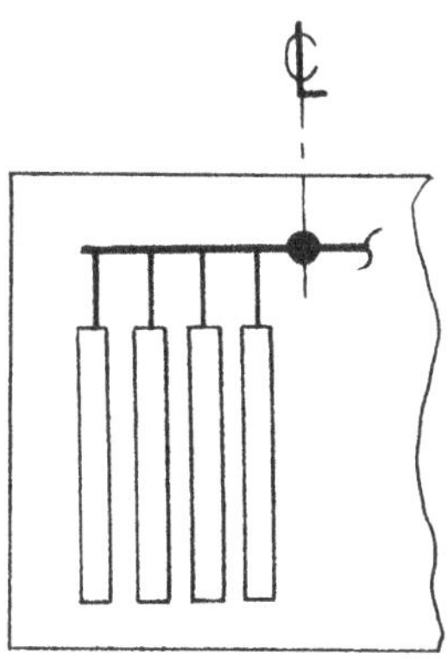

FIGURE 8.13. Unbalanced multicavity mold.

Except for family molds, multicavity molds should mold parts to the same shape, and all parts must be gated at the identical location. When locating cavities into a mold, the flow length of the runner feeding the part cavities must be minimized. Heat loss and pressure drops in the resin in the runner system can seriously effect the dimensions of the outer cavities in unbalanced molds. Excessive amounts of regrind are also generated.

RUNNER SYSTEMS

Cavity Runner Layout

The material reaches the mold cavity through the runner system. In all cases, route the runner system so that it is as nearly equal to all cavities as possible from the central sprue. A shorter flow distance and a fewer number of bends reduces variances in part dimensions. Longer flow lengths and more bends dramatically reduces the cavity fill pressure that controls part tolerances. Therefore, the runner should be as small as possible, but still able to supply sufficient melt to fill and packout the cavities. This will also help to minimize the amount of regrind generated. The optimum runner layout will have the runner filling uniformly and material flowing through each part's gate at the same time and under equal pressure. This promotes uniform cavity filling, part packout, and equal dimensions.

Runner System Design

The runner system should be shaped and sized for maximum flow of the plastic resin, but with a minimum cross section to keep its weight low. This statement is intentionally redundant. In most tools, the runners are too big and the layout of parts so poorly placed that long oversized runners are required to feed the cavities. Your suppliers can recommend the runner shape and size best suited for their materials.

Suppliers can also provide data on the melt flow length of their resins. With this information, you can determine if the runner is sized correctly to properly fill and pack out the cavity. Suppliers generated these data from spiral flow molding evaluations of their material, using standard processing temperatures, injection pressures, and their experience working with customers. They may recommend shielding the runner system from the mold's cooling system, which may inhibit flow in the runner system. There are also software programs that can assist in runner sizing and provide data on pressure drops in the runner system. However, many of these programs are still evolving, and experienced mold and process engineers should review the output and qualify the results.

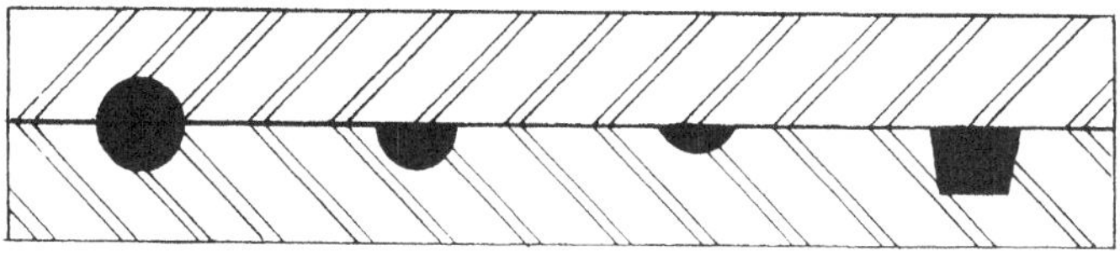

FIGURE 8.14. *Examples of four types of runner designs. (Adapted from Ref. [1].)*

To minimize the pressure drop in the runner system, the number of bends should also be minimized. Each time the runner turns a corner, the pressure drops. While designing part layout in the cavity, plan to minimize the number of runner bends. This will ensure that the maximum injection pressure is available to fill and pack out the cavity.

Resin flows from the runner in a manner similar to lava flowing from a volcano. The molten core of the resin flows through and out the material first laid down in the runner channel. When the polymer comes into contact with the cooler mold surface, it solidifies and remains relatively stable as the molten resin flows out ahead of it. For this reason, a texturized cavity surface only adds marginal resistance to resin flow during filling. Only very deeply texturized surfaces add appreciable resistance to flow. Runners must be neither rough nor highly polished. The finish should be smooth to ensure ease of flow and without any surface undercuts that might restrict removal during ejection.

The full round runner, followed by the trapezoidal, are considered the best designs. The half and quarter round runners are less desirable, as the flow area is minimal and rapid cooling of the melt can result in premature gate freeze-off. Remember that the center of the runner section is the main feed channel to the part, and it will always be circular. Figure 8.14 shows these in profile, relative to the mold parting line. The only problem with the full round runner is that it must be machined into each mold section.

Table 8.6A shows recommended round runner sizes for typical resins. The size of the runner is also dictated by resin flow length and cavity part thickness. Table 8.6B shows the minimum runner diameter based on part thickness. The longer the flow and thicker the part, the larger the runner. It is always best to undersize a runner if you are unsure of its required size, as it can always be increased if part fill and pack out problems occur.

The runner system should also have cold slug wells (traps) at each bend to trap any cooled resin before it reaches the gate (see Figure 8.15. Runner cold slug well traps, whose lengths equal the diameter of the runner, are required for the sucker pin style of sprue puller, which has no provisions for trapping the initial cold slug injected into the mold at the start of the molding cycle.

TABLE 8.6A. Recommended Material Runner Sizes.

Material	Diameter (in)
ABS, SAN	0.187–0.375
Acetal	0.125–0.375
Acrylic	0.312–0.375
Impact acrylic	0.312–0.500
Cellulose acetate	0.187–0.437
Cellulose acetate butyrate	0.187–0.375
Cellulose propionate	0.187–0.375
Ionomer	0.090–0.375
Nylon	0.062–0.375
Polyallomers	0.187–0.375
Polycarbonate	0.187–0.375
Polyethylene	0.062–0.375
Polypropylene	0.187–0.375
Polyphenylene oxide	0.250–0.375
Polysulfone	0.250–0.375
Polystyrene	0.125–0.375
Polyvinylidene fluoride	0.125–0.312
Polyvinyl chloride (plasticized)	0.125–0.375

ABS, acrylonitrile butadiene styrene; SAN, styrene acronitrile.

Source: Adapted from Ref. [1].

TABLE 8.6B. Minimum Runner Diameter.

Part Thickness (in)	Runner Length (in)	Minimum Runner Diameter (in)
0.020–0.060	Up to 2	1/16
0.020–0.060	Over 2	1/8
0.060–0.150	Up to 4	1/8
0.060–0.150	Over 4	3/16
0.150–0.250	Up to 4	1/4
0.150–0.250	Over 4	5/16

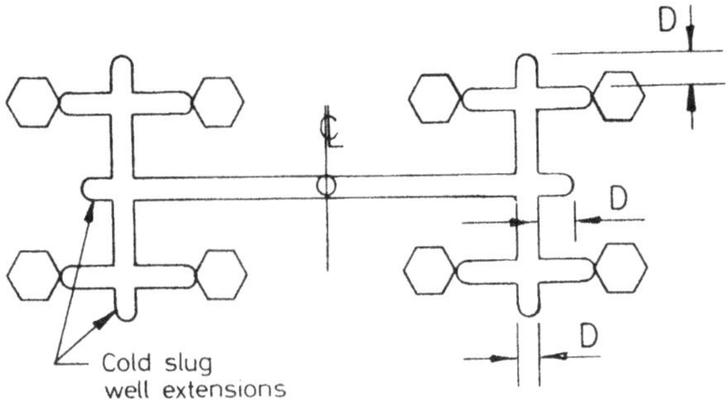

FIGURE 8.15. *Runner cold well slug trap.*

GATING THE PART

The part's gate location, size, and type control the flow of resin into the cavity from we runner system. The runner system, in turn, provides the melt to the cavity's gate, which feeds, directs, and controls the filling of the part cavity. The gate acts like a dam. When it solidifies or freezes, it restricts any molten resin in the part cavity from flowing back into the runner. The gate also controls part weight and, thereby, the dimensional quality of the molded part. During filling, it controls the molecular orientation of the resin and any fillers or reinforcements in the polymer. Glass fibers in molten materials will line up in the flow direction of the melt. When injected into thin-walled parts (0.10 inch thick or less), they will orient in the flow direction. In thicker sections, because of more turbulent resin flow, they will tumble and become more random in their orientation. The latter will result in a more uniform resin shrinkage in the thicker glass—reinforced molded parts.

The gate location, type, and opening size affect part quality. If possible, the gate should be located at the thickest part section to ensure that the melt flow fills the cavity before freeze-off occurs at the gate. Some parts may require more than one gate; this is determined by part shape, section thickness, flow length, and resin type. A. good uniform flow of resin into the cavity is necessary to ensure uniform filling and good packout pressure for uniform dimensions throughout the part. For tight tolerance, a part's gate design and location is critical.

All parts require adequate and. uniform fill, as well as uniform packing pressure distribution. For most parts, one gate serves these purposes. For tight-toleranced parts, the number of gates is usually doubled to insure adequate fill, reduce resin flow length, and create a uniform packing pressure gradient in the cavity. To obtain uniform shrinkage and dimensions, the packing pressure should be as uniform as possible throughout the part. As packing pressure decreases, the further the resin flows from the gate, the greater he shrinkage. This will cause more part-to-part shrinkage variances, resulting in varying dimensions in the part. Multiple gates reduce the flow length, yielding parts with tighter dimensional control.

Multigated parts should not be opposite each other in the part layout. If opposite each other, they will have converging melt fronts that will fight each other in filling out the cavity. The trapped air in the cavity will also cool and slow down the meeting of the melt fronts. This can result in a weak weld line, voids, and porosity in thicker parts.

To avoid weld line formation, multiple gates should not be located at critical part sections, such as flex, impact, high-load, and visible points on the part. The cooling channels must also be located near or at these junctions, so that correct mold cavity temperature can be controlled to insure good part properties and appearance.

Adequate venting at melt front meeting points is required to ensure strong material junctions and eliminate burning of the resin. This is covered in greater detail in the cavity venting section.

For tight tolerance control, a balanced runner system for locating gates may not always be possible. Therefore, with multiple gates, an unbalanced runner system is used to obtain the tolerances required for the part. The examples shown in Figure 8.16 illustrate typical flow and pressure gradient lines in a single- and multiple-gated mold.

When molding reinforced resins, particularly glass fibers, additional gates will result in a more random fiber orientation. Fiber orientation will have a definite affect on part shrinkage and the control of finished dimensions. For parts with section thicknesses of less than 0.100 inch, the fibers will line up in the material flow direction, with the fibers' length restricting shrinkage in the flow direction. But in the transverse direction across the fibers' length, the base resin's shrinkage will control the part shrinkage. It is also affected by the volume of glass fibers occupying this space. In thicker parts, the glass fibers' orientation is multidirectional because the fibers tumble and randomize during filling. This results in more uniform material shrinkage at these sections.

Material Shrinkage

A resin's mold shrinkage is developed by the resin supplier from molding flex bars (½ × ⅛ × 4 inches long), which are end gated and not typical of any molded part. These flex bars are molded under standard molding conditions.

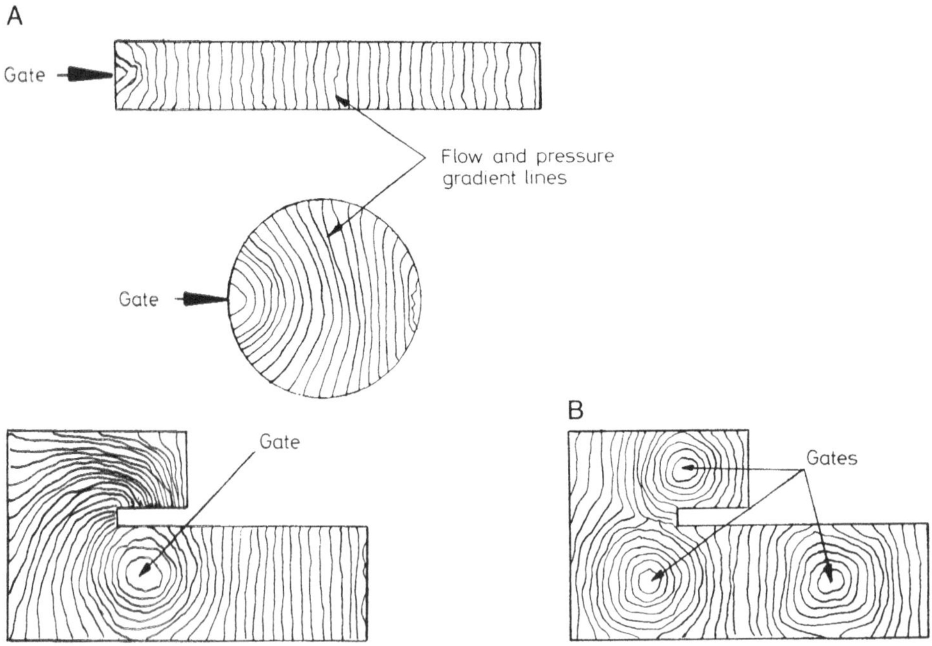

FIGURE 8.16. Flow and pressure gradient lines in (A) single- and (B) multiple-gated molds. (Adapted from Ref. [16].)

Their shrinkage is measured based on the mold cavity dimensions. These shrinkage dimensions are only guidelines to be used when calculating the cavity dimensions for the resin to be used for the part.

A resin's shrinkage varies based on the many factors in the molding cycle. Experienced tool designers and processing personnel have come to rely on their experience in calculating shrinkage values. Calculating resin shrinkage will be discussed in greater detail in the Mold Shrinkage section later on in the chapter.

If cut too small, the gate size can influence part shrinkage as much as 50 percent. Injection pressure to fill the part cavity should never exceed 1200 PSI. If this occurs, the gate opening is too small. Too small a gate results in increased resin shear heat build up and resin decomposition. If this occurs, open the gate dimensions.

Too small a gate will also result in premature resin freeze-off at the gate. This restricts packing out the cavity to obtain maximum part weight, resulting in more shrinkage and an undersized part. This varies with amorphous and crystalline polymers; crystalline polymers show larger variances because of their higher shrinkage rates.

Gate Location

The gate's location at the cavity is positioned so that the resin is injected against a side wall or opposing surface, or in a controlled filling process that results in a uniform melt front that will flow uniformly to fill the cavity. This is illustrated in Figure 8.17 which avoid the resin jetting shown in Figure 8.18. Resin jetting results in uneven fill uneven fill and causes dimensional part warpage and surface problems. By relocating the gate, or changing its type and size, or by controlling the injection rate, uniform fill can be achieved. If the gate's location cannot be changed, the cavity fill rate (programming the screw and injection pressure profile) can be used to obtain a uniform fill rate. First, use a slow fill rate to obtain a uniform melt front of resin in front of the gate. The part will begin to fill uniformly if the fill rate is increased by additional injection pressure. Production department personnel help to provide the necessary injection profile. They can also best determine whether the modification of process conditions will produce quality parts with the gating specified. If not, determine whether the gate can be moved or modified to produce good parts.

When deciding gate location, always gate the part at its thickest section. This ensures that the gate remains open as long as possible, so that the maximum amount of resin is injected into the cavity and held under pressure until gate freeze off. Only by completely filling and packing out the cavity can dimensions be held consistently to meet customer tolerances.

Gate Terminology

The influence of gate type with the four basic runner configurations—half round, full round, trapezoidal, and U-shaped—will now be discussed. Good

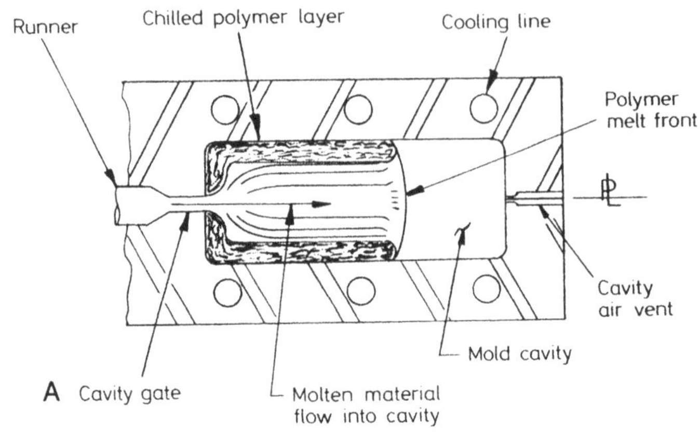

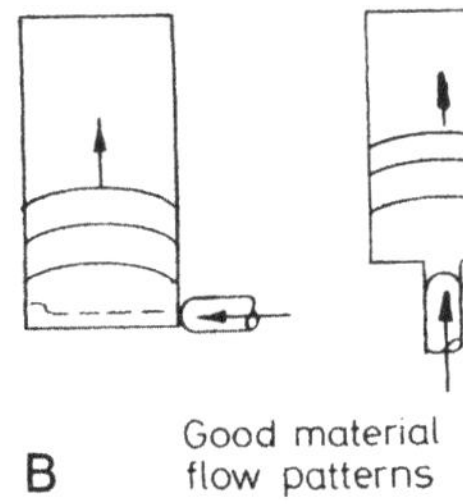

FIGURE 8.17. *Correct resin flow into mold cavity: (A) slow even fill to establish melt front; (B) gate design and placement to obtain uniform melt flow.*

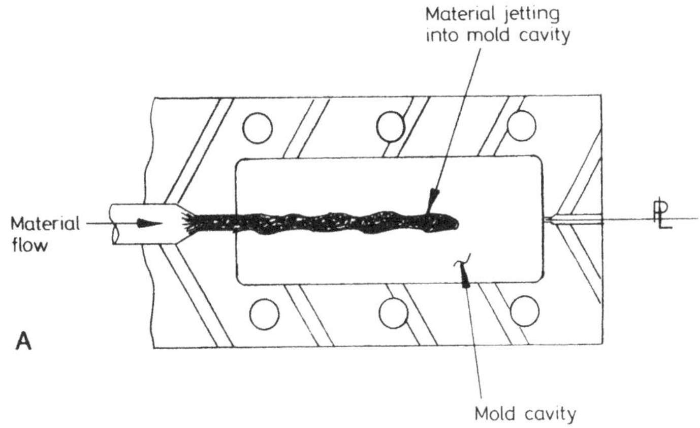

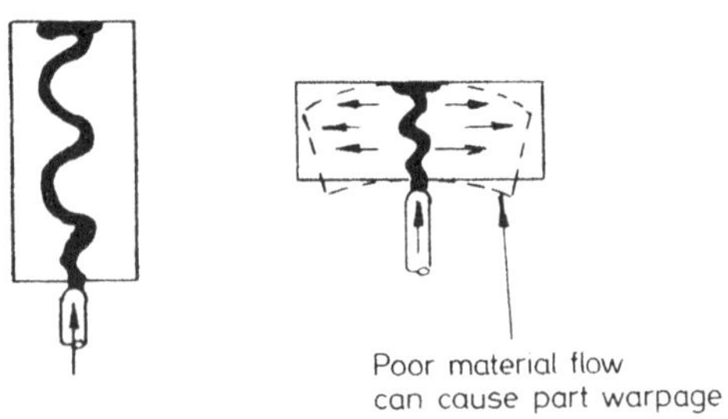

FIGURE 8.18. *Poor resin flow patterns due to resin jetting: (A) high injection pressure and poor gate design; (B) reduced mold fill rate and relocated gate impinge on side wall.*

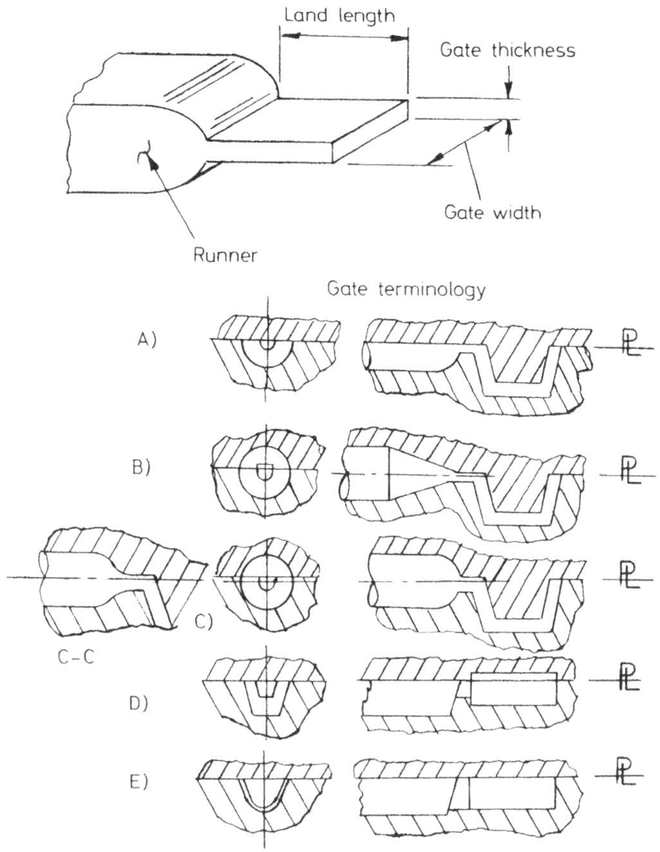

FIGURE 8.19. Runner and gate design combinations: (A) half round; (B) full round; (C) full round; (D) trapezoidal; and (E) U-shaped. (Adapted from Ref. [8].)

and bad runner/gate combinations are shown in Figure 8.19, with gate terminology explained for better understanding.

1. The half-round runner and pin gate are poor because of premature cooling and gate freeze-off before the cavity can be filled and packed out.
2. The full runner with taper to semicircular gate is better, but the taper interferes with proper cavity filling.
3. The full circular runner with slight taper (CC) is a good design that reduces pressure loss, improves flow, and minimizes glass breakage if resin is reinforced. It has a recessed semicircular gate with runner. The gate's centerline is in the cavity side of the mold parting line. This provides continued polymer flow. It is generally used for pin-point gates with the land length determined by the minimum metal strength of the mold.

4. The trapezoidal runner and sprue gate or standard style gate is used if the runner must be located in only one half of the mold. It is used due to core pulls in the runner area or if the cavity is restricted to one side of the mold parting line. Polymer flow is restricted to a circular section drawn within the trapezoid due to polymer drage at the corners.
5. The U-shaped runner with full gate for very thick section parts (over 0.120 inches thick) provides good part pack out. The problem with such a thick runner and gate combination is that the cycle is dependent on the rate of resin freeze off at the gate or entire runner in this case.

The trapezoidal and full-round runner system are recessed below the parting line. They require an ejector system to ensure positive ejection of the runner from the mold during operation.

With these styles of runner systems, there are many gate designs to choose from. Each must be selected based on location, part configuration, fill pattern, tolerances, tool type, and degating operation. The type and size of each part's gate will determine how a part fills and is packed out. The final part tolerances are the controlling factor for cycle time. Not until gate freeze-off can the screw be retracted to build up the melt for the next cycle.

Gate Type

The different types of gates with their good and bad features are shown in Figure 8.20. Your material supplier, mold designer, and production personnel can assist in selection of the correct type of gate for each part. Remember that if part aesthetics are important, the gate location must not be visible after the part is assembled.

1. The standard gate is most often chosen. It is easily degated with trim fixture or hand nips.
2. The full-edge gate, with a full wall thickness, is used in thick-walled parts for good pack out.
3. The fan gate, which is relatively wide but has a thin edge, spreads out the material flow. It therefore minimizes flow lines. Its parts have thin flat sections.
4. The flash or film gate, which is used for intricate parts, has uniform filling and parallel direction of flow. It may taper from thinner section at the center to larger thickness at the edges to compensate for the pressure differential in runner feed to part. It is used to control warpage in reinforced parts. It should be degated as soon as possible to avoid warping the part.
5. The edge gate is a pin-point gate used for flat, thin-walled parts.
6. The tab gate is used to obtain good part pack out in thick parts and out-of-the-way of mold cam operation. It uses a machine operation to remove the gate.

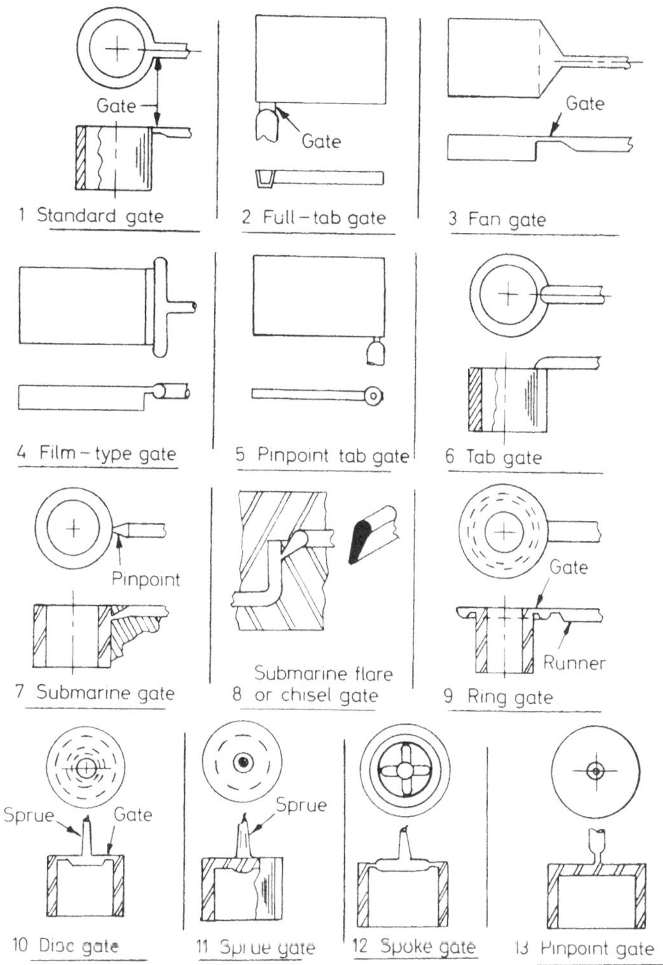

FIGURE 8.20. *Types of cavity gating. (Adapted from Ref. [1].)*

7. The pinpoint submarine gate has automatic degating below the parting line, which usually has an 0.020 inch diameter. It is used on small parts. The diameter of a pin gate should be 50 to 70 percent of the nominal wall thickness, with a minimum diameter of 0.050 inch and a maximum of 0.100 inch. Gates smaller than 0.050 inches can shear the material and cause premature freeze off before the part is packed out. Gates larger than 0.100 inches can ause extended cycle time, difficulty in cutting the gate, a higher stressed area, and a large visible gate blemish on the part.
8. The submarine flare or chisel gate is similar to the gate just discussed. Its chisel edge is longer and used on larger parts. It has automatic degating. Examples of the submarine gate's operation are illustrated in Figure 8.21. As the mold opens, the gate shears and the runner and

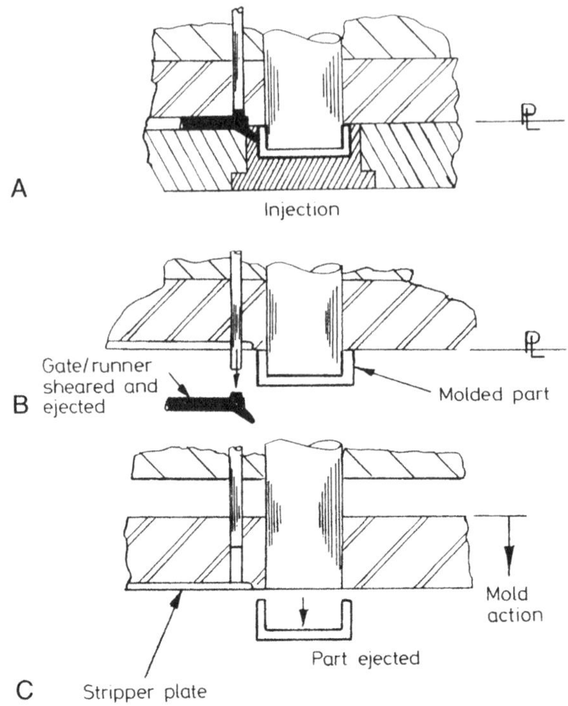

FIGURE 8.21. Automatic submarine gate shearing: (A) submarine shear gate into part; (B) gate is sheared on mold opening; and (C) part is stripped off core by stripper plate. (Adapted from Ref. [8].)

sprue are ejected. The core then retracts and the parts separate from the tool core.

9. The ring gate is used with external, multicavity tools suitable for circular parts. It prevents weld lines and gives good part concentricity.
10. The ring or disk gate is used with internal, single-cavity, or three-plate tools suitable for circular parts. It has good concentricity.
11. The sprue gate is for single-cavity tools. It has good packing pressure, excellent concentricity, rapid fill, and can be used on large semiflat parts in three plate design. It requires mechanical degating.
12. The spider gate is used with multiple injection sites. A weld line will occur between gates and its concentricity is poor, but it is easily degated.
13. The center gate, pin point, has good fill but poor packing pressure. It is easily degated and can be used with a three-plate tools, but has a slight pinch point that creates a flush break off.

There are many ways of gating the part, and each is selected to obtain the desired part and processing effects. These are the basic styles of gates one can select. The exact style should be chosen to suit the requirements and quality of the molded part.

Gate Control of Weld Lines

Gating controls many factors of a part's strength. But once past the gate and inside the cavity, parts with cored-out sections or multiple gates present another issue. The bond strengths of the two opposing melt fronts must be addressed. Wherever resin melt flows meet, meld lines occur. Weld lines occur when the melt flows around core pins or obstructions and then meet. They also form where multiple gate melt fronts meet and at cored-out ribbed sections. In the most severe case, the melt fronts meet and poor low results (see Figure 8.22). This may cause lower part strength at these weld lines. While the cavity is filling, the melt front is cooled continually by pushing the air ahead of it out of the cavity. When the melt fronts meet, they may not obtain a good fusion bond at the weld line. Weld-line strength can be controlled and improved by a number of factors, including the following:

1. Adequate venting at the weld line site. If poorly vented, trapped air retards the flows from meeting. The melt flows only touch and do not fuse properly. Note: Glass-reinforced polymers at these sites will only exhibit the weld-line strengths of the parent resin. The fibers will not flow across the melt junction and mingle with the opposing melt front.
2. Use high-speed injection pressure to keep the material hotter when the melt fronts meet. Venting is critical if this technique is employed.
3. Correct resin and mold temperatures. Material melt temperature may have to be raised and/or the tool heated in these local areas to promote melt flow bonding.
4. Provide a weld-line overflow tab, as illustrated in Figure 8.23. This allows the fronts to meet and flow together a short distance to promote bonding. The tab will later be removed. This is not true for glass-reinforced materials.

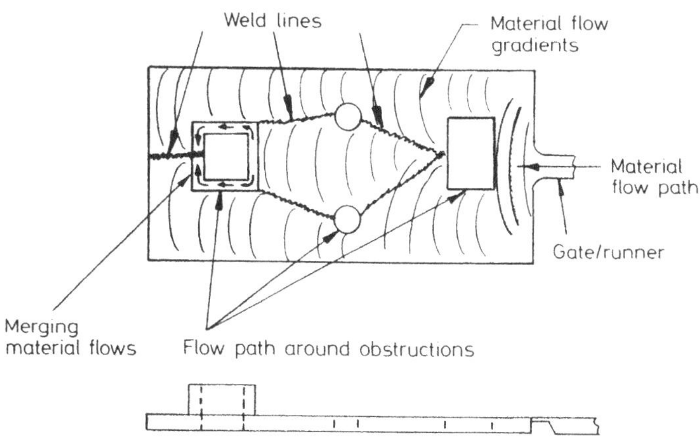

FIGURE 8.22. Weld line formation in molded parts.

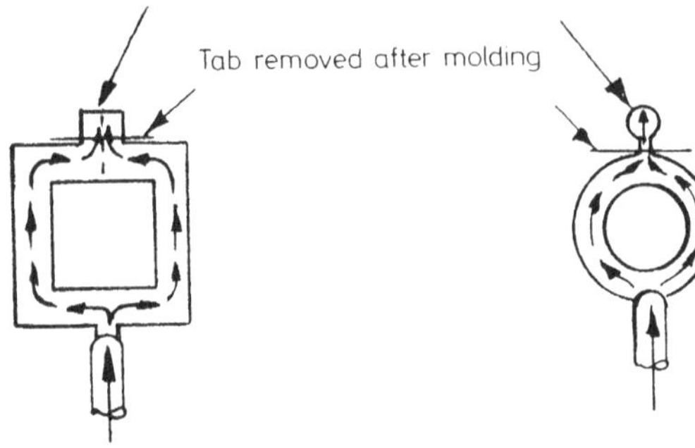

FIGURE 8.23. *Flow tabs improve weld line strength.*

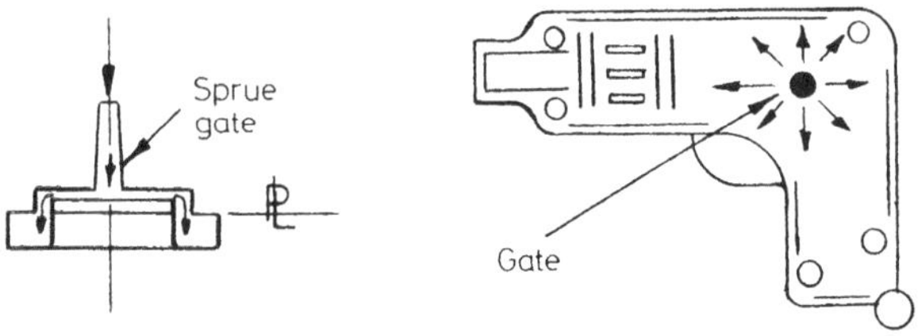

FIGURE 8.24. *Gate location that minimizes weld lines and improves dimensions.*

If weld lines form at critical areas, such as impact points or high-load bearing areas, then the gate location will have to be changed to minimize or move the weld line effects at these areas.

In a critical section, where weld lines are not permitted or their effects must be minimized, a different type of gating may be required. This may require a different type mold, such as a three plate versus a two plate, so that the gate can be located within the part's internal shape (see Figure 8.24). This also reduces flow length and aids part pack out and dimensions. Another alternative is the use of multiple gates to move the weld line(s) to a less critical section of the part.

Multiple gating is also used for parts requiring a living hinge, if the material cannot flow sufficiently through the hinge area to fill the other side. A second gate is then located in this section, as shown in Figure 8.25. By using a longer runner to feed this section, the fill pattern is changed, filling out the rest of the part cavity and moving the weld line into the thicker section of the part.

Wherever weld lines are formed, localized cavity heating can aid an improving their strength. In grid-shaped and cored-out parts, weld lines are inevitable. By correctly locating the gate, the resin melt flows can be directed to meet at the grid intersections. By controlling how the melt flows meet and flow together, instead of allowing them to meet head on at these intersections, weld line and part strength can be greatly increased. By reconfiguring the shape of the grid patterns in a part, the melt flow and resulting weld-line strength can also be improved (see Figure 8.26).

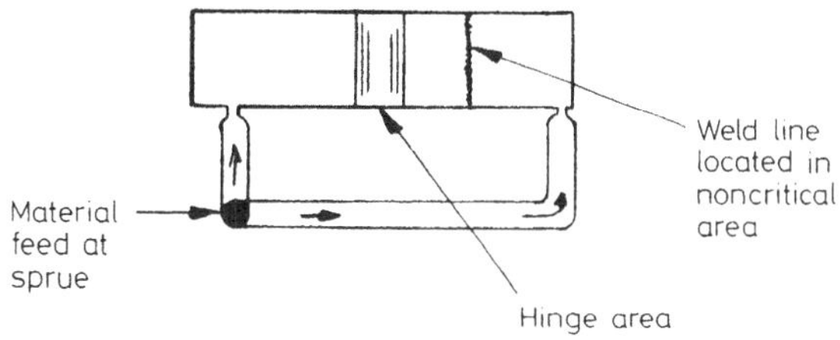

FIGURE 8.25. *A longer runner shifts the weld line into the thicker section of the part.*

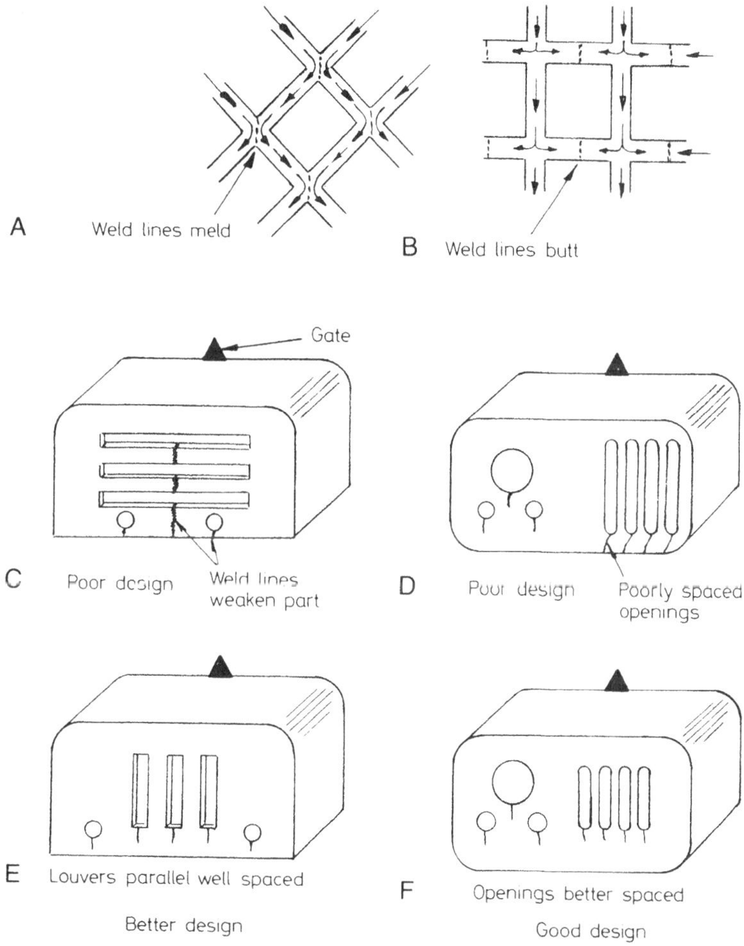

FIGURE 8.26. *Weld line improvement (Adapted from Ref. [1].)*

SPRUES AND NOZZLES

Sprue Pullers

The mold's sprue puller is designed to pull and clean out the mold's sprue bushing and the tip of the injection molding machine's nozzle of solidified polymer when the mold opens. It is located in the moving half of the mold. It is also used as a cold-slug well to trap any solidified resin remaining in the nozzle on the next injection cycle. All sprue pullers, except the sucker-pin style, have a cold-slug well located in line with the sprue opening. An exception to this is when the sprue is the runner and gate in a single-cavity mold. Examples of different styles of sprue pullers are shown in Figure 8.27. The runner system should also have cold-slug traps at each bend of the system, especially for the sucker-pin sprue puller that has no trap for the nozzle's cold slug.

Sprue Bushing and Nozzle Seating

The mold's sprue bushing must mate with the injection molding machine's nozzle to create a tight seal against the injection pressure developed to fill the mold. The sprue bushing, which has a minimum draft of three degrees for easy

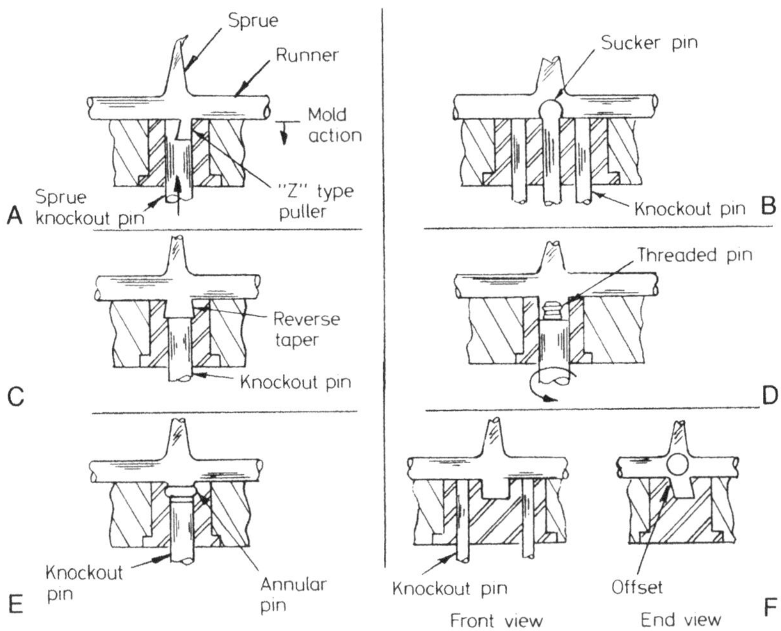

FIGURE 8.27. Mold sprue pullers: (A) Z type; (B) sucker pin; (C) reverse taper; (D) treaded pin; (E) annular pin; and (F) offset. (Adapted from Ref. [1].)

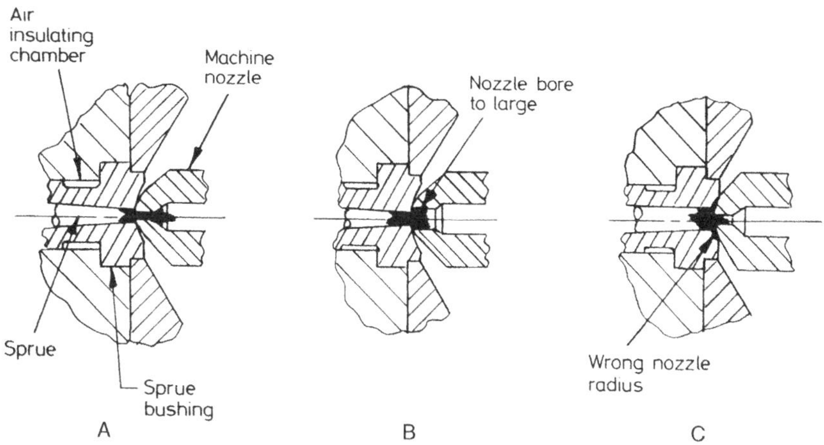

FIGURE 8.28. *Nozzle and sprue seating. (Adapted from Ref. [8].)*

release during pulling, is surrounded by an insulating air chamber that prevents the polymer from cooling down too rapidly. Nozzle and sprue configurations are shown in Figure 8.28. They illustrate (A) the proper fit of the nozzle to sprue bushing versus problems (B) a nozzle bore too large, and (C) an incorrect nozzle radius not matching the sprue. The radius of the nozzle seat in the sprue bushing must be identical to the radius of the nozzle. This is required to achieve correct pressure and material seal off of the nozzle to the sprue bushing. If sized incorrectly, the resin will become trapped, as shown in (B) and (C), and the sprue will not pull properly. A line contact seal is used to minimize heat loss from the nozzle to the tool. If the nozzle temperature control becomes a problem during molding because of heat loss to the mold, the nozzle heater band temperature must be increased. To avoid drool at the nozzle, use a stainless steel sprue bushing with its lower heat conductivity. To verify that this is the problem and to avoid increasing nozzle temperature, use an insulating material between the sprue bushing and the nozzle. This could be several sheets of paper, thin cardboard, a high-temperature thin plastic film, such as teflon or any other low-heat conductivity material. This thin layer of insulating material will help to prevent or retard the heat transfer from the nozzle to the cold sprue bushing.

The mold's sprue bushing should also be slightly larger in opening diameter than the nozzle's opening, to prevent the problem of resin damage resisting sprue pull illustrated in Figure 8.28B. Standard size sprue bushings are available from mold suppliers to match machine nozzles, so this should never be a problem. But because of constant butting against tools, machine nozzles eventually become flattened. They should be remachined to match the sprue bushing radius or be replaced. This will be apparent just by examining the end of the sprue.

PARTING LINES

Cavity Parting Line Location

The decision as to where to part the tool or to place the part's parting line depends on the following factors:

1. Size, shape, and complexity of part.
2. Number of cavities.
3. Layout of runner system and type of gating.
4. Tolerane requirements and cooling system layout.
5. Resin selection.
6. Molding machine size.
7. Venting requirements—any blind cavities?

The part's parting line should be located to maximize part release and minimize deformation, damage, or surface defects during ejection of the part from the cavity. Automatic operation should also be considered for degating and ejecting parts from the mold, so they can be conveyed to their next station.

Part removal from the mold cavity must have ease of release, with uniform ejection and maintenance-free operation of the ejection system. Otherwise, cycle irregularities may occur and can cause dimension problems, poor economics, and variable part quality. In some cases, poor part release on ejection can cause parts to hang up in the tool. If the mold closes to them, they will damage or destroy the tool. Always strive for full part and runner release from the mold. The operator must not be required to pull the part off the core or out of the cavity.

When determining the part's position in the mold, try to use a single-plane parting line for both simple and complex parts. The operations that occur on the parting line are shown in Figure 8.29, along with mold component identification. They include resin injection through the runner system, followed by part ejection by the ejector system. When the tool opens, the ejector pins push the part from the cavity along with the runner from the movable half of the tool.

Complex Parting Line

With more complicated parts used to form complex closed structures and undercuts, these molds may require a multiplane parting line. For these parts, various methods have been developed to aid manufacture and ease of part removal. These complex parts are formed in the cavity with the following items:

1. Side core pulls.
2. Wedges to move mold steel away from the part's vertical sections.

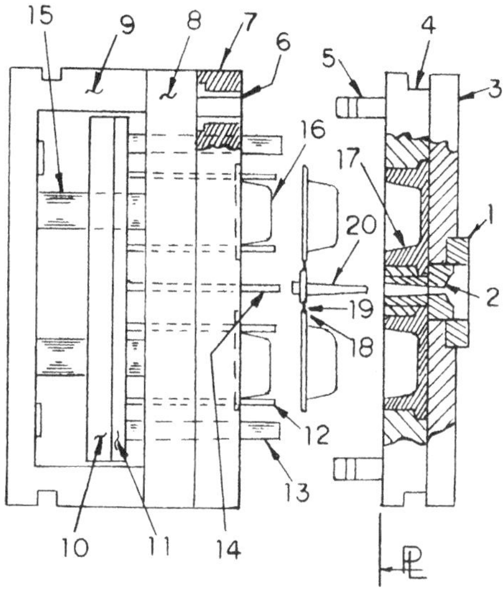

1. Locating ring
2. Sprue bushing
3. Clamp plate
4. Cavity retainer plate
5. Leader pin
6. Leader pin bushing
7. Core retainer plate
8. Support plate
9. Ejector housing
10. Ejector plate
11. Ejector retainer plate
12. Ejector pin
13. Return pin
14. Sprue puller pin
15. Support pillar
16. Core
17. Cavity
18. Gate
19. Runner
20. Sprue

FIGURE 8.29. *Parting-line operations and mold components. (Adapted from Ref. [14].)*

3. Rotating and collapsing cores for forming external and internal threads.
4. Inserts to be removed after the part cools. These are more expensive and are labor/operator intensive.

The selection of one or more of the above is governed by the shape and complexity of the part, the resin selected (it influences part flexibility, rigidity, and shrinkage), and by the end-use requirements and part quality.

For parts with external sections or undercuts methods 1, 2, 3 can be used. For parts with internal threads or part section variations, item 3 can be used. The method depends entirely on part design and end-use requirements.

These systems can be operated by manual, hydraulic, mechanical, pneumatic, or electromechanical systems. The selection depends on what services are available at the molding machine, the toolmaker's preference, tool-cost guidelines, and the degree of automation required in the tool.

SIDE CORE PULLS

There are various side core pull mold designs. Some are positive acting, operating as the mold opens; others are delayed action, operating only after the tool has opened and unlocked the caming action.

Side-Action Core Pull

Side-action core pulls are used to form complex part shapes where mold steel must be removed from the part before it can be ejected from the mold. Figure 8.30 shows a side operating mechanical core pull in the closed and opened positions. Side wedges on the fixed half lock up the core pulls operated by the cam bars or dowels. The cam bars are mounted at not more than 25 degrees to keep their bending stress low. This is because of the high forces exerted on them during opening and closing. The side core pulls ride up and down in "T"-slot on the moving half of the mold. At the end of their operating stroke, they are captured either by a spring operated cam block locking pin (Figure 8.30), or a slide retainer mounted on the mold surface (Figure 8.31). The slide retainer holds the slides in the open position during part ejection and then releases the slide during mold closing. The angle on the locking side wedges should differ by 5 degrees from the core pulls' locking surface. This prevents jamming and galling of the wear surfaces during mold closing.

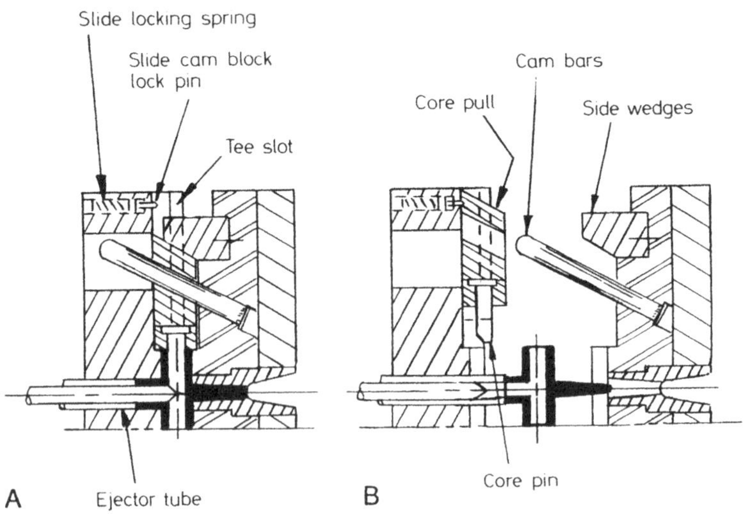

FIGURE 8.30. *Side-action core pull: (A) closed position and (B) opened position. (Adapted from Ref. [8].)*

A

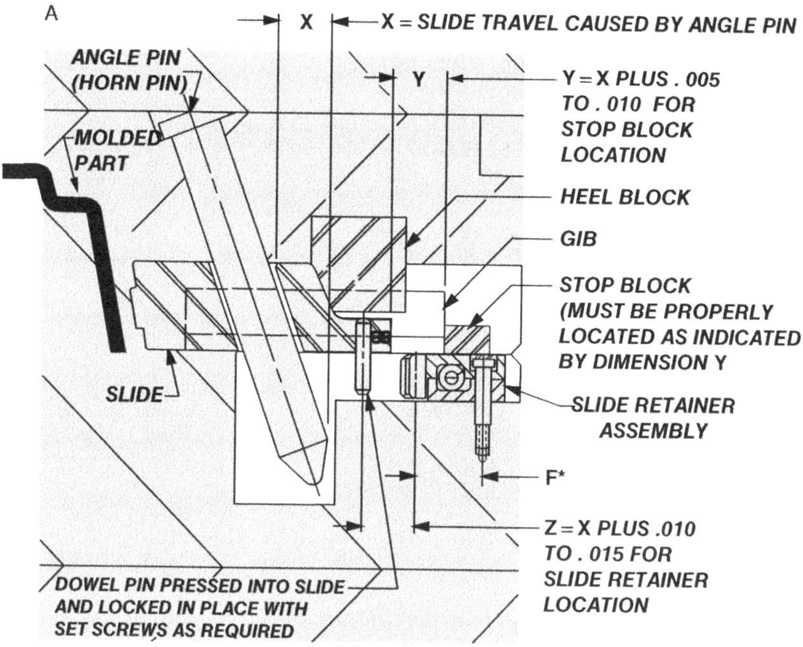

B

FIGURE 8.31. *Mold slide retainer mechanism: (A) typical set-up dimensions; (B) retainer to suit application; (C) conventional slide retention—when mold opens, dowel pin in slide is engaged in retainer jaw; and (D) holding or pulling of floating plates—retainers keep third parting line closed until first two have full opened. (Courtesy of D-M-E Co.)*

Delayed Side-Action Core Pull

Delayed side-action core pulls, as shown in Figure 8.32, are similar in operation. There is, however, a delay built into the caming operation that allows other internal mold operating functions to occur before the side cores move.

With hydraulic or pneumatic-actuated side cores, the operation is much simpler and can be timed to operate when desired. These operations require external equipment, which may cause a fit or mounting problem on the molding machine. But, they are also easier to maintain, experience less wear, and yield greater maintenance savings.

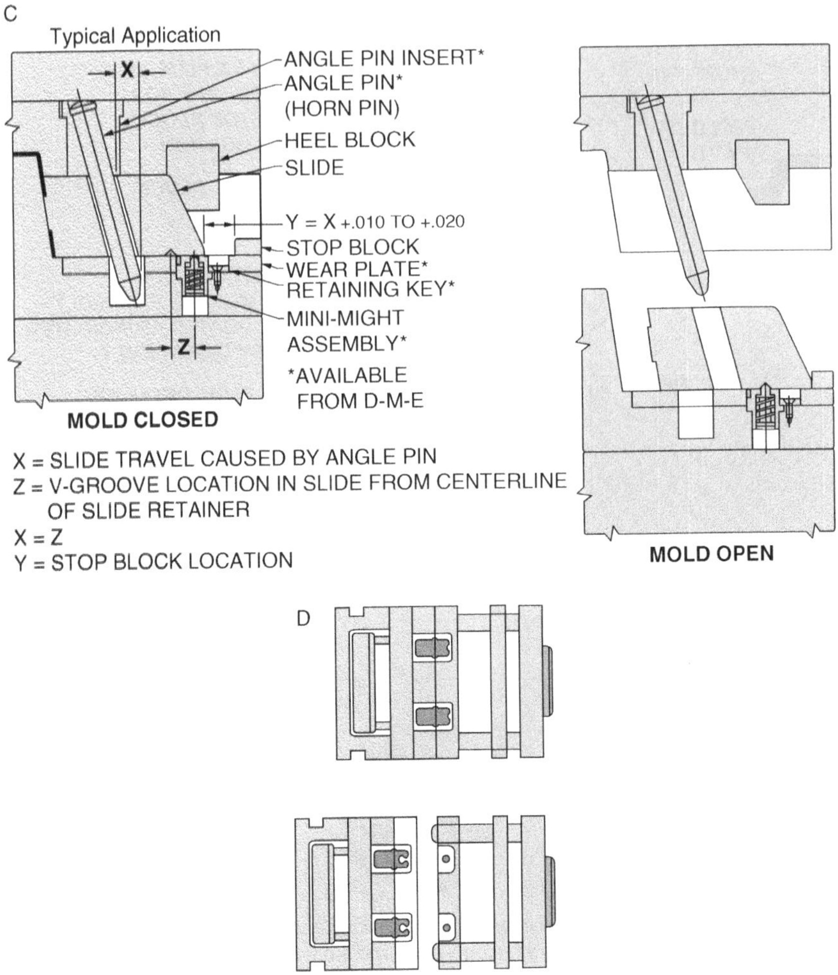

FIGURE 8.31. *(Continued)*

Slide Retainers

In cam action molds, there must be a positive locking mechanism to retain the slide cores after the mold is opened. This prevents the slide cores from dislodging or operating out of sequence, both of which may damage the mold during closing. Several styles are used, such as internal or external spring-retainer pins or electrical/hydraulic mechanisms. The method selected is based on the mold and part design requirements. The preferred method incorporates slide retainers within the confines of the mold's dimensions, where it is easily accessible if service is required. Slide spring retainer systems, as illustrated in Figure 8.30, is, if the mold is heated, subjected to thermal stresses and shorter operational life. An alternate attachment for slide retainers is illustrated in Figure 8.31. These retainers are mold mounted, external to the cooling system,

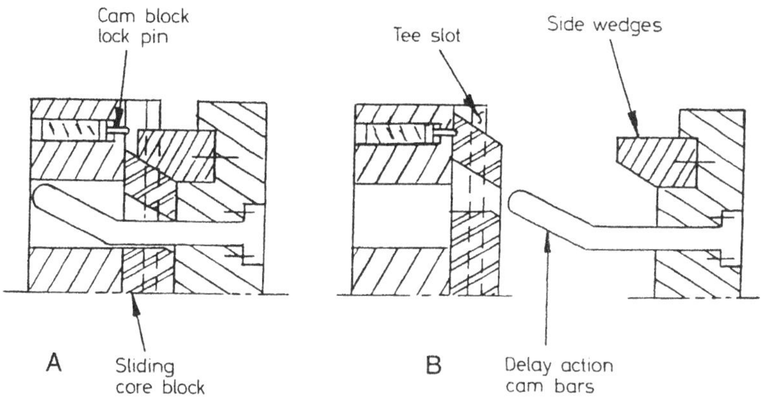

FIGURE 8.32. *Delayed side action core pull: (A) delay in operating core block controlled by cam bar and mold opening times, and (B) angled cam bar operates cam block, forming undercut to allow ejection of part. (Adapted from Ref. [8].)*

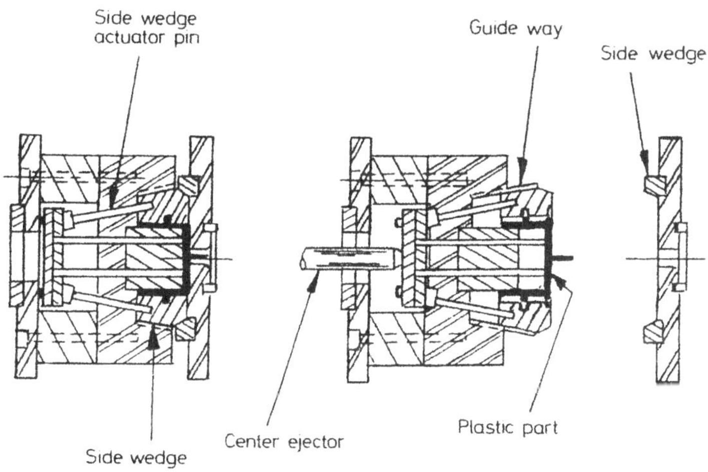

FIGURE 8.33. *Wedge action core pull. (Adapted from Ref. [8].)*

and can be insulated from the mold's surface. They are available in different holding force ratings, easily serviced, and are less expensive than in-mold slide retainers that are mounted.

Quality and dependable operation are most important in any slide action mold operation. This area, often neglected during mold design, is an important factor in the life of the mold and in part quality.

Wedge Action Core Pull

A wedge mold is shown in Figure 8.33, with a center ejector simultaneously activating the side wedges that release the part as it is ejected from the tool.

The wedges are on guideways that control their movement in and out of the cavity surfaces. Temperature control of the operating components and matching the mold materials' thermal expansions is critical to ensuring that the sections seal but still slide freely during operation.

CORE SELECTION

Collapsible Cores

Internal threads, cutouts, and undercuts are made by several different methods. Internal threads are produced by molds with rotating cores. They are expensive and slow down the cycle. In addition, collapsible cores are used and require more maintenance and are size limiting. Another method used is stripping, but it is more material sensitive. The resin must be rigid enough to be ejected off the core pin and still have enough elongation to stretch over the major thread diameter of the core pin without fracture and permanent part deformation.

Collapsible cores consist of a segregated, perfectly matched collapsible sleeve that has the shape to be made machined into its outer surface. The internal core pin keeps the core expanded when the mold is closed. When the mold opens, it withdraws inward causing the core to collapse (see Figure 8.34). The degree of undercut is limited by the amount of collapse. The collapsible core is a very precise tool that is often difficult to cool. Materials for its construction must be selected for sliding wear and heat transfer. Collapsing cores can reduce cycle time by as much as 30 percent over conventional unscrewing or other complex actuating mechanisms.

Unscrewing Cores

Rotating or unscrewing cores are very precise in dimensional capabilities. They operate as the tool opens, if combined with gearing, or can be operated independently with pneumatic, hydraulic, or electrically driven toothed racks. They can activate before, during, or after the tool opens. The latter is preferred, as long as the part is supported in the mold and cannot rotate during unscrewing. Figure 8.35 illustrates a worm shaft combined with gearing that operates as the mold opens.

During the mold's opening an closing, the drive gear is rotated by a coarse-pitch worm shaft that is fixed in the stationary mold half. The drive gear transmits rotation to the geared core, which unscrews from the part by means of a screw thread bushing. The pitch of each screw's threads must match perfectly for smooth operation. The speed of operation is based on mold-opening speed, which is kept slow to reduce the likelihood of damaging screw threads, reducing wear, and damaging the part. Ejection of the part from the cavity proceeds as the core is unscrewed. In this case, an external unscrewing

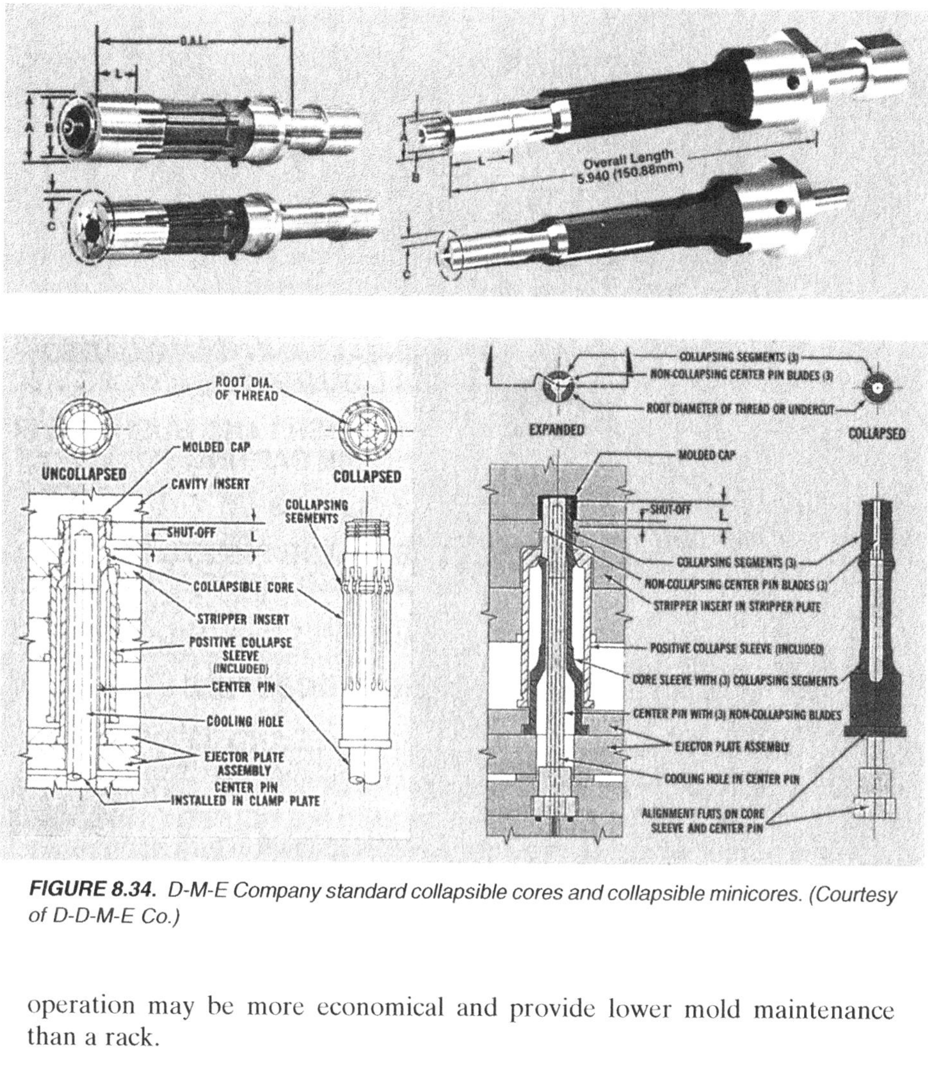

FIGURE 8.34. *D-M-E Company standard collapsible cores and collapsible minicores. (Courtesy of D-D-M-E Co.)*

operation may be more economical and provide lower mold maintenance than a rack.

PART EJECTION

A smooth clean release of the part from the cavity must occur on every cycle. The ejection of the part is critical to preventing distortion, warping, scuff marks, or any marring of the part. After the mold is filled and both pack out and gate seal-off has occurred, the part must remain in the mold to cool. This hold time is required for the part to solidify, thereby enabling part ejection to proceed. The hold time must be long enough so that the ejector pins do not penetrate or distort the part and not so long as to increase the overall cycle time. Part ejection from the cavity is performed by the mold's ejector system,

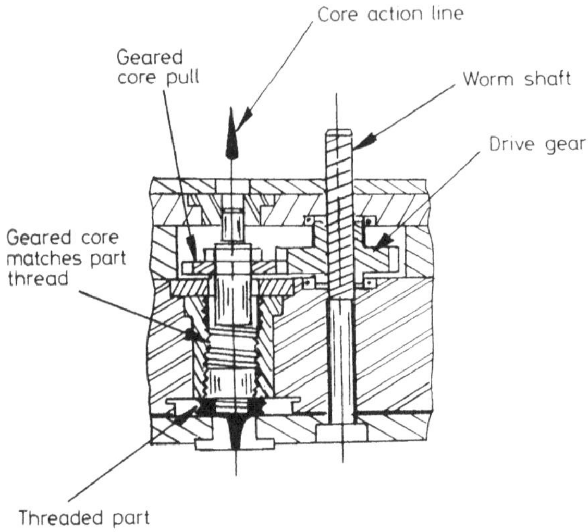

FIGURE 8.35. *Unscrewing cores with a worm shaft and gear combination. Slow mold opening/closing operates drive gear mechanism. (Adapted from Ref. [8].)*

using knockout pins, strips, bushings, plates, or rings. The choice of the ejector system and the number of ejector sites on the part are governed by the shape of the part, its complexity, and the resin used. The cavity ejectors are located at key points on the part in the cavity to push the part out of the mold. The ejector system is always in the moving half of the mold and is operated off the molding machine's ejector system.

In 90 percent of all mold designs, the part is positioned in the cavity so that it shrinks down on an internal core in the moving half of the mold. This ensures that the part is removed from the fixed half of the mold and then ejected from the moving half. This means that on cooling, the part shrinks away from the female cavity side and shrinks down on the core side. The ejector system then pushes the part off the core, releasing it from the mold.

Each molding machine has, for its size of platens, an ejection pin system located on either 3½ or 4 inch off-center spacings. The placement of the ejector pins must coincide with the mold's knockout plate, which operate the knockout system. The molding machine's ejector pins are attached to the knockout plate to provide positive action of the knockout system. This is used primarily for positive retraction of the knockout pins into the mold during mold close to ensure that the mold does not try to close on an unretracted pin.

Older molds or even some new ones may have springs mounted within the mold to push the knockout plate back. This retracts the knockout pins when the mold closes. This system is not recommended, because the springs will fatigue over extended time of use and may break. Serious damage can result if the mold closes on an unretracted knockout pin.

Positive Early Ejector Return

To insure positive return of the ejector system, there is an external mounted device that returns the ejectors to their closed position. It is mounted externally, so that it does not interfere with cavity and cooling channel layout, slides, or other required mold components. The "Toggle-Lok" system, designed by the D-M-E Company, uses a lever to activate cam linkages to positively return and lock the ejector plate in its home position prior to final mold closing. This is illustrated in Figure 8.36 in both the open and almost closed positions. This system does away with springs, pneumatics, hydraulics, and other required mold components.

Accelerated Ejectors

Accelerated ejectors are used to increase the speed and stroke of ejector pins, sleeves, or entire assemblies. The two styles are shown in Figures 8.37 and 8.38. The first operates on a rotating rack that has an activating stud for ejection and a bumper stud for retraction. The second uses a pivot-type motion with a stud that activates the knockout pin's motion and travel. The knockout pin return is controlled by the ejector plate during mold closing.

With heated molds that reach above 150°F, provisions must be made for the expansion of the mold plates and operating system. There is often an unequal expansion between the mold's main cavity plates, which are heated, and the knockout and pin plates, which hold the knockout pins to the knockout plate. Because the knockout plate only receives radiated heat from the main mold sections, it expands at a slower rate. This results in uneven expansion. The knockout pins and pin plate must have clearance to move and compensate for these center-to-center variations and changes as the mold heats up during operation.

If the mold is heated, it should be insulated from the molding machine's platens. They act as a huge heat sink extracting heat from the mold. In fact, all molds should be insulated from the machine's platens, because uniform and consistent mold temperature is required for quality molding. This will be discussed in this chapter under "Insulating the Mold for Temperature Control."

The ejector system can also serve a dual purpose by acting as cavity venting points, especially in blind (nonventable) areas of the part cavity. In some cases, an ejector pin may not be used for ejection but primarily for venting. In all cases, the pin should still move a small amount to avoid having material plug the air space around the pin.

VENTING THE CAVITY

When the polymer is injected into the cavity, which only takes a few seconds or less, the trapped air in the cavity must be able to escape or be vented from the cavity at the same rate as the entering polymer. The venting of cavity air

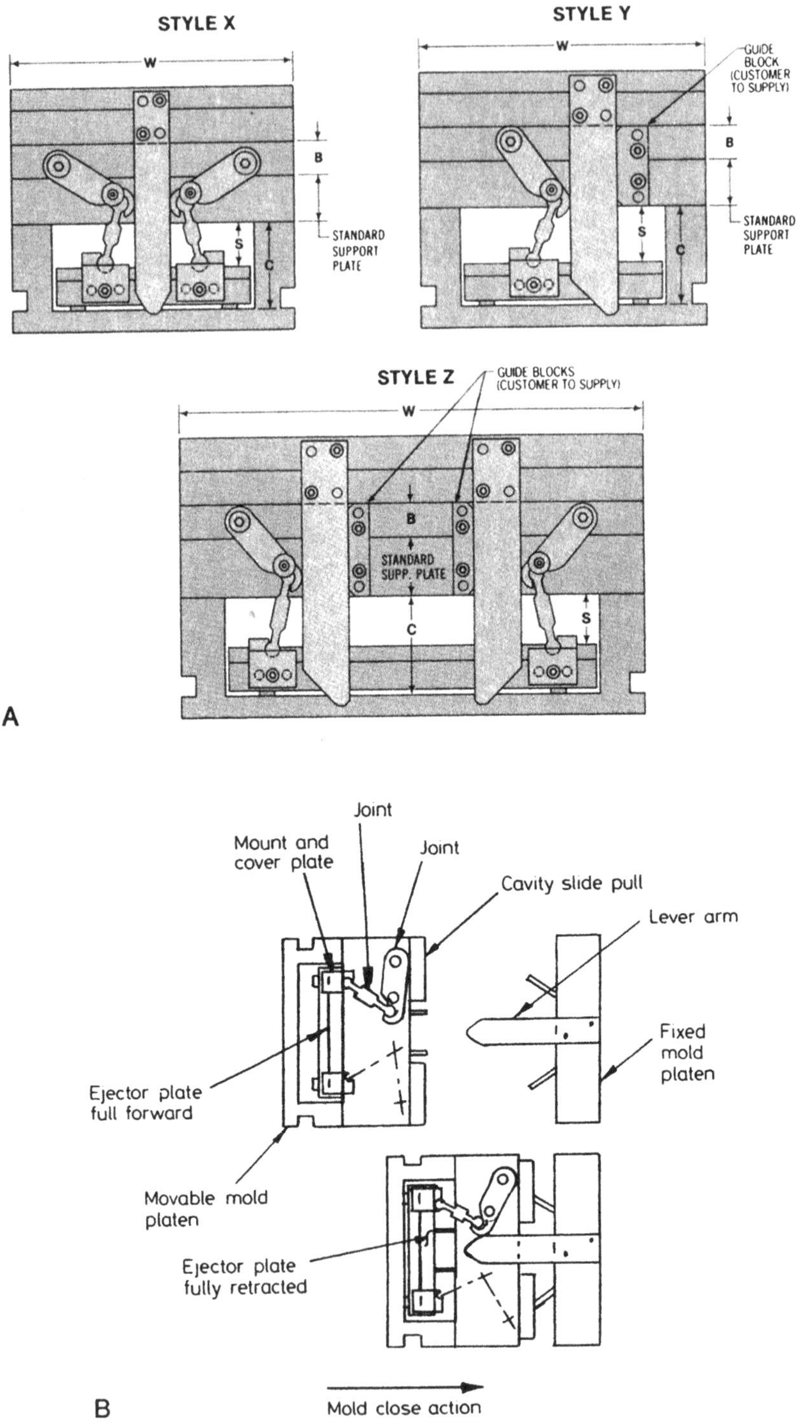

FIGURE 8.36. *(A) Three positive early ejector-pin return mechanisms. (B) Lever arms ensure positive action of cavity side pulls. (Courtesy of D-M-E Co.)*

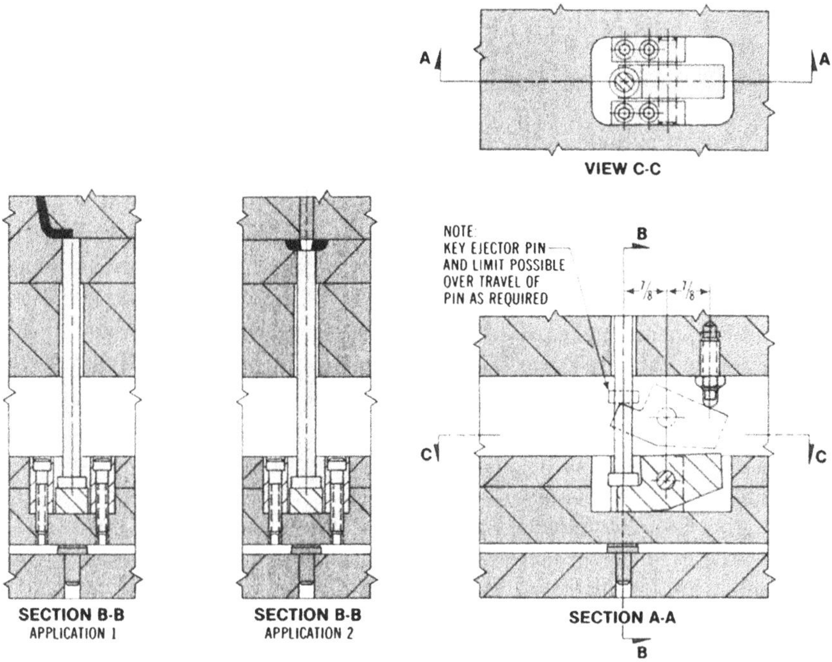

FIGURE 8.37. *Accelerated pivot-type knockouts. (Courtesy of D-M-E Co.)*

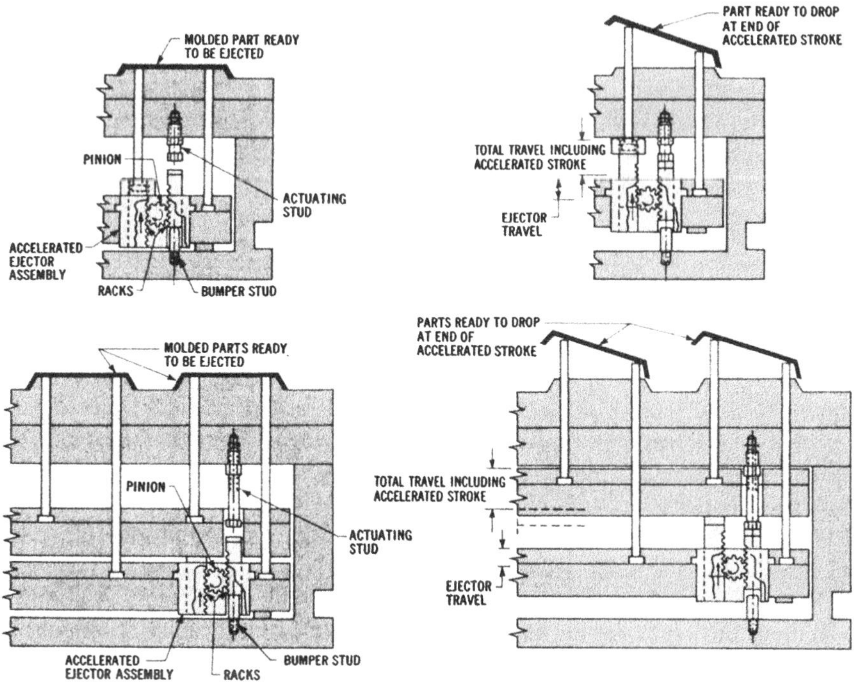

FIGURE 8.38. *Accelerated rack-and-pinion ejector. (Courtesy of D-M-E Co.)*

and polymer gases is critical for producing quality parts. It affects dimensional control, part strength, and surface aesthetics.

Adequate cavity venting is necessary so that the flow of resin into the cavity is not restricted. This would cause the melt front to cool and inhibit filling the cavity. If venting is inadequate, premature polymer pressure buildup occurs. In extreme cases, burning of the material occurs. As the hot polymer is injected into the mold cavity, the air must escape through a carefully thought out and designed venting system. The size and location of the venting system depends on part shape, layout in the tool, number of gates, and resin selected.

In simple parts, the vents are usually located opposite the gate and in the direction of resin flow when filling the mold cavity. Vents are usually located on the parting line. However, with complicated, multiflow fill patterns and cavities with varying section depths and potentially blind pockets, the ejector system may also be used for cavity venting sites. The ejector system can assist in part release by providing a path for air to enter around the cores to break the seal of the material to the cavity surface.

When processing polymers known to exhibit mold deposit problems, the correct sizing of the venting system is essential for allowing trapped air and gasses to escape rapidly from the cavity so they will not form on the cavity surfaces or at the vents. Vents should not be so large as to permit flash to form at these points, thereby creating a post-trimming operation.

Mold deposits consist of the smoke, oils, lubricants, plasticizers, and other vaporized components of the resin that form during heating. They collect and plate out on the mold surfaces over time as a varnish that builds up and plugs the vents. These deposits will vary from resin to resin. There are sprays to assist in their removal during molding, but if build-up is excessive, the tool should be pulled and the vent system restored to its original state. This can often be detected by burn marks at the plugged vents in natural parts, but it is very hard to detect in dark or black parts. When the air cannot escape and an ignition of gasses occurs at the trapped area, the process is called dieseling.

Typical vent size for a simple edge gated part is shown in Figure 8.39. The higher flow resins, such as nylon, acetal, polyester (generally the engineering crystalline polymers), require shallower vents, because their flow in the mold cavity is more fluid and they will flash easier.

The land length of the cavity's vent dimension (B) should be minimal. The longer the land length, the greater the air release pressure that develops in the cavity and restricts rapid venting. Vent size is important because air flowing across and through a restriction, such as the vent, suffers a pressure drop. This drop is governed by the cross-sectional area of the vent and its land flow length. Therefore, a short land length with the same cross-sectional area will experience a smaller pressure drop than a longer land length. Therefore, the land length should be minimal and the cavity air vent channel at least 8 to 10 times the vent land depth in order to rapidly vent the air out of the mold cavity plate.

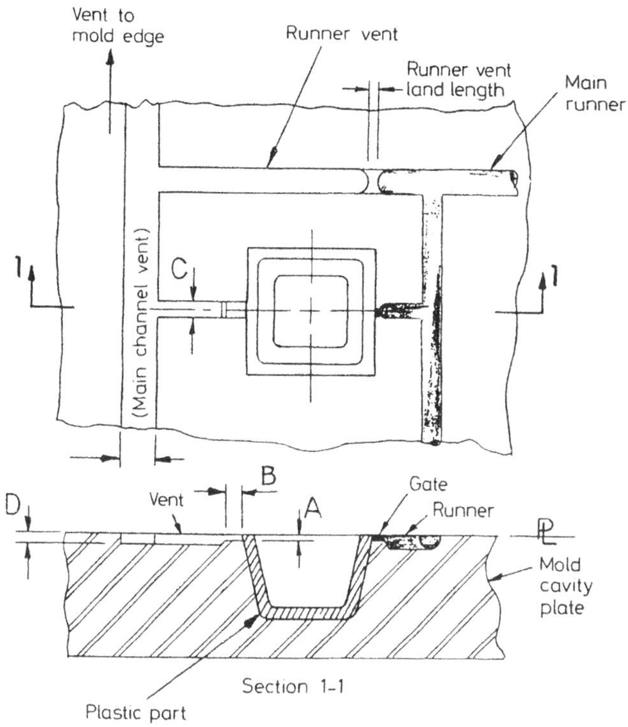

	Most Thermoplastics	Acetal/Nylon	Polyester
A.	0.003	0.0015	0.0010
B.	0.125	0.1250	0.1250
C.	0.500	0.5000	0.1250
D.	0.010	0.0100	0.0100

FIGURE 8.39. *Simple mold edge vents and sizes.*

Cavity venting is usually cut in the runner or movable section of the tool. Before starting up a tool, check the vent opening for clearance by applying toolmaker's blueing to one surface of the tool and clamping it in the machine under full pressure. If blueing is transferred to the mating steel, the vent is not big enough and must be recut.

If venting problems are suspected during molding, apply a single piece of masking tape to the mold plate at the problem area. This keeps the mold from closing tightly during the next cycle and allows any trapped air to escape at this localized spot. If this solves the problem, then additional or deeper/wider vents should be added to the cavity parting line at this location.

One method for quickly adding a vent if the tool must stay in production is to tape a piece of small diameter nylon fishing line from the cavity to the outer mold block. Then, slowly close the tool several times under full clamp pressure, without injecting resin. On removal, notice the fish line has coined a small channel vent in the mold plates even in the hardest steels. Then, when the tool can be removed from production, a suitable vent can be installed.

Cavity Shutoff

A good shutoff land around the cavity parting line concentrates the mold's clamping force so that the cavity does not flash under injection pressure. A shorter land length maximizes the clamping force at the cavity shutoff perimeter. The wider the land length, the more the clamping force is dissipated over the tool face. By providing additional clearance between the mold halves and reducing clamping surface area, venting becomes more efficient.

Cavity Considerations

To ensure that the part stays on the core or in the cavity on the movable side of the tool when the mold opens is important. Suction plays a major part in determining to which half of the mold the part sticks. To have the part stay in the movable half of the tool, the designer uses material shrinkage, draft relief, and draw polishing on the cavity surfaces. Other means, such as molded-in undercuts, help achieve this result. If the part sticks to the mold's fixed half, it is usually because of suction, insufficient draft, hidden undercuts, or insufficient part pullers in the moving half of the mold. By adding perimeter venting on the fixed-mold half installing passive vents on the sprue bushing, part suction is broken and the part remains with the movable half of the mold when it opens. If this fails, then check the draft or the fixed cavity half for undercuts. Polish these out to attain part release.

On a large surface area, flat parts often experience high, failure rates because ejection stresses pop the parts off the ejector side of the tool. This occurs when the ejector pins operate before air can creep in behind the part, thereby allowing full uniform section release. Vented core pins at the ejector sites are a solution to this problem. They admit air to relieve the effects of suction. For tacky resins, a positive air injection at the ejector pins can aid in part release if necessary.

Core or ejector pins are self-cleaning natural vents. One way to increase the efficiency of core pin venting, illustrated in Figure 8.40, is to remove 0.001 inch of material on the diameter of the top half of the pin 0.060 inch deep and to add a vent groove to the outside. This is more positive and effective than the method of polishing multiple 0.001-inch flats on the side of a core pin.

With the use of EDM, mold makers have lost the natural venting that occurred when the cavity was machined in several pieces and bolted

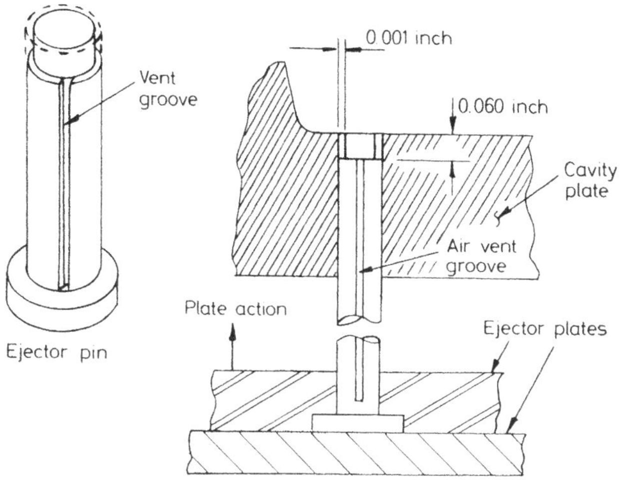

FIGURE 8.40. Vented core pin (self-cleaning from action). (Adapted from Ref. [21].)

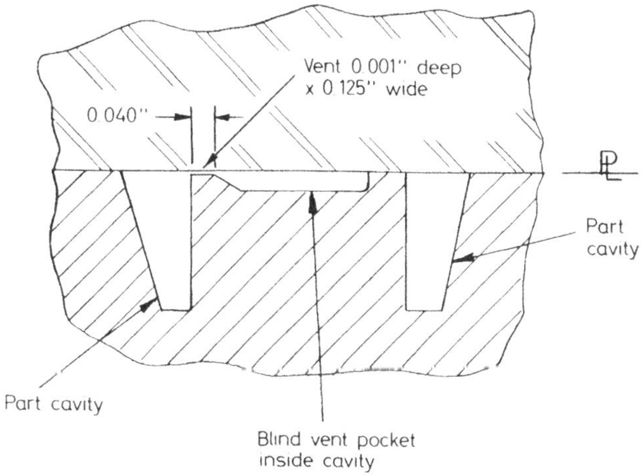

FIGURE 8.41. Blind passive vents. (Adapted from Ref. [21].)

to the mold sections. Therefore, the use of multiple action, moving, self-cleaning core pins is recommended for easy venting and part ejection from the cavity.

Passive Vents

When active vents are not possible, passive vents are used. Passive vents can be placed around parts with internal windows or cutouts, as shown in Figure 8.41. They are similar to parting line or perimeter vents, usually with a smaller cross-sectional area but a similar land length. They are blind pockets positioned in the flow path of the material and sized to capture the anticipated captive air vented into them. If any mold deposits fill up the pockets, they can be cleaned similarly to the parting line vents. Inserts can also serve as vents as long as there is a passage through the core plate. The disadvantage is that they easily become fouled and must be disassembled for cleaning.

Porous Metal Vents

There is another way, not often thought of, to vent blind or dead-ended cavity pockets. This uses a metal insert that is porous enough to allow trapped air to pass through but dense enough to restrict the resin from plugging up the air passages. This technique requires a cooling line behind it, so that escaping air can be bled off and out of the mold. The porous metal insert can be placed directly into the cavity surface and then machined and polished or texturized to match the cavity surface. The insert size is determined by the quantity of trapped air to be evacuated from the blind pocket.

CORE VENTING

Poppet valves are sometimes used to assist in relieving suction from parts on the cavity or core side. They can also assist in ejecting parts off the core. They are usually mounted in conjunction with core pins. Conical seating of the poppet valve is recommended, as shown in Figure 8.42. It requires minimal travel and part distortion to relieve the suction as it injects air behind the part. Figure 8.43 allows too much travel before the air can be injected. This can cause part distortion and failure.

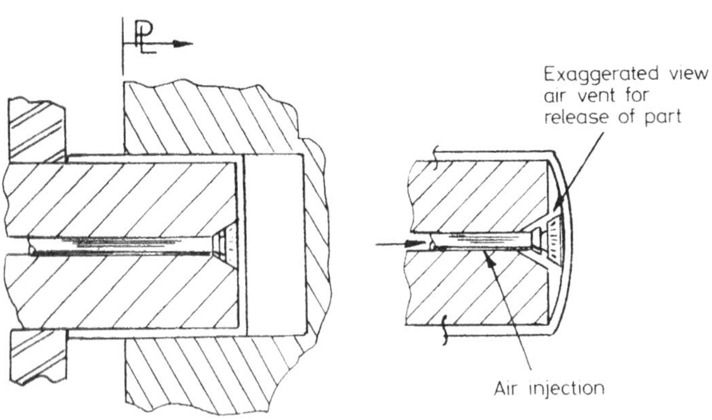

FIGURE 8.42. Poppet valve (correct design). (Adapted from Ref. [8].)

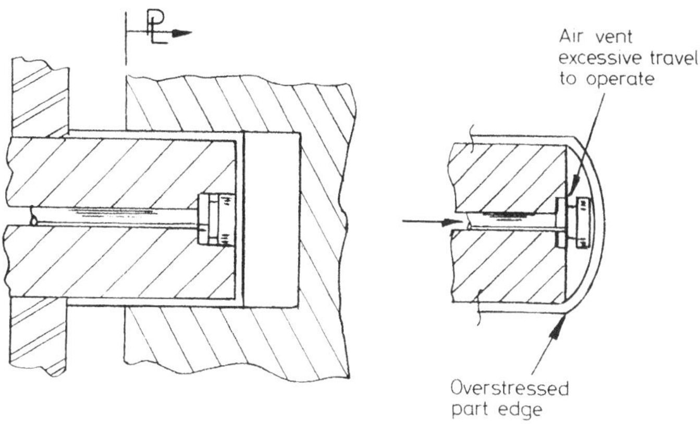

FIGURE 8.43. Poppet valve (incorrect design). (Adapted from Ref. [8].)

POSITIVE CAVITY VENTING

The positive cavity venting system is gaining more recognition for use in molds with deep blind pockets and parts with multiple weld lines. The system functions after the mold closes by evacuating all trapped air in the tool. During the injection phase, it also evacuates gas from the molding resin as it flows into the cavity. Positive venting can help to eliminate burning, voids, and short shots, as well as improve weld-line strength by eliminating the air cushion formed when the two melt fronts meet. The seals at the cavity and mold parting lines using the air evacuating system are very important. Maintenance of the seal area must be continuously monitored for the system to function as designed.

Blowback System

Another system, similar to positive venting but working in reverse, is the blowback system. The blowback system can assist in part ejection by injecting positive clean air into the mold. The air can be injected at the knockout system or along selected points on the mold cavity's parting line. The air is also used to blow residual polymer gases from the tool, thereby reducing the distillate's effect on the cavity surfaces before the next cycle. It has recorded higher part quality, lower mold maintenance, lower injection pressures, and faster cycles.

TEMPERATURE CONTROL

Insulating the Mold for Temperature Control

Too often, insufficient concern is given to the mold and machine interface. The metal platens of the molding machine, where the mold is clamped, are

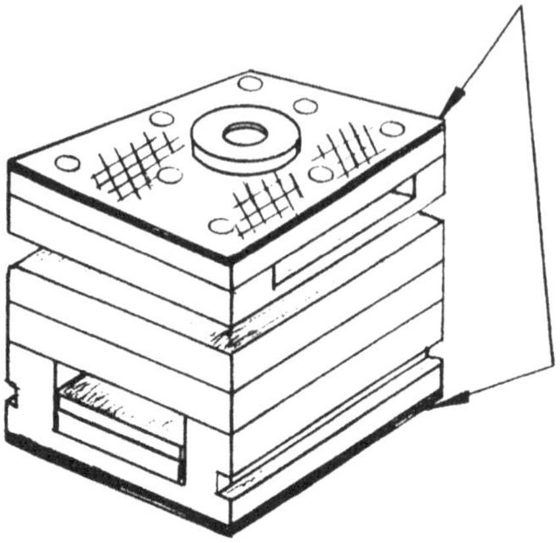

FIGURE 8.44. Mold insulator sheet.

huge heat sinks. They may either help or hinder control of the mold's temperature. When a mold is heated and not insulated from the machine's platens, there is a high loss of heat from the mold to these platens. As a result, more time and energy are needed to bring the mold steel and cavities up to the required temperature and to maintain it. This may take hours, since the mold will never truly reach equilibrium until the platens reach the same temperature. There will always be a loss of heat, which causes the mold temperature controller to constantly supply heat to compensate for the loss. This is less critical when the mold is cooled, but still will affect mold temperature and cooling equipment output.

There are high-temperature insulating sheets in varying thicknesses that can be used on the mold's clamping surfaces and the machine's movable and stationary platens. These sheets isolate the mold from the machine, thereby permitting more accurate and consistent mold and cavity temperature control. This is essential for the control of part dimensions. An example of this sheet on a mold is shown in Figure 8.44.

Depending on the thickness of these sheets, the mold's locating ring must be long enough to compensate for the loss in the depth of the ring because of the thickness of the insulation. This is shown in Figure 8.45 which assures more than the required $\frac{5}{16}$-inch pilot engagement for correct mold and machine nozzle mating.

Mold Temperature Control

To produce parts to the correct dimensions and tolerances, the cavity temperature must be controlled during the molding process. With increasing part quality requirements, tighter dimensional tolerance, and reduced rejection rates, tighter temperature control of the mold cavity is paramount. It affects

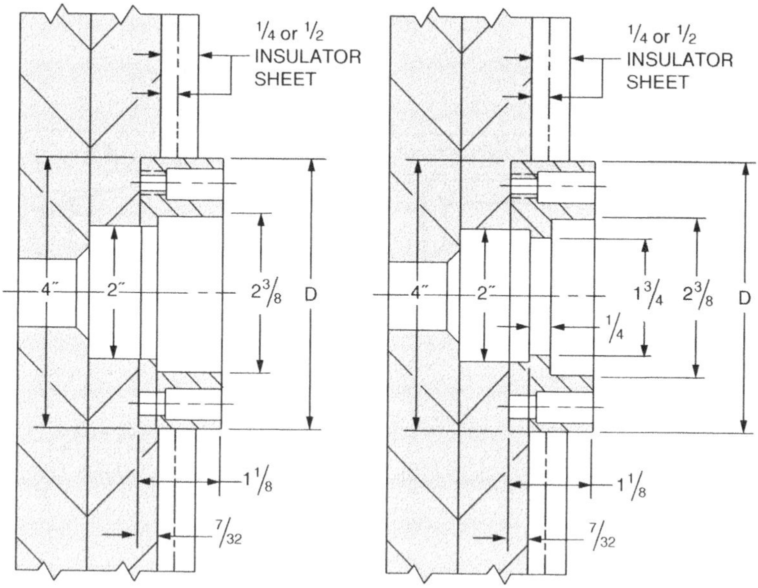

FIGURE 8.45. *Locating rings for use with insulator sheets. (Courtesy of D-M-E Co.)*

molding cycle rates and the part's physical properties, appearance, and dimensions. Uneven cavity temperatures result in parts with molded-in stresses, warped sections, sink marks, poor surface appearance, and varying part dimensions from cycle to cycle and even cavity to cavity.

Typically, the cooling system is the last area to be considered when designing and building a mold. As a result, the cooling channels must be routed around the cavity as best they can, and the number and size of the cooling circuits are often limited by the part's ejector system. Too often, cooling is not even considered until it is too late in the tool design stage to change the mold so that it provides the proper temperature control required for the cavity.

Computer programs can now analyze a mold's temperature control requirements and recommend a cooling circuit layout that provides uniform temperature control for all cavities. The filling and cooling phase of the molding process are the crucial stages for maintaining uniform temperature control from cycle to cycle. This involves monitoring the cavity and core temperature to decide on improvements in part quality, repeatability of the process, and productivity.

Until recently, molds were usually built with single in-and-out cooling systems. The results were temperature gradients varying as much as 50° from the inlet to the outlet and across the tool surface. Clearly, the coolant picked up heat while passing through the mold. This caused varying part dimensions and, as the tool warmed up, uncontrolled temperature drifts and part size variations. Examples of temperature variance across a mold's surface are

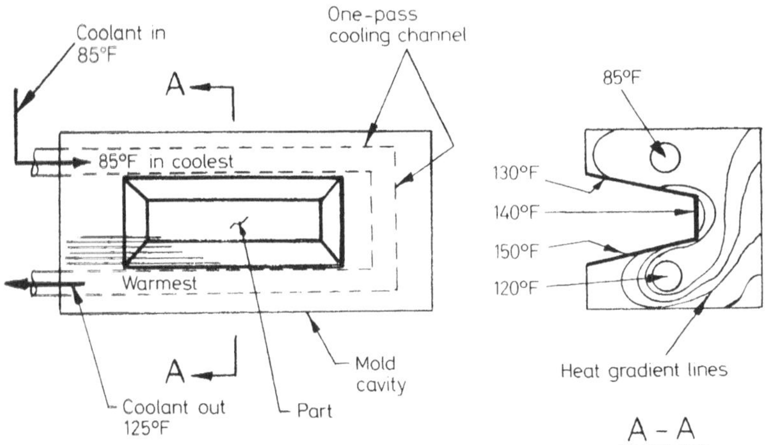

FIGURE 8.46. *Restricted mold temperature control. (Adapted from Ref. [21].)*

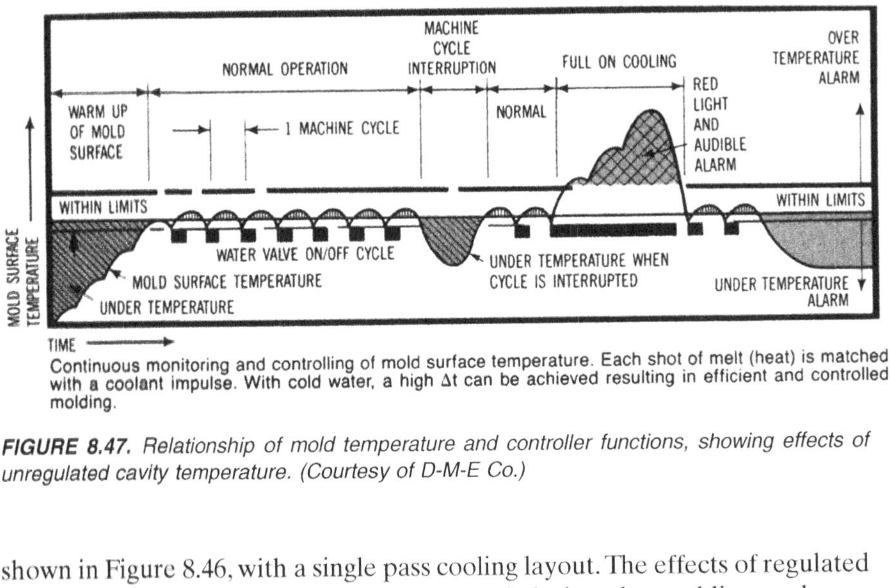

Continuous monitoring and controlling of mold surface temperature. Each shot of melt (heat) is matched with a coolant impulse. With cold water, a high Δt can be achieved resulting in efficient and controlled molding.

FIGURE 8.47. *Relationship of mold temperature and controller functions, showing effects of unregulated cavity temperature. (Courtesy of D-M-E Co.)*

shown in Figure 8.46, with a single pass cooling layout. The effects of regulated and unregulated cavity temperature control during the molding cycle are shown in Figure 8.47.

The addition of more cooling circuits dramatically helped to maintain correct temperature control. But temperature gradient problems still occurred, and other mold temperature control methods were instituted to provide even better cavity temperature regulation.

Pulse-Modulated Cooling. Pulse-modulated cooling is one method now used for maintaining more uniform cavity temperature gradients across single and multicavity mold surfaces. This method responds to increases in cavity temperature during the molding cycle by injecting a coolant only for the time necessary to maintain a uniform cavity temperature. Sensors, which are located in specified locations in the mold, are hooked up to a controller that monitors these temperatures and operates the valves that inject coolant at timed intervals. This pulsing of coolant at a minimum of 40 PSI across the mold face (incoming coolant pressure over outgoing coolant back pressure) creates a turbulent flow that is more effective in heat transfer than a streamlined laminar flow, because of the transverse movement of the liquid particles.

The temperature sensor hookup and cooling channel layouts are separate for the core and cavity. This separation controls coolant flow and pulse time for each circuit, as the parts shrink to the core and pull away from the cavity, which affects each area's heat transfer differently. Thus core cooling stays on longer with a shorter cavity time to maintain thermal balance in the mold cavity. These features are illustrated in Figures 8.48 and 8.49, where the cooling pulse duration is timed cycle to cycle to maintain cavity temperature.

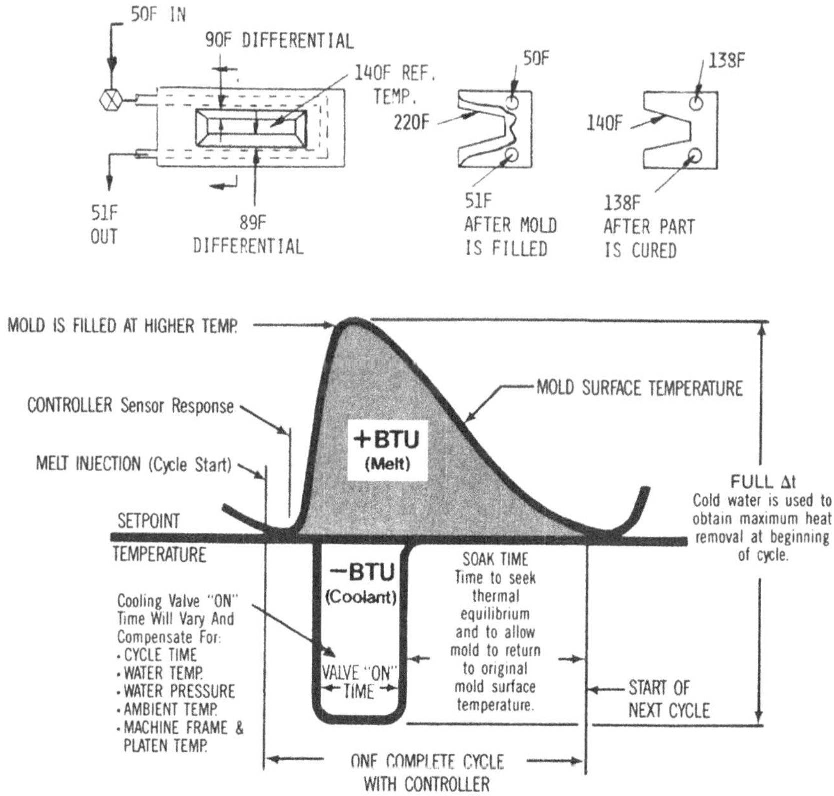

FIGURE 8.48. *(A) Pulse system for thermal equilibrium in the mold. (B) Improved temperature control through a complete molding cycle. (Courtesy of D-M-E Co. and adapted from Ref. [23].)*

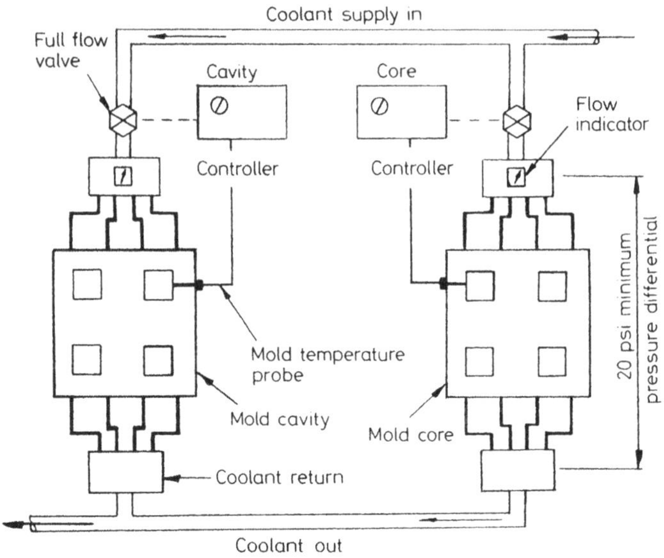

FIGURE 8.49 *Pulse surge layout. (Adapted from Ref. [23].)*

The temperature sensor location is critical to temperature control and must be located within 0.10 inch of the part surface. The steel used for the cavity and core must also have good heat conductivity for rapid sensor response time.

The engineering resins usually require hot molds (150°F and higher) with cartridge heaters, hot water, or oil as the heating medium. If the pulse system is used for these materials, an electrical heating system is preferred for heating the mold. The cooling lines should be installed in the normal layout to allow the pulse system to control the thermal balance in the tool. The reason for this is obvious—the tool must be heated independently of the cooling system, which is used only to control the thermal balance of the mold cavity.

The advantages of pulse modulated cooling are (1) no temperature gradients the melt must overcome when entering the cavity, (2) uniform temperature control of each cavity, resulting in consistent cavity-to-cavity dimension, and (3) uniform part cooling and reduced cycle times, resulting in more consistent part manufacture. Pulse cooling will allow zero-defect parts to be produced cycle to cycle, because the mold is consistently operating under tight temperature control.

Cavity Temperature Control

To obtain uniform mold cavity temperature control, the cooling system must be designed to remove the heat from the melt as it enters the cavity. It must also provide rapid cooling of the part so that a repetitive and economical cycle can be attained. The mold cavity(ies) requires a balanced temperature gradient across its surfaces so that residual molded in material stresses are minimized. This occurs when the melt enters the cavity at one temperature and flows into the cavity to meet an increasingly hotter cavity surface. This would be caused by the coolant, which initially enters at the mold's cooling channels—the gate area. As the coolant flows around the cavity, it loses cooling capability as the resin heat is transferred into the cooling medium. This temperature differential in the cavity can cause a resin skin effect on the part, which, if severe enough, could lock-in differential surface stresses. After the part is ejected, this could cause the part to warp or bend. This is often seen with large flat parts, both with and without side walls and ribs for support. Therefore, a well thought out and routed cooling system is very important for attaining the desired part dimensional stability and quality.

COOLING SYSTEMS

Cooling System Layout

The layout of the cooling systems depends on the part geometry, number of cavities, ejector and cam systems in the mold, part quality and dimensions required, part surface appearance, resin used to make the part, and the required cavity temperature profile. The sizing of the cooling channels is dependent on the rate of cooling and temperature control needed for controlling the part's quality.

Cavity temperature control is regulated by the flow of the cooling or heating media—water, water and glycol mixture, oil, etc.—through the mold's cooling channels. It also depends on the rate of heat transfer through the mold cavity steel to the coolant. Some mold steels have better heat transfer rates than others, and their selection must be made to provide the temperature control required to meet the part's dimensional requirements, cycle time, and quality. The coolant medium flow through the cooling system must also be turbulent. The maximum heat transfer results with this type of flow, as compared with laminar flow (see Figure 8.50)

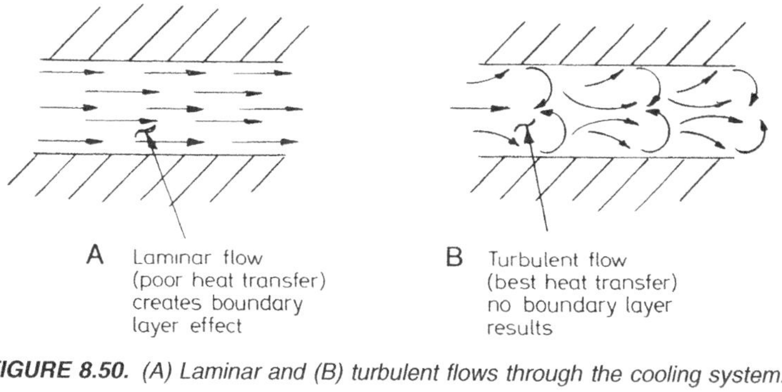

FIGURE 8.50. (A) Laminar and (B) turbulent flows through the cooling system.

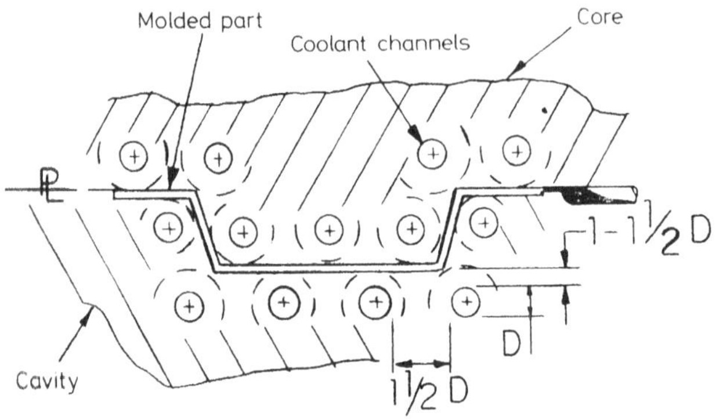

FIGURE 8.51. *Proper location of coolant channels.*

Molders are now using pure ethylene glycol as the primary fluid transfer medium in closed-loop cooling systems to both heat and cool the mold. Ethylene glycol eliminates rust and mineral deposits in the cooling system; it also has a higher heat transfer rate than water or mixtures of water and glycol. Quick disconnected fittings or mechanical shutoffs are used to avoid losing the cooling medium during mold changes. This will be discussed under "Mold Cooling Line Connections" in this chapter.

Uniform cavity temperature control is achieved by selecting the correct number of cooling circuits and the correct placement of the cooling channels around each cavity. The cooling circuits should be routed around the part cavity as consistently as the part shape will allow and as close to the cavity walls as possible. This is determined by the strength and rigidity of the mold's materials. The channels should be located one to one-and-a-half times the diameter of the cooling channel, away from the cavity surface (see Figure 8.51). The cooling channels should be short and independent routings to yield the best temperature control performance.

In multicavity molds, this may not always be possible for each part cavity. In these cases, as many independent circuits as possible should be used with a parallel circuit as opposed to a series cooling circuit (see Figure 8.52). The cooling circuits should also avoid dead spots and obstructions, as they are potential hot spots and inhibit cooling. Coolant pressure should be no lower than 40 PSI with a known CFM (cubic feet per minute) flow rate through the cooling channels; this affects turbulent flow. Remember that if channels become too large, turbulent flow is lost because of lower coolant velocities. Conversely, too small a channel yields insufficient coolant flow for correct temperature control.

Core Cooling

The cavity cores used to form the part are the most difficult to cool, based on their geometry. As they extend out from the cavity surface, special care and

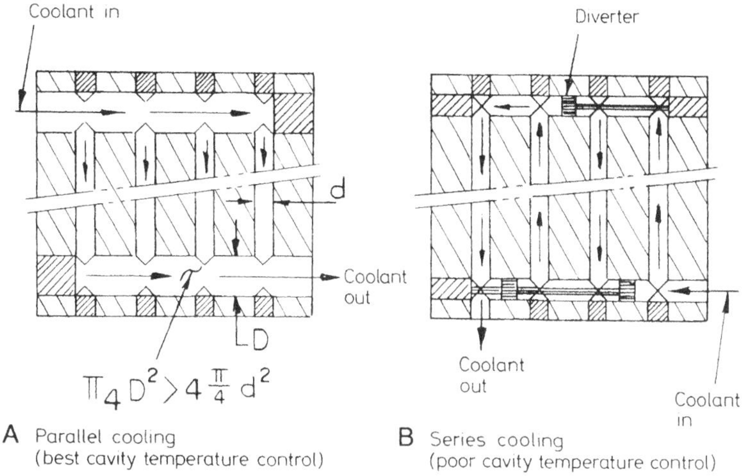

FIGURE 8.52. Mold cooling layouts: (A) parallel and (B) series. (Adapted from Ref. [8].)

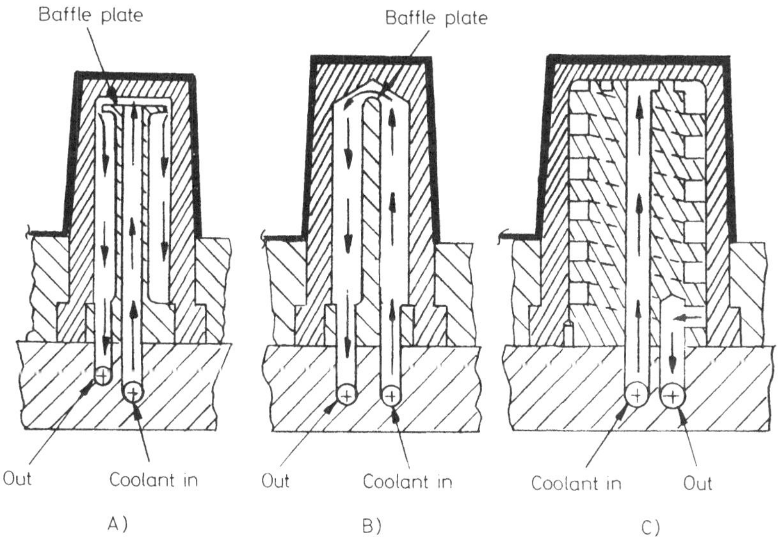

FIGURE 8.53. Three examples of core cooling systems. (Adapted from Ref. [8].)

design must be used to regulate their temperature. For small cavity cores or core pins, where either insufficient space or their size does not permit coolant flow within their geometry, a change of core material is recommended. Beryllium copper is often used; it has five times the heat transfer rate of typical tool steels to drain away the heat. These cores and pins should be located near cooling channels to transfer the heat to a source of coolant. In cores and pins of sufficient size to allow coolant flow, see Figure 8.53 for

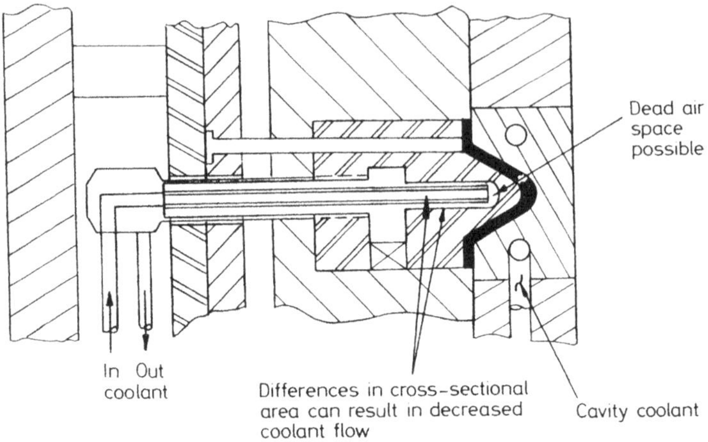

FIGURE 8.54. *Core bubbler. (Adapted from Ref. [18].)*

examples of cooling systems that can be used. In all of these cases, adequate coolant flow is achieved, flow is turbulent, and no dead spots occur in the coolant flow area.

A typical core bubbler, or fountain arrangement, is shown in Figure 8.54. The coolant is directed at the top of the cavity core, which requires cooling in a restricted area. If the coolant flow is restricted at the end of the tube and internal coolant channel, then turbulent flow ceases and inefficient cooling occurs. Adequate turbulent flow and the return of the coolant down and around the bubbler tube is required to provide temperature control of the cavity surface.

In this core-cooling example, assume that an 0.44-inch diameter hole is drilled in the center of the core and an 0.25-inch bubbler tube is installed. The tube must be cut to the correct length to eliminate a dead air space occurring at the apex of the drilled hole.

Calculating the coolant flow through the tube and its return down the cavity hole yields the following cooling efficiency for this example.

Diameter of cavity hole	= 0.44 inch
Area of cavity hole	= 0.150 square inch
Tube outer diameter (OD)	= 0.25 inch
Area of tube OD	= 0.049

Annular area between tube OD and inner diameter (ID) of hole:

(Ah) = 0.150 − 0.049	= 0.101 square inch
Wall thickness of tube	= 0.040 inch
ID of tube (0.250 − (2 × 0.040))	= 0.170
Area of tube ID (At)	= 0.023 square inch

The velocity of the coolant in the two channels—the tube ID and annular area between the tube and wall of the cavity hole—is inversely proportional to their areas and can be expressed as:

$$\frac{Vh}{Vt} = \frac{At}{Ah} \quad \text{or} \quad Vh = \frac{0.023}{0.101} Vt = 22.8\% Vt$$

where,

V = velocity
A = area
h = cavity hole
t = tube

With such a restriction in flow around the top of the cooling tube and down the cavity exit hole, this arrangement would restrict flow and not provide adequate cooling. Therefore, the cavity bubbler hole must be enlarged to obtain nearly 100 percent of the coolant volume flow from the bubbler tube down the cooling channel.

Similar considerations must be employed with the system shown in Figure 8.53A ,which uses a baffle plate when calculating the flow over the baffle. This example is less efficient than that shown in Figure 8.53B, as in that system an air bubble could form at the top creating a dead air space and hot spot.

The example in Figure 8.53C uses a helical channel, which is more complicated and subject to leaks. It also requires more mold maintenance than the first two examples. A temperature gradient that will also form down the core may affect part tolerances, depending on the length of the core and the flow distance.

Coolant Channel Seals

The seals in the coolant system must be leakproof and not affected by the mold's operation. They must always seal, even when the mold expands and contracts during thermal cycling. The seals and O-rings should be positioned so that there is no chance of them being damaged or improperly seated during mold assembly. This would cause the mold to leak and, in rare cases, restrict the coolant flow. Seal and O-ring grooves should be machined to match closely the contour of the seal. This ensures that the seals will have the metal support in the mold to provide the required sealing. Examples of good and bad designs are shown in Figure 8.55, with A being the preferred layout for both seals and cooling.

The main items for controlling cavity temperature with the mold's cooling system are as follows:

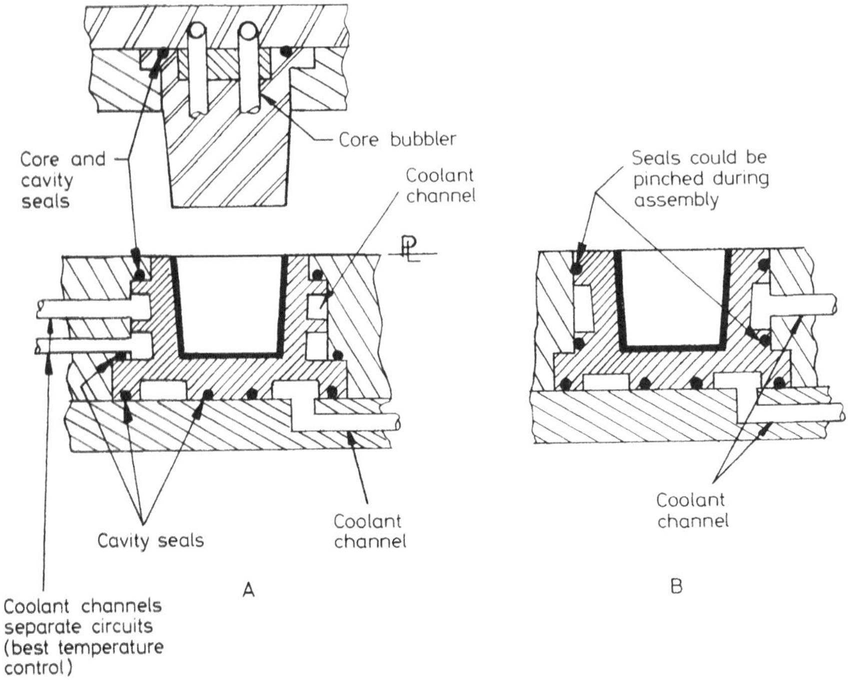

FIGURE 8.55. *Cooling circuit and seals: (A) good design and (B) poor design. (Adapted from Ref. [8].)*

1. Design the cooling system with an adequate number of cooling channels at equal distances from each other and from the cavity walls. Size the channels to carry a coolant volume that will maintain the correct cavity temperature.
2. Position the cooling circuits symmetrically around the cavity(ies), providing as many independent circuits as necessary to maintain a uniform temperature gradient across the cavity.
3. Design the cooling system for turbulent flow in the channels, with sufficient coolant velocity to keep the tool temperature in balance.
4. Eliminate long cooling lines and dead air spaces at corners and other areas.
5. In hard-to-cool areas with small cores and pins, use metals such as beryllium copper alloys with high heat conductivity connected to cooling channels.
6. Provide temperature sensors in each mold half to monitor and control cavity temperatures. Most molds never have temperature sensors installed and rely on mold heaters/chiller temperature controllers, which estimate cavity temperature based on entering water tempera-

ture. The actual cavity temperature can be as much as 20 to 30 degrees higher than the coolant temperature controller registers.

7. When running hot tools (about 150°F), always insulate the mold from the machine platens to reduce heat transfer loss from the mold into these huge heat sinks. Use compression resistant heat insulating plates for this function.
8. Provide easy access to mold cooling channels for periodic cleaning (removal of lime deposits, etc.), even if the system already uses closed loop chemically treated water. Consider ethylene glycol for closed-loop systems as a better heat transfer medium.
9. Use mold temperature control units to regulate mold temperatures. Do not rely on city or tower water for temperature control to the mold. This water temperature will vary with the seasons.
10. Use efficient sealing methods and materials to eliminate coolant leaks.

Mold Cooling Line Connections

Additional mold cooling or heating problems occur when the set-up person connects the wrong cooling line to the mold's cooling connections. All connections should be plainly marked and identified. Male plug connectors are mounted to the mold's cooling line connections and must be protected to avoid damage during installation or when working on the mold during maintenance. A damaged plug will not mate correctly with its socket connector, and fluid and pressure loss will occur. The flexible, cooling hookup lines, from the cooling or heating sources, have socket connectors that attach to the mold's male plug connectors for ease of attachment.

When installing or disconnecting cooling lines, exercise extreme care because of all the electrical lines around the machine. Carelessness can result in electrical shock or serious burns if heated water and oil are used in the mold. For these reasons alone, all connections must be carefully performed and all plug and socket connectors inspected and kept in perfect operating condition.

Mold Connection Types

There are basically three types of plug and socket connectors: no shut-off, one-way shut-off, and two-way shut-off. The one- and two-way shut-off styles are shown in Figure 8.56. When using city supply line water, the no shut-off style is often used. However, it is not recommended because of water quality questions, temperature variance, cost, and wastefulness. The closed, controlled fluid cooling system approach is preferred, with the shut-off connector on the cooling hookup line using a one-way shut-off socket. For systems that use glycohol/water mixtures for heating or cooling or especially oil heated

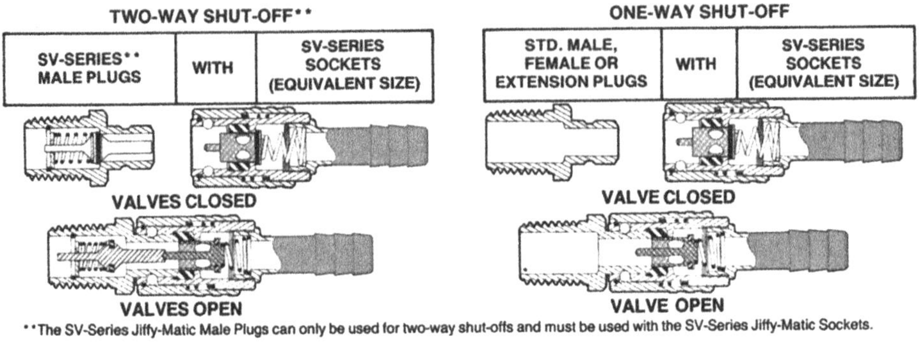

FIGURE 8.56. *Cooling plug and socket connections. (Courtesy of D-M-E Co.)*

systems, two-way shut-off systems should always be used for both the mold and hookup lines. This system eliminates loss of coolant, ensures safety, and reduces costs when making the connection to and from the mold. It also protects the mold set-up operators when electrical connections are close to the mold or adjacent equipment and ensures that the coolant/heating system is not contaminated. Safety is always first, followed by quality temperature control of the mold's surfaces and cavity.

The plug and socket must be a matched set to operate correctly. To avoid the possibility of incorrectly hooking up any cooling line connections to the mold, different sized sockets and plugs for the "in" and "out" lines should be used. Failure to do this is the most common set-up error. The external connections for "in" and "out" should also be "color coded" to help identify and reduce setup errors.

All trapped air should be bled out of the mold and hookup lines before starting up the cooling system. This ensures fluid flow and proper heat transfer of the fluid's medium at the mold's surfaces. Trapped air in any system will greatly reduce efficiency. In some cases, an air pocket may inhibit fluid flow into a critical area affecting the quality of the part.

In the case of long cycling molds, a cooling water manifold is often attached to the mold to facilitate, identify, and correctly ensure that supply lines are sized and hooked up correctly. The manifold will have two or more sizes of fittings for "in" and "out" connections. The manifold and fittings can also guarantee that the size of the feed-and-return supply line hoses are correct. For correct heat transfer, the volume, pressure, and rate of fluid flow must be maintained. This will be discussed in greater detail in the Auxiliary Equipment section.

Cooling Time

A major factor in controlling part quality and reducing the molding cycle is how long it takes a part to cool before it can be ejected. Mold hold or

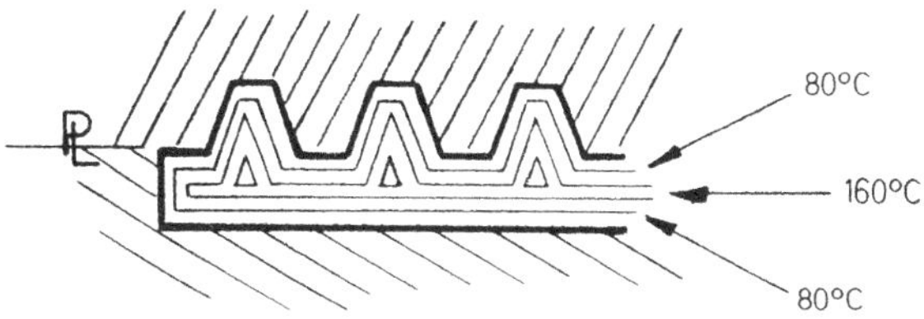

FIGURE 8.57. Temperature gradient within a molded part. (Adapted from Ref. [8].)

cooling time depends on coolant temperature, melt temperature, and how quickly the heat can be removed from the part. Theoretically, as soon as the gate freezes off during the molding cycle and a skin forms on the part, it can now be ejected from the tool. But if the center of the part has not had sufficient time to cool, the knockout pins may punch through the part or the part may not be rigid enough to hold its shape. Therefore, the minimum cycle time is the time required for the thickest section of the part to cool sufficiently to be ejected without distortion. Because plastics are excellent insulators, the cooling system must continue to draw out this trapped heat to reach the part's ejection temperature. For the part to become rigid, this rate of heat removal is determined by the type of resin—crystalline or amorphous—used. The amorphous polymers require more time for heat removal from the part. This is illustrated in Figure 8.57, which shows the center at 160°, with the outer surfaces in contact with the mold surfaces at 80°F. In contrast, the crystalline polymers only have to be cooled to just below their melt temperature for parts to become rigid enough for ejection. The elevated mold temperatures also assist the crystalline structure to develop faster; this also aids in improving part set-up time.

Therefore, the minimum cooling time required for a part to reach mold release temperature is based on the following:

1. Wall thickness of the part
2. Temperature differential between resin and mold temperature
3. Temperature differential between part ejection temperature and mold temperature
4. Efficiency of the cooling system

MOLD SHRINKAGE

To obtain the correct part size, the mold cavity must be cut to compensate for the plastic's shrinkage during the molding operation. Calculating the shrinkage of a resin in a mold cavity to obtain the part dimensions required is more of an art than a science. There are many variables in this process. The part designer must specify the required part dimensions with tolerances. Then,

TABLE 8.7. **Nominal Mold Shrinkage Rates for Thermoplastics (inch/inch).**

	Average Rate per ASTM D-955		Directional Rates 0.062 Inch Sample	
Material	0.125 inch	0.250 inch	Flow	Transverse
ABS				
unreinforced	0.004	0.007	0.005	0.005
30% glass fiber	0.001	0.0015	0.001	0.002
Acetal, copoly				
unreinforced	0.017	0.021	0.022	0.018
30% glass fiber	0.003	NA	0.003	0.016
HDPE, homo				
unreinforced	0.015	0.030	NA	NA
30% glass fiber	0.003	0.004	0.003	0.009
Nylon 6				
unreinforced	0.013	0.016	0.014	0.014
30% glass fiber	0.0035	0.0045	0.003	0.004
Nylon 6/6				
unreinforced	0.016	0.022	0.021	0.021
15% glass fiber + 25% mineral	0.006	0.008	0.006	0.007
15% glass fiber + 25% beads	0.006	0.008	0.006	0.008
30% glass fiber	0.005	0.0055	0.003	0.005
PET polyester				
unreinforced	0.012	0.018	0.018	0.015+/–
30% glass fiber	0.003	0.0045	0.003	0.007
Polycarbonate				
unreinforced	0.005	0.007	0.006	0.006
10% glass fiber	0.003	0.004	0.003	0.004
30% glass fiber	0.001	0.002	0.001	0.002
Polyether sulfone				
unreinforced	0.006	0.007	0.006	0.006
30% glass fiber	0.002	0.003	0.001	0.002
Polyether etherketone				
unreinforced	0.011	0.013	0.009	0.011
30% glass fiber	0.002	0.003	0.002	0.004
Polyetherimide				
unreinforced	0.005	0.007	0.006	0.006
30% glass fiber	0.002	0.004	0.001	0.002
Polyphenylene oxide/PS alloy				
unreinforced	0.005	0.008	0.005	0.005
30% glass fiber	0.001	0.002	0.001	0.002
Polyphenylene sulfide				
unreinforced	0.011	0.004	0.009	0.011
40% glass fiber	0.002	NA	0.001	0.003
Polypropylene, homo				
unreinforced	0.015	0.025	NA	NA
30% glass fiber	0.0035	0.004	0.003	0.009
Polystyrene				
unreinforced	0.004	0.006	0.005	0.005
30% glass fiber	0.0005	0.001	0.001	0.002

Shink rates vary considerably between amorphous and crystalline resins. Note also the effects of variations in flow direction, wall thickness, and the presence of reinforcing fiber.

Notes: Combined flow and transverse values from end-gated 5 × 0.5-inch bars 0.125 or 0.250 inch thick. Measured from molded 4-inch edge-gated disks or tab-gated 2.50 × 1.75-inch chips. Homopolymer values are about 5% higher.

HDPE, high-density polyethylene; PET, polyethylene terephthalate.

Source: LNP Engineering Plastics, division of ICI Advanced Materials.

the tool designer computes the cavity dimensions based on the material's shrinkage rate and tolerances. The tool is built to meet these requirements. Production personnel are then tasked with making the part to meet the customer's requirements.

Many factors affect a part's shrinkage in a mold cavity. Each resin has its own particular mold shrinkage values that are provided by the material suppliers. The difficult task is determining how the material will shrink in the mold cavity based on its part geometry, cooling rate, gating location, and resin flow in the cavity. The task of calculating cavity dimensions is based on these factors and the tolerances expected for the part.

The tool designer begins by selecting the mold and cavity steels and determining the cavity layout in the mold, gate and runner size, and cooling layout and ejector systems. Cavity dimensions are calculated based on the material supplier's resin shrinkage values per ASTM D-955 procedures. These procedures determine the material's shrinkage for a flex bar but often will vary considerably for the part being tooled. The typical shrink rates of polymers are shown in Table 8.7, which does not take into effect such other variables as gate size, mold temperature, and processing conditions. More examples, shown in Figures 8.58 and 8.59 illustrate how material, mold, and processing conditions affect part shrinkage. Additional examples of how part geometry and gate location affect shrinkage are shown inn Figure 8.60.

The tool designer often calculates a plastic's shrinkage rate based on experience with other molds and similar parts in the same resin. If the designer is unsure of the actual shrinkage to be obtained during molding, the cavity is cut to the minimum shrinkage value. Then, before fine tuning the cavity's final dimensions, the mold is sampled and part measurements are taken. If shrinkage adjustments are required, the cavity dimensions can be adjusted by metal removal. This is initially more expensive but less than making adjustments when the tool is in production. The result is a tool that can produce parts that meet customer requirements.

Post-Mold Shrinkage

Post-mold shrinkage of the finished part must also be considered before predicting a resin's mold shrinkage and calculating mold dimensions. This is often overlooked but is critical if the part must fit into another assembly.

All plastic parts continue to shrink to some degree after molding. The amount depends on the resin used (amorphous or crystalline), the mold's temperature, the amount of molded-in stresses, processing conditions, and the part's end-use service. All materials want to reach their lowest or most stable energy condition, which is a stress-free state. This is achieved by exposure to elevated temperatures that unlock molded-in stresses and result in further part shrinkage. Crystalline polymers, when exposed to

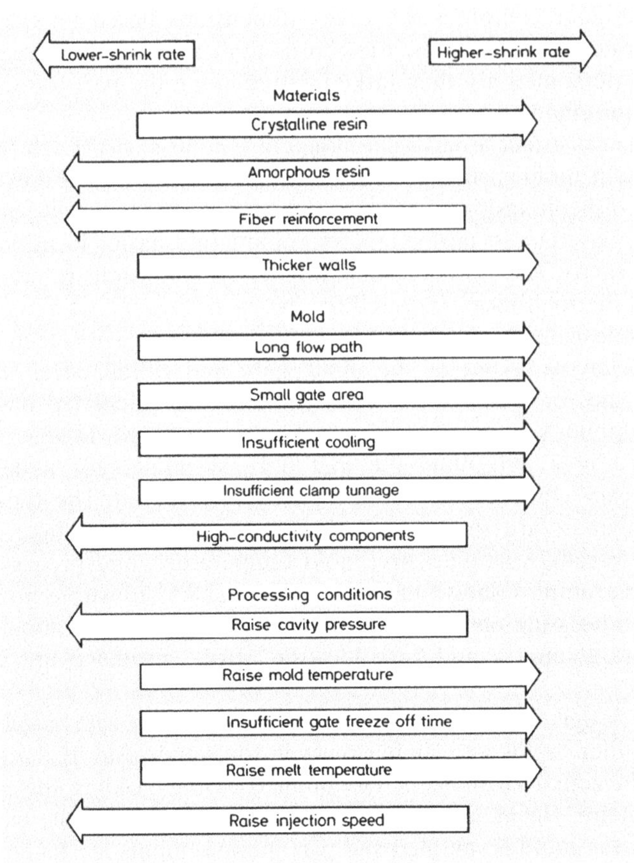

FIGURE 8.58. *Chart illustrating how design and processing affect shrink rates. (Adapted from Ref. [12].)*

temperatures above their mold cavity temperature, will continue to shrink as their structure becomes more crystalline or ordered. Amorphous resins, which do not experience this change, are more stable after molding, but may still experience some post-mold shrinkage. Too often after parts are made, they are stored or shipped at temperatures above their mold temperature. As a result, they change dimensions or warp. Some parts will not reach stable conditions and dimensions until a few hours or even several days after molding. In addition, environmental conditions can affect the degree of a part's post-mold shrinkage. Nylon, a crystalline polymer, grows as it picks up moisture, but the moisture has an annealing effect that unlocks molded-in stresses. This results in some minor additional shrinkage. The dimensional growth for nylon as it absorbs moisture is shown in Figure 8.61B.

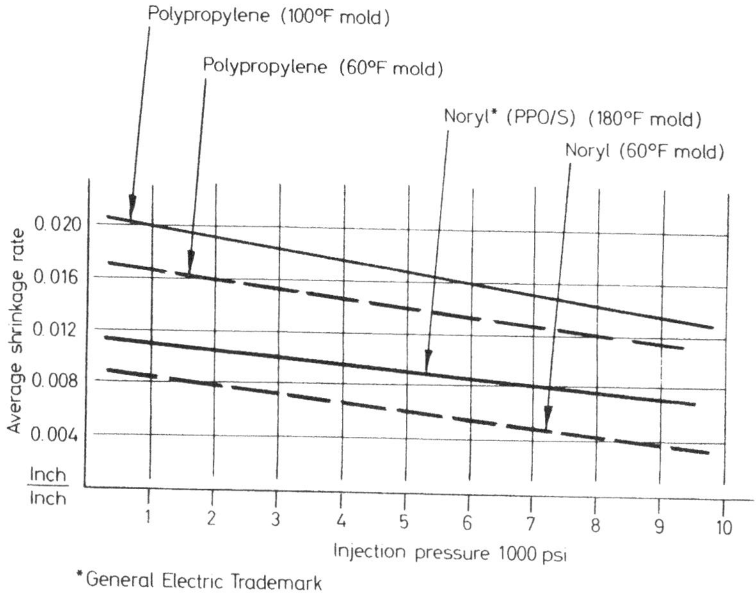

FIGURE 8.59. *Factors affecting material shrinkage. (Adapted from Ref. [12].)*

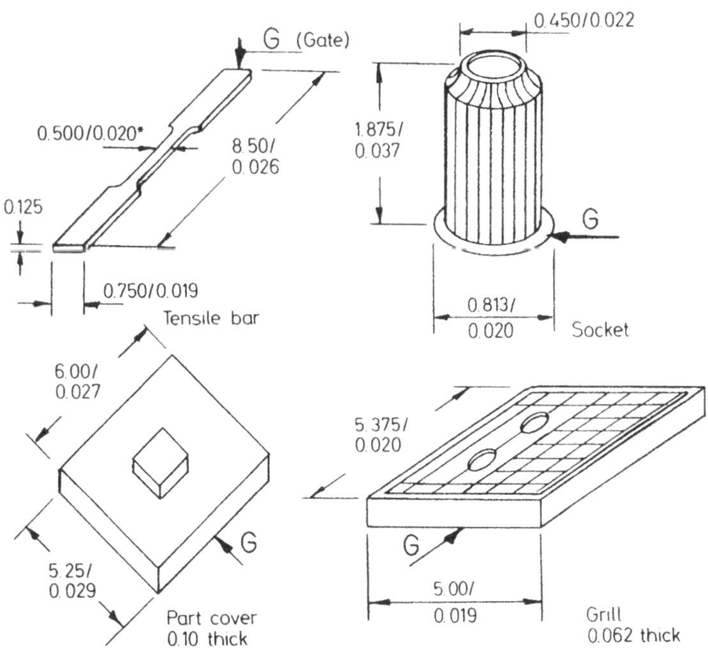

FIGURE 8.60. *Shrinkage effects due to part geometry and gate location. Values: dimension/shrinkage in inches with acetal copolymer. (From Hoechst Celanese and adapted from Ref. [12].)*

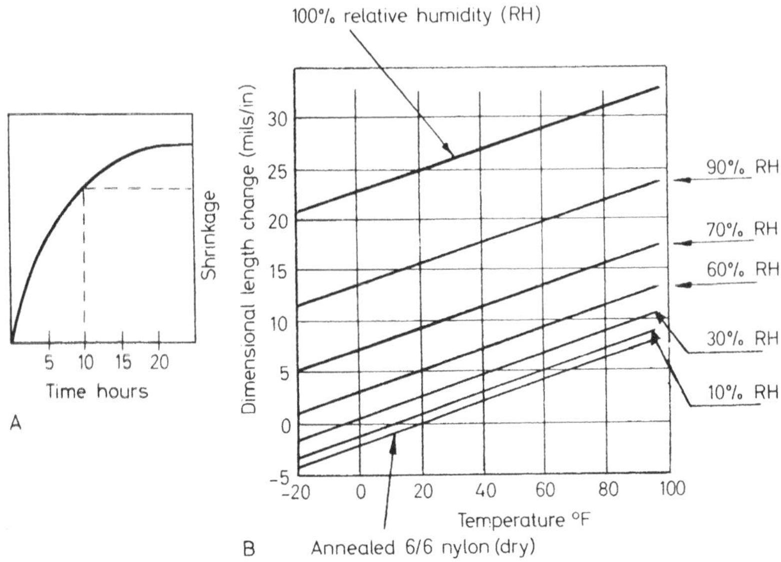

FIGURE 8.61. *(A) post-mold shrinkage for polypropylene, an amorphous plastic and (B) dimensional growth for nylon as it absorbs moisture. (Adapted from Refs. [12] and [25].)*

Amorphous plastics may also have some post-mold shrinkage, as illustrated in Figure 8.61A for polypropylene. This amorphous polymer experienced a 20 percent part shrinkage rate 10 hours after the part was molded. Dimensional part checks at the machine must allow for continuing size change, as most materials do not reach equilibrium until 24 hours after molding. For this reason, most parts that are exposed to elevated service temperatures are molded in hot molds to maximize shrinkage. Dimensions are set so that the part's post-mold shrinkage and dimensional changes are minimized. Processing conditions as well as the mold's cooling system are important in producing parts with the correct shrinkage.

Calculating and Estimating Part Shrinkage

To manufacture the part to the correct size, the mold's dimensions must be calculated based on the material's shrinkage values under standard or preselected mold and processing conditions. The material suppliers provide shrinkage values for their resins based on ASTM D-955 procedures. The SPI has also developed, for the generic polymers, recommended part tolerance guidelines. Molders and customers can use these in determining the degree of tolerances anticipated under commercial and fine-tolerance molding conditions for these generic resins. This information is available in their publication *Standards and Practices of Plastics Molders*. A review of the normal, accurate, and precise requirements are listed as a guide for understanding tolerance requirements.

Tolerances for Parts

Normal	Accurate	Precise
The required dimensions of the mold can be achieved using normal tool making techniques.	Accurately dimensioned mold cavities are necessary.	High-precision molds are required, depending on a product often of sophisticated design; the final sizes of the mold cavities are established after making a series of injection molding trial runs.
Product dimensions can be achieved in multiple cavity molds.	Multiple cavity molds can be used occasionally.	In many cases, only single-cavity molds can be used.
Injection molding conditions can be adapted for low-cost manufacture.	Molding conditions are more critical.	Molding conditions must be carefully controlled.
Regrind can be reused if required mechanical properties allow to do so.	Reuse of regrind is only possible to a limited extent and with due observance of required mechanical properties.	Processing of regrind not allowed.
No special inspection of final product size is necessary.	Random inspection of final products is usually required.	Thorough inspections of final product size is of vital importance.
Normal cost level.	Moderate additional costs.	High additional cost.

An example of the SPI tolerance standards for molding ABS in shown in Figure 8.62. The part illustrated is of uniform wall section with generous radii at the corners to maximize material flow and avoid pressure drops in the tool. Gating is not indicated, as its type and placement will have a dramatic effect on part dimensions and tolerances attainable between the fine and commercial ranges of manufacture.

The design of the part and the mold cavity dimensions selected will determine the dimensions and tolerances that can be held on the molded part. For very simple part geometries, tooling and resin advances have measurably increased tolerance repeatability with resin shrinkage more predictable and uniform from lot to lot of the same material. Table 8.8 lists tolerances for very simple parts that may be held, but never for the total part. Processing consistency also has a major bearing on holding these tolerances. However, complex parts will never achieve this degree of tolerance repeatability. What must also be understood is what part dimensions in the mold cavity are controlled by the mold and which ones are not due to shrinkage of the resin. The dimensions can be controlled in the mold in certain areas, as shown in Figure 8.63 A, B, and C. View A shows which part dimensions are controlled by the

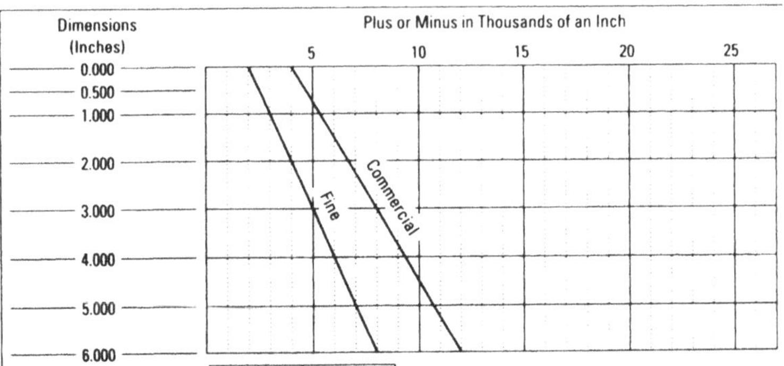

Drawing Code	Dimensions (Inches)	Comm. ±	Fine ±
A = Diameter (See note #1) B = Depth (See note #3) C = Height (See note #3)	6.000 to 12.000 for each additional inch add (inches)	0.003	0.002
D = Bottom wall	(See note #3)	0.004	0.002
E = Side wall	(See note #4)	0.002	0.003
F = Hole size diameter (See note #1)	0.000 to 0.125	0.002	0.001
	0.126 to 0.250	0.002	0.001
	0.251 to 0.500	0.003	0.002
	0.501 and over	0.004	0.002
G = Hole size depth (See note #5)	0.000 to 0.250	0.003	0.002
	0.251 to 0.5000	0.004	0.002
	0.501–1.000	0.005	0.003
H = Corners, ribs, fillets	(See note #6)	0.027	0.017
Flatness	0.000 to 3.000	0.015	0.010
(See note #4)	3.001 to 6.000	0.030	0.020
Thread size (class)	Internal	1	2
	External	1	2
Concentricity	(See note #4) (F.I.M.)	0.009	0.005
Draft allowa nce per side	(See note #5)	2.0°	1.0°
Surface finish	(See note #7)		
Color Stability	(See note #7)		

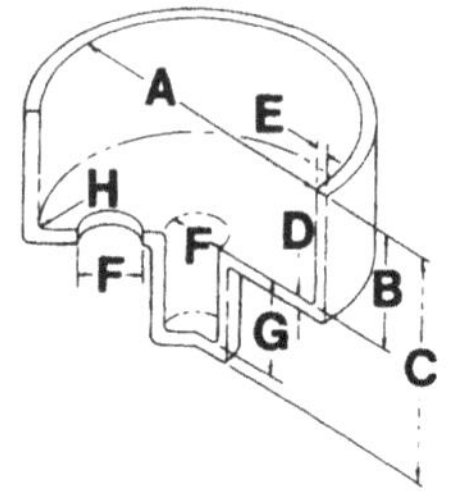

Reference Notes

1. These tolerances do not include allowance for aging characteristics of material.
2. Tolerances are based on 0.125 inch wall section.
3. Parting line must be taken into consideration.
4. Part design should maintain a wall thickness as nearly constant as possible. Complete uniformity in this dimension is sometimes impossible to achieve. Walls of nonuniform thickness should be gradually blended from thick to thin.
5. Care must be taken that the ratio of the depth of a cored hole to its diameter does not reach a point that will result in excessive pin damage.
6. These values should be increased whenever compatible with desired design and good molding techniques.
7. Customer-Molder understanding is necessary prior to tooling.

Note: The *Commercial* values shown below represent common production tolerances at the most economical level. The *Fine* values represent closer tolerances that can be held but at a greater cost. Any addition of fillers will compromise physical properties and alter dimensional stability. Please consult the manufacturer.

FIGURE 8.62. *Standards and practices of plastics molders. Material: ABS. (Reprinted from* Standards & Practices of Plastics Molders, *courtesy of the Society of the Plastics Industry, Inc.)*

mold cavity and View B the dimensions not controlled by the mold cavity. Dimensional control of the part can be enhanced in the mold cavity by the use of extending side walls and by adding ribs and forcing material to cool around cores. This will increase the molder's tolerance capability, as in View

TABLE 8.8. Tolerances for Simple Parts.

Dimension		Tolerance
1.0000 inch	+/–	0.0005 inch
2.0000 inch	+/–	0.0010 inch
4.0000 inch	+/–	0.0020 inch
6.0000 inch	+/–	0.0030 inch

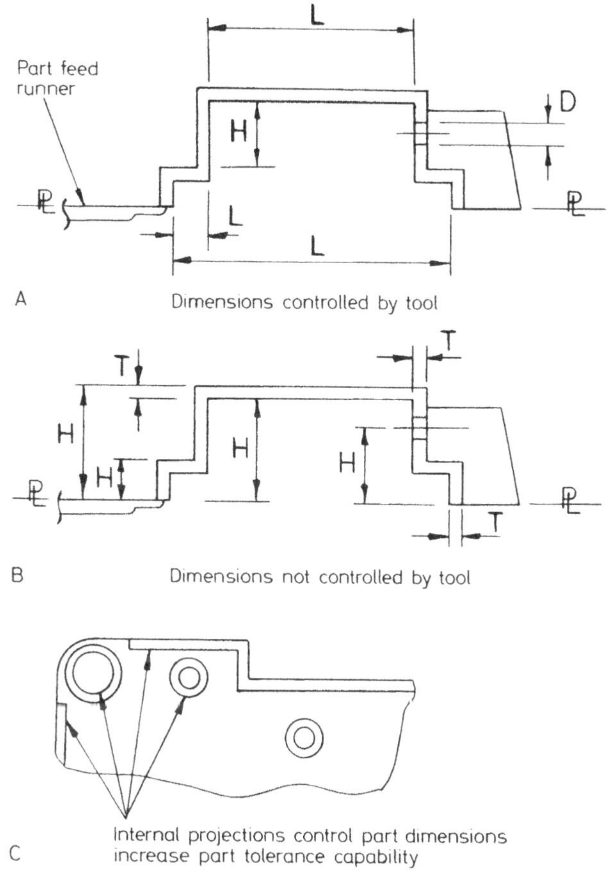

FIGURE 8.63. The relationship of the mold cavity to part dimension control. (Adapted from Ref. [22].)

C. These features will lock in key dimensions and hold dimensions as built into the mold.

Determining Cavity Dimensions

In determining the part cavity dimensions based on the material's shrinkage rate, two basic formulas are used. The current formula is not as accurate as

TABLE 8.9. Error in Mold Size*.

Part Size (in)	0.004	0.008	0.012	0.016	0.020	0.030	0.040	0.050	0.060	0.070	0.080	0.090	0.100	0.200	0.300	0.400	0.500
1.0	−0.02	−0.06	−0.1	−0.3	−0.4	−0.9	−1.7	−3	−4	−5	−7	−9	−11	−50	−129	−267	−500
3.0	−0.05	−0.19	−0.4	−0.8	−1.2	−2.8	−5.0	−8	−11	−16	−21	−27	−33	−150	−386	−800	−1500
5.0	−0.08	−0.32	−0.7	−1.3	−2.0	−4.6	−8.3	−13	−19	−26	−35	−45	−56	−250	−643	−1333	−2500
7.0	−0.11	−0.45	−1.0	−1.8	−2.9	−6.5	−11.7	−18	−27	−37	−49	−62	−78	−350	−900	−1867	−3500
9.0	−0.14	−0.58	−1.3	−2.3	−3.7	−8.4	−15.0	−24	−34	−47	−63	−80	−100	−450	−1157	−2400	−4500
11.0	−0.18	−0.71	−1.6	−2.9	−4.5	−10.2	−18.3	−29	−42	−58	−77	−98	−122	−550	−1414	−2933	−5500
13.0	−0.21	−0.84	−1.9	−3.4	−5.3	−12.1	−21.7	−34	−50	−68	−90	−116	−144	−650	−1671	−3467	−6500
15.0	−0.24	−0.97	−2.2	−3.9	−6.1	−13.9	−25.0	−39	−57	−79	−104	−134	−167	−750	−1929	−4000	−7500
17.0	−0.27	−1.10	−2.5	−4.4	−6.9	−15.8	−28.3	−45	−65	−90	−118	−151	−189	−850	−2186	−4533	−8500
19.0	−0.31	−1.23	−2.8	−4.9	−7.8	−17.6	−31.7	−50	−73	−100	−132	−169	−211	−950	−2443	−5067	−9500
21.0	−0.34	−1.35	−3.1	−5.5	−8.6	−19.5	−35.0	−55	−80	−111	−146	−187	−233	−1050	−2700	−5600	−10,500
23.0	−0.37	−1.48	−3.4	−6.0	−9.4	−21.3	−38.3	−61	−88	−121	−160	−205	−256	−1150	−2957	−6133	−11,500
25.0	−0.40	−1.61	−3.6	−6.5	−10.2	−23.2	−41.7	−66	−96	−132	−174	−223	−278	−1250	−3214	−6667	−12,500
27.0	−0.43	−1.74	−3.9	−7.0	−11.0	−25.1	−45.0	−71	−103	−142	−188	−240	−300	−1350	−3471	−7200	−13,500
29.0	−0.47	−1.87	−4.2	−7.5	−11.8	−26.9	−48.3	−76	−111	−153	−202	−258	−322	−1450	−3729	−7733	−14,500
31.0	−0.50	−2.00	−4.5	−8.1	−12.7	−28.8	−51.7	−82	−119	−163	−216	−276	−344	−1550	−3986	−8267	−15,500
33.0	−0.53	−2.13	−4.8	−8.6	−13.5	−30.6	−55.0	−87	−126	−174	−230	−294	−367	−1650	−4243	−8800	−16,500
35.0	−0.56	−2.26	−5.1	−9.1	−14.3	−32.5	−58.3	−92	−134	−184	−243	−312	−389	−1750	−4500	−9333	−17,500
37.0	−0.59	−2.39	−5.4	−9.6	−15.1	−34.3	−61.7	−97	−142	−195	−257	−329	−411	−1850	−4757	−9867	−18,500
39.0	−0.63	−2.52	−5.7	−10.1	−15.9	−36.2	−65.0	−103	−149	−205	−271	−347	−433	−1950	−5014	−10,400	−19,500
41.0	−0.66	−2.65	−6.0	−10.7	−16.7	−38.0	−68.3	−108	−157	−216	−285	−365	−456	−2050	−5271	−10,933	−26,500
43.0	−0.69	−2.77	−6.3	−11.2	−17.6	−39.9	−71.7	−113	−165	−227	−299	−383	−478	−2150	−5529	−11,467	−21,500
45.0	−0.72	−2.90	−6.6	−11.7	−18.4	−41.8	−75.0	−118	−172	−237	−313	−401	−500	−2250	−5786	−12,000	−22,500
47.0	−0.76	−3.03	−6.9	−12.2	−19.2	−43.6	−78.3	−124	−180	−248	−327	−418	−522	−2350	−6043	−12,533	−23,500
49.0	−0.79	−3.16	−7.1	−12.7	−20.0	−45.5	−81.7	−129	−188	−258	−341	−436	−544	−2450	−6300	−13,067	−24,500

*As a result of using incorrect shinkage formula (mils).

Source: Adapted from Ref. [4].

desired; as part size increases, so does the resultant dimensional error. This formula currently used is:

$$SDc = FL(1 + SR)$$

where

SDc = cavity or core dimension
FL = finish part dimension
SR = material shrink rate

This formula errs by not accurately determining the cavity dimension required to mold parts under standard molding conditions.

The formula that more accurately predicts the correct cavity dimension is a slight variation. Instead of adding one to the shrinkage rate and multiplying it by the finished part dimension, the shrinkage rate is subtracted from one and then divided into the finished part dimension.

$$SDn = FL/(1 - SR)$$

As an example, consider the following that uses a crystalline resin molded into a flat bar 7 inches long, 1 inch wide, and 0.300 inches thick. The part is edge gated at one end, with a gate opening area of 0.030 inches squared. Using high-injection pressure of 16,000 PSI, a hot mold of 200°F, and a melt temperature of 410°F, a shrink rate of 0.030 inch/inch over the length can be expected.

Using the current formula, SDc yields a cavity length dimension of 7.210 inches, whereas the new formula, SDn, yields a cavity length of 7.2165 inches. Molded length yielded an SDc of 6.9937 inches and an SD_N of 7.000 inches. The difference of .0063 is minor, unless the designer had a (+/–) 0.002-inch tolerance on the length dimension.

Care must also be exercised when calculating shrink rate using the SD_N formula with respect to varying shrink rates and part sizes. Table 8.9 shows the variance, or calculating error, between the two formulas and at what part size and material shrinkage rate the error will be significant enough to affect cavity and part dimensions. The tool designer can use the table or calculate the difference using the following error equation, E = SDc – SDn. The error is E = (FL × SR)/(SR – 1).

The calculation using the table would show that a material with a shrink rate of 0.008 inch/inch would have to be 15 inches in length to have a cavity dimension resulting part error of 0.001 inch. This implies that the new SDn formula is more accurate as part size and shrinkage rate increase. Since greater accuracy is the key, the SDn formula should be used. The more accurately the part's cavity shrinkage can be estimated, the

TABLE 8.10. Thermoplastic Injection Temperatures°.

Thermoplastic Short Forms	Average Melting Temperature*	Material Structure**	Mold Temp.	Hot Runner Process Temp.	Glass Fiber Weight (%)	Mold Temp.	Hot Runner Process Temp.
PE	140	SEMI	25	250	30	40	230
PP	170	SEMI	35	255	30	40	245
PS	110	AMO	45	225	30	65	245
SAN	115	AMO	80	255	30	90	260
ABS	110	AMO	75	250	30	90	260
PMMA	100	AMO	70	245			
POM	181	CRYS	100	200	30	105	210
CA	227	AMO	75	235			
CAB	140	AMO	55	225			
CAP	190	AMO	65	280			
PETP	225	CRYS	140	255	30	140	285
PBTP	225	CRYS	35	300	30	90	243
PC	150	AMO	90	250	40	120	310
PA 6	220	CRYS	90	285	40	110	280
PA 6/6	255	CRYS	90	250	30	110	300
PA 6/10	215	CRYS	90	230			
PA 11	175	CRYS	60	230			
PA 12	175	CRYS	60	230			
PPO	120	SEMI	80	300	30	105	325
PVC	100	AMO	35	195			
PUR	160	AMO	35	205	30	50	215
PSU	200	AMO	150	315	30	160	385
PPS	290	CRYS	110	330	40	120	315
PES	230	AMO	150	350	40	150	360
FEP	275	CRYS	150	315	30	230	355
PAI	300	AMO	230	365			
PEI	215	AMO	100	370	30	150	420
PEEK	334	CRYS	160	370	30	180	380
LCP	330	CRYS	175	400	30	180	400

*Temperature in degrees centigrade.
**AMO = amorphous; CRYS = crystalline; SEMI = semicrystalline.
Source: Adapted from Ref. [17].

easier it will be for production personnel to mold parts to customer requirements.

HOT-RUNNER MOLDS

Hot-runner molds are rapidly becoming the new tooling wave; they make up 15 to 20 percent of all tools now being built. Hot-runner molds, with their more precise temperature controllers, new tip and nozzle materials, and varying types of gating systems, are able to process all of the polymers currently available. Customers are requiring higher part quality, no regrind usage, and tighter control of tolerances on molded parts. The hot runner molds are better at satisfying these requirements. They operate best when designed for automatic operation.

The trends in hot runner equipment include the following:

1. Miniaturization—more cavities per square surface area and some nozzles with multiple gates.
2. Standardization of mold components for reducing mold design time, ease of replacement as parts wear, and lower mold cost.
3. Better mold temperature control to eliminate processing problems. Process control feedback should be able to lower mold temperatures if machine cycle interruptions occur.
4. More knowledge and experience in correctly designing tools, melt flow paths, mold steels, gating, and maintaining even pressure and temperature gradients in the tools.

Because the melting and softening characteristics of crystalline and amorphous materials differ, there are distinct ways of processing each and selecting the mold materials and components. The typical processing temperatures for most plastics are shown in Table 8.10. Figure 8.64 illustrates the temperature processing curve for amorphous polymers, which have a much broader processing window than crystalline polymers, which have a very sharp melting point.

Hot-runner molds must accurately control the polymer's temperature in the mold's feed system, from the injection machine's nozzle to the mold cavity's hot tip bushing. Material hold-up points or dead spots are not allowed in the feed system and flushing of resin through the feed channels is required to avoid material degradation during each injection cycle. Maintaining low shear stress at the gate is also required, because excessive gate shear will degrade many polymers and lower the physical properties of the molded part.

The typical hot-runner layout, shown in Figure 8.65, points out the critical areas of the system. A thermocouple cartridge heater and a typical hot-runner

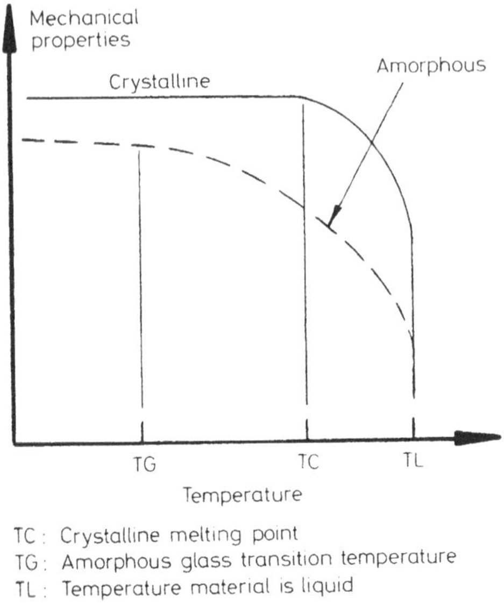

FIGURE 8.64. Temperature processing curve. (Adapted from Ref. [17].)

gate are illustrated in Figure 8.66. Various hot-runner gate types are shown in Figure 8.67.

Processing for Hot-Runner Molds

Tighter control of the injection molding process is required for hot-runner molding. The processing diagram in Figure 8.68 illustrates the critical nature of temperature control on the polymer during the molding cycle, as related to processing and gate temperature. Too high a polymer melt temperature increases chances of degrading the resin, especially with the more heat sensitive resins, such as the PVC compounds. As with conventional tooling, the cavity's gate must freeze-off. This can be accomplished in three ways: 1) lowering of heater gate temperature with a temperature controller, 2) using the mold's cooling system to cause freeze-off, and 3) using a positive value shut-off so that after injection, a plunger moves forward and physically shuts off the material flow while maintaining cavity pressure.

All three gate types work well with the amorphous plastics. With the crystalline resins, the shut-off is preferred because it prevents nozzle gate drool. Temperature control of the tip can also work, but it is very sensitive, more difficult to adjust, and must operate within 5° of the melt temperature to be effective.

For hot-runner molding to be successful and controlled, the hot-tip bushings and mold manifold system are isolated from the main sections of the mold. This assists in maintaining even temperature control, as shown in

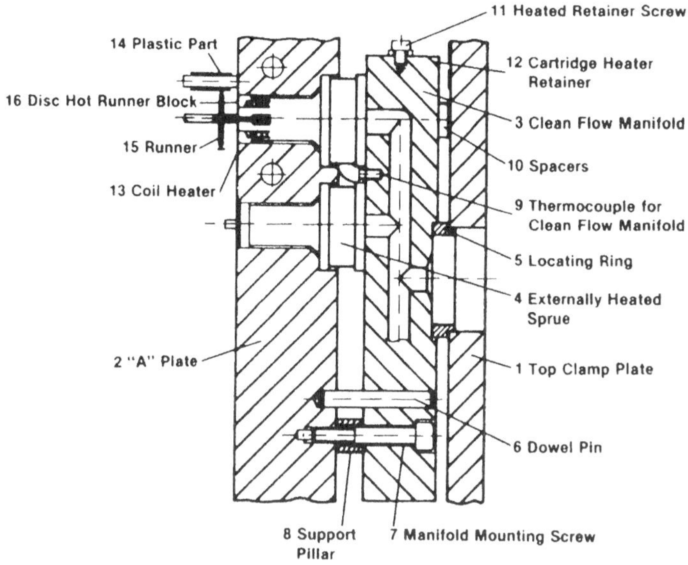

Cutaway drawing of hot runner mold system producing polycarbonate parts

FIGURE 8.65. Hot-runner mold layout. (Courtesy of INCOE Corp.)

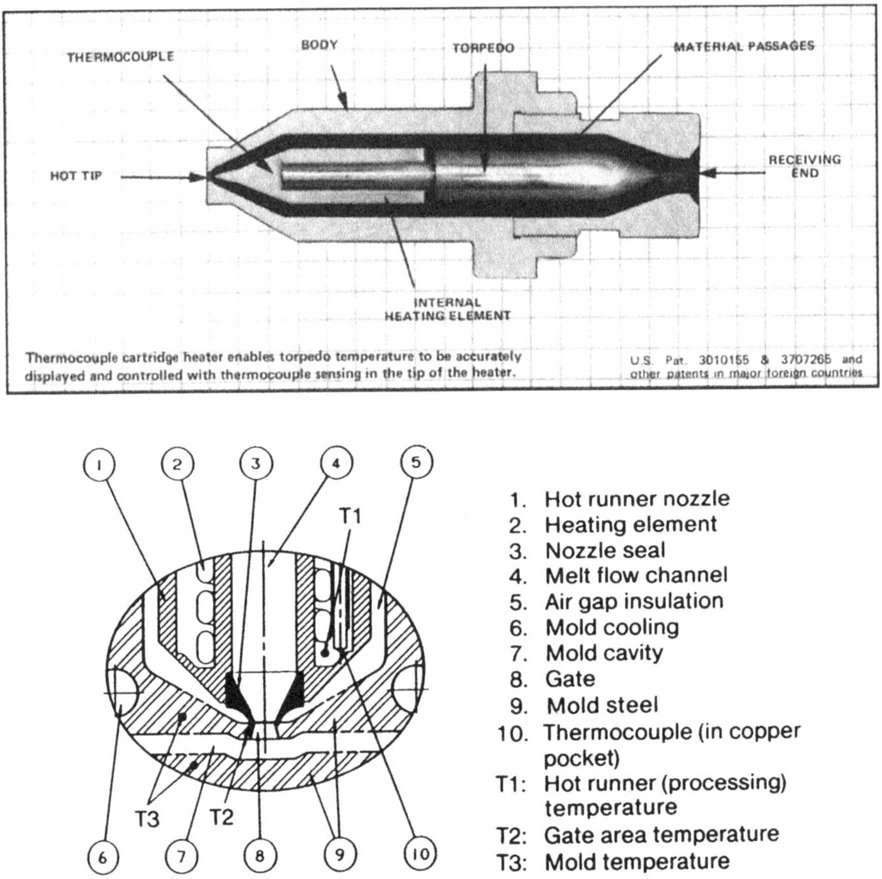

FIGURE 8.66. *Hot-runner gate: (A) thermocouple cartridge heater and (B) typical hot-runner gate. (A: courtesy of INCOE Corp. and B: courtesy of Mold-Masters Limited, Georgetown, Ontario, Canada.)*

Figures 8.69 and 8.70, for single and multicavity molds. The feed system to the hot tips is of a large enough diameter to reduce pressure drops in the feed channel. It should be well polished and without dead spots or sharp corners. The hot tips must be thermally insulated from the mold cavity and cooling channels to attain uniform temperature control during operation.

The correct selection of the hot-tip gate design is necessary for part filling, pack out, and cosmetics, if the circular witness mark left by full body construction is not desirable. Examples of available hot-runner gating for these types are illustrated in Figure 8.71, which shows examples of full body gate/part contact.

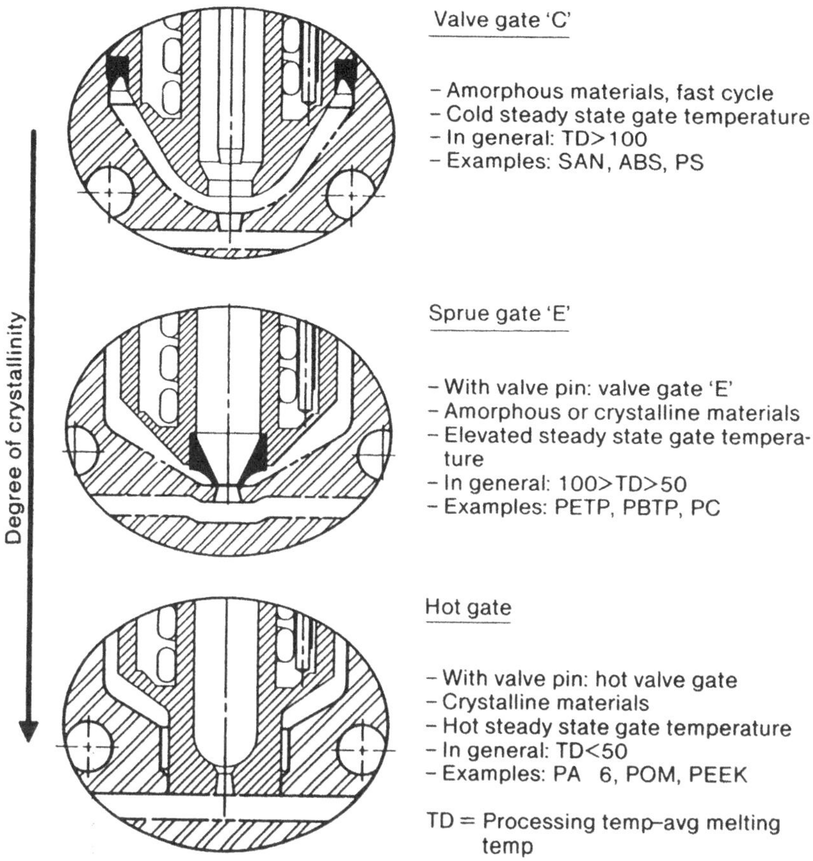

FIGURE 8.67. *Various hot-runner gate types. The hot gate for crystalline polymers transfers heat directly from its large nozzle end to the gate area for high temperatures, while the valve gate "C" for amorphous polymers has a plastic film around the nozzle tip to aid in rapid heat dissipation. (Courtesy of Mold-Masters Limited, Georgetown, Ontario, Canada.)*

An example of resin flows within these types of gates and how they are sized for specific hot-tip bushings is shown in Figure 8.72. Exact temperature control is essential with a hot-runner tool, and controllers must always be in calibration and well maintained. Operator training of the hot-runner system is paramount to proper operation, startup, and molding of quality parts. There are many good hot-runner mold designers, who can assist in selecting the best method and parts to build the desired tool.

The main advantage of runnerless molds are as follows:

1. No runner system and no regrind
2. Use of virgin material for all parts, higher part quality, and no material variations

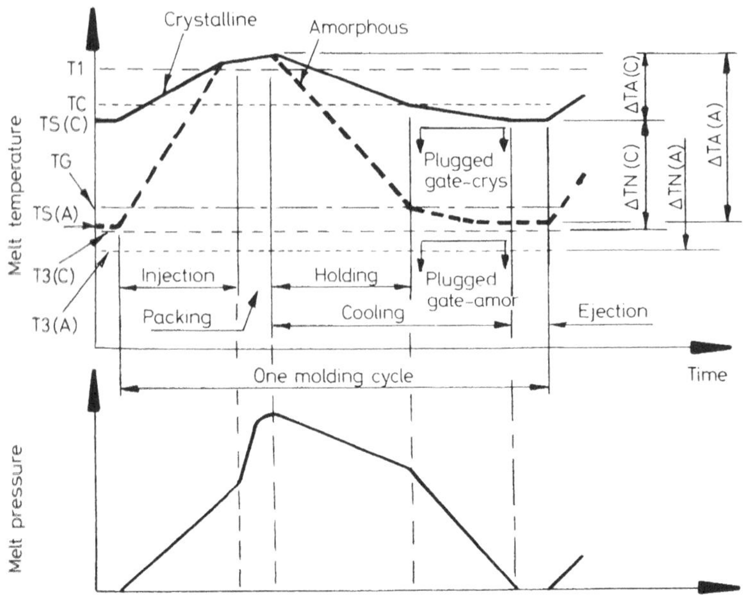

----T2: Gate temp. (crys ex: PA 6)
----T2: Gate temp. (amor ex: ABS)
T1: Hot runner (processing temp.)
TC: Crystalline melting temp.
TG: Amorphous glass transition temp.
T3(C): Mold temp. (crystalline)
T3(A): Mold temp. (amorphous)
ΔTN(C): Gate temp. elevation(crys)
ΔTN(A): Gate temp. elevation(amor)
ΔTA(C): Gate temp. addition (crystalline)
ΔTA(A): Gate temp. addition (amorphous)
TS(C): Steady state gate temp. (crystalline)
TS(A): Steady state gate temp. (amorphous)

FIGURE 8.68. *Temperature control for hot-runner molding. (Adapted from Ref. [17].)*

3. Reduced material loss
4. Less stress in molded part
5. More uniform part weight and dimensions
6. Reduced cycle time
7. Automatic operation
8. Fast startup of system

Selecting the correct size and type of gating to be used for filling the part is the most important item to consider in hot-runner molding. The items that must be considered are as follows:

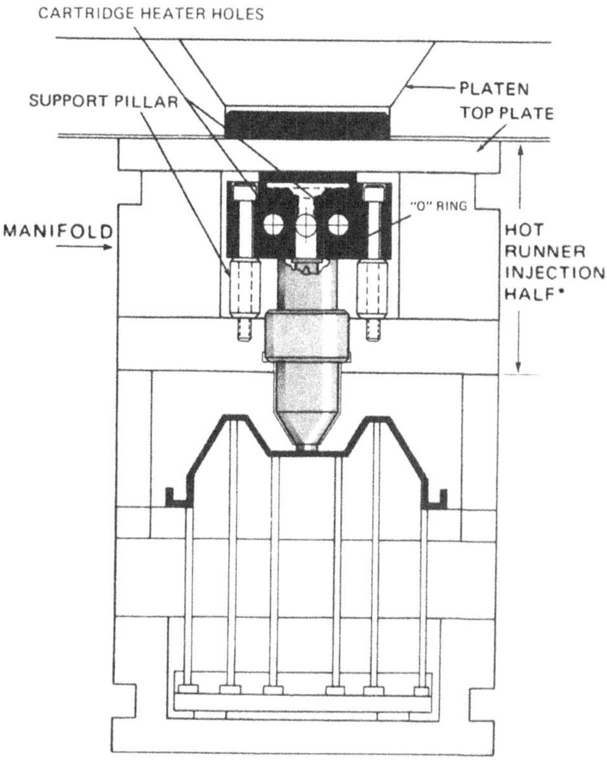

FIGURE 8.69. *Hot-runner single cavity mold. (Courtesy of INCOE Corp.)*

1. Type of resin and its shear rate
2. Injection rate required to fill the cavity
3. Part size and geometry, including wall thickness

Gate drool is controlled by precision temperature control. This permits the use of large gates when necessary to obtain good part fill and pack out with amorphous and crystalline polymers.

The use of closed-loop voltage proportioning controllers is required to maintain continuous control of the resin's melt temperature in the hot-runner system, from the feeder manifold to the hot tips. Autotransformers and time-proportioning type controllers are not recommended. They require constant operator monitoring and adjustments that will vary constantly throughout the run as conditions vary in the molding room.

Training in correct start-up procedure for hot-runner systems is very important. This is due to the type of heater systems used and the specific procedure that must be followed to eliminate shorts and burnouts during start up. Operators must also be instructed in proper operating temperatures

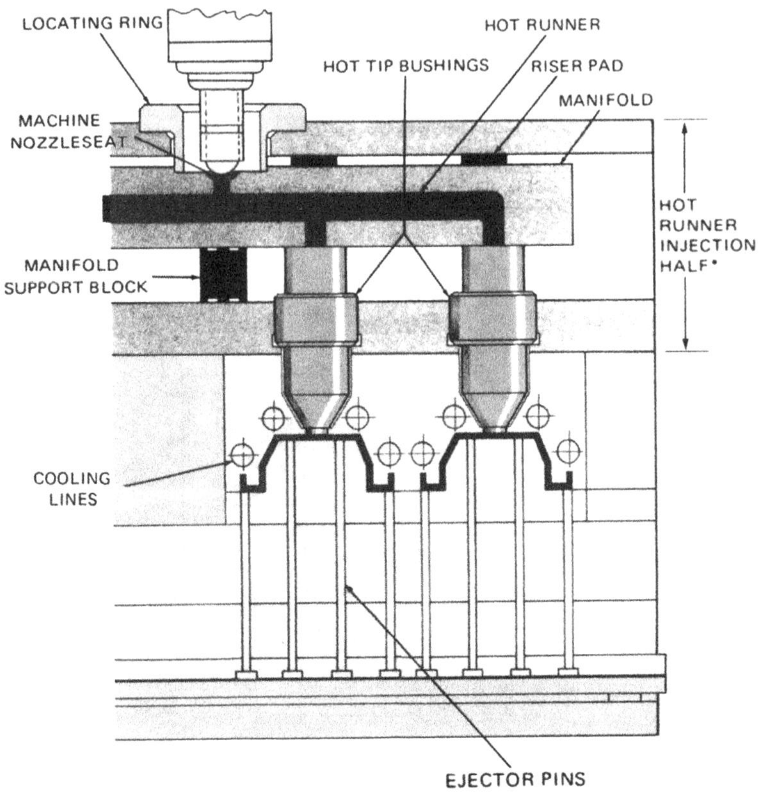

FIGURE 8.70. *Hot-runner multicavity mold. (Courtesy of INCOE Corp.)*

and preventive maintenance. Hot-runner molds are proving their worth in controlling material and labor costs, while improving part quality and profitability.

MOLD MAINTENANCE

To produce quality parts the mold must always be kept in excellent condition. If the mold and its components are not kept clean and in good repair, it will not be able to produce quality parts each time it is operated.

The mold should be scheduled for both preventive and quality maintenance after each run to ensure that the dimensions, gates, operating mechanisms, cooling system, and cavity surface and vents are within specifications and functioning properly. Production records should be kept of the number of cycles and parts produced from the mold so that maintenance will know what items in the tool should be checked for wear and possible replacement.

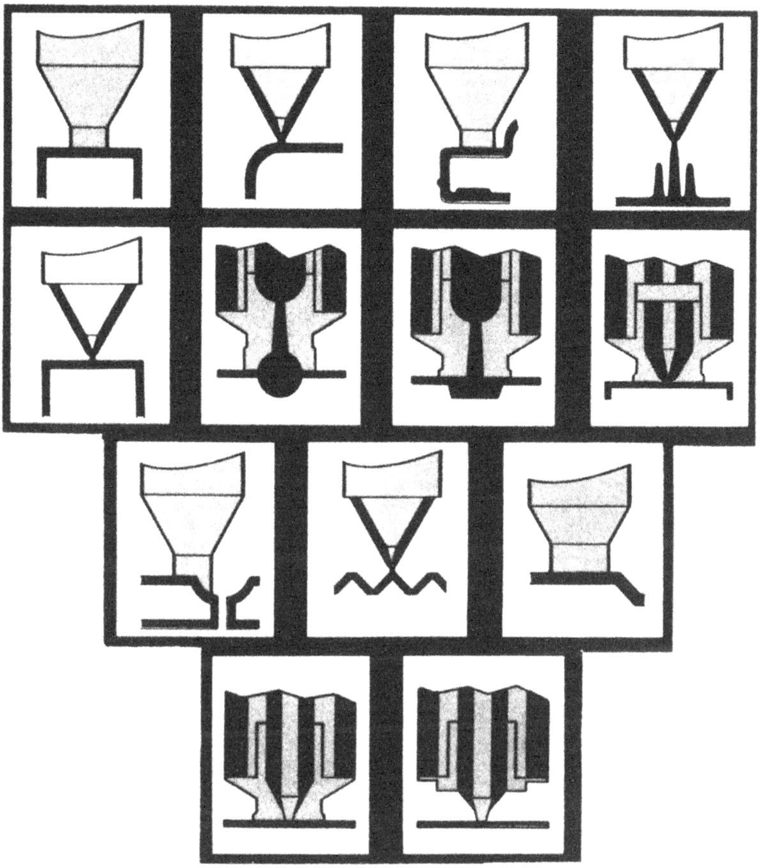

FIGURE 8.71. *Hot-runner gating. (Courtesy of INCOE Corp.)*

Depending on the materials used in a mold's construction, it is estimated that annual mold maintenance costs should range from four to eight percent of the original cost of the tool. Keeping accurate records on each mold's maintenance will help to extend the life of the tool.

A tool's predicted, as opposed to expected, life is based on historical data for the number of cycles it is to operate, material molded, tool steels used in its construction, and the amount of wear experienced on bushings, guide pins, etc. This information is necessary for the maintenance department to determine when a mold is losing efficiency so that repairs can be made before part quality drifts. This can vary in multicavity molds from cavity to cavity and should be corrected at once. If a problem occurs in one or more cavities, then some molders shut off the failing cavities. This unbalances the tool and can cause problems with the remaining cavities' part quality. This should be avoided by correcting problems with the tool as they occur.

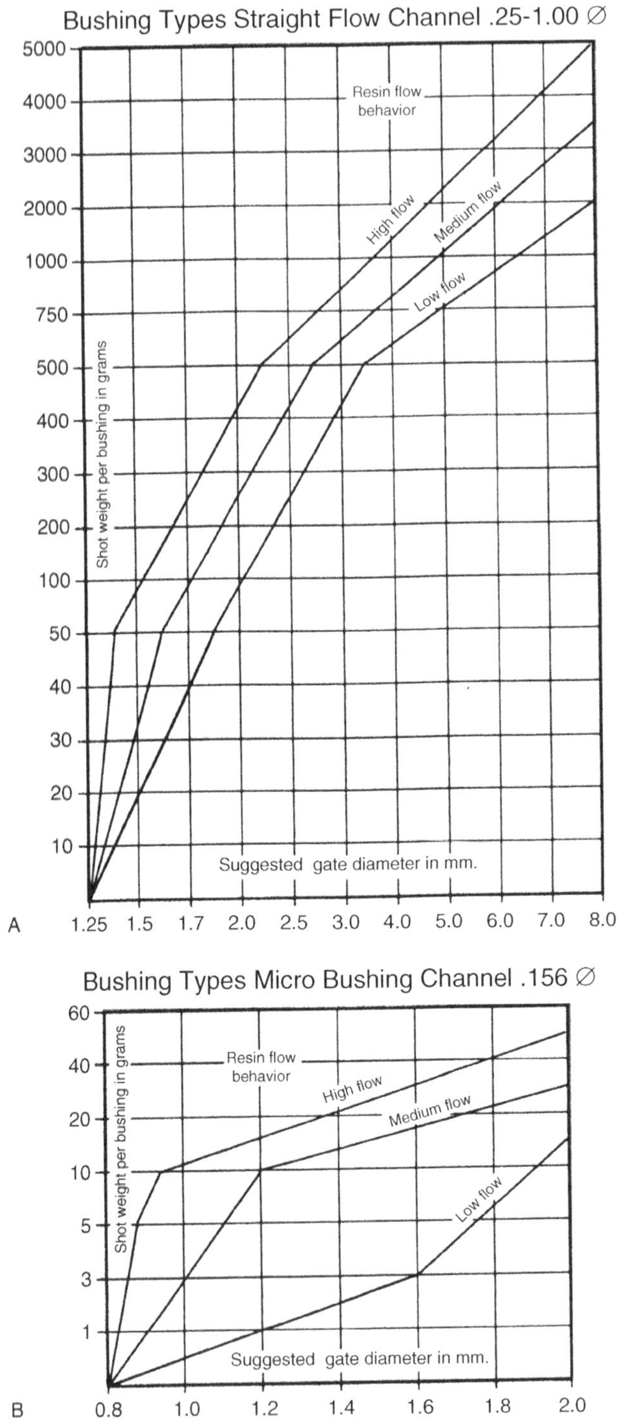

FIGURE 8.72. *Hot-runner gate selection guide for bushing types: (A) straight flow channel, and (B) microbushing channel. (Courtesy of INCOE Corp.)*

Each mold should have setup procedures and the work order should itemize the specifics for each mold's requirements. This information is used to identify the details of the mold and the mold's inspection and repair history. The setup procedure should list all special tools and equipment required to operate the mold.

A mold work order should include the following information:

1. Customer name
2. Job number
3. Mold number
4. Part identification
5. Type of machine, including press number, minimum press size, and shot size
6. Quantity of parts to run and the number of mold cavities
7. Type of material, including lot number and material identification information
8. Start date, end date, and due date
9. Set-up time and procedures, including machine settings
10. Auxiliary equipment required to run the job
11. Auxiliary equipment settings

The mold inspection and maintenance record should also be a part of this document and includes the following (see Figure 8.73):

1. Maintenance requirement notes during operation
2. Last and previous maintenance dates
3. Problems encountered with tool during operations
4. Cycles to date in hours and parts produced
5. Cycles since last maintenance
6. Next scheduled maintenance date
7. Areas to be checked—complete tear down or periodic inspection of cavities, gates, etc

Some of the newer machine process control units will collect the information on hours of operation and number of cycles the mold operates. This information must be recorded at the end of each run to determine correctly the amount of wear the mold experiences over its operational life and to schedule maintenance. There are now computer systems to organize and calculate this information and provide a basis for mold maintenance schedules. Some maintenance operations are coded with an identification number to keep track of when each specific maintenance operation is required. A typical mold-tracking schedule is shown in Table 8.11. This record tracks the mold's operating history for part quality listing problem areas, a preventive

MOLD NAME
MOLD DRAWINGS

DATE LAST SERVICED
MOLD NUMBER
MOLD MATERIAL/PARTS LIST
MOLD SIZE: WEIGHT

_______ × _______ × _______
length width height

PART DRAWING______FOR CAVITY DIMENSION CHECK

RUN HOURS/NO. PARTS TO DATE________________
DATE RECEIVED________________

FITS FOLLOWING MACHINES________________
MATERIAL RUN-IN-MOLD________________
SPECIAL SET-UP EQUIPMENT REQUIRED________

Mark (X) if item is o.k. for next run. Mark (R) if item must be repaired before next run. Give brief description.

MOLD CONDITION											Describe Maintenance Requested	Date	Describe Maintenance Completed	Date
Insp. Date	Sprue Bushing	Runner	Puller (Sprue)	"A" Cavity & Cores	"B" Cavity & Cores	Ejectors	Cooling Flow	Side Action	Vents	Wiring & Heating				

MAINTENANCE PERFORMED BY________________
PERFORMED MAINTENANCE INSPECTION AND APPROVAL________ DATE______

FIGURE 8.73. *Mold inspection and maintenance record.*

TABLE 8.11. Mold Maintenance Due Forecast (Next 30 Days).

Mold	Code	Date Last Performed	Cycles Since PM	Mold 59762 Cycles to PM	Due Date	Overdue	Description
59762	429	06/04/90	0	5804	06/05/90		Check eject pins
	430	06/01/90	58357	157643	07/01/90		Polish
	431	04/02/90	-39553	0	00/00/91	X	Change guide bushings
	432	02/01/90	379704	1120296	00/00/91		Replace core
	433	05/28/90	10653	0	00/00/91	X	Check spring
	435	06/03/90	1917	483	06/04/91		Inspect gates

(PM) Preventative Maintenance.

Source: Adapted from Ref. [19].

maintenance schedule, and the job requirements for the tool. This is shown on a 30-day moving forecast schedule. Specific items in the mold maintenance program are identified by a code number for the specific operation. They should be as specific or general as required for the individual mold maintenance program.

The mold maintenance-due forecast schedule is based on mold history, preventive maintenance, and the job-run schedule. When an incomplete date is shown in the "Due Date" column for a specific maintenance code item, it indicates insufficient run time for the tool to set a date for maintenance. But the number of hours or cycles until the next maintenance period are still noted. This will assist in maintaining part quality and eliminate unscheduled mold maintenance.

Once the tool is put into the production schedule, these items may be scheduled depending on anticipated run time. Any item not completed at time of forecast generation is noted by an asterisk in the overdue column. Attention will be drawn to them and corrective action will be taken.

Mold operation and maintenance reports should also indicate problems associated with the mold during operation to document problem areas. As soon as the job ends, the problem can be corrected. These data can be inputted by an operator at the press or by using the process controller setup that keeps track of part quality. The number and type of defects can thus be identified and corrections made to eliminate them in the future.

When maintenance is performed on a mold, the technician enters the respective maintenance code of work to be performed, its identification number, hours spent, parts replaced, and part cost. The cost of maintaining a tool should be built into the molder's part quotation, so that the toolowner pays for these repairs as a function of piece-part price.

A number of computer software programs provide mold maintenance scheduling assistance. For example, Microsoft Excel and/or Lotus 1-2-3 can be used to track these functions. No matter what system is used, there must be management, personnel, materials, and experience to maintain the degree of part quality required for the customer's products.

9

Manufacturing Equipment

The injection molding machine must be selected and sized to produce parts with the tooling supplied within the cycle and cost objectives of the customer. The prerequisites for equipment selection go beyond the old norm of selecting any machine that met the mold's shot size, clamp force, and platens. With today's higher part quality requirements, the production department must be more selective. The best designed part, a capable tool, and lot-to-lot material uniformity will be squandered if the injection molding machine cannot consistently make quality parts.

MACHINERY SELECTION

All machinery manufacturers compete to offer customers the best price for the best machine, often noted as their basic model, with proven mechanical, hydraulic, and electrical systems. Special features can be added to improve its efficiency and quality. These include the following:

1. Computer-integrated process controls, with varying output software systems to control the processing cycle. Closed and open loop information feedback systems for total integration of the machine and support equipment.
2. Selection of proportional valves in combination with variable displacement pumps or servovalves.

Total Quality Process Control for Injection Molding, by M. Joseph Gordon

3. Complex core sequencers.
4. Ceramic heater bands.
5. Special wear-resistant adaptors, barrels, and screws.
6. Specially designed processing screws.
7. Mold clamp selection—toggle or hydraulic.
8. Other machine features as required for the job, including oversized platens, external connections for auxiliary equipment support, heavy duty pumps, electric drive, etc.

Many machinery suppliers are designing and building special machines to produce complex parts for the medical and electronics fields. These require a very high degree of quality and tight tolerances. To achieve this, machinery suppliers are becoming involved with their customers at the design stage. They discuss machinery with options and special features to produce the desired part.

For this to happen, machinery suppliers are concentrating on improved mechanical performance of their equipment, including electrohydraulic machine operations, the speed and accuracy of hydraulically actuated machine functions, cycle time reductions, energy conservation with lower maintenance costs, and long-term equipment and component reliability under vigorous production conditions. The goal is to produce quality parts right out of the mold, cycle-to-cycle, ready for assembly or packaging without costly inspection.

PROCESS CONTROL

To control and monitor these machines, equipment suppliers are using state-of-the-art computer process control units. These units record and store multiple processing parameters for many molds; the exact processing parameters can be recalled each time the tool is run. The controllers monitor, compare, adjust, and maintain processing conditions with preselected processing windows, and they can be connected with mainframe computers to control and monitor the manufacturing process accurately.

These control units constantly compare and adjust the machines' processing variables through closed and open loop process control systems. They can also be networked with the auxiliary equipment to provide a total work cell control system. They can signal problem areas when the process variables go beyond their capability, so that an operator may solve the problem. Thus, they are excellent controllers and monitors of the processing cycle.

These machines have built-in sensors to monitor certain aspects of the mechanical, hydraulic, and electrical equipment components. Such aspects include how the components respond and function while providing feedback to the process controller. These machines can now monitor processing parameters for material set-up in the mold to yield the minimum cycle production times for producing quality parts. The process controllers operate in real-time parameters to control part quality and respond to processing problems as they occur so they can be solved immediately. Besides monitoring all electrohydraulic system responses, these machines can also gather, store, and output statistical process/quality control information that proves that the part's manufacture meets quality standards. This saves an operator from gathering these data, which must be used in a real-time mode to maintain the production process within manufacturing control limits. These control systems also have alarms that warn operators if the system is going out of specification. With some systems, this will shut down operations until necessary corrections are made. There are also systems that will check part weight, as fine as 0.001 gram, to verify the molding process is functioning as required, using part weight as a quality parameter.

Some molders have considered upgrading their existing injection molding machines with the new process-control units. The cost to upgrade as opposed to purchasing new machines must be evaluated, along with the degree of process control attainable by upgrading an older machine. Many older machines were not built to the standards now required for the computer process systems. To just add the computer process control systems, without upgrading the injection molding machine's internal valving, pumps, motors, and mechanical components, will not yield the required results.

This does not mean it is impracticable to upgrade an existing machine. But, if a machine is upgraded, its mechanical, electrical, and hydraulic systems and components must also be upgraded.

ELECTRIC INJECTION MOLDING MACHINES

All-electric injection molding machines have gained acceptance as their price approaches the initial costs of hydraulic machines. The electric machines claim a 30 percent faster molding cycle, along with 50 to 75 percent lower power consumptions, when compared with conventional hydraulic units. These quieter machines realize these savings because their electric motors only operate when required; hydraulic system pumps run continually. The machine's mold toggle clamping unit, with four servomotors, also provides high-speed, smooth clamping. The electric machines also have add-on options that, when molding small thin-walled parts, can double the injection pressure to minimize part shrinkage and distortion. These machines can also come with auxiliary outlets, hot-runner control, automatic heat-up control, bump ejection, two-stage melt generation and injection chambers of greater size and stroke, automatic shutdown sequencers, and cavity pressure transfer controls. Today, machine sizes vary from 30 to 950 tons. As advances occur, presses with higher tonnage are predicted. Their faster response is attributed to the higher efficiencies of electric motors over hydraulic pressure systems. Many problems associated with hydraulics, such as hydraulic fluid leaks, contamination cleaning and/or replacement of hydraulic fluids, are also eliminated. Finally, the variable of hydraulic fluid temperature/viscosity control is eliminated. This, if not regulated, contributes to molding cycle operation inefficiencies.

INJECTION MOLDING MACHINE NOMENCLATURE AND OPERATION

The typical injection molding machine is a reciprocating screw hydraulically operated machine, as illustrated in Figure 9.1. There are older versions of injection molding machines, particularly the plunger or ram machine (see Figure 9.2).

The plunger or ram machine is still used to obtain special two- or three-color molding effects, such as mottled or wood grain patterns on parts. The ram machine uses barrel heat and pellet-compression shear heat to create

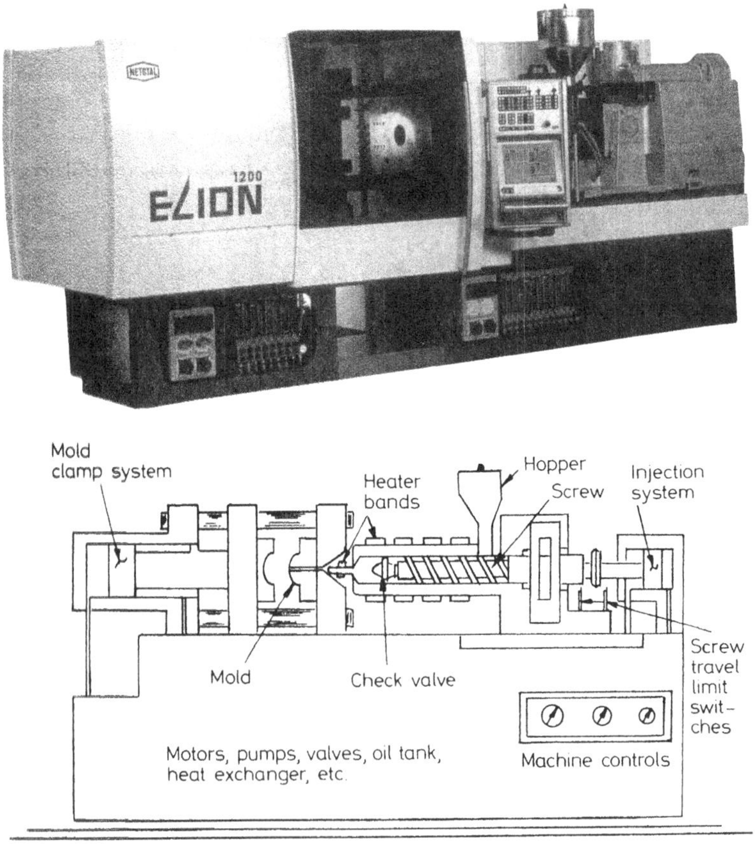

FIGURE 9.1. *Injection molding machine layout.*

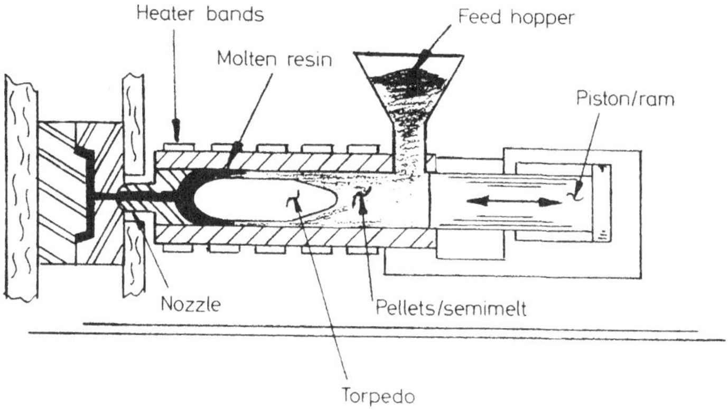

FIGURE 9.2. Hydraulic ram injection molding machine. (Adapted from Ref. [1].)

melting as the resin flows around an obstruction in the barrel called the torpedo or spreader. As the plastic pellets are forced down the heated barrel, they are compressed and forced around the torpedo and up against the barrel's surface where there, they are melted.

Once past the torpedo, they form a pool of melt. During the injection cycle, this melt and any additional melt is created as the mold is filled.

RECIPROCATING SCREW INJECTION MOLDING MACHINE

The reciprocating screw injection molding machine is efficient, reliable, and able to process all of the thermoplastic polymers to create a high-quality melt for the manufacture of plastic parts. It is basically an extruder with a screw that retracts by melt pressure as it conveys plastic pellets forward from the feed hopper. The screw's action compresses and forces the pellets against the heated barrel surface. Then, through barrel and shear heat, the pellets are melted and conveyed down the barrel through progressively shallower screw flight depths. The primary heat for melting the plastic pellets is supplied by the external heater bands, which heat the barrel's inner surfaces. The melt created in the forward sections of the screw is then forced forward by the screw's pumping action through a check ring at the end of the screw. This forms a supply of melt in the barrel in front of the screw. As the supply of resin builds up in front of the screw's check ring, it creates pressure that forces the screw back until a preset volume of resin is built up ahead of the screw. This is the material to be injected into the mold during the next injection cycle.

The reciprocating screw injection molding machine, offers many advantages over the ram systems, such as the following:

1. Improved melt homogeneity, with colors/additives thoroughly mixed
2. Precise temperature control of the resin
3. Better and more consistent pressure transmission during injection, and the ability to profile injection speed and pressures
4. Faster and consistent molding cycles
5. No resin hang-up in barrel
6. Faster purging and turnaround for material changes
7. Resin handling versatility and screw design versatility

Injection Molding Cycle Operations

The basic operations of the injection molding cycle begin by closing the mold and applying clamping pressure to keep the mold closed and locked during the molding cycle. The basic operations are as follows:

1. Material in the hopper is fed to the screw that conveys resin forward, melting or softening it by barrel heat and screw shear action.
2. The check ring at the end of the screw is forced forward by the screw's pumping action. This forces the melt to flow through the openings in the screw tip. As a result of the screw's pumping action a molten supply of resin is created ahead of the screw. The screw then retracts due to the pressure build up by the melt at the front of the screw.
3. Once the selected amount of melt is built up, as determined by the selected screw-retraction distance, the screw is triggered to stop by a preset switch.
4. The injection cycle then occurs. Hydraulic first-stage injection pressure forces the screw forward. The check ring is forced rearward by the injection pressure, thereby creating a seal that restricts resin from flowing back over the screw. The mold is then filled. The fill rate is determined by the hydraulic pressure forcing the screw forward.
5. The screw remains forward while the hydraulic pressure is lowered to second-stage packing pressure. This continues until the cavity gate freezes off. Based on timer settings, the screw begins to rotate to build up a new quantity of melt for the next injection cycle. This occurs while the mold is closed and locked.
6. After a present holding time and while the part is cooling and solidifying, the mold opens and the ejection system pushes the part out of the cavity. The mold then closes for the next cycle to begin.

The repeatability of the cycle as well as the melt temperature of the resin, injection speed and fill profile, and packing and hold pressures must be uniform from cycle to cycle. This is a function of the quality of the machine and its controllers.

Machine Selection for the Molding Cycle

The correct sizing of the injection molding machine is necessary, as many variables affect part quality.

Every injection molding machine is sized for its resin melting and shot-size capability, its potential mold clamp tonnage, and its type of clamp operation—hydraulic or toggle action. A 4-ounce, 75-ton hydraulic action reciprocating screw injection molding system describes the main features of the machine. Machines have basic clamp tonnage capabilities and types, hydraulic or toggle action, with interchangeable barrels and screws. As the machine's clamp tonnage increases, so does the melting and shot-size, or volume, of melt that it generates for injection into the mold. This operation is dictated by the barrel and screw size. But molders can specify different size barrels and screws to suit the molding jobs and tool sizes they prefer to handle. These features will vary with each machine manufacturer's decision to compete by supplying machines in the various size ranges. These features, plus others, and their operations will be developed in this section.

These versatile machine controls that regulate the molding cycle are becoming standard on newer machines. There are options or add-on features that can be selected, but in general the standard operational controls remain the same. They are, however, improved in terms of reliability, functionability, and reaction time.

Resin Melt Shot Capacity

Each injection molding machine is rated on its capability to create a maximize shot weight of plastic. Polystyrene (PS), a general purpose resin, is used as the standard. A rating of 4 ounces means that the machine can inject this amount of PS resin during each cycle. This is the ideal situation, but it is not recommended as it implies that the machine is 100 percent efficient. It would also tax the machine to its full limits, which would jeopardize part quality.

Using polystyrene with a specific gravity of 1.06, the shot capacity of other resins in a 4-ounce or other size machine will vary more or less, depending on their specific gravity ratio to polystyrene. For example, a type 6 nylon has a specific gravity of 1.13; the maximum shot weight for nylon would be $4 \times 1.13/1.06$, or 4.26 ounces. But the full shot capacity of a machine is rarely used, as unmelt may occur because of machine inefficiencies. This taxes the machine to its full limits. Therefore, when calculating the shot weight required for a tool, the molder uses only 80 percent of the machine's ounce capacity rating. The typical shot weight selected for a machine is based on 20 to 70 percent of the melting capacity in ounces, adjusted for the resin's specific gravity. Shot weight is calculated and based on the total part and runner's resin weight necessary to fill the cavity during injection. There must also be some residual resin in the barrel, so that the packing pressure can be maintained on the melt until the cavity gate freezes off.

When highly lubricated materials are molded, it may be necessary to use at least 75 to 80 percent of the machine's shot capacity. This prevents the material from slipping back over the check ring and screw flights during the injection cycle. It is mainly typical of highly lubricated materials and is discussed in greater detail in the machine melt output and check ring section in this chapter.

If too large an ounce capacity machine is selected, then heat sensitive resins may have too long a residence or hold-up time in the barrel. The resin could degrade, causing poor quality and lower physical properties in the part. Therefore, the machine's ounce capacity must be sized to the mold's shot weight requirements and anticipated cycle time. These determine the polymer's residence time in the barrel. Under ideal conditions, two to three shot weights in the barrel, under typical cycle times, will not result in resin degradation. If in doubt, consult the supplier about the effects of residence time on the resin's properties and thermal stability. For general polymers [not polyvinyl chloride (PVC) compounds that are not thermally stable], under controlled supplier recommended barrel temperature settings and minimum screw retraction with rotations per minute (RPMs) up to 5 to 10 minutes will normally not affect a resin's physical properties. This relates to five to six shot weights of resin in the barrel's length as a standard, as long as the hold-up time does not exceed 10 minutes. Too often, molders select a machine with too high an ounce rating for small parts; material degradation can then occur because of long residence time in the barrel. In these cases, barrel heats may have to be reduced.

Each machine is also rated on the screw's L/D ratio—length of screw (L) over screw diameter (D). A machine with a 1-inch diameter screw that is 20 inches long would have an L/D of 20 to 1. Certain resins will require a minimum L/D ratio to properly plasticize or melt. These include the crystalline engineering resins. If colors are to be blended by the machine's screw, a screw with a higher L/D ratio may be required to ensure thorough mixing in the barrel to obtain uniform resin color on each cycle. The correct screw L/D ratio recommended for a resin should be confirmed by the supplier.

Machine Melt Plasticizing Capability

The ability of a machine to produce a uniform and consistent melt is a process requirement. This should be obtained without excessive screw RPMs and back pressure to eliminate unmelt in the resin, keep resin screw shear heat under control, and avoid all forms of resin degradation during molding.

The molding machine's plasticizing capacity or melt generation relates to production cycle time requirements and its capability to supply the necessary heat to produce a quality supply of melt within the time frame. Each reciprocating screw injection molding machine is basically an extruder, with modifications to allow injection of resin into a mold. Under standard conditions, barrel heats, and screw RPMs, its rating is based on the number of pounds of polystyrene resin it is capable of melting per hour.

Based on the resin to be processed, screw design and speed, taking into consideration that the screw turns about $\frac{1}{4}$ to $\frac{1}{3}$ of the time, processing conditions must be adjusted to insure a good melt supply. This is accomplished by adjusting the barrel's temperature profile, selection of the screw's design, varying screw speed, and using back pressure on the screw to make it work harder. This creates more shear heat in the resin.

Based on mold shot weight and cycle time, the machine's melting capacity should be within 90 to 95 percent of its ability to meet the job's resin capacity requirements to supply a quality melt to manufacture quality parts.

Injection Rate and Pressure

The filling of the mold cavity is controlled by the machine's timers and hydraulic pressure settings, which are preselected to push the screw forward and inject resin into the mold. These pressure and time settings control the forward motion of the screw, and thus the time required to fill the cavity in seconds. This is commonly called the cycle's injection or fill rate. With the newer machines, the operator can select one or more high and low pressure settings to obtain a cavity injection profile that suits filling the mold's runner system and cavity. Injection pressure profiling can assist in controlling the cavity's filling rate and can be adjusted to compensate for the filling requirements of complex parts. Injection profiling can solve such molding problems as jetting and flashing, as well as other complicated fill problems required for a part, by ensuring that a uniform melt front fills the part under the necessary injection pressure.

The crystalline resins, and their many alloys, usually require fast fill rates because of their sharper melting and freeze-off temperatures and higher fluidity. With these resins, in 1 to 3 seconds you want to inject and fill the cavity to 90 to 95 percent of its capacity with first-stage injection pressure of 500 to 1800 PSI. The higher the injection pressure, the faster the screw will move to fill the cavity. The amorphous resins, with their higher melt viscosity, will typically have slower fill rates and pressures. This is especially true of the more shear sensitive resins.

Higher injection pressure creates excessive shear heat in the melt, particularly if the gates are too small for the material to easily fill the cavity. Jetting and other cavity fill problems may also occur. If too high an injection pressure is used with too small a gate opening, increases of 200 to 300° in material temperatures are not uncommon. These lead to material and possibly part degradation and failure. Opening the cavity gate, slowing the fill time, and reducing cavity injection pressure may be the solution.

The use of full injection pressure to fill a mold's cavities is seldom required or desired. A general rule used by most molders is that if more than 1200 PSI gauge pressure is required to fill a mold, the gate and/or runner size is too small.

Packing Pressure

Packing or second-stage injection pressure is required once 90 to 95 percent of the cavity is filled. It uses one half the injection pressure to avoid flashing the mold cavity and to finish packing out the cavity. At the same time, it maintains the packing pressure necessary to obtain the correct cavity part dimensions. Once the cavity is filled and packed out, the machine maintains this packing or holding pressure on the cavity. This holding pressure prevents the cavity from depressurizing (i.e., allowing melt to flow back out through the gate into the runner system) until the gate area solidifies or freezes off. After gate freeze-off, the holding pressure is released and the screw can start pumping to build up the next shot. With amorphous resins, the cavity hold pressure should be about 60 percent of injection pressure and held until gate freeze off occurs.

Back Pressure

Back pressure on an injection molding machine makes the screw pump harder to build up the melt in front of the barrel. It produces shear heat in the resin, which raises its temperature to promote more rapid melting of the pellets and produce a homogenous melt. It is used in situations where additional heat is required to melt the plastic, as when the screw's rearward travel and material plasticating time may not be sufficient to create the melt supply for the next shot. Back pressure is then applied with often-increased screw RPMs to ensure a good melt. It is also used when time is not a factor to reduce the formation of unmelt in the resin supply and to promote more complete mixing of colors in the barrel.

Back pressure should be used prudently as it creates a very rapid temperature increase in a resin. This can cause less thermally stable resin and color pigment systems to decompose faster, and it creates more glass fiber breakage in reinforced polymers.

Back pressure on the screw is created by restricting the oil flow from the hydraulic reciprocating screw injection chamber. This occurs as the screw pumps back, while building up melt for the next shot. Restricting the flow of oil through these valves determines the ease with which the screw can be pushed back by pumping melted resin forward. This creates pressure on the screw tip, thereby forcing the screw back and the oil out of the injection chamber. Back pressure is also used to aid in the removal of trapped air, gases, and moisture in the pellets by compacting them in the screw flights before melting. The use of back pressure can increase melting efficiency in a machine with a low L/D screw ratio, but it is not recommended. However, if it is necessary to obtain a good melt supply, the back pressure should be no higher than 25 to 50 PSI, depending on the resin and its additive system.

Time Variables and Controls

The molding machine operates on timer settings preselected for controlling the molding cycle. Selecting the right times for each function can mean the difference in obtaining the correct part shrinkage, maintaining part dimensional control, eliminating warpage and molded-in stresses, and proper filling of the part. In essence, it determines total part quality.

Injection Molding Cycle

The injection molding cycle is controlled by very precise timers. These time controllers are either of analog or digital design and are used to control the machine's operations precisely to within fractions of seconds. Most timers are digital, as they can be set more precisely. As an example, the control of time for one complete cycle of a molding operation is shown in Figure 9.3.

The three phases of cycle time control are injection time, cooling time, and ejection time. Because precise control of the molding cycle and part quality go hand in hand, select the most accurate timers for repeatable control of each cycle. Profitability is also closely tied to the molding cycle. Piece-part price is based on the overall cycle time required to make a part, and the molder wants to produce the parts in a minimum cycle that ensures part quality and a positive return for labor.

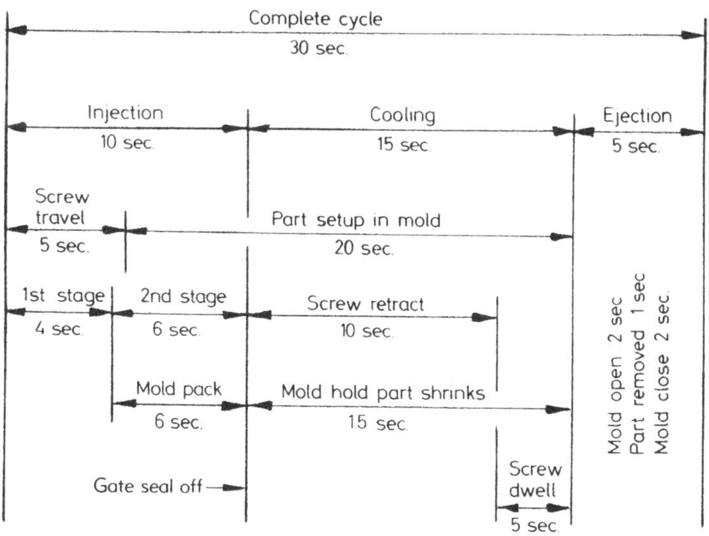

FIGURE 9.3. The injection molding cycle.

Injection Time. Injection or mold-fill time is broken down into filling the mold cavity and maintaining packing pressure until the gate freezes off. During this period, the timers control the first- and second-stage pressure settings while the mold is filled and packed. During the injection stage, timers control the high-volume hydraulic injection pump or electric motors (if an electric drive machine is used) to create the pressure that forces the screw forward to fill the mold cavity. At the start of the injection cycle, the first-stage timer will control the forward motion of the screw to almost fill the mold cavity. It switches to second-stage packing pressure just before the screw stroke reaches the preset resin cushion in the barrel. The resin cushion is necessary to ensure there is always a positive packing pressure on the resin in the mold cavity before gate freeze-off. This cushion or pad, as it is often called, is usually a half-inch minimum of screw travel. If the screw should bottom in the barrel, the question always exists—is the cavity fully packed out? Packing pressure, along with totally filling the mold cavity, is one of the controls for obtaining part dimensions and tolerances.

Packing Time. Once the second-stage packing pressure begins (typically 40 to 60 percent of injection pressure), the part's final dimensions are being defined. Packing pressure must keep the screw against the resin cushion and continue to force material into the mold cavity as the part shrinks and solidifies. Packing pressure is held on the part cavity until gate freeze-off. If the packing time is not long enough, the molten resin in the part's center will flow back through the gate into the runner. This causes part depressurization that results in varying dimensions and part quality problems. The packing time is determined by the resin, its melt temperature, gate size, and the mold's cooling system. The larger the gate size for an amorphous resin, the longer the packing time. Conversely, if a crystalline polymer is used, then the packing time may be less. This is because of the material's sharper melting point; less heat needs to be removed before it solidifies. Packing time is variable dependent.

Determining Packing Time. To obtain the required part dimensions, the molder should always select the "pack time," which is critical for yielding maximum cavity part weight. The easiest and accepted way to determine "pack time" is to check the weight of the molded part by using the established molding cycle. Select one cavity and carefully degate the part, recording its part weight and pack time. This should be done for three cycles to determine whether the part weight is remaining constant or varying. On the next cycle, increase mold pack time by 1 second and determine whether the part weight increase. Continue this procedure until the part weight stabilizes and does not increase with the next additional second of pack time. Use the stabilized pack time for the new cycle's "pack time." While this test is performed, the molding cycle must be in thermal and process control. The melt and mold temperatures must be at their operating temperature settings, with injection pressure and other cycle times at their optimum cycle settings. If changes are made to any process variable other than pack time, this procedure must be repeated. This procedure is for a balanced or single-cavity mold, where packing pressure is the same for all cavities.

It is more difficult to establish "pack time" for an unbalanced tool, as the part weight varies slightly from cavity to cavity. For an unbalanced tool, select the cavity farthest from the sprue, because it will probably be the last to fill and obtain gate freeze-off. Of course, this may not always be true; it depends on the routing of the cooling system. Conduct a similar test to that used with the balanced mold system to establish pack time. Verify the results by also testing a cavity midway between the sprue and test part. For the tests to be meaningful, the part's dimensions should be measured to be sure they are within specification. With maximum part weight established and dimensions within tolerance, this will be the minimum mold "pack time" for the cycle.

Hold Time. Once gate freeze-off occurs, the mold hold timer begins. This is the length of time required for the part to cool sufficiently to be ejected from the tool. The time chosen must be sufficient for the part not to distort, have knockout pin penetration into its thick sections, and be rigid enough for removal from the mold. Similar tests can be conducted to determine the minimum hold time, but part weight has no bearing here. This test must determine the part's ability to be ejected without damage.

Some molders may use mold hold time to maximize a part's shrinkage. By using mold temperatures that are above the part's anticipated service-use temperature, maximum part shrinkage results. In crystalline or semicrystalline parts this will reduce postmold shrinkage and relieve molded-in stresses. This is an expensive way to anneal parts. It increases cycle time, but it is most effective if the cycle time increase is not too great. If postannealing is required, consider holding fixtures in a hot air oven or heated oil to minimize air degradation of the surface. Generally, a hot tool with a slightly extended cycle time will do this job without postannealing. The amount of holding time for each specific part must be planned when determining the cycle times to achieve the required dimensions for the part.

During the mold hold cycle, allow sufficient time for the screw to retract and build up resin for the next shot. When running on very fast overall cycle times, this may require the use of back pressure to ensure a uniform melt supply.

THE INJECTION MOLDING MACHINE

The reciprocating screw injection molding machine is available in many ranges of size, ounces of material capacity, tons of clamping force, hydraulic or toggle clamp lockup, platen size, and tiebar spacings, as well as hydraulic or electrical drive systems. The main section of an injection molding machine is the barrel

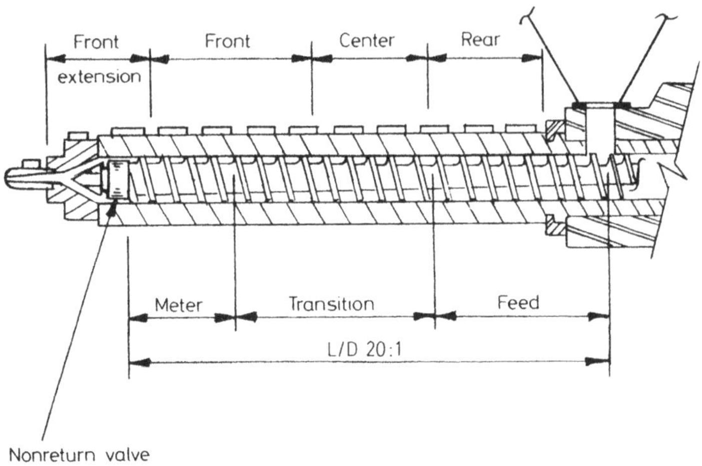

FIGURE 9.4. *Barrel and screw assembly.*

and reciprocating screw section. This is shown in Figure 9.4 with the hopper feed section, heater band sections on the barrel, nonreturn valve, front extension, nozzle, and screw.

The Barrel and Screw Assembly

The barrel and screw are a matching set. Special steels with hardened wear-resistant surfaces are preselected for each to ensure good wear and part life. They are designed with specific clearances and tolerances to ensure that the resin pellets will be conveyed along the screw and then compressed and melted. This is accomplished by screw shear heat and barrel heater bands.

A typical molding machine has three barrel heating sections—the rear, center, and front zones. Most machines also have a nozzle heater, with an optional adaptor or front extension zone. Each section or zone heater band has a separate temperature controller. Temperature settings for each zone are recommended by the resin suppliers. The temperature profile set on the barrel is normally a low-to-high profile, from the rear zone to the front zone. If a front extension or adaptor section is used, its temperature setting will match the front zone. The nozzle attached to the front of the barrel, or adaptor, is set at the same temperature or at 5 to 10° lower for crystalline polymers to prevent resin drool from the nozzle during mold open time. The correct nozzle selection will minimize this problem.

Screw Decompression. If unable to control nozzle drool with temperature or nozzle design, the molder can use the machine's decompression or suckback feature, which can be programmed to operate as the tool opens. Decompression is a slight backward movement of the screw—a few tenths of an inch—to pull the molten resin back into the nozzle; it is timed to occur when the mold opens. This method is effective in drool prevention, except that it draws air into the nozzle. This may cool the melt slightly and cause formation of a skin that must be captured by the cold slug wells in the sprue and runner system during the next cycle. If drool persists, then a reverse taper nozzle should be used. The nozzle heater band and its controller should also be checked to discover whether a problem exists with maintaining the correct temperature.

Hydraulic Oil Temperature Control. With hydraulic machines, temperature control of the oil should be maintained to within ±5° of its recommended operating temperature. This will ensure consistent operation of the machine's hydraulic and pressure transfer systems. As oil temperature increases, viscosity is lowered. This results in less efficient pumping, pressure variances, and increases in cycle times. This affects both the clamp and injection phases of the system and can lead to variances in injection and mold clamp pressure and their operating speeds. Therefore maintaining the correct hydraulic oil temperature is important to uniform molding cycles and process quality control.

The Reciprocating Screw

The reciprocating screw and its design are the main component of an injection molding machine for the production of a quality melt. It is aided by the heat input from the barrel heater bands, but without a properly designed screw, a good quality melt of resin will be difficult to produce consistently.

With the many different generic resins available and their intermingled alloys, it is often difficult to decide what type of screw should be used for each resin. Unless specified, most injection molding machines will come with a general purpose (GP) screw, as illustrated in Figure 9.5. The GP screw has a gradually shallowing flight depth, as it approaches its front end. The difference in flight depth at the rear, as compared to the front, is the screw's compression ratio.

The action of melting the resin with a GP screw is similar in all screw designs, but may not be as efficient for all types of resins. The screw's length is divided into three main sections: feed, transition, and metering. Within these

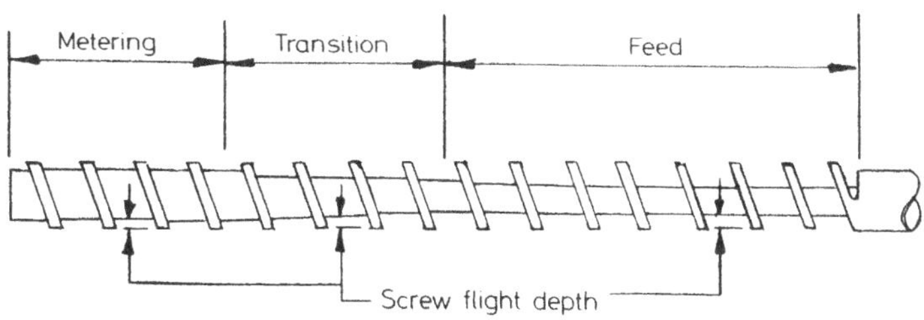

FIGURE 9.5. *General-purpose screw configuration.*

sections, the resin pellets are conveyed and melted through barrel and screw shear heat.

Resin Size and Regrind. Resins are produced in the shape of pellets. Most are cylindrical, measuring about 0.05 to 0.10 inches in diameter by 0.125 inches in length. Others are spherical; their shape and size depends on how they were manufactured. Others are extruded as long strands and die cut into pellets. They may also be pelletized underwater, as hot melt is pushed through a die face and when cut form a spherical pellet.

Besides using virgin resins, regrind from sprues and runners is fed back into the system at varying percentages based on the quality level required by the part's end-use function. The regrind particle size must be controlled, so as to avoid feed or processing problems. The particle size is determined by the granulator system that is used to chop up the sprues and runners, as well as by the size of the holes in the screen of the granulator. The regrind particles should match the size of the resin feed supply as closely as possible to obtain uniform feed by the machine's screw.

All fines, dust, and very small particles should be discarded; they can cause feed and processing problems during the molding process. How regrind should be handled, dried, stored, blended, and fed back into the system will be discussed in later chapters.

Poor regrind handling has caused more processing and part quality problems than material saving benefits. This is why regrind must be properly recycled back into the system. The resin feed supply—virgin pellets or mixtures of virgin and regrind—must be uniform to continuously produce quality parts.

Melt Generation. The melting process begins as the resin pellets are gravity fed from the machine's feed hopper into the screw's feed section (see Figure 9.6). The screw's feed section fills with pellets and, on turning, conveys them

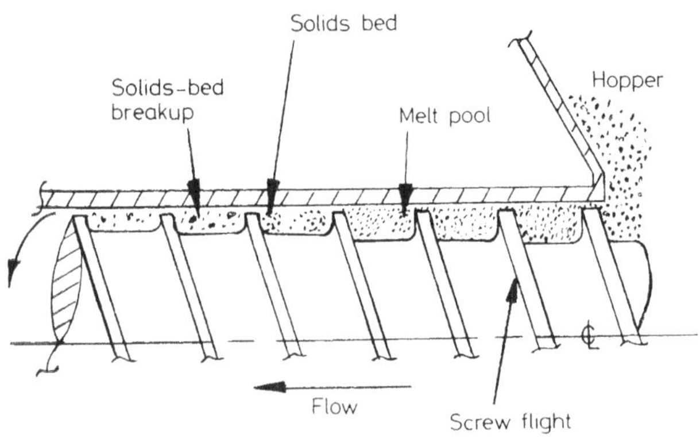

FIGURE 9.6. Standard screw melt model. (Adapted from Ref. [2].)

forward by compression that results from the screw's shallow flight depths. The barrel's heat soaks into the pellets, which begin to melt and adhere to the barrel's walls. This adherence is the main melting mechanism—shear heating of pellets at the screw flight/barrel interface.

As the screw turns, the pellets are conveyed along the screw's decreasing flight depth. The pellets are compressed and forced up against the heated barrel surface, to which they adhere. The shearing action of their conveyance along the barrel's surface produces a film of melt that is shaved off by the rear screw flight. When the pellets reach the transition zone, they become a mixture of melted and unmelted pellets. This semifluid mixture is compressed and melted even more. By the time it reaches the metering section, it is a homogenous melt. Amorphous resins will soften and melt very easily with a GP screw, which permits a lower L/D compression ratio. They process well on a machine with an L/D ratio of 16:1 or less and a compression ratio of 2:1 to 2.5:1.

Crystalline polymers have sharp melting points similar to ice—a solid at 32°F and a liquid at 33°F. Crystalline polymers begin to melt when the outer surface of the pellet is raised above its melting point. An increase in temperature is required to continue the melting process.

Crystalline resins require higher barrel heat inputs to continue melting as they are conveyed forward by the screw. They also require higher screw L/D compression ratios to provide more time for heat to soak into the resin to ensure complete melting. Compression ratios of 3:1 to 5:1, with L/D ratios of 16:1 to 25:1, are recommended to produce a quality melt.

The temperature on the barrel's heaters must be high enough to transfer sufficient heat to the pellets in the feed section of the barrel. The screw must not be allowed to stall when conveying the pellets and initial melt into the transition section.

Many molders use GP screws for all polymers, except those with very high melting points. GP screws can process crystalline resins, if the screw has an L/D ratio of 16:1 or greater. But when using a GP screw with crystalline polymers, its melting and shot capacity may be limited by up to 25 percent of its rated capacity because of inefficiency in the screw's melt capacity.

If the screw's L/D compression ratio is too small, some resins may not totally melt and unmelt may occur. Depending on the degree of unmelt, difficulties such as gate blockage, dimensional problems, and surface and part defects may occur. The only cure is to try raising the barrel's temperature, increasing screw RPMs, or adding backpressure. If not successful, use the right screw design for the resin. The solution may also be to use a machine with a more efficient melt capacity that does not overwork the polymer and cause it to degrade.

Some resins, like the crystalline polymers and newer alloys, require a special screw design as shown in Figure 9.7 for improved melt generation. The design has definite screw-section transitions, where the screw's flight depths are suddenly decreased. This creates greater pellet/melt compression and more shear

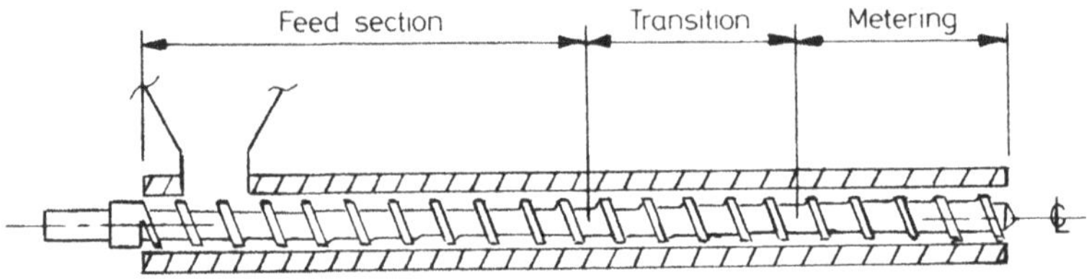

FIGURE 9.7. Nylon screw.

melting at the flight/barrel-wall interface. It is often called a nylon screw, as it was first designed to produce a uniform melt for this polymer.

Other screw designs, shown in Figure 9.8 are used for the more difficult-to-melt resins. Some are designed for special conditions, such as color mixing, additional additive disbursing, and more heat-sensitive materials. The barrier screw design, which uses melt-pool channels separated by solid-bed channels that grow progressively smaller as they proceed to the end of the screw from the feed section, is one such design. The barrier screw captures the solids until fully melted by allowing the melt to flow out into unrestricted channels (see Figure 9.9). Material suppliers can determine which type of screw is best suited for processing their resins.

Machine Melt Output. When selecting the molding machine, examine the machine's output of melt-per-hour capability, as compared with its shot-weight capacity. The molding machine's best melt-generating performance range will be less than 50 percent of its shot capacity. If this range is exceeded, the machine's melt production performance is lowered. This is because of the machine's reduced screw shear heat input, which is produced only when the screw is retracting, and there is insufficient barrel heat soak time for the pellets to continue melting efficiently. Therefore, screw retraction time should be a minimum of 75 to 85 percent of the mold cool time to ensure that enough shear heat is generated to produce a uniform melt. The barrel's heating profile must also be adjusted to have sufficient time for heat to soak into the resin.

Most machinery manufacturers rate their screws according to Society of the Plastics Industry, Inc. SPI's Screw Plasticating Code. The code lists screw recovery time in ounces per second, using polystyrene as the base resin, and the number of seconds required in the molding cycle for screw recovery. For example, if the shot size is six ounces and the machine's screw recovery rate is one ounce per second, the minimum screw recovery time will be 6 seconds for polystyrene.

Screw rotation speed is also very important. The ratio of setting screw retraction RPMs assumes that the screw's retraction requires 75 to 85 percent of the mold hold time to build up the melt for the next shot. Care must be exercised to adjust the screw's RPM to the barrel's temperature heat setting to ensure that the excessive shear heat generated by high screw RPMs will not degrade the polymer or induce a heat spike in the material. This heat spike

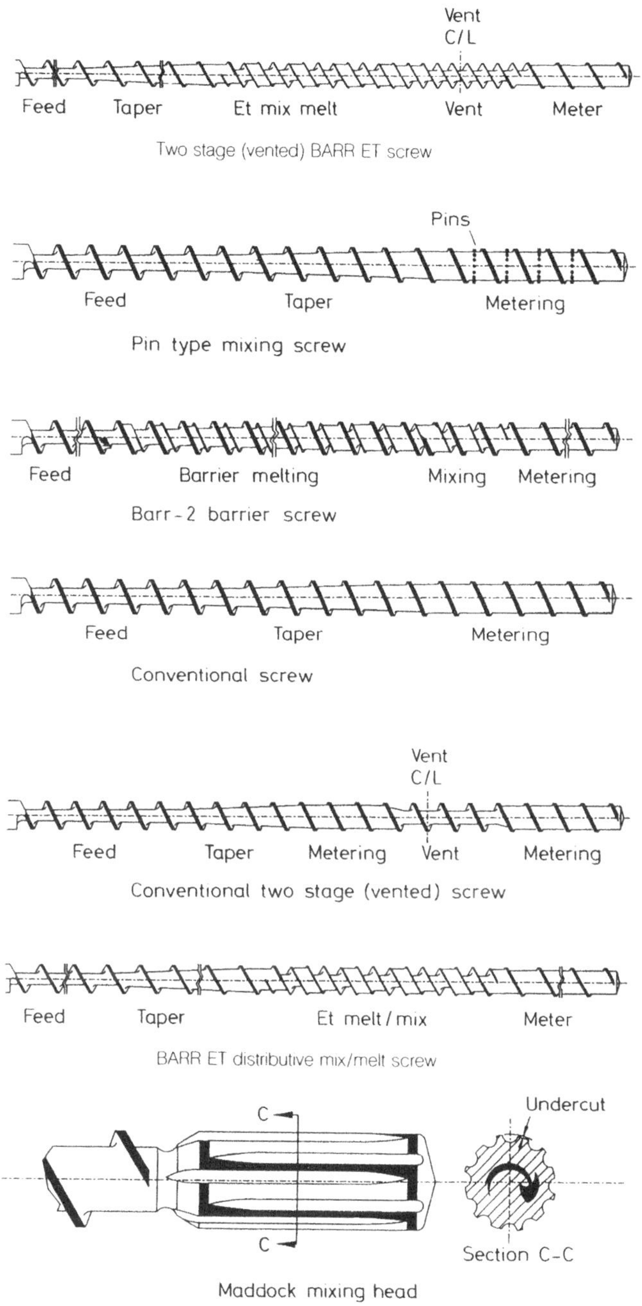

FIGURE 9.8. *Screw designs. (Courtesy of Robert BARR, Inc., Virginia Beach, VA.)*

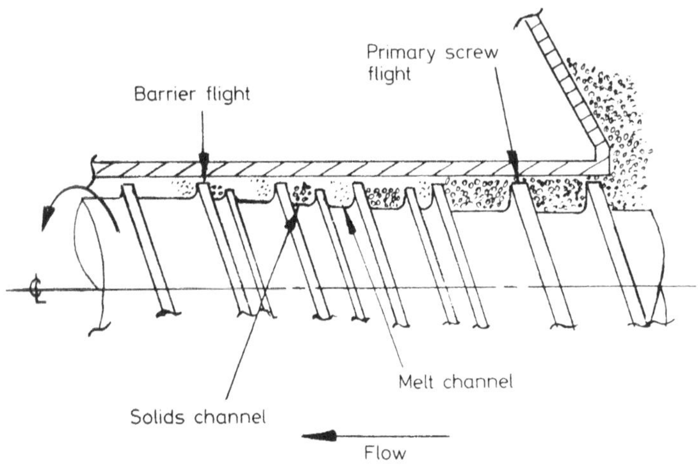

FIGURE 9.9. Barrier screw melt model. (Adapted from Ref. [2].)

can raise the melt temperature so high that it causes viscosity problems in the resin when filling the mold.

The length of the feed, transition, and metering sections will vary according to screw design and barrel length. If the correct L/D and compression ratio for the polymer are selected, the screw section ratios can vary as noted and still produce a uniform melt.

Screw Section Lengths (in Percent).

Feed	Transition	Meter
50	25	25
40	30	30
30	35	35
30	30	40

Each molding machine's barrel and screw combination, or melt generation capability, is rated in pounds per hour. This figure is based on the screw running continuously, which never occurs in injection molding. Therefore, a machine with a 200-pound-per-hour melt output should yield 100 pounds of melt product per hour based on a 50 percent screw turning utilization factor based on cycling time. When a machine has cycle problems and the screw cannot produce the quantity or quality of melt required, the problem is the screw's lack of performance. Therefore, to produce a homogeneous temperature-controlled melt, the correct size and type of screw is critical.

Some molders prefer a large diameter screw machine for processing heat-sensitive resins. The larger screw, with its higher L/D ratio and greater capacity for creating melt in the barrel, can use lower screw RPMs and barrel temperatures to more gradually bring the resin up to molding temperature, thereby reducing thermal stress on the resin. This technique can also be used for heat-sensitive pigment systems, where the longer barrel length will aid in promoting better mixing and more uniform colors. But residence time at temperature must be considered to avoid degrading the pigment system and causing part color problems.

Nonreturn Valves

Nonreturn valves control the flow of a resin at the end of the screw. They are used to contain the melt in the barrel and to force it into the mold during injection. There are three basic types of nonreturn valves or check rings, as shown in Figure 9.10. These are the sliding ring, ball check, and smear-type valves.

The check valve serves two functions. When the screw is turning and pumping resin forward to build up the melt supply for the next shot, the check valve is pushed open allowing the melt to flow around it and the screw tip. At the start of molding, with the barrel full of material, the adaptor and nozzle section from a material seal at the head of the machine. The screw can now retract to build up the next cycle. This restriction in the nozzle is sufficient to overcome the melt-head pressure created by resin flowing through the check ring by the screw's pumping action to push the screw back. The check valve is also used to create a seal at the end of the screw that restricts resin from flowing back down the screw flights during the injection stroke.

The nonreturn sliding-ring valve is used for amorphous and crystalline resins, as is the ball-check valve and smear-tip. The latter is preferred for the more shear sensitive resins, such as PVC. The smear-tip design produces less resin shear heat as the polymer flows past the screw tip.

Sliding Check Ring. The sliding check ring, shown in Figure 9.10A, is made of a wear-resistant steel of nominal Rockwell RC-44 hardness. The screw tip and check-ring seal are of RC-52 or higher. The screw tip is made of harder steel so that as wear occurs, the less expensive ring will wear first and be replaced. This is preferable to replacing the screw tip, with its check-ring seal.

There should be a clearance of 0.001 to 0.002 inches between the outer diameter (OD) of the check ring and the inner diameter (ID) of the barrel, which is the typical manufacturing clearance. As the check ring wears, this gap becomes larger and will eventually allow the material to backflow over the screw during the screw-injection stroke. This can be observed at the screw coupling to the gear box, as the screw will rotate counterclockwise because of pressure on the screw flights from the backflow of resin over the worn check ring. This can also be noticed if the packing cushion or pad cannot be held during packing time. Check-ring life can vary from 6 months to 1 year or longer, depending on the corrosive action and wear of the resins it processes. Glass-and-mineral reinforced or highly filled resin will create much faster ring wear.

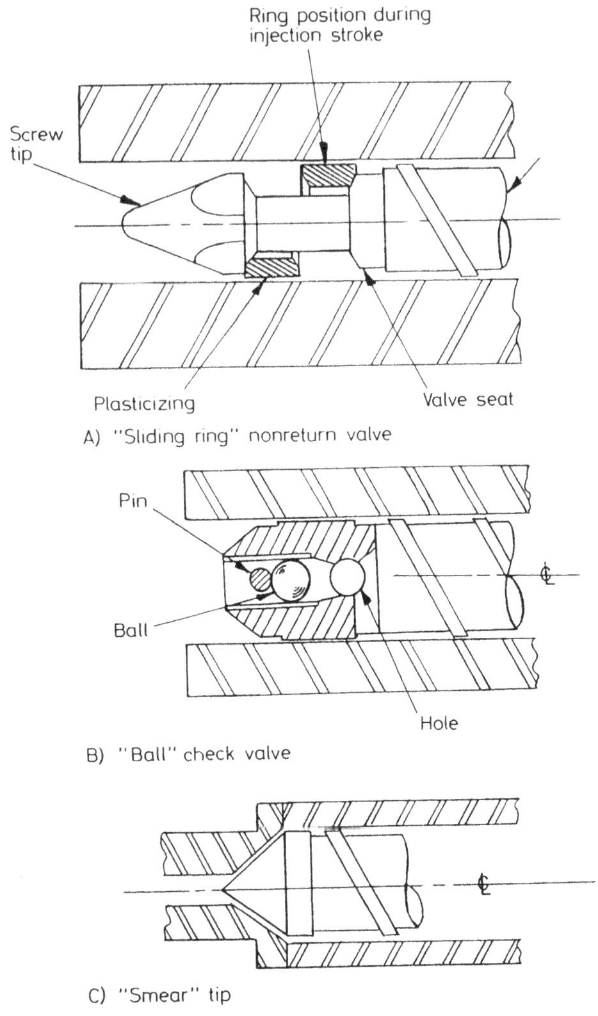

FIGURE 9.10. *Screw shutoff valves.*

Ball Check. The ball-check valve is used to a lesser degree because when it wears out, the entire screw tip must be relaced. The major problem is that the ball may develop a flat spot on one or more surfaces and provide only intermittent material sealoff as it rotates. This makes it much harder to know when replacement is required and can therefore affect part quality. Production personnel must be aware of this potential problem and must constantly be monitoring part weight, injection, and packing pressures. You should always know the type and condition of the nonreturn valve on the screw.

For the check-ring and ball-check valves to function correctly they must meet the following requirements:

1. No hold-up spots
2. No flow restrictions
3. Good seal
4. Monitoring of wear

Smear Tip. The smear-tip design is used for heat-sensitive resins and relies on the material in the nozzle's sprue tip to create a seal. This occurs as it retracts during melt buildup, while pumping melt into the adaptor zone. Let us use PVC as an example. You do not want any holdup spots in this area to trap the polymer and cause degradation. When PVC degrades, it creates hydrochloric acid, resulting in chlorine gas formation and corrosion in the adaptor area. Both are dangerous and must be avoided. If a prolonged hold-up time—5 to 10 minutes—is encountered when molding PVC or other heat-sensitive resins, the molding machine's nozzle should be retracted from the sprue bushing and low RPMs put on the screw to slowly purge the barrel and bring fresh melt into the system to avoid degradation.

The typical screw-tip adaptor clearance for the smear tip is 0.015 inches. The adaptor's temperature must be controlled when processing PVC to within 10 to 15°F of the recommended processing temperature to have good flow and yet avoid degradation of the polymer.

Barrel Adaptor

The adaptor, as shown in Figure 9.11, is designed to attach the nozzle to the end of the barrel. It must fit exactly to avoid holdup areas at mating surfaces (A and B). The mating surfaces must have short cylindrical sections to maintain accurate diameter matching if it becomes necessary to machine or reface the mating surfaces. The seal surfaces (C) should be narrow enough to develop a good seal, yet wide enough to not peen over when torqued tight to the barrel.

The adaptor is the mechanical reducing mechanism for the barrel to the nozzle. It also provides a thermal isolator from the nozzle to the barrel for better nozzle temperature control. In most cases, heater bands are used on the adaptor to assist in melt temperature control.

Screw Tip

The screw tip shown in Figure 9.11 performs other functions. It is the support for and the sealoff point of the check ring and provides a streamlined flow path for resin as well as the pressure head used by the melt to push the screw back in conjunction with the check ring. The screw-tip section (D) is slotted or fluted to allow unimpeded resin flow past the tip with the check ring in the forward sealoff position. The space (E) must be of sufficient depth to allow easy melt flow access to the screw-tip slots. The screw tip must mate perfectly with the end of the screw (F) and the check-ring seal (G) to avoid any resin

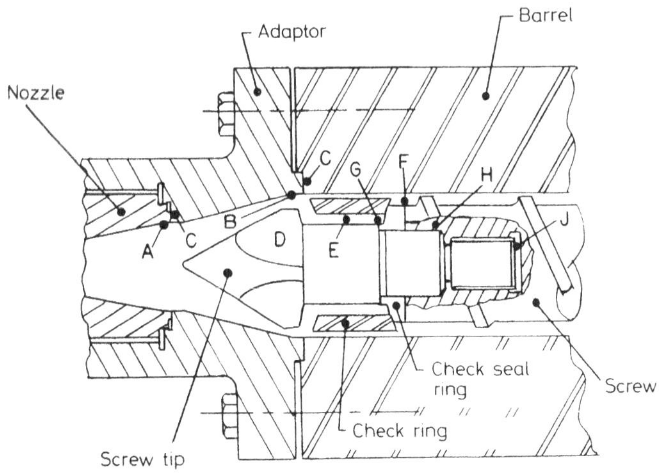

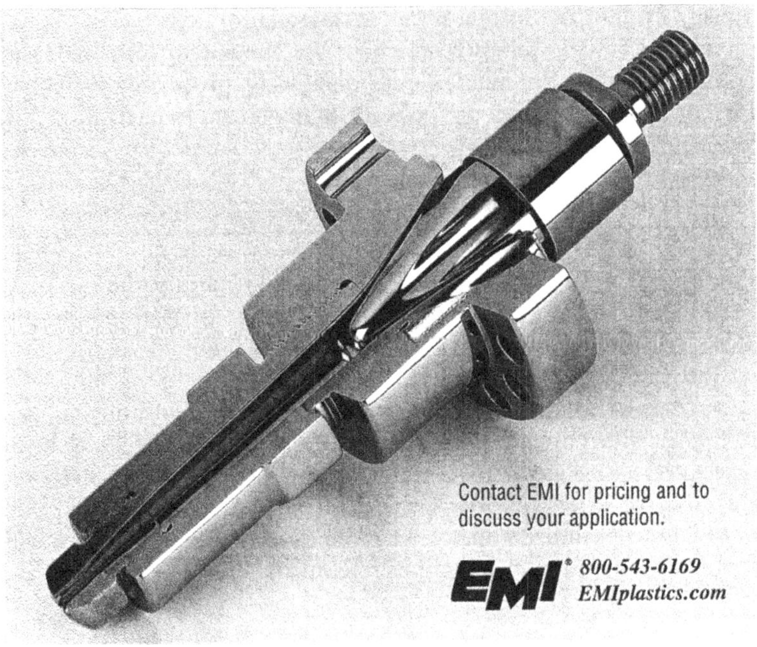

FIGURE 9.11. *Check-ring screw-tip adaptor assembly. (Adapted from Ref. [3].)*

holdup areas. The screw tip also has a matching section (H) ahead of the screw tip threads to match the screw's counterbore to align and support the screw tip and seal ring. The screw tip threaded counterbore (J) should be minimized to avoid the possibility of molten resin being trapped and degrading. This is done to avoid any possibility of degraded polymer outgasing and violently expelling the screw tip during equipment disassembly.

Nozzles

The machine's nozzle provides the path for the melt to the mold through the sprue bushing. It creates a pressure seal at this point. There are three basic types of nozzles (see Figure 9.12)—the straight bore or general purpose, the shut-off, and reverse taper—plus the positive shut-off type. The interior taper of the nozzle can be smooth or transitioned. Some material suppliers recommend different nozzle designs to suit their polymers. Because nozzles can be

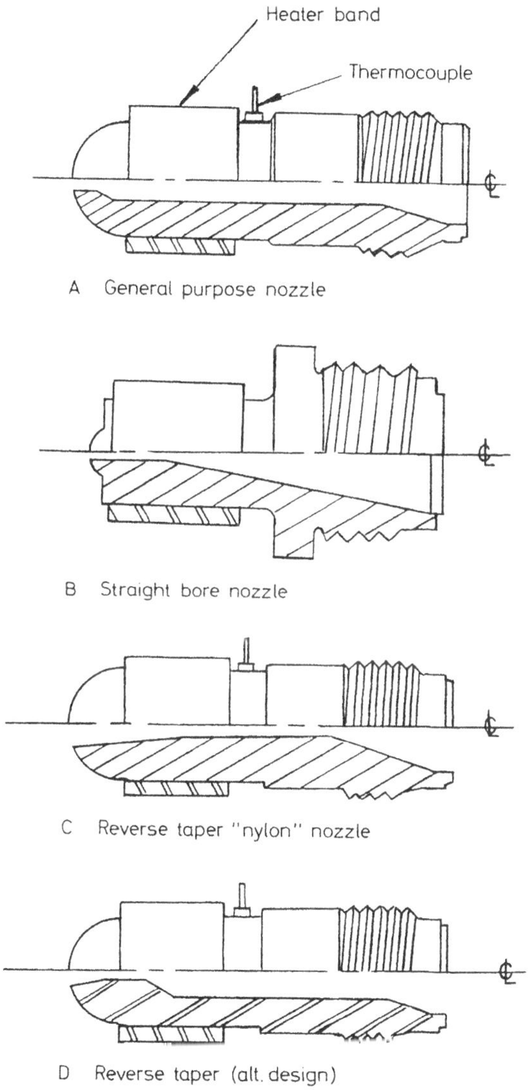

FIGURE 9.12. *Nozzle designs.*

easily interchanged, the correct nozzle design for each resin type should be used.

All nozzles require heater bands for temperature control and must have a thermocouple either centrally located for the straight and GP nozzles or over the straight section of the reverse taper nozzle. Thermocouples can be located through or next to heater bands depending on nozzle length and band size but must never contact the band if erroneous temperature readings are to be avoided.

The heater bands must extend as close as possible to the rounded nozzle end and cover as much of the nozzle as possible. This is required to counteract any heat loss to the mold's sprue bushing. If this becomes a problem, especially when running very cold molds, a thin insulated material may be placed between the nozzle and sprue bushing to limit heat transfer to the mold.

Straight-Bore and General-Purpose Nozzles. The straight-bore and general-purpose nozzles (see Figure 9.12A and B) are used for amorphous polymers, which have higher viscosities. Good nozzle temperature control can limit drool. The GP type has a slightly deeper inner channel that acts as a sloping dam to restrict resin flow through the straight diameter land opening. The length of the land is usually five to ten times the diameter of the nozzle opening, but it can vary for different polymers. These nozzles can be used with crystalline resins but require tighter temperature control because of the more fluid viscosity of these resins and their greater tendency for drool.

Reverse-Taper Nozzle. The reverse-taper nozzle (Figure 9.12C and D) was developed for crystalline polymers, with their more fluid melt properties. Drool with the crystalline resins was a problem with straight-bore nozzles. The reverse-taper nozzle was developed to pull a longer sprue length when the mold opened. This controllable sprue length within the nozzle resulted in a longer flow path for the resin before drooling occurred at the tip. Combining tighter temperature control with the use of reverse-taper nozzles has solved the drool problem with crystalline resins.

Shutoff Nozzles. Shutoff nozzles are used when screw retraction is still required after the mold has opened or with vertical cylinder molding machines. They provide a positive shutoff and seal, but are susceptible to wear and must be closely monitored. Resin erosion of the mating surfaces not only causes seal problems, but can create holdup spots that result in polymer burning and rejection of parts. Each case must be evaluated before using this type of nozzle. These nozzles are operated by compressed air, electromechanical mechanisms, or spring-loaded pressure valves, with the shutoff nozzle shown in Figure 9.13 operated by air.

The electromechanical system is timed to the cycle functions, opening during injection and closing when gate freeze-off occurs. The pressure-sensitive system uses springs that, under injection pressure, are compressed

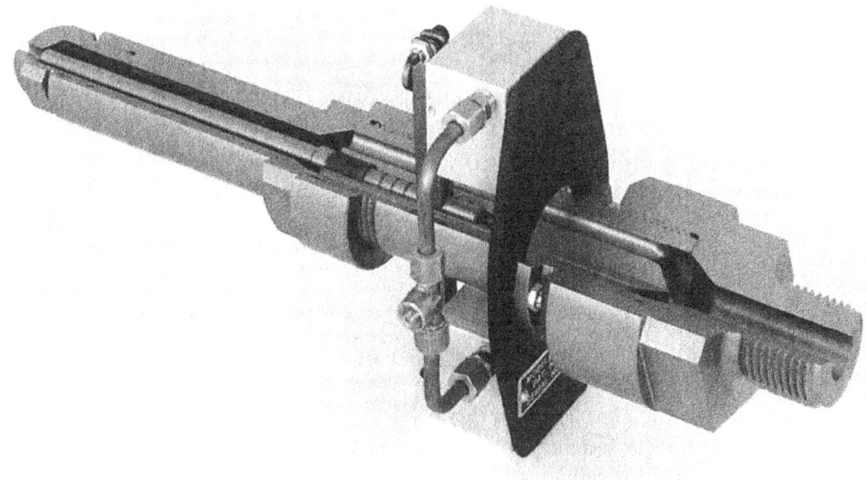

FIGURE 9.13. Air operated shutoff nozzle. (Courtesy of EMI Corp.)

by the valve's movement. When the cycle times the screw back, it closes and pressure is reduced. Because of thermal aging and the number of cycles, the springs may break or relax their tension with time. This results in poor performance.

SELECTING BARREL HEATER CONDITIONS

The barrel heater bands are the primary external source of heat for melting the resin to create a homogenous melt. Each resin requires a preselected range of temperature settings for the barrel's three-heated zones to produce a uniform resin melt temperature. Besides barrel heating, other factors must be taken into account. These factors have a bearing on the temperature profile set on the barrel's heater bands. The other factors are the size of the machine, screw L/D ratio, type and compression ratio, and the machine's rated pounds-of-melt capacity per hour. Additional considerations are the size of the shot to fill the mold, barrel melting capacity in ounces of resin, screw travel and time to build up the next charge of melt, and polymer residence time in the barrel.

As an example, use a GP screw. Calculations have shown that the volume of resin in a 20:1 L/D screw in pellets and melt is 3.5 times its rated shot size for the screw. Therefore, a 12-ounce machine screw would hold 42 ounces of material if all screw flights are full. If a 2-ounce shot size is used in a 15-second cycle, this means 21 shots are in the barrel. The calculated barrel residence time for the polymer is 5.25 minutes. This is the minimum time the polymer experiences elevated temperatures, not counting the drying time.

$$\text{Residence time} = \frac{\text{Ounces in barrel}}{\text{Shot size ounces}} \times \frac{\text{Cycle time}}{60}$$

If the shot size is doubled to 4 ounces, then 10.5 shots are in the barrel. However, only 2.625 minutes are available to reach the desired melt temperature. If a 6-ounce shot is used, only 1.75 minutes are available.

The temperature profile will have to be adjusted for each specific resin to obtain the correct melt temperature. The best time to determine this is before seating the nozzle to the mold. Apply steady-state heating on the barrel's zones. Then begin cycling and ejecting the melt under the chosen cycle's molding conditions. Be sure to "reduce" the injection pressure to avoid injury. Do this for a minimum of three cycles and then take the melt temperature of the resin with a pyrometer. If the resin's melt temperature is within ±5 degrees of the recommended melt temperature, bring the nozzle forward, adjust injection pressure, and begin molding. If it is not within those parameters, adjust barrel heats, back pressure, and screw RPMs to bring the melt temperature within the desired range.

Heater band temperatures should also be monitored during startup and molding to be sure that their controllers are regulating the electrical input properly. This can be done with a pyrometer, which must be in calibration. By using the pyrometer, you can also tell when heater bands are going bad or the controllers are out of calibration. It can also be used to verify whether the barrel's thermocouples are operating correctly.

PYROMETER

A pyrometer is a temperature measuring device with two interchangeable temperature tips. The one for melt temperature is needlelike; the other is for flat surfaces. It can be an analog or digital type, but analog responds much faster. The digital pyrometer can, however, be forced to respond faster by preheating the temperature probe to the anticipated temperature to be measured. This will save time by not allowing the melt to cool off while rising to the temperature reading. Molders prefer analog style as it records temperature changes faster, but it is not nearly as precise for recording the temperature. The analog needle swings to a point on a scale, whereas the digital has a numerical output.

THERMOCOUPLES

You should also know the depth of the thermocouples in the barrel for each machine. Thermocouple depth determines the barrel's temperature exactly at that point and not the actual temperature experienced by the resin at the inner barrel surface. The deeper the sensor, the more accurate the inner barrel temperature will be. The sensor's depth has a definite connection to the selection of barrel temperature zone settings, particularly if the tool is set up on another machine with different thermocouple depth recordings. The goal is a steady-state barrel temperature profile based on the resin processed and indicated by the material's melt temperature.

But factors change with time, season, temperature in the molding plant, air flow around the machine, and power fluctuations in voltage feed to the machine. To maintain melt quality, the barrel temperature profile must be tightly controlled and monitored. Some shops wrap insulating pads or binders around the barrel to isolate external effects and improve heat-band efficiency.

MOLD FIT AND SUPPORT

When selecting the molding machine, the mold must fit between the machine's tie bars and be fully supported by the machine's platens. Any portion of a mold extending beyond the platens during clamp up will not experience maximum and mold-clamp force. As a result, the mold can become warped, flash, and result in varying dimensions and subquality parts. Extreme care must be exercised when mounting and dismounting the mold. It is very heavy, and if not properly clamped in place, it could fall and damage the tool or cause personal harm. The mold's sprue bushing must match the opening on the fixed platen to obtain correct alignment of the tool to the machine's platens.

MACHINE AND MOLD CLAMPING SYSTEMS

The hydraulic clamp and toggle clamp are the two basic machine-mold operating systems used to open and close the mold and to apply pressure to keep the mold closed during molding.

Hydraulic Clamp

The hydraulic-clamp system used a large hydraulic cylinder and piston to open and close the mold, as shown in Figure 9.14. It also provides the clamp tonnage required to hold the mold closed against injection and packing pressure. By transferring oil from the reservoir to the hydraulic cylinders, hydraulic pumps provide the pressure necessary for operation.

To close the mold, the hydraulic pumps use low pressure to the hydraulic cylinder. The cylinder operates the piston to provide a smooth start that overcomes the weight of the machine's platens and mold. It also overcomes friction from the bearings on the machine's tie bars. Reduced hydraulic pressure is used to avoid jerking and violent starts. The slow start can be very brief and is controlled by a limit switch or potentiometer. Once motion is obtained,

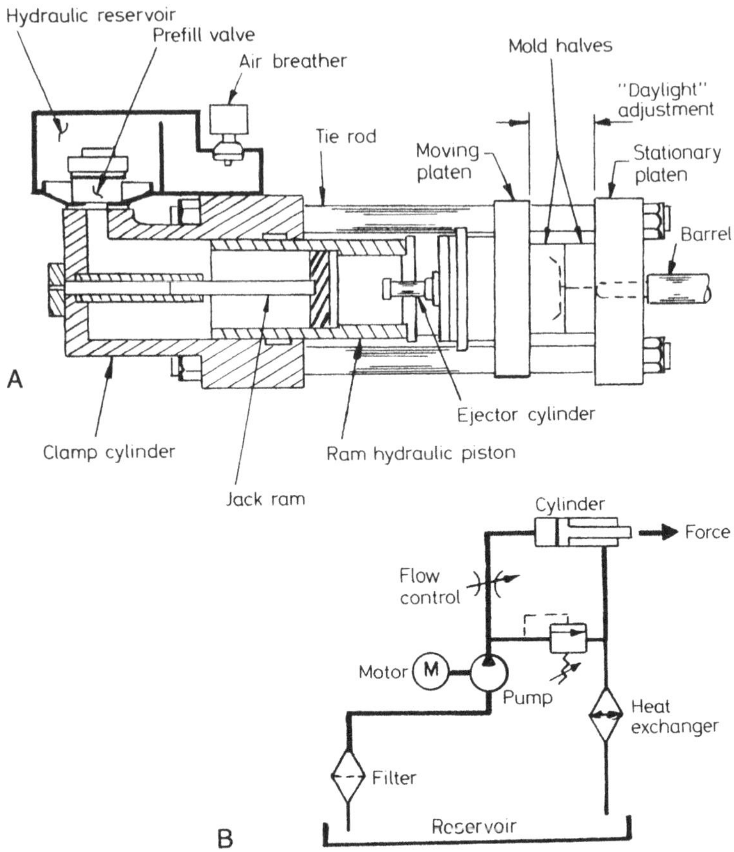

FIGURE 9.14. *Hydraulic machine clamping system. (Adapted from Ref. [5].)*

hydraulic pressure is cut back as mold closing may only require 500 PSI hydraulic pressure. The mold then continues to close at its normal speed until its faces almost come together at about ¼ to ½ inch. Limit switches or potentiometers then reduce cylinder pressure to its minimum as the mold faces come together.

Most machines also have an additional safeguard to prevent mold damage from parts not fully ejected from the mold. This low-pressure closing uses just enough pressure to bring the mold faces together. If an obstruction exists, then a pressure sensor will trigger and stop the cylinder from closing the mold and damaging the tool. If no obstruction is sensed, then the low pressure slows the mold closing enough for a smooth touch of surfaces to occur. The hydraulic control system then switches to high pressure to develop the full mold-clamp pressure required to keep the tool closed. On a hydraulic machine, the mold-clamp pressure can be adjusted as necessary. The hydraulic press is even more adjustable. As more oil is pumped into the hydraulic cylinder, the piston pressure is raised up to the machine's rated clamp tonnage capacity. If full machine rated clamp pressure was used on a small mold, it could crush the tool.

This slow start and slow down at the mold's closing is necessary to avoid a rapid starting jolt to the machine. It also slows down the velocity of the closing tool, so that it does not smash against itself or a stuck part during closing. With the correct pressure settings, limit switches, and potentiometers, the closing-and-opening cycle will be smooth and within standard time limits. Trying to speed up the cycle with high pressures and limit switches will not save enough time to warrant the damage that can occur. It is not practical to gain cycle time at the expense of mold damage.

The mold opening and unlocking action with the hydraulic clamp is controlled by low pump pressure, which develops a smooth opening without shocking the system. After overcoming the mold-opening force, interlocks are tripped or potentiometer settings used to speed up the pumps to open the mold faster. The mold-opening force is based on parting-line geometric resistance (depth and draft), material shrinkage, and part-packing pressure. These factors could require a force equal to 10 to 20 percent of the nominal clamp force to unlock the mold. Once the mold is moving, similar controls are used to slow and then stop the mold's backward motion to eliminate bumping the mold and the moving platens on their stops.

Once the mold has opened, the jack ram (see Figure 9.14) moves forward to operate the mold's ejector plate on which the knockout pins are mounted, thereby stripping the part from the cavity. When the mold closes, the jack ram retracts to its preset position.

Cycle time can be gained, but it should not be done at the expense of the machine or mold. By adjusting the mold's opening stroke to the minimum dimension required for part removal—known as the daylight dimension—valuable cycle time can be saved, as compared with using the entire opening-stroke distance.

Three plate molds may require an additional limit switch to slow down mold closing so that the center plate is smoothly brought into contact with the movable plate.

Toggle Clamp

The toggle-clamp system uses a smaller hydraulic cylinder to open and close the mold, as shown in Figure 9.15. Based on the toggle-clamp system's geometry at mold lockup, a series of mechanical toggle arms are moved off center to lock the mold. The off-center action develops a mechanical advantage of between 25:1 and 50:1 for each 100 PSI of hydraulic pressure applied. This equals 2500 to 5000 pounds of clamping force. The use of toggle clamps as compared with hydraulic clamping systems is the same up to 500 tons. But after 500 tons, the hydraulic system is more efficient and versatile in setup, adjustment, and productivity.

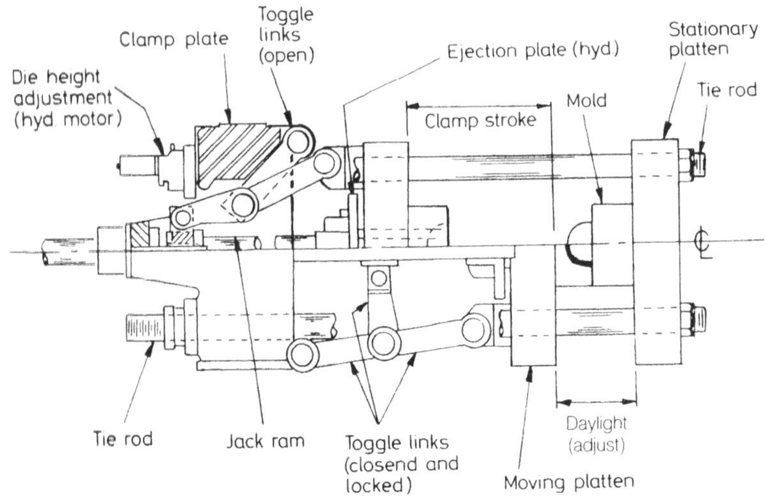

FIGURE 9.15. *Toggle machine clamping system. (Adapted from Ref. [5].)*

Setting the toggle clamp for mold lockup is more complicated than setting the hydraulic system, which relies on cylinder pressure. Care must be exercised in developing the required mold-clamp pressure to hold the mold closed, but not so high as to crush and peen over the mold's mating surfaces.

Depending on the age of the machine, the mold height (daylight) adjustment or the length of the closed mold between the platens, must be set manually or controlled by a hydraulic motor. To allow for variable mold heights, the tie-rod end-nuts, which are usually located on the mechanical end of the machine, are adjustable. They have sufficient thread length to cover the advertised range of minimum-to-maximum mold-closed height that the machine was designed to cover.

The mold-height adjustment must be set carefully to avoid crushing the mold and causing excessive strain in the tie bars and platens. But it must also be tight enough to develop sufficient mold-clamping force to prevent the cavity from flashing and losing dimensional control of parts. When properly set, the toggle machine will always develop its full mold-clamping capability because of the mechanical linkage setup.

Mold-height adjustment is set to lock up the mold and put stress on the tie bars but not use all the available clamp force, which will, with time, cause machine or mold damage. This is best explained by looking at Figure 9.16. The solid curve is the last one percent of the mold's closing clamp-force curve. This force is reached when the toggle action locks up after the mold halves touch.

To avoid overstressing the system, the correct mold-height adjustment is necessary. At mold kiss, the toggle action should occur and the clamp force, a linear straight line on the curve, should develop as all elements of machine and mold deflect to allow for toggle action and lock up. The slope of this solid

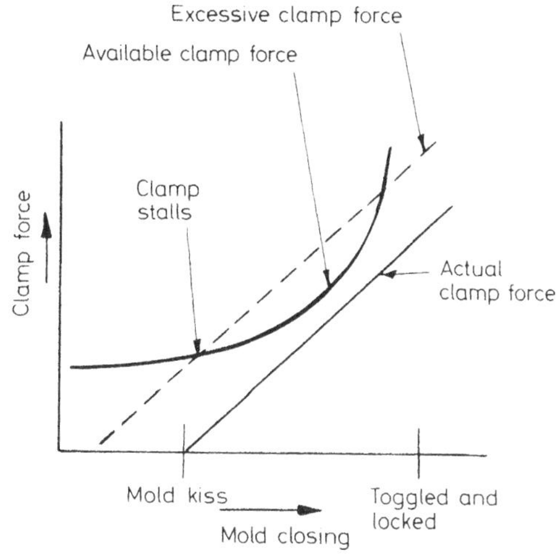

FIGURE 9.16. Excessive, available, and actual toggle clamp force. (Adapted from Ref. [5].)

line—the actual clamp force—is known as the spring rate. The distances from the kiss point to the righthand axis (toggled) represents the total deflection of all the clamp elements and mold.

The intercept of the actual clamp force with the right hand axis is the actual clamp farce when the clamp is toggled. The actual clamp force should always be kept below the available clamp force curve. Otherwise, the toggle will not function correctly. If the kiss point or mold height adjustment is moved to the left—the dashed line—or set too tight, the clamp-and-toggle action will stall as the available clamp-force curve falls below the actual clamp-force curve. If when setting the mold-height adjustment, the actual clamp force falls just below the available clamp-force maximum, the lock-up force will be developed without damage to mold or machine.

The mechanical advantage of the toggle is also used during the opening action. Only a few hundredths of an inch of movement is required to initiate mold opening. This then requires only one-to-two percent of the nominal clamp force to move the movable half of the mold back for the part ejector system to function.

The cost of a typical mold is about one third the cost of the injection molding machine. The machine and mold must be handled carefully by setting the control switches to maximize running conditions while protecting the machine and the mold.

VENTED-BARREL MACHINES

The vented-barrel (VB) injection molding machine offers specific advantages over conventional closed-barrel machines. Part quality improvements have been achieved when processing hygroscopic resins. These improvements range from better surface appearance to reduction or elimination of porosity and voids. The latter resulted from moisture and gases in the melt that were not released during initial pellet/melt compression by the screw.

Vented-barrel machines can be used on almost every resin where moisture or other nonsolid contaminants create quality problems. They also release residual monomer and other volatile materials from the melt that could, over time, build up on the mold face and block the vents.

The list of resins used in vented-barrel machines includes nylon [acrylonitrile butadiene styrene (ABS), polystyrene (PS)] cellulosics, acrylics [polycarbonate (PC), styrene acrylonitride (SAN), polyphylene sulfide (PPS), polypropylene (PP), polyethylene (PE), polyphylene oxide (PPO)], Acetal (PS), and high-flew PVC and PVC powders. The alloys of these polymers are also included. These resins are usually dry enough to process right out of their shipping packages. But old wet regrind or packaged material that has been opened for a long time should be dried to reduce the effects of moisture.

Two resins never run in VBs are polybutadine terephthalate (PBT) and polyethylene terephthalate (PET). These resins are so moisture sensitive they can pick up moisture at the vent. PBT and PET must always be dried prior to processing and always run in closed-barrel machines. Moisture levels above 0.02 percent reduce these polymer's molecular chain lengths by breaking the chemical bonds. The result is lower melt viscosity, which produces good-looking, but brittle, parts that have greatly reduced physical properties.

Vented-barrel machines are essentially two screws in a series, with special features to drive off volatiles from the melt before the material is injected into the mold cavity. VB machines have their screw divided into two stages, as shown in Figure 9.17.

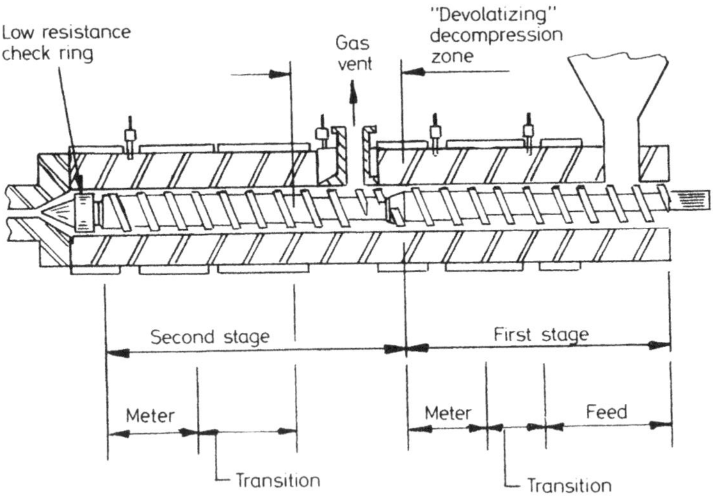

FIGURE 9.17. Vented-barrel and screw assembly. (Adapted from Ref. [4].)

The first stage is similar to a regular GP screw, except shorter, with the screw flight depth. This creates a large free-volume space between the barrel and the screw. This causes decompression of the melt. The moisture and volatiles, because of their high vapor pressure, escape from the melt and air vent area.

The second-stage screw sections begin here with a rapid deepening of the screw flight depth. This creates a large free-volume space between the barrel and the screw, which causes decompression of the melt. The moisture and volatiles, because of their high vapor pressure, escape from the melt and are vented out of the barrel through the vent located on the barrel's surface. This is similar to carbon dioxide gas being released from a can of soda when opened, except that it is continuous while the screw turns.

Depending on the polymer being processed, the exhaust gases are either vented to the atmosphere or first trapped and run through a conversion or scrubber unit, often with a vacuum assist at the vent. Consult with your material supplier as to the makeup of the volatiles that are exhausted and how they should be controlled and vented. The material safety data sheets may also answer these questions.

After the molten resin enters the decompression section of the screw and venting occurs, the melt enters the second stage. It is again compressed in the transition and metering sections of the screw, while being pumped forward to build up the next shot or charge of resin for the next injection cycle.

Decompression screws all operate on the same principle, but there are proprietary differences in L/D ratio, flight geometry, shear and mixing components, and other features. The various combinations provide a different balance of devolatilization rate, melting efficiency, recovery rate, residence time, and cost. VB screws can be designed for a specific resin or left as a general purpose screw for a variety of resins. These screws can also be equipped with mixing and barrier sections.

For the VB screw to function correctly, the second-stage screw design must be able to handle all the melt it receives from the first stage without generating back pressure. Any resistance to the forward motion of the melt into and through the second stage will push the melt back along the screw and out the vent. To reduce this tendency, the second-stage screw design will have slightly deeper screw flight depths. If back pressure is used, it may have to be adjusted to a lower level or may require the use of a low-flow resistance check ring in front of the screw. Some suppliers recommend using a starve feeder to control the feed of resin to the first stage of the screw. Unless starve feeding is correctly controlled, however, other problems can occur that cause the machine to work harder to produce the quantity of melt demanded by the cycle and shot size.

Other items to consider for each type of resin are the barrel temperature profile, screw speed, feed rate, and back pressure. With amorphous resins, the barrel heaters straddling the vent are often 20°F lower to inhibit the melt from back flowing at the vent.

Vented-barrel machines will not produce the same pounds of melt per hour as their closed-barrel counterparts. The typical drop off is at 15 to 25 percent. This occurs because the vented screw section takes up the equivalent of eight screw flights for devolatilization, thereby reducing melting capacity. The shorter melting length in the first stage, plus the lower or nonexistent back pressure, reduces the machine's melt output capacity.

The melt production drop off depends on the L/D of the screw, the size of the shot it must prepare, and the cycle time. Raising barrel heat on the first stage to induce more heat into the melt may cause other problems with melt quality, such as resin decomposition, burning, and color changes. Therefore, most machine manufacturers recommend a minimum L/D of 26:1, especially if a general purpose screw is used.

If lower L/D ratio screws are used, the screw sections may also need a barrier section, with larger presses, to produce a better quality melt. This assumes that recovery rates are minimal. This is part dependent, as cycle time is based on resin set-up time along with the machine's recovery and melt generation rate. These must allow sufficient time for screw recovery and good melt to be produced without these added screw sections.

There is always some controversy over the ability of VB machines to process resins, with or without predrying. The concern centers on the part appearance and physical properties of the molded parts. Controlled tests have been run on predried and saturated resins in vented-barrel machines to try and answer these questions. Optimum process conditions were used, along with starve feed and barrier sections. On the screw in a 26:1 L/D machine, these yield equal and in some cases superior properties for the saturated resin as compared with the resin dried in a desiccant drier. One caution is that some material properties are severely affected if processed while too wet. Most resins can be processed directly from their package as they are dried prior to packaging. Wet regrind could change these results. Impact tests on molded parts will usually answer these questions, as part toughness is usually an indicator of lower part physical properties caused by excessive moisture in the resin during processing.

Vented barrel machines come in L/D ratios of from 20:1 to 32:1. The main reasons for using vented barrels are as follows:

1. Predrying of resins is reduced or eliminated, resulting in cost and time savings.
2. Productivity rates are equivalent with those of closed-barrel machines, although screw modifications and additions may be required. A 26:1 vented-barrel machine with double barrier sections performs the same as a 20:1 closed-barrel machine.
3. Utilization of same floor space for equipment. Retrofit kits are available if the job size allows the same L/D ratio.

All of the above can be attained if the machine size, screw design, and resin are complementary. The vented-barrel machines can produce high-quality parts. More molding problems occur from lot-to-lot resin viscosity differences than from all the melt operation problems of vented-barrel machines.

MAINTENANCE OF MACHINERY

To mold quality parts, the machine and its controls must be in top condition and in calibration. As equipment is run, it wears; problems must be quickly diagnosed and repaired.

As with a mold, when a machine is taken out of production, it should be broken down, inspected, repaired, and put back into calibration. Although this is often not the norm with injection molding machines, it should be. Once a failure occurs and the machine goes down for repair, part quality suffers, and production time is lost.

All components in a machine have a life span based on usage, cycles of operation or hours, and temperature history. Therefore, a preventive maintenance schedule should be prepared for each machine that records the number of hours of operations for motors and pumps, number of cycles for bearing, and moving part wear and materials processed so that wear on the barrel and screw assembly can be checked. It is too late to switch the mold if subquality parts are being produced by a machine with a problem and if another press is not available to meet the schedule and part requirements. It is expensive and time consuming to keep good machine records and do preventive maintenance, but it costs more to lose a customer.

Preventive Maintenance

Your equipment supplier can provide machine preventive maintenance schedules based on experience with the specific items of equipment. This may include lubrication points to be serviced, filters changed, oil replaced, wear points measured, gauges calibrated, heater bands replaced, screw and barrel measured for wear and corrosion, thermocouples calibrated, hydraulic hoses replaced, and pressure valves checked. This is not an easy task, but if monitored and performed on a regular schedule, it will prove to be a cost saving. The production department will always have a process capable machine. By performing preventative maintenance checks, one can determine whether components are wearing out, going out of calibration, or need replacement. Parts can then be ordered and downtime planned so that repairs do not impact negatively on the production schedule.

There are tests that can be performed to monitor the machine's functions and indicate potential problem areas before a failure occurs or molded part quality suffers. Using an oscilloscope that is hooked up to monitor the machine's hydraulic and timer functions, the uniformity of operations from cycle to cycle can be monitored. The scope displays a trace of the pressure and time curve that represents the initial cycle monitored. Each succeeding cycle is then compared with the original trace to detect inconsistencies. If any are noted, their degree of variance can be compared. Adjustments in machine conditions can then be made and monitored to determine whether they can be maintained. If not, there is a problem that needs to be isolated and the part replaced or repaired as necessary.

The replacement of systems and parts should be done in units or sets. If one item of a set needs replacing, the others in the set are also worn. Always replace the entire set. If only the worn part is replaced, it could cause the others in its set to wear more rapidly. The new part, now in calibration, puts additional and uneven stress on the other parts, causing them to wear faster. This may result in the need to replace all parts sooner than anticipated.

The screw-and-barrel and nonreturn valves are usually the first major items that require repair and replacement. Each screw and barrel is sized to have a set clearance between components. As wear occurs because of resins and their fillers, corrosive additives, metal-to-metal contact, and resin flow along and through its parts, the screw flights wear down and the barrel lining thins, or grows larger. It does not take more than 0.002 to 0.003 inches of barrel and screw wear to have problems. The screw flights, where the pellets and melt are conveyed forward, will suffer wear and corrosion. As this wear progresses, pits and rough surface areas occur that inhibit the resin from being smoothly pumped forward. This can cause the resin to adhere and decompose.

As a result of these problems, the machine will lose its plasticizing capability and the cycle will very slowly start to lengthen. Unless attention is paid to and accurate records kept of the machine's cycle, no one in production control will be aware of the problem until part quality goes out of specification. Warning signs that indicate wear at the check valve, nonreturn, and screw and barrel are as follows:

1. Machine cannot hold packing pressure or cushion.
2. Excessive screw recovery time.
3. Screw rotates backward during injection stroke.

Other symptoms of wear that affect part quality and performance are as follows:

1. Decreasing part weight and varying tolerances.
2. Nonuniform surface appearance, causing aesthetic problems.
3. Sink at thick sections and part warpage.
4. Dark or varying colored specks or streaks in natural and colored parts as a result of degraded resin.

5. Contamination from previous resin run if barrel is hard to clean because of resin hangup in worn areas.
6. Metal particles in the part and flakes from plating on the barrel and screw surfaces.

Metal flake contamination can occur in many ways.

1. Abrasion from reinforced materials.
2. Screw sag from a long L/D screw and metal fatigue.
3. Mechanical misalignment due to worn screw drive and support bushings.
4. Supplier contamination that has nothing to do with your machine, but can be caused by the production extruder or metal in the package from wear in the conveying and packing out system.

It is recommended that a metal magnet for trapping these fine particles be placed in the throat of the machine feed hopper. This protects the quality of the part and prevents a large metal object-nut/bolt from damaging the screw and barrel if it is fed into the system.

The barrel, screw, and check ring should be periodically inspected. If problems occur, inspect the screw for worn areas and polish up areas showing wear. Check rings can be replaced and screws refurbished if not worn beyond repair and resurfacing. Barrels can also be relined, but it is expensive and time consuming. Benefits from resurfacing the screw and/or relining the barrel must be weighed against the cost of total replacement.

Screw and barrels running reinforced materials, flame retarded grades, and other corrosive resins, such as PVC, can have long life and minimum wear if common sense approaches to their processing and cleaning are followed. When processing flame retarded (FR) grades of resin, be sure that they are thoroughly dried. The FR grades contain brominated and chlorinated compounds that are corrosive in their own dry state, but if overheated and containing a little moisture, they form strong acids from the breakdown of sulfur. These acids contain additives that attack metal surfaces. Therefore, predry these resins and check the moisture after every run. Pull the screw and clean and polish it and the barrel to neutralize any effects of the FR compounds on the metal, especially if acid was formed during the run. A good poly purge may negate this requirement, but checks should be run to prove screw and barrel damage are not occurring.

Excessive melt temperatures should be avoided with fluoropolymers and PVC resins; they decompose easily and form harmful acids. Periodically check melt temperatures and do not allow the resin to cook in the barrel if a long cycle interruption occurs. If there is a holdup time of more than 5 minutes, back the screw from the mold and, with low RPMs, pump resin through the barrel. If a long holdup time is contemplated, reduce barrel heat—not below the melt point—and keep the screw turning.

When starting a new job, a machine check-off procedure sheet should be used to ensure that all hydraulic lines, seals, controls, and other components are within specification and not showing wear which can lead to a problem.

Monitoring hydraulic fluid levels and temperature is also critical. A deviation in these can cause pressure and cycle variations that affect part quality and output. If the equipment supplier did not furnish a checklist, you can develop your own and use it daily to ensure that the machine is capable and within processing limits to continue making good quality parts.

Maintenance Checklist

To establish a good checklist, the following items need to be considered:

1. Are all safety features operating correctly and in calibration? Machine manuals list these and their inspection procedure. Never, in any case, should a safety feature or interlock be wired shut to complete an operating safety circuit. Any cycle savings here could result in serious injury.
2. When was the last hydraulic oil change? Changes are recommended every 2000 hours or once a year, if equipment runs only 8 hours per day.
3. Replace the hydraulic filter on the pressure side at the time of an oil change. The suction side is similar, but it can be cleaned and replaced. Some machines are equipped with water traps that should be inspected at more frequent intervals and emptied to insure a uniform oil viscosity.
4. What are the name brand hydraulic oils recommended by the supplier? They should only be used with good antiwear and corrosive additives. Hydraulic oil filter systems that can remove even the smallest micron of foreign material, are now available. If used continually, or at quarterly or semiannual intervals, they will almost guarantee, like new, oil in the hydraulic system. A filter is an inexpensive item to replace or clean; more frequent inspection and replacement is a small investment in time and money.
5. Is the hydraulic oil temperature within manufacturer's specifications? This is usually between 100 and 110°F. The oil heat exchanger should be maintained to allow this to occur. Try not to overdrive the system and overheat the oil, as unnecessary wear will occur and shorten its life.
6. Is the hydraulic pressure system properly adjusted? Your manual will give pressure settings and adjustment procedures. Calibrating the gauges at this time is also suggested.
7. Have you checked the minimum and maximum screw rotation speeds under a no-load condition?
8. Are all hydraulic lines checked for worn spots and hydraulic fittings? These should be replaced as required.

9. Has the system been checked for inoperative temperature controllers, barrel heater bands, and thermocouples? A calibrated pyrometer can assist in these tests.
10. Has power been disconnected so that all or any loose electrical connections can be tightened?
11. Is the malfunction light in good operation order?
12. Has the nozzle alignment been inspected for good nozzle to sprue bushing contact? Blue dye on the nozzle can indicate contact points. If the barrel nozzle bounces on its travel into the bushing, alignment is required.
13. Have past processing records been reviewed to determine whether the nonreturn valve and screw need inspection? If required, remove the nozzle and adaptor, pull the screw, and inspect it.
14. Are all machine functions checked in manual, semiautomatic, and automatic operations?

There may be other points to inspect periodically. If not noted in your equipment manual, your supplier's service department can alert you to them.

It is essential to train maintenance personnel to keep equipment in top running order. Equipment suppliers run training and maintenance seminars to make your personnel knowledgeable in these areas. In serious cases, contact your machine supplier's technical staff for support.

Machines requiring the most maintenance are in the 300-ton and lower classes, mainly because there are more of them. They are used more frequently for complex parts, often with tough resins, with faster cycles, and with more mold changes per machine. This results in greater wear and more frequent breakdowns. Surveys indicate hydraulic and toggle clamp presses require the same amount of service. The major machine components requiring service are electronic controls (52 percent), hydraulics (34 percent), mechanical (12 percent), other components (2 percent). Therefore, your machine maintenance personnel should follow these guidelines.

Maintenance of your machine to keep it in a high state of quality control requires an investment by management. They need to supply the assets and motivation necessary to train your operators and maintenance personnel to inspect, calibrate, repair, and replace machine components as required. They also need to monitor and record all running times, service, and machine operating parameters to ensure that only quality parts are produced.

10

Auxiliary Equipment

To make quality parts, the auxiliary equipment must consistently supply and monitor the support services required by the molding cycle. This equipment, which includes dryers, mold temperature controllers, granulators, feeders, part removal systems, robots, conveyor systems, part handling, and plant support equipment, should be networked into a local or plant control system (see Figure 10.1). They should also have diagnostic and trouble shooting capabilities that signal operators when systems are failing or going out of control.

Without this auxiliary equipment, it is impossible to make quality parts consistently. All of these items must be able to operate independently and have the capability to be tied into a local or central microprocessor network. The network then controls, monitors, and provides operator and plant feedback in real time. The data stored in these systems can be used to plan and better manage the assets and equipment.

Careful thought should be given to the most effective and efficient use of the auxiliary equipment. This may include sizing auxiliary equipment to handle more than one machine, as with dryers, material feeders, and granulators that use closed-loop feedback systems to control material feed systems, eliminate contamination, and reduce material drying time.

In plants with long production runs, using the same materials, position the auxiliary equipment in job cells that support more than one molding machine. This reduces the number of variables to monitor if the auxiliary equipment

Total Quality Process Control for Injection Molding, by M. Joseph Gordon

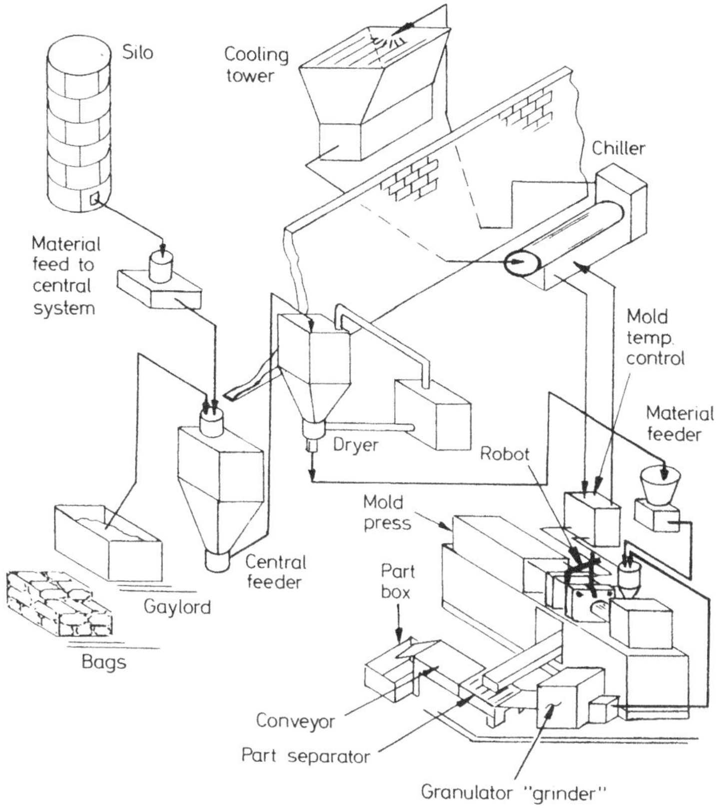

FIGURE 10.1. Auxiliary equipment layout.

supports the same molding conditions, drying requirements, material blending, and feed systems. In plants running a more diverse material mix and molding conditions, individually controlled auxiliary equipment is required. The right combination of auxiliary equipment promotes automation, reduces the amount of manual labor, and frees up operators to monitor the manufacturing system and develop methods to improve the quality of the finished parts.

No single piece of auxiliary equipment is more critical than another. Each machine must repeatedly perform its assigned task and function within its set processing parameters. The three primary pieces of auxiliary equipment are the material supply system, material dryers, and mold temperature controllers.

MATERIAL FEEDERS AND BLENDERS

After the incoming resins have been inspected for quality, they must be properly stored. When needed, these materials are conveyed to the machine's feed hopper in two ways—automatically (pneumatic or vacuum loading systems) or loaded by material handlers. Each is acceptable as long as quality material handling procedures are followed. Automatic handling is preferable, because it lessens the possibility of contaminating the resin and keeps the material dry. If regrind is permitted, it should be fed back into the system immediately through a proportional feeder, where the correct resin-to-regrind ratios can be maintained. A greater number of problems with quality occur from the improper control of regrind than from any other material problem.

AUTOMATIC SYSTEM

Automatic material handling systems can be centrally located. Smaller systems, fixed or mobile, can also be used (see Figure 10.2). The smaller systems are usually installed on the molding machine's feed hopper or positioned adjacent to the machine. Central material handling systems are used in plants that are molding the same resins on one or a battery of machines. Material feed lines are routed to dedicated machines from the central drying system. Material feed is usually controlled by resin level sensors in the feed hoppers. When the amount of material in the feed hopper drops below a specified level or weight, the feed system is triggered to replenish the hopper.

The material feed system is basically a closed-loop system that uses filtered dried air to convey the material to each machine. As an air-tight system, it uses either positive air flow to push the resin through its piping or a vacuum to suck the resin through the system. The piping should be of stainless steel, or another low-wear, nonrusting material. The number of 90-degree bends in the system should be minimized to reduce pressure loss and abrasive wear from the pellets and to eliminate any potential blockage in the system.

Automatic hopper loaders use a vacuum to lift the resin to the hopper from boxes or drums. A siphon tube on a flexible hose is inserted through the package liner to lift the resin to the feed hopper. The tube should be sealed tightly to the liner in the shipping container to keep the material dry and avoid contaminating the resin.

A protective drum or gaylord cover should also be used to protect the material from contamination. Most covers have windows that enable operators to verify that siphon tube is kept in the resin supply. There are also material package tilters that automatically lean the drum or gaylord at increasing angles as the package is emptied by the siphon tube. Each feed system should have a dust and fine filter system to trap any small resin particles. If allowed to build up in the hopper, they can cause machine feeding problems.

Automatic resin loading systems are preferred over hand loading systems. Climbing up on a machine to feed the hopper manually is a slow, dangerous process that can result in approximately a 2 percent resin loss because of spillage. Hand loading also requires an extra person to service the machine and,

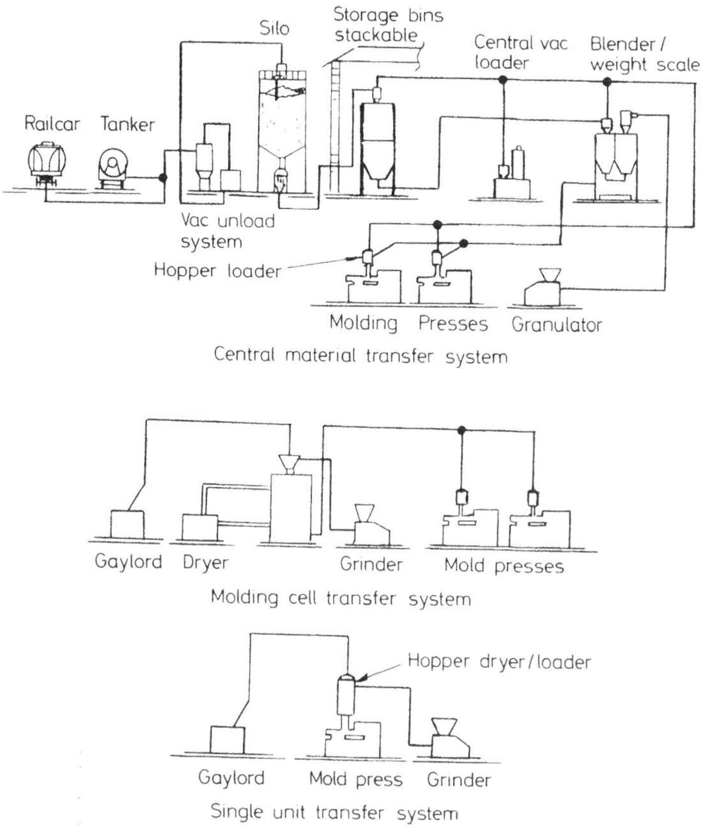

FIGURE 10.2. Material transfer and feeding systems. (Adapted from Ref. [20].)

if hygroscopic resins are being used, the material package is usually left open to facilitate frequent loading. This allows the material to pick up excessive moisture, which leads to longer drying times, or it becomes contaminated in other ways that can increase the risks of quality problems. If regrind is used, hand loading greatly reduces the probability of correct ratio mixing.

CENTRAL SYSTEMS

In larger plants, bulk material silos or large air-tight containers are used to store the resin received in bulk shipments. Hopper railcars or bulk tankers can deliver 45,000 to 200,000 pounds in a single shipment. Depending on the plant's location, type of resin, and size of the storage system, the tanks may be located in or outside of the main building. These storage tanks are dedicated to specific resins and designed for the storage of each specific type, from powders to pellets.

When outside storage is used and the containers are not insulated, care must be taken to reduce condensation from forming inside the storage system. Resin is often warm when loaded into the tanker at the supplier's plant. This can cause condensation to form in the tanker. When discharged into the uninsulated silo, it may also attract moisture that condenses on the inside of the tank walls when temperature conditions vary outside the tank. Tanks should be made of corrosion-resistant materials to avoid material contamination and have air-tight sealing systems with vents to control the inside temperature. Some storage systems even have the capability of putting inert nitrogen blankets of air over the entire system to retard moisture pickup by the resin.

Always perform incoming resin quality control checks before any new lots of material are loaded into your bulk storage system. It is important to verify that the material is within specification limits. Perform these checks as a precaution even if the supplier's certification has been received. Once the resin is in the silo, it is too late to return offspec or wrongly graded material that has contaminated good material. Develop tests with your supplier that will accurately evaluate the resin's quality and not delay unloading.

During cold weather, material stored outside the plant or in an unheated warehouse should warm up in the plant before use, possibly in an auxiliary holding system. This brief warm-up reduces system shock and increases the auxiliary drying equipment's efficiency.

The quality of the air or vacuum used to convey the resin to the machine must also be considered. There are two basic transfer mechanisms—one is open to the atmosphere; the other is a closed-loop transfer. Avoid using non-dehumidified plant air to transfer material after it is dried. Moisture-sensitive resins can easily become contaminated, causing molding and part quality problems. The closed-loop system, which uses dehumidified air from the drying system is preferred. An inert gas system that protects the resin's hygroscopic properties is also acceptable. The intent is to protect the material in the handling system to keep it as dry as it was when packaged and shipped to your plant.

All material conveying equipment should have dust and fine filter traps to capture any resin dust and material fines generated during the supplier's loading, while in transit, and when off-loaded at the plant. The harder and more abrasive resins, as well as those reinforced with glass and minerals, create particle dust and fines during transfer. The fines collect in the material transport piping in vertical bends or low-flow areas, thereby reducing resin flow to the machines, causing air flow to decrease by plugging up the filters on the dryers. This results in improperly dried material which can cause part and process quality problems. If these resin particles are not removed, they will clog the material handling system, causing poor material conveyance and feed problems at the molding machine. The accumulation of fines also inhibits uniform resin feeding by building up in the feed section of the screw. Such an accumulation in a low-compression screw will prevent material from feeding and causes the material in the screw—fines and pellets—to churn in place.

This starves the machine, leading to longer cycle times and variable shot melt quality. Ultimately, it can lead to a shut down of the machine so that it can be cleaned out.

All parts of the system must be monitored and cleaned regularly. The system must also be designed to avoid any dead spots where pellets can hang up. With the more abrasive reinforced resins, the piping at bends will thin, wear out, and require replacement. The speed used to convey pellets and the pressure or vacuum in the system should be adequate to move, not blast, the resin to its next point of use. For each type of resin, there are recommended pipe sizes and bend radii. The system should also be periodically cleaned and wire brushed at piping bends. At these choke points, the softer resins may form polymer skins on the piping surface. This builds up, flakes off, and can lead to feed problems.

Material level indicators or weight sensors are used in bulk storage tanks. These recorders assist purchasing in tracking material resin inventory levels and usage so that orders can be placed and deliveries made within production schedule requirements.

It is very important to safeguard resin quality by avoiding dust, moisture, and cross-resin contamination. The plant's storage bins and silos, as well as the material transfer lines dedicated to specific resins, should be well labeled and tagged, so that designated resin is conveyed into the system. A verification and sign-off system should be followed before loading bulk resins into your central storage system. It first verifies the quality of the resin and then its storage in the correct unit. If the plant consumes more than one type of bulk resin, use different size resin loading fittings or color code the fittings to ensure that the resin is put in the correct storage container. With the proliferation of resin types, many look alike and have similar coding or material identification nomenclatures. They can easily be misidentified and put into the wrong storage container.

All material transfer systems are not designed to handle all types of materials. Therefore, consult with your equipment or material supplier for the system best suited to your requirements.

MATERIAL FEED TO THE INJECTION MOLDING MACHINE

Each injection molding machine has a feed hopper that holds various amounts of resins. These feed directly into the feed throat of the barrel. With cylindrically cut or spherical pellets, the feed hopper is designed for gravity fall of the pellets into the throat. With fine resin or powders, like Polyvinyl Chloride (PVC), a screw auger may be installed in the hopper to provide a positive feed of compound. The auger, often called a crammer, is timed to operate as the machine's screw turns and forces the material into the throat.

The problem of resin bridging (pellets or powders packing together and inhibiting free flow) is minimal when running virgin resins. The feed hoppers are designed and sized to inhibit pellet/powder compaction. However, when feeding regrind with virgin resins, particle size differences may cause bridging at the feed throat. The solutions may be improving the control of regrind particle size (i.e., bringing it closer to the pellet size) or mounting a vibrator on the hopper to ensure better flow. The typical molding machine's vibration during operation is usually enough to keep pellets flowing freely. But if pellets are surface coated with oils or lubricants, specially mounted hopper vibrators are often necessary to ensure constant uniform resin feed. With large-volume material hoppers, reducing the volume of resin stored has a direct relationship to the compressive load on the resin as it narrows down at the hopper feed throat. The resin must feed freely into the barrel to ensure a constant supply to the screw generating the melt. The screw can only meter and stay efficient if the resin supply is constant. Intermittent feeding can lead to short shots, cycle interruptions, and problems with sufficient melt for the next injection cycle. The machine's operations are based on timers and screw position indicators. All must function in a prescribed manner or the cycle gets out of sequence, produces poor quality parts, and loses efficiency.

MATERIAL BLENDING AT THE HOPPER

Often additional material enhancers or modifiers are required or desired at the hopper or at the machine's feed throat. These additives may take the form of color concentrates in liquid, powder or pellet concentrates, or other additives, such as blowing agents and lubricants. These modifiers are usually fed directly at the hopper/barrel feed throat interface by the use of a mounted auger or other type of proportional feeder. Uniform control is attained by metering these additives with a proportional feeder at the barrel's feed throat.

A new multiplate dry concentrate feeder is now available. It operates on the pulsed multishuttle plate principle, with varying sized feeder holes. This feeder controls additive levels very accurately. It operates only when the screw turns and is driven off the machine's microprocessor control unit.

Top feeding of additives into the hopper system is not desirable if different additive particle sizes or fine powders are involved. The feed mechanism uses a vortex action, with the center section feeding more directly into the throat. As a result, top additive feeders must disperse only free flowing additives over the entire surface of the hopper to achieve uniform blending of ingredients. Top feeders are used, however, for adding regrind and blending resin by using proportional vacuum feeders timed to convey the correct proportions into the top of the hopper. A deflector or director in the top disperses the material evenly over the entire surface of the hopper. Samples should be taken periodically at the hopper feed throat to verify that the correct ratio is being obtained. Because most feed hoppers have a clean-out gate, the sample can be obtained here. If not, the hopper may have to be unbolted and moved aside to obtain the sample.

BLENDING QUALITY CHECKS

To check whether the additive percentages are correct, collect approximately one cup of material and separate the elements in the sample. Usually the additives are dissimilar enough for this to be easily accomplished. Then, weigh the components on a digital scale to two decimal places. By comparing the individual weight ratios to the total sample weight, the percentage of each additive can be determined. If the additive percentage is off, adjustments must be made to produce the correct mixture proportions. Control of regrind and resin blends is important to ensure that the part's color, physical properties, and processing parameters are maintained from cycle to cycle.

Some reasons for using hopper blending are small runs of special colors, material cost savings using color concentrates as opposed to compounded colors, and the ability to make special resin blend combinations and to add special additives to enhance a material's properties. These are important because the major resin suppliers resist supplying small quantities of colors or specialty blends unless it is advantageous to business. Although some suppliers specialize in these types of compounds, they are expensive and may require long lead times. If special control is exercised and testing proves that the blend works for the part, hopper blending can be successful. Tighter quality control of materials, blending, and processing are needed to ensure that this method of manufacture will yield the desired results.

COLOR CONCENTRATE BLENDING

Blending color concentrates at the hopper for injection molding is becoming more common. The feeders are usually the auger or volumetric type set to run during the screw retraction or feed cycle. They meter set amounts of colorants in the form of pellet concentrates directly into the feed hopper with virgin resin. The accuracy of the feed system for volumetric blenders is about 1 percent, which means that color uniformity is usually undetectable to the eye. This method is preferred over weighing out set amounts of materials, blending by mixing, and then feeding the hopper. Poor material ratios and blending can make this an expensive undertaking, so it must be properly controlled and checked.

When uniform part color is critical, the material supplier usually specifies a compounded colored resin. The use of regrind with colored parts is questionable, especially if tight color control is to be maintained. On each pass through the barrel, the pigments degrade. Based on the part's color and material's pigment system, they probably will fade if too high a regrind percentage is used. With some colors, a 5 percent regrind may be tolerable without causing a color shift, but each program must be evaluated separately. With the shift to organic pigment systems, with their lower heat stability, this becomes even more questionable. Most molders do not feed regrind back for colored parts because additional color concentrate must usually be fed into the system to maintain exact color control. The cost of these concentrates, their metering, and verification of part color control does not usually justify any cost savings that could be realized by the use of regrind.

REGRIND USAGE

Regrind can usually be fed back into the system at 20 to 25 percent without affecting a part's properties. It should be fed back as soon as it is generated. After the parts are separated, the sprue and runners are fed to the granulator, chopped to a uniform particle size with fines separated, and then fed in timed cycles back into the hopper. The material feeder system is tied to the virgin resin feed system to meter correctly the percentage of regrind allowed back into the system. Regrind can be conveyed directly into the top of the hopper or to a gravimetric or volumetric feeder that uniformly feeds it back into the hopper at the desired ratio. Tight timing control of the feeders is necessary to ensure the correct ratios are maintained in the feed hopper.

To determine the accuracy of the feeder for either type, use the following equation (IT = ingredient tolerance):

$$\text{IT } (+/-\ \%) \text{ in final blend} = \text{Ingredient feeder accuracy } (+/-\ \%) \times \frac{\text{Ingredient proportion}}{100}$$

The quality specifications for the final blend are usually expressed in terms of the allowable percentage variation of each ingredient. Therefore, if you are feeding a color concentrate at a 2 percent level and the feeder accuracy is +/– 0.50 percent, then the variation in the final blend is:

$$\text{IT} = +/-\ 0.5\% \times 2/100 \text{ or } +/-\ 0.01\%$$

Similar calculations can be made for any additive fed into the system to determine whether the correct additive level is being maintained in the hopper. The quality of the molded parts is related to how accurately the feeder system operates.

MATERIAL DRYING

To ensure part quality, almost all plastic processors dry resins prior to molding. This includes even the nonhygroscopic resins, where removal of any surface moisture is all that is required. A material is dried by passing super dry-hot air through the pellets in the hopper. This process evaporates the moisture and removes it from the system. For the hygroscopic resins, such as nylon, polycarbonates, and polyesters, the pellets must be heated sufficiently for a set period of time to extract the moisture that is absorbed within them. This operation requires dehumidified hot air that is dried to a dew point as low as –40°. The dew point is important for successful drying of the material. The dew point is the temperature to which the air must be cooled before water vapor condenses. It is therefore the temperature at which air becomes saturated and produces dew. The lower the dew point, the dryer the air.

Failure to dry the materials properly before processing can result in surface and interior part defects as well as a drop in the material's physical properties. If a material is processed with too high a moisture level, the molecular chains of some resins will fracture from hydrolysis in the melt stage. This can cause processing problems and irreversible deterioration of the polymer's physical properties. With some polymers, excessive moisture will show up as surface effects on the part such as splay. But with other resins, it may not show at all as with polyesters. Polyester resins that are molded when too wet become brittle; only physical testing or viscosity measurements will reveal this damage. Therefore, the question of using vented-barrel machines without knowing the resin's moisture level or predrying the material becomes a factor for hygroscopic materials.

The moisture content of all resins should be known before drying. Otherwise, the time and temperature for drying cannot be accurately determined. However, most resins require a minimum moisture level for proper processing. Nylon, for example, requires a minimum moisture level of between 0.05 to 0.10 percent, depending on the type of filler and its level in the resin. Drying some resins too long or at too high an air temperature may drive out friendly and required additives. This may be noticed by a color change in the molded part, if the material processes poorly, or if the parts are brittle. If materials are to be left in the dryer for long periods of time, the temperature should be reduced. By knowing the moisture level of the resin, production personnel can reduce drying time, thereby saving energy and money. An item often forgotten concerns filled materials. The more filler a resin contains, the lower the allowable moisture level for processing. If in doubt, consult the material supplier. The typical recommended resin drying times are shown in Table 10.1. For materials using regrind that was exposed to wet air, a longer drying time is required.

Material Drying Systems

There are many types of material drying systems. They vary from the old resin pan-loaded baking style ovens to hot air convection ovens and hot-air desiccant dehumidifying closed-loop drying systems to microwave, radio frequency, and refrigerant styles that use hot dehumidified air to extract moisture from the resins. The systems use desiccant beds or mechanical means to dry the air, which in turn dries the resin.

TABLE 10.1. Material Drying Time and Temperature.

Material Type	Drying Temp °F	Drying Time (hrs)	Virgin Resin (Permissible Moisture Content)
ABS	180–200	3–4	0.10–.15
Acetal	185	2–4	0.1
Acrylic	160–180	2–4	0.02–0.10
Cellulosics	160	3–4	0.4 max
Ionomer	150	8	—
Nylon	180	4–6	0.05–0.20
Polycarbonate	250	3–4	0.02 max
Polyethylene	195	3	0.02 max
PET (high heat)	250	4–6	0.02
PBT	250	2–3	0.05–0.20
PET	160	3–4	0.10
Polyamide	250	2	0.10
PP	195	1	0.10
PS (GP)	180	1	0.02
PS (HI)	180	1.5	0.02
Polysulfone	250	4	0.02
Polyurethane	180	3	—
PPO	255	2	0.1
Styrene (SAN)	180	2	0.10
Vinyls (PVC)	160	1	0.08

ABS, acrylonitride butadiene styrene; GP, general purpose; HI, high impact; PBT, polybutylene terephthalate; PET, polyethylene terephthalate; PP, polypropylene; PPO, polyphylene oxide; PS, polystyrene; PVC, polyvinyl chloride; SAN, styrene acrylonitrile.

There are basically three major types of drying systems. Each uses the desiccant bed, hot-air drying system, which is the most economical and reliable for drying amorphous and crystalline resins. These are all central, multimachine, and single-machine systems. The central dryer system is usually permanently located near the resin storage area close to the molding room. The multimachine unit may be mobile or permanently positioned in a molding cell of two or more machines. Its drying hopper is sized to suit the resin's drying requirements for the machines it serves. Individual feed lines are routed to a smaller, insulated, closed hopper for feeding to the barrel. The volume of resin in the machine's feed hopper is usually restricted to what can be processed in 15 to 30 minutes or less. This reduces cooling and moisture pick-up time if the system is not equipped with its own hopper dryer maintenance unit.

Many custom molders have dedicated dryers mounted on the machine's feed hopper. They are individually controlled and can dry the many different grades of resins as long as sufficient drying time is allowed. Hopper resin capacity, resin throughput, and drying time at temperature must all be considered when sizing the drying equipment.

The advent of quick mold change has required adjustments. Unless the same resin is processed in the new mold, the hopper dryer arrangement may not always be satisfactory and productive. To solve this problem, the mobile dryer system loads and dries the resin before the mold change. Then, when the mold is ready to run, it is rolled up to the machine and pumps the dried material to the feed hopper. Each plant will have to select the drying system best suited for its operation. But with more just-in-time (JIT) manufacturing, which requires more frequent mold and resin changes, the latter, with its faster cleanout and turn around times, is becoming more popular for short runs.

An important area often overlooked in sizing the dryer is the system's drying capability, expressed in pounds-per-hour of resin output. If sized incorrectly, the resin may dry improperly or overheat in the dryer. This causes degradation and discoloration of heat-sensitive polymers.

The biggest problems with dryer systems occur with capacity selection, efficiency, and the lack of adequate controls, air flow, and monitoring systems. Consider also the maintenance of the unit, as well as whether operations are automatic and accurate.

There are two basic dryer types—conventional and high temperature. The first is used to remove surface moisture from most resins and to dry the hygroscopic type resins [e.g., polycarbonate (PC) and nylon]. The second is used for resins that require extremely high temperatures and very low moisture levels [e.g., PBT, PET, liquid crystal polymer (LCP), and polyphylene sulfide (PPS)]. Equipment suppliers can recommend the best system to suit your requirements. The dehumidifying hot-air dryer system is efficient and capable of drying all resins (see Figure 10.3). This dryer system can dry air down to a −40°F dew point. In contrast, the refrigerant dryer system can only dry air down to a +40°F dew point.

Dryer Analysis

To reduce air temperature loss, locate the dryer as close to the molding machines as possible and insulate the material transfer line and feed hopper. As much as 10°F of drying air temperature per foot can be lost in an uninsulated material transfer line.

The humidifying drying system operates in the following manner (see Figure 10.3). Hot, dehumidified dry air is injected at the hopper inlet (A), where it circulates around the base of the hopper into and through the resin. The temperature of the entering air must be monitored by a thermocouple or thermometer, as it is a control point for the system. After the drying air passes through and extracts moisture from the resin, the now wet air exits the hopper then through a filter (6), which removes fines and any volatiles released during drying. The air is then cooled by passing it through a heat exchanger (4) that extracts the accumulated moisture from the air. When the returning air is cooled, it aids the dryer's desiccant beds in removing a higher percent of moisture from the air (see Figure 10.4). A trap (5), which is located below the aftercooler, collcets any volatiles from the condensate formed on the cooling coils. The cooled air then passes through a filter system (7) of adequate size to remove any fines or contaminants that could poison the desiccant beds (8).

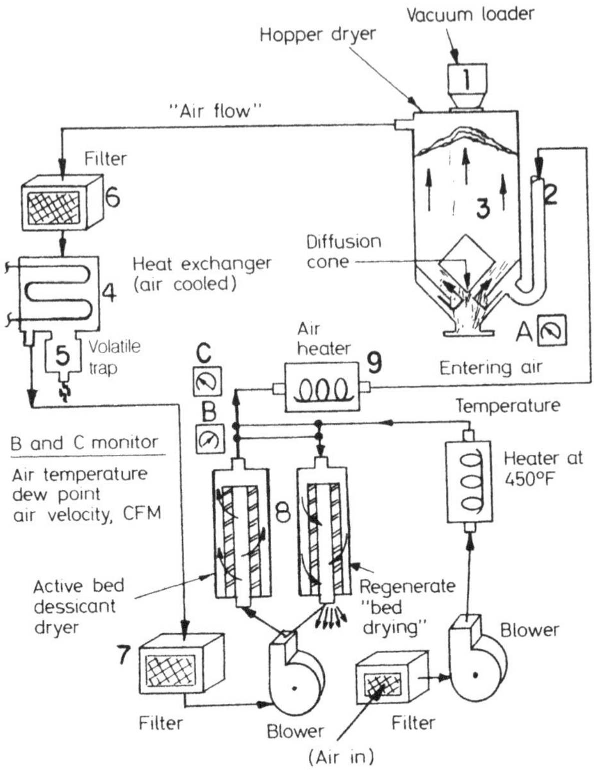

FIGURE 10.3. Hopper dryer/dehumidifier schematic. (Adapted from Ref. [16].)

After the air is conditioned by the desiccant beds, which extract the remaining moisture from the air, it exits where the dew point (B) of the air is continuously monitored. This measurement is very important, as it is a quality checkpoint for how efficiently the exiting air is fried. Dryers are equipped with continuous readout, or red and green lights, which indicate air dryness quality—red is above the setting and green is at or below. Some have audible alarms that signal personnel when the air is above its set point. The operator uses these warnings to determine when a bed's drying efficiency is decreasing and to switch to the alternate dry bed. With the automatic and process control units, the beds are switched automatically to maintain the dryness required for the air.

An air flow meter (C) should be installed with an alarm to indicate whether process air filter plugging is occurring. This can supplement the dew point alarm and hopper inlet temperature control. The air is heated to the necessary drying temperature by the process heater (9) and injected into (2) the hopper inlet feed system.

The air flow rate through the drying system ensures adequate drying of the resin. Each pound per hour of resin to be dried requires 0.8 to 1.0 cubic feet

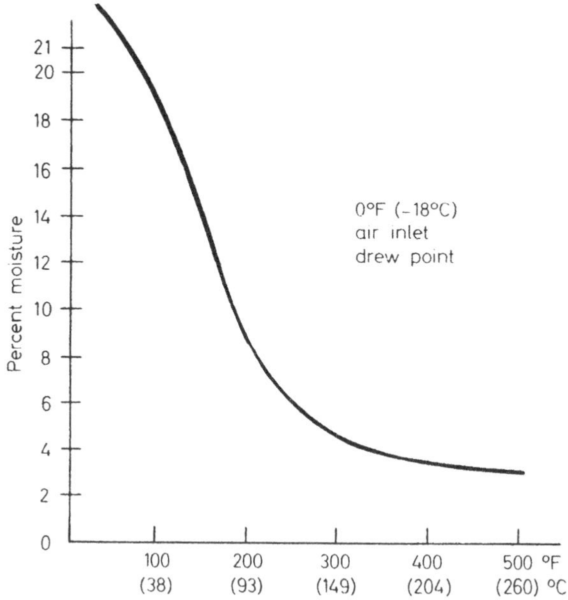

FIGURE 10.4. Air temperature moisture absorption. (Adapted from Ref. [16].)

per minute (CFM) of air. The capacity of the dryer must be known to calculate the air flow required for drying accurately. For example, if 60 pounds per hour of resin is to be molded, 48 to 60 CFM of dry air will be needed. Lower airflow rates will significantly reduce the resin temperature in the drying chamber and either prolong drying time or not dry the resin properly.

The hopper must always be kept full to ensure adequate drying time for the resin. This is determined by hopper capacity and resin usage, which is based on machine throughput rates and drying time. If the resin is too wet and the hopper drying time and temperature too low, the material will never reach the required dryness. With materials requiring long drying times or using wet regrind, separate dryers may be needed to bring the resin to the correct dryness.

Each resin has a recommended drying time and temperature, which should be closely followed. If residence time in the hopper exceeds supplier recommendations or if overnight or weekend drying is required, then the drying temperature must be reduced so as not to cause degradation of the resin.

Material Drying

After the material has been dried for the recommended period of time and before molding begins, open the hopper cleanout gate and remove the initial charge of resin below the air inlet connection. This resin should be considered wet or inadequately dried, because no dried air flowed through it. This resin should then be fed back into the system and redried as normal resin. In most cases, it is purged through the machine during the startup procedure. This is expensive, as there is often 5 to 6 pounds of resin in this zone of the feed hopper.

The equipment necessary to dry a resin adequately includes the following:

1. Automatic hopper loader (preferred)
2. High temperature dryer that is resin dependent
3. Insulation of inlet air line
4. Insulation of hopper
5. Aftercooler/volatiles trap
6. Temperature monitor at hopper inlet
7. Continuous dew point analyzer
8. Airflow meter
9. Desiccant beds and automatic bed changer

For any drying system to function properly, there must be no external air leaks into the system. All seals at joints and switching points to the second desiccant drying bed must be air tight. Remember that the drying system is a closed-loop system.

The major problems associated with drying systems relate to the care, maintenance, and activity of the desiccant bed, and how the filter system protects the drying desiccant bed from contamination.

Dryer Bed Analysis

The desiccant bed is a "molecular sieve" made of a moisture-absorbing material manufactured into round spheres or cylindrical extrusions that contain internal pores of various diameters. The material is slightly charged so that water molecules, with a slightly negative charge on the oxygen side, are readily drawn into the sieve and absorbed into the desiccant. Whether loose or in cartridge form, these beds should provide service for 1 to 3 years if cared for properly.

Within a desiccant bed (see Figure 10.5), almost all moisture extraction occurs during initial contact with the bed and very little toward the end (curve A). As the bed goes through its extraction and regeneration cycle (curve B), it loses some of its capacity and efficiency to absorb moisture. Eventually the bed must be replaced (curve C).

The key to proper drying of resins is maintenance of the drying system. A procedure should be followed for daily and periodic checks of the system. The

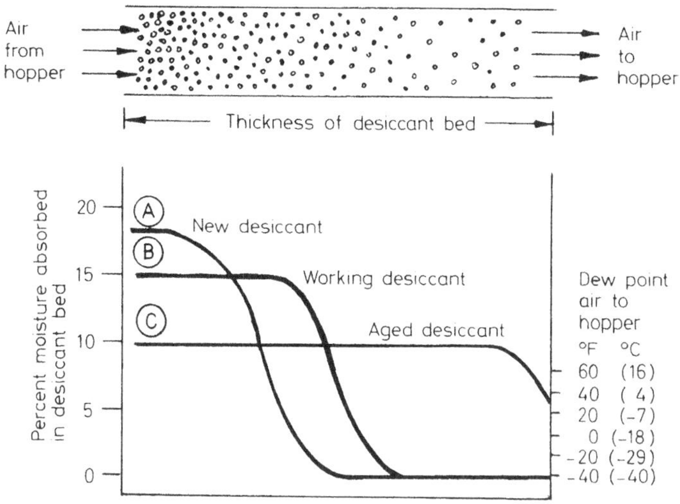

FIGURE 10.5. *Desiccant bed service life. (Adapted from Ref. [16].)*

following items are essential to maintain the system and not blame the desiccant bed as the cause of dryer problems:

1. The key elements (particularly the filters) should be checked frequently. Dew point, hopper inlet temperature, and air flow should be continuously monitored and recorded. If these elements show variations, then the following items should be investigated over and beyond routine maintenance: process and regeneration heating elements, bed transfer system, O-rings, gaskets and hoses, blowers, thermocouples, voltages, and meters. Poor filter maintenance can contribute to early desiccant bed failure by contaminating the bed surfaces with resin fines, dirt, or dust. Even when filters are operating properly, they can plug up and reduce air flow to unacceptable levels. All three filters, before the aftercooler, desiccant bed, and regeneration heater, should be checked once per shift to determine necessary cleaning and replacement frequency. Depending on the type of resin dried, cleaning and replacement may be required as often as twice a week or every 2 weeks. Turn off the dryer when doing these inspections.
2. Check for leaks. Seals and O-rings age and any outside air drawn into the closed-loop system will affect capacity and drying quality. The dehumidifier dries the air, the air dries the material, and the air stream must be a closed-loop from the dehumidifier to the hopper and back again. Humid outside air ruins the system.
3. In dryers left idle, the desiccant bed absorbs moisture from the atmosphere. If a dryer has been idle, do not introduce resin for drying until

you run the system through several complete cycles. This "dry-cycle" regenerates the desiccant beds. Checks on dew points and the inlet air temperature will indicate when the system is ready to be charged with resin.

4. To check the activity of the desiccant bed, measure the dew point of the air at the inlet and outlet of the bed. If the temperature differential is 30° or higher, the desiccant is working well. The differential between the inlet and outlet temperatures is critical. The outlet temperature by itself is not a sufficient measurement because an outlet dew point that is too high does not mean the bed is not drying properly. For example, if the outlet temperature is 0°F, but the inlet is 50°F, the desiccant is working fine; there is just too much moisture in the system and the cause must be found.

If the differential is less than 30°, the regeneration cycle needs to be checked. Examine the timer, bed heater, and bed thermostat. If these are functioning correctly, then check the desiccant bed.

A visual check will sometimes yield the following clues to the condition of the desiccant bed:

1. Color changes in the desiccant. A purple desiccant turns pink as it absorbs moisture. Not all change color, so check with your supplier. If your bed is not supposed to change color and it does, it is probably due to volatiles in the resin being deposited on the surface of the desiccant. This can be determined by examining the surface under a 20× or 50× microscope. An active desiccant surface will resemble a sponge with visible peaks and valleys. Use a replacement sieve as a guide during the examining. If you cannot see the valleys in your sample, then it is highly likely that the pores are clogged with waxes, lubricants, or other contaminants, and should be replaced.
2. If you see powder, chips, or broken beads around the sieve container, the cause may be bed disintegration. These pieces will filter down, causing the air to flow around instead of through the bed without coming in full contact with the desiccant. If this occurs with desiccant beads, the bed level drops. However, these bead beds can often be regenerated. Dump out all of the desiccant, filter and replace it by adding new desiccant, and then vibrate and pack the cartridge tightly to bring the bed level to its correct height.
3. If the dew point is correct at the start of the cycle, but declines too rapidly, the equilibrium capacity of the desiccant has slipped. This could be caused by desiccant aging or acid damage. Some resins containing halocarbons that volatilize during drying will attack the sieve and shorten its drying life. Check with your material supplier to see if the resin releases acidic volatiles during drying. If so, then check with your dryer

supplier to see if a special filter could be used to trap these volatile s before the bed is damaged.

Desiccant Bed Analysis

To check the activity of a bead desiccant bed to see if it needs replacement, run the following test. Because desiccant beds give off heat when moisture is absorbed, you can take its temperature.

1. Obtain a 3 to 4 ounce sample of the desiccant from the active area of the bed about 4 to 6 inches dowm.
2. Dry the sample at 350°F for a minimum of 2 hours. Place in a dry air-tight jar and allow to cool to room temperature.
3. Pour 1.5 to 2 ounces of water into a 6-ounce glass and record its temperature. Use this temperature as the reference point.
4. Pour in the desiccant into a similar size glass to a point 10 percent above the level of the water in the first glass. Place the thermometer in this glass and pour in the water from the first glass.
5. Stir the mixture with the thermometer and note the increase in temperature. Record the maximum peak reading that occurs in 10 to 20 seconds.
6. Subtract the peak temperature from the water reference temperature and record the difference. If the difference is 40°F or greater, the bed is active and the sieve in excellent condition. If less than 40°, the desiccant should be replaced.

To avoid snap decisions, do this test a minimum of three times and average the results to make a final determination.

The desiccant is checked periodically (e.g., quarterly). As the temperature differential approaches 40°F the bed's drying capacity slowly decreases. This means longer drying times are necessary to reach the material's correct moisture level. Raising the dry-air temperature does not increase dryer efficiency, as the bed is losing its ability to absorb moisture from the air. Only by replacing the desiccant will the dryer's efficiency be restored. When regenerating the desiccant bed, refer to the supplier's drying temperature recommendations. Drying the bed above 550 to 600°F will shorten its life by causing thermal breakdown of the desiccant sieve and bed. A desiccant bed with properly maintained filter systems should dry the typical generic resins for 1 to 2 years. But with the newer alloys and their additive packages, these checks will prove helpful in determining the useful life of the desiccant bed.

A problem solving checklist is provided in Table 10.2. Your dryer supplier can also provide a checklist and a recommended item maintenance schedule that should be performed at least once every 2 months.

TABLE 10.2. Solving Common Dryer Desiccant Problems.

Problem	Cause	Solution
Desiccant picking up environmental moisture	Leak in system	Check hopper lid and all hose connections.
	Dryer sitting idle	"Dry cycle" system for several cycles to purge of excess moisture.
Dew point too high, desiccant active	Malfunction in cycle timer, bed heater, bed heater thermostat, air control valves	Check and repair/replace or adjust.
	Desiccant assembly not transferring	If valve system, check and repair valve/drive assembly. If rotational system, adjust drive assembly. Check electrical connections on motor; replace motor if needed.
Poor airflow	Dew point meter incorrect	Check meter and recalibrate.
	Filter dirty or damaged Powder/ chipped desiccant channeling airflow Contaminated desiccant	Clean or replace. Inspect bed bottom. Sift and replenish. Replace.
Dew point cycling from high to low	Electrical malfunctions	Check electrical connections on heater/controller.
	Desiccant contaminated	Replace.
Contamination	Damaged filter	Replace filter, replace desiccant.

Alarm Display	Malfunctions	Solutions
High temperature limit Regeneration process	Either the regeneration or process temperature limit has been exceeded because of an abnormally high temperature in the heater box.	Check both heater boxes for signs of excessive heat. Clear restricted lines. Check for proper air flow in both circuits.
Overload process blower	The overload on process blower has tripped. This could be caused by not having the overload properly adjusted or the motor may be drawing excessive current.	Check setting on overload and adjust if necessary. Reset overload. Check motor current against name plate current to insure motor is not drawing excessive amperage.
Index cycle too long Check limit switch motor	A. The carousel has rotated too far during indexing. B. Carousel has not rotated due to drive motor malfunction or gear-gate drive is defective.	Check position of limit switch relative to bed plate. A. Make sure limit switch is properly adjusted. B. Test motor for proper operation. C. Check motor control circuit including fuse.

TABLE 10.2. (*Continued*)

Alarm Display	Malfunctions	Solutions
Delivery air probe malfunction	Either the temperature sensors have not been properly connected to control box or the sensor is defective.	Check connection. Check probe for obvious damage and replace if necessary.
Regeneration probe malfunction	Same as above.	Same as above.
Process protection probe Malfunction	Same as above.	Same as above.
Process temperature unsatisfactorily low	The process temperature is below an acceptable level. A. Process temperature setting too high for dryer to maintain temperature at set point. B. Dryer may be too far from hopper. C. Air flow may be reversed. D. Heaters may be defective. E. Process temperature sensor not properly positioned at inlet of hopper. F. Supply voltage different from nameplate voltage.	Check heater amperage for defective heaters. Check air flow direction. Check location of process temperature sensor. Check supply voltage against nameplate voltage.
Process temperature unsatisfactorily high	The process temperature is above an acceptable level. Heater contactor may have failed.	Check contactor for damage.

Source: Adapted from Ref. [22].

Dryer Problem Checklist

In developing your own checklist, the following areas should be monitored:

1. Check thermocouples and their calibration.
2. Amp meters at each heating element should be calibrated and checked. Low amperage results in lower heater-output temperatures.
3. Clean traps periodically to remove contamination.
4. Dew meters should be checked and calibrated.
5. Temperature monitors should be checked and calibrated.
6. Air-flow meters should be checked and calibrated, noting if pitot tube is plugged with contamination.
7. Use portable calibrated dew, temperature, amperage, and air-flow meters to verify instrumentation on the dryer.
8. Clean and flush aftercooler coils to remove resin deposits.

Because a dryer only measures the dew point of drying air, there is no way to determine how dry the resin is without taking a sample and running a moisture analysis in the lab. But there is a new, closed-loop microprocessor system available that can monitor the moisture level in the pellets as they exit the drying hopper.

Dielectric Closed-loop Moisture Analysis

The system detects changes in resin dielectric properties caused by variations in its moisture content and converts the readings into measurements. If the resin exceeds preset limits, the system's microprocessor corrects the deviation. The system thus makes possible automatic, online, closed-loop control of moisture in the resin. Benefits include no underdried or overdried material, less scrap, improved part quality, and energy savings. When operated in conjunction with a properly maintained dryer system, this system greatly improves part quality and machine resin processing.

Microwave Dryers

There are also newer microwave or ultrahigh frequency dryers on the market that can shorten resin drying times to 30 to 60 minutes. They use dehumidified air to extract moisture from the system. But, when properly maintained and serviced, any drying system will improve the quality of molded plastic parts and increase JIT techniques.

PLANT EQUIPMENT COOLING SYSTEMS

In a molding plant, temperature control is very important for efficient operation of equipment and part quality. The cooling system must remove the heat generated by the machinery. There are different ways to ensure the cooling requirements for the plant's equipment from dedicated central systems, such as cooling towers and refrigeration units, to portable stand-alone chillers.

Cooling towers or evaporative coolers circulate water at around 80°F to cool such plant equipment as chiller compressors, heat exchangers, hydraulic units, and air compressors. Tower water is cooled by evaporation to within 7°F of the ambient wet-bulb air temperature. Water lost through evaporation, windage, and the bleeding of tower water to maintain mineral concentration at a proper level, must be replaced by treated water. Cooling tower capacity in tons has a heat transfer rate of 15,000 BTU/hr. Towers are rated at 78°F wet-bulb temperature. For each degree above 78°F, towers lose 10 percent of their cooling capacity.

Chiller Systems

Some plants that require large amounts of chilled cooling water have central refrigeration units supplying cooling water to portable chillers. These in turn are cooled by a water tower. A schematic of a system incorporating all the above mentioned is shown in Figure 10.6.

The molding machine's hydraulic fluid system must be maintained at a recommended temperature to obtain uniform pressure outputs during its operation. During the molding cycle, the machine's hydraulic oil temperature increases. As a result, the oil's viscosity is lowered. This change may create cycle variations and varying part quality. Therefore, hydraulic oil temperature must be controlled by the machine's oil coolers. Otherwise, thinning and premature breakdown of the oil may occur and cause burning or tar formation that can trigger hydraulic system failures and erratic pressure outputs. Figure 10.7 shows an oil-to-water heat exchanger that uses cooling water to carry off the excess heat.

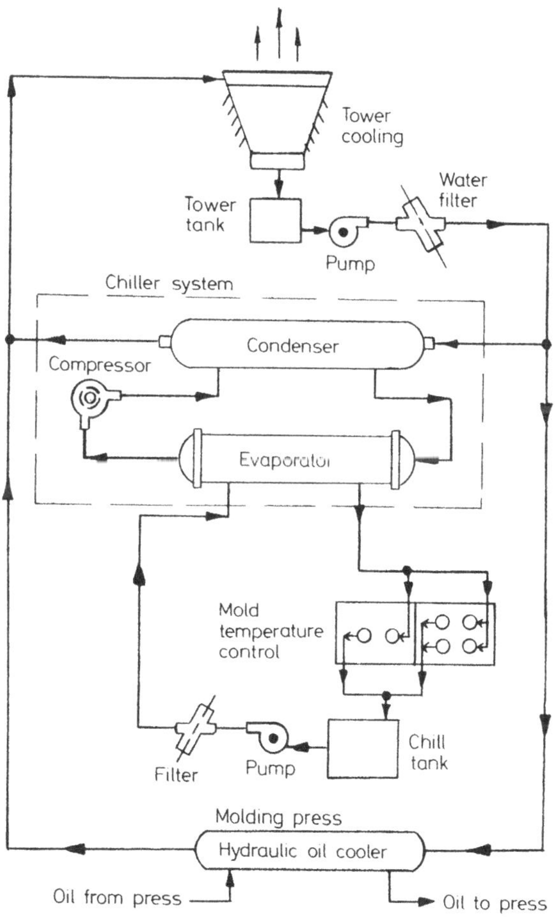

FIGURE 10.6. Plant cooling system. (Adapted from Ref. [15].)

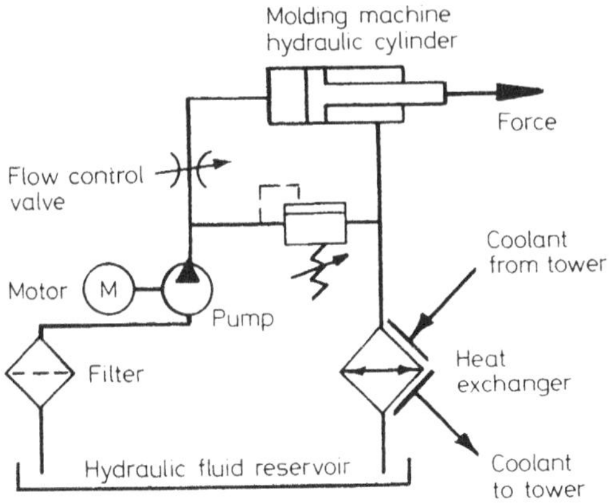

FIGURE 10.7. *Hydraulic cooling system. (Adapted from Ref. [14].)*

There are basically two types of temperature control systems used for molding equipment and cooling. These are the chiller systems and mold temperature controllers.

Chillers. Chillers provide cooling water under controlled temperature conditions from 20° to 60°F. They may be central units using closed-loop refrigeration or portable systems that can be stacked to obtain the desired coolant volume and temperature. Central cooling systems can tie into these portable heat exchangers to remove the heat from the closed-loop system and increase the chiller's efficiency. Water can be used for a leaving water temperature (LWT) of 45°F or greater; a water-ethylene glycol mixture should be used for 45° to 20°F. For more efficient cooling with greater heat transfer at low temperatures, pure ethylene glycol is often used.

The chiller transfers heat from the source to the coolant fluid. The heat is removed from the coolant by a refrigerant system that is in turn cooled by air or water from its condenser. Water-cooled chillers are used when a cooling tower is available. The heat extracted from these units is often used to heat the plant in the winter months, thus making them very cost effective. In the warmer months, the heat is exhausted from the plant.

A chiller's rated capacity is expressed in tons of refrigeration. One ton of refrigeration capacity is equivalent to a heat transfer rate of 12,000 BTU/hr. Chillers are considered 100 percent efficient, when LWT is 50°F. Each increase of 10°F LWT reduces a chiller's capacity by about 20 percent.

Within any cooling system, the coolant water or water/glycol must be maintained, tested, and treated to avoid rust, mineral deposits, and bacteria from developing. If the system is not monitored and treated, these contaminants will destroy it and, over time, plug up your condensers and molds. No water system should be considered harmless until samples have been analyzed. If necessary, an ongoing treatment and testing plan should be implemented.

Chillers require rust inhibitors and water/glycol requires an antifoaming agent when operating above 45°F.

Water treatment companies can test and recommend appropriate chemicals for your system. The results should be recorded to determine whether adjustments are necessary. Plant maintenance should test the system weekly and the water treatment contractor monthly.

All closed-loop cooling systems, from portable chillers to central systems, should have a filter system to remove any contaminants in the liquid system. There are filter systems to remove airborne dust, debris, sand, algae, bacteria, and other solids. The filter systems are designed to handle 10 to 10,000 gallons of flow per minute at pressures up to 250 PSI. They can remove micron-size particles automatically and operate online pressure. Rinse cycles to flush the screens can be short, with a minimal loss of make-up water. The cost of a filter system is minimal, especially when compared with the down time of the cooling system if pollutants that precipitate on the warm surfaces of heat exchanger tubes must be removed. If maintenance is not regular, cooling efficiency falls, part quality suffers, and extensive down time to clean the system will result in lost production time and profits.

It is not advisable to use city tapwater as the cooling medium, because the chemical treatment of this water is questionable and feed temperature will vary from week to week. To get the maximum results from your cooling system, the temperature of the coolant must be tightly regulated. In all cooling systems, a turbulent flow is needed to obtain efficient heat transfer. A minimum flow of 10 gallons per minute is required for turbulent flow. A pump capacity chart to obtain turbulent flow is shown in Figure 10.8 for varying horsepower pumps.

Chiller Selection. When selecting a chiller system you must consider the reliability of the unit and the system. The type of chiller system—air versus water cooled—needs to be decided. Air-cooled units can aid in plant

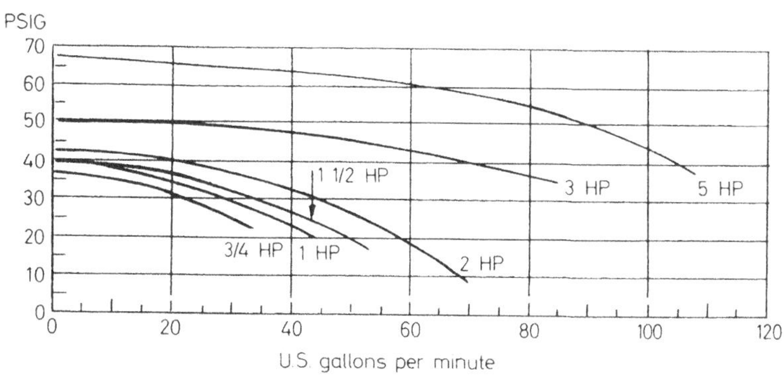

FIGURE 10.8. Pump capacity chart. (Adapted from Ref. [3].)

ventilation and, if located indoors with heating in the winter. If located outdoors, they may have to be modified to operate in ambient temperatures below 60°F. The only drawback is that air-cooled units have about a 15 percent lower cooling capacity than water-cooled equivalents. Water-cooled units should always be installed indoors if the danger of freezing exists. This is why molders located in Southern climates prefer water-cooled units, and those in Northern climates prefer the air-cooled units. Most coolant lines have quick disconnect shut-off fittings to eliminate cooling loss when attaching and disconnecting the line from the mold. The use of standard open fittings is discouraged, because of the chance of electrical shock and loss of coolant during mold change.

The following questions will assist in determining which cooling system to select and what results to expect.

1. What equipment do you want to cool?
2. What temperatures will you require?
3. What degree of accuracy is required?
4. What are the operating cost factors?
5. How energy efficient is the system?
6. How is the system to be sized in relation to anticipated load requirements?
7. Is future plant expansion considered?
8. Can the system be used to maintain plant temperature control?

The answers to these questions can lead to the proper selection and sizing of current cooling needs and assist in planning for the future.

Mold Temperature Controllers

Each mold usually has its own dedicated temperature controller. It regulates the mold's cavity temperature to obtain consistent cycle-to-cycle part dimensions and to produce parts at a specified rate. The amorphous resins require more cooling water than crystalline resins to remove heat from the mold in order to increase part solidification or setup so that cycle time is minimized. The crystalline resins usually use warm-to-hot molds to aid part setup, thereby reducing the mold's cooling requirements.

Most mold temperature controllers have a dual-purpose design—heat or cool the heat transfer fluids in a temperature range of 20° to 250°F. They have built-in refrigeration and heater units to provide the temperatures required for the molding cycle. Some are stand-alone, self-contained units, whereas others are equipped with fittings to be supported by cooling tower or central chiller units. They come with different pump output capacity as well as compressor cooling and heating element capacity ranges to suit different requirements. These controllers are closed-loop systems to and from the mold for uniform temperature control. They can operate as single units or in tandem to support more rigorous molding demands. Running units in tandem will increase cooling efficiency when multiple mold-cooling circuits are required and temperature control is critical. Because some molds may require different cooling temperatures (spot cooling) in different sections of the same tool, more than one temperature controller may be necessary for a single tool.

Chiller Types

Air-Condensed Chillers. There are two basic types of chillers—air-condensed and water-condensed units. The air-condensed units rely on ambient air temperatures (85° to 95°F maximum) supplied by fans or centrifugal blowers. The air is moved across a Freon-to-air coil to remove heat from the chiller coolants. They require only electricity—no cooling water—and are very versatile in the plant's production environment. Although not as efficient as water-cooled systems, they can supplement plant heating and ventilation systems.

Water-Condensed Chillers. The water-condensed systems are more efficient and less expensive. They transfer heat through a Freon-to-water tube and shell condenser to the primary tower, tap, or central chiller water. The primary water temperature should not be above 85°F for efficient heat transfer.

Chillers should have a minimum capacity that is five percent greater than the required load, so that the compressor runs continuously. When loads fall below 50 percent of a chiller's rated capacity, the compressor cycles continuously and this shortens its working life. The chiller capacity listed in Table 10.3 for a 1-ton unit shows the anticipated resin throughput rates, expressed in pounds per hour, for some standard resins. This information is used to

TABLE 10.3. Chilling Capacity of a 1-Ton Unit for Injection Molding.

Resin	Resin Capability (lb/hr)
HDPE	30
LDPE	35
PS	30
PVC	75
PET	45
Nylon	45
ABS	50
PC	60
PPO Alloy	50
Acetal	45

HDPE, high-density polyethylene; LDPE, low-density polyethylene.
Source: Adapted from Ref. [3].

calculate the chiller's unit load requirements based on resin type and pounds molded in each mold per hour. Most chillers can control water temperature within +/– 2°F, if operated at close to load capacity. Most chillers have a capacity control system to regulate the temperature of the cooling water supply and are efficient even at 50 percent of rated capacity.

Chillers are sized to provide a LWT of 50°F. For lower temperatures, an ethylene glycol/water mixture is required. Since low-temperature operation reduces a chiller's ability to remove heat, any temperature below 50°F LWT derates the chiller's capacity by 2 percent per degree. For example, a 5-ton capacity chiller operating at 40°F LWT will have an actual capacity of 4 tons.

$$50°\text{F} - 40°\text{F LWT} = 10°\text{F (2 percent derated per degree F)} = 2.0$$

$$5 \text{ tons} \times 2.0 = 1 \text{ ton derated capacity}$$

$$5 \text{ tons} - 1 \text{ ton derated} = 4 \text{ tons rated for } 40°\text{F LWT}$$

Mold Heaters

Mold temperature controllers may also incorporate a heating function using water, water/glycol, or oil heat-transfer media. A heater coil or plate system is used to heat the fluid for transfer to the mold. Most engineering plastics require a hot mold to maximize part shrinkage in the tool, trigger polymer crystallization, reduce molded-in stress, create better material flow, reduce warpage, and improve surface finish and weld line strength. Mold heaters control water temperatures up to 250°F; special heaters using synthetic heat transfer fluids (oil) can control up to 600°F. High-temperature mold heaters require special hoses to deal with the pressure and temperatures of the heat transfer fluids. Armored lines are recommended to protect plant personnel from hose failures. These units also have additional safety features for rapid shut down in case temperature or pressure override.

Malfunctioning heaters should be checked with an amperage probe. A valve or solenoid may be the only problem. As with chillers, fluid treatment and maintenance of filters are required.

Temperature sensors should be checked with a pyrometer on a periodic basis and replaced as required. Knowing where temperature differentials should occur and measuring them will let you know whether the system is operating properly.

Temperature Setting

One important item to remember when using a chiller is the "cavity" temperature required for the mold. There will be a difference between LWT and actual cavity temperature, which is what the unit controls.

Because of heat buildup in the mold's steel and the efficiency of heat transfer from the steel to the coolant, the chiller temperature will have to be adjusted to obtain the right cavity temperature. Too often, operators accept LWT for actual cavity temperature. The LWT is never the temperature of the mold's cavity surface.

All cooling and heating units, but especially central units, should be located where they can be easily serviced and inspected. They should be in dust-free areas, with adequate ventilation to insure proper heat transfer. Units placed in non-vented or out-of-the-way areas are soon forgotten and inadequately serviced.

A written preventive maintenance procedure should be adhered to for reliable operation and output. The following minimum maintenance schedule, if adhered to, will ensure that the mold temperature-control system will always supply the mold with the correct temperature medium for cycle control and part quality.

1. Monthly service for a plant that operates 24 hours a day, 6 or 7 days per week.
2. Service every 2 months for a plant that operates two shifts, 5 to 6 days per week.
3. Service quarterly for a plant that operates one shift, 5 days per week.

Filters on chillers should be checked daily and always kept in good, clean condition. All inspection results should be recorded, along with any required maintenance, to show trends and prevent future breakdowns.

Maintenance Checks

A maintenance check should involve the following:

1. Lubricate all motors and bearings and adjust or replace drive belts, pulleys, or couplings. Check for unusual noise or vibration.
2. Test the refrigerant circuit for operating pressure, valve settings, refrigerant charge, compressor oil pressure and level, expansion valve operation, water flow through the chiller unit, and heat transfer rate.
3. Test the operation of all water pumps and check seals.
4. Verify that all operating and safety controls are functional.
5. Record voltage and amperage readings of motors under load at motor terminals.
6. Perform chemical tests of the water circuits for air-cooled compressors.
7. Clean and brush the condenser coils. Cleaning is required if head pressure is higher than normal or if a visual inspection indicates foreign material on the face of the condenser.

When chiller problems occur, the solution needs to be quickly implemented to reduce lost production time. Table 10.4 lists the common problems and how to identify and fix them.

TABLE 10.4. Chiller Troubleshooting Guide.

Problem	How to Identify Problem	Solutions
Pump seal failure	Leakage between the pump and motor.	Pump seal life is usually 18 to 24 months. If premature failures occur consistently, install a filter on the water supply.
Cooling valve failure	Unit will not heat or cool on demand from controller input.	The solenoid valve diaphragms from some suppliers rated for 2.5 to 3 million cycles. Install a filter on the water supply if solids contaminate the valve.
Control hardware problems	Temperatures swing beyond control tolerances.	Check flow through the mold. Increase flow if possible. Check sensor devices and remove scale if insulating effect is slowing sensor response time.
Control software problems	Odd displays. Control will not respond to input commands.	Remove the power rack and reactivate the controller. If the controller responds, check the unit's grounding and/or move the unit's power supply away from dc drive motors and resistance welding operations.
Heater failure	Unit will not heat up or fuses blow when heat is energized.	Check continuity and resistance to confirm the failure. Water or scale build-up on sheaths causes heater failure.
Burned-out heaters	No heat.	Check for adequate water flow across the heater and for proper water conditions—fouling of the heater will cause premature failure.
Stuck solenoid valves	The unit will not come up to heat. Grit particles may cause the solenoid valve to stick in a partially open position, causing leakage.	Flush out the solenoid valve. If leakage continues, disassemble and clean, or replace the solenoid valve.
Leaking pump	Leaking pump seals.	Replace seal if water has a considerable amount of particulate contamination. Install a filter or a heavy duty seal. Check seal flush line.
Unit will not start	Check incoming water pressure. It should be 30 to 50 PSI.	Check hand valves and pumping system.
Unit will not cool	Temperature keeps climbing, heater off, cooling on.	Check for restrictions in the cooling water "out" line. Also, check for back pressure on cooling water "out" line.
Undersized water lines	Slow water flow will create a long reaction time, which causes overheating and overcooling, as indicated by temperature fluctuations.	Contact the manufacturer's representative for proper sizing and connections of water passages.
No water treatment	Immersion heaters are scaled.	Ensure water is properly treated. If malfunction continues after scale removal, notify a qualified electrician for service.
Pump rotation is in the wrong direction	Unit will not operate properly. Pump operates backward.	If the unit is unwired for any reason, recheck the proper motor rotation before the motor is rewired.

Source: Adapted from Ref. [4].

The manufacturers of chillers and mold temperature controllers are now installing computer-integrated manufacture (CIM) control systems. These have feedback loops for temperature sensor feed from the mold to hold cavity temperatures to +/– 0.5°F. For the higher quality parts now being demanded, these precision temperature settings are necessary.

GRANULATORS OR GRINDERS

Until all injection molding converts to hot-runner molds, granulators will be required. Granulators are machines with rotating knives and nibblers that reduce sprues and runners and rejected parts to small particles that can be fed back into the system as usable regrind (see Figure 10.9). Most plastic parts allow some percentage of regrind in their manufacture. The percentage is based on the part's specifications, tolerances, and end-use requirements. Any contaminated runners or parts should never be reused. Regrind may also be saved and used for less critical parts or sold as scrap to material reclaimers and reprocessors.

The most important things about regrind are to keep it clean, dry, segregated, and identified. The cost of ignoring these factors exceeds any anticipated savings the use of regrind could have generated.

Granulators are designed for handling specific applications, sprues and runners, parts of varying sizes, different types of materials, and feeding procedures. Granulators can be located by the press, in network molding cells, or centrally located to be fed from multiple machines. They can come with different cutting and dicing systems as well as varying motor output drives and control systems for tying into computer-integrated manufacturing systems. Most regrind occurs from sprue and runner systems. Part and runner size reduction occurs in the granulator's cutting chamber, where a series of rotating knives catch the plastic between stationary bed knives and slice or chop the material into small particles. There are varying designs, such as the scissor cut, which provides a shearing action across the length of the knife as a result of the skewed position of the blades; the straight cut, in which the blades rotate in the same plane as the fixed stationary blades, thereby causing a chopping action; or the rotor design, which can be opened or closed to adjust material flow or particle size, or can direct consistently the parts through the bed knife gap into the screen area. These machines also have special options, such as hardened blades and cutting chambers for the more abrasive materials and nibbling action for finer particle sizes.

To produce consistently sized particles, a screen is mounted on the lower portion of the cutting chamber under the rotating knives. Screen openings can vary in size from ⅛ to ½ inch depending on the ultimate use of the granulated material. The screen retains the material in the cutting chamber until it is small enough to pass through the screen. Screen size and the type of cutting action

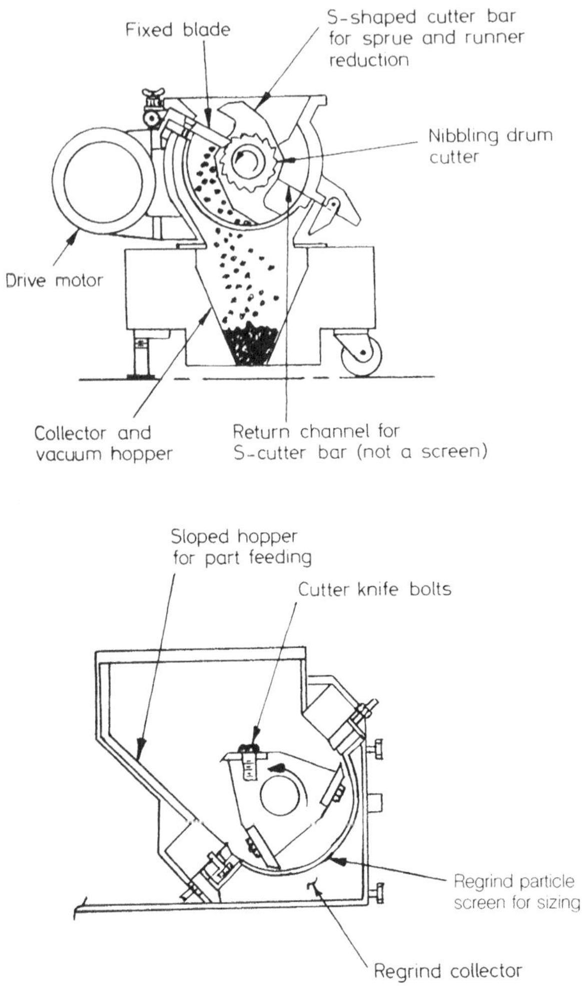

FIGURE 10.9. *Granulators. (Adapted from Refs. [5] and [23].)*

will determine the throughput to be expected with gravity and air-vacuum assist promoting flow into the catch bin.

Granulator specifications are based on the dimensions of the cutting chamber where material is reduced and on the capacity to produce regrind which is expressed in pounds per hour based on the horse power output of the motor. The machine size also depends on the type of resin to be reduced, its heat content if ground hot, and its future use in the plant's operation. When using air or vacuum assist during grinding, a dust and fine particle separator will result in higher regrind quality. Fines and small particles, particularly if hygroscopic, will absorb moisture much faster when ground hot, right out of the mold.

The newer grinders use low RPM grinding systems to reduce noise and fines as well as to produce more uniformly sized particles and conserve energy. Noise levels below 80dB are preferred and can be achieved with the newer, more efficient sound-deadening systems. These low RPM grinders can improve the quality of molded parts by fine reduction, which decreases cutter blade heat that causes burning and black spots, and by producing a more consistent particle size for uniform drying.

Some grinders now have controls that assist in problem diagnosis, obtaining longer granulator blade life, and producing more consistent particle size and output. These sensors need to be monitored. When used in conjunction with alarm and shut-down systems, they protect the granulator and ensure quality regrind.

These controls include ammeters, cutting chamber heat sensors, and hopper and bin level detectors. These sensors record hopper, infeed, cutting chamber, and rotor-bearing temperatures, as well as motor amperage, blade RPM, and granulate level values. As amperage or temperature rises in the cutting chamber, these instruments can determine whether a backup is occurring or if blades are dulling. This could melt the granulate and clog he system. These sensors can also trigger an alarm and shut down the machine for service or repair. This feature can prevent the burning up of motors, bearings, or belts, while preserving the quality of the regrind.

Other sensors can monitor grinder operation to produce quality regrind consistently. For example, when molding cycle speed increases, more hot and semisolid parts are fed to the grinder. Such a situation requires cooling the air or using water coolers in the cutting chamber to reduce heat. For heat-sensitive resins, these cooled cutting chambers or low RPM cutters may be necessary to avoid degrading the resin regrind.

Granulator Selection

Several types of granulators are available; their selection should be based on the following considerations:

1. Scrap-part size, resin type, and the size of the runner to determine throat and feed opening size.
2. Thickness and hardness of the part and its temperature when reduced to determine blade type and configuration, as well as power requirement.
3. Throughput requirements to maintain acceptable online process speed.
4. Particle size required and screen selection size available.
5. Fine and dust collector, with associated air-vacuum assist feature connection.
6. Sound deadening below 80dB.
7. Machine size and floor space available.

8. Ability to add or automate feed system both to and from the granulator.
9. Control and monitor the sensor system to protect regrind quality and machinery.
10. Adaptability for easy cleanout if multiple machine/resin use is planned to enable quick change over from one dedicated application to another.
11. Ease of maintenance and availability of spare parts.

Press-Side Granulator

The most popular granulator is the press-side unit shown in Figure 10.10. It is portable and can be placed in a position by the machine for hand, automated (robot), or conveyor feeding from one or more molding machines. The throughput rate with most resins and parts is up to 800 pounds of regrind per hour. Special hopper designs are available to allow for both manual and automatic feed. The press-side unit is also more adaptable for rapid cleanout and change over to granulate other materials.

FIGURE 10.10. Press-side granulator.

The change over or cleanout is very important for protecting the quality of the next molded material. Too often, an air hose is used to blow out the granulator and collector bin. More quality problems occur at this point than any other in the material handling system. When change over occurs, the granulator should be opened up to inspect the cutting chamber, knives, and screen, and all material should be brushed, wiped, and vacuumed out. The dust filter should be changed, the bin vacuumed and wiped clean, and the conveyor system to the hopper thoroughly cleaned and inspected. Failure to do this may leave small, foreign resin particles in the system. These can contaminate the new resin in many future cycles. A written procedure should be followed for this change over with each machine setup.

For sensitive resins and critical parts, a grinder is often dedicated to a particular setup and production run. Cross contamination is expensive to reverse; therefore, cleanup and change over must be handled correctly.

Central Granulator

Central granulators are more powerful than press-side granulators. With 30 to 400+ horse power (HP), they are located in secluded and soundproof areas. They are fed by hand or by conveyor systems using only one or two main resins. Often, they are connected to a central dryer and proportional feed system for their injection molding machines. The centrally fed granulators are usually designed for automatic sprue and runner feed, using conveyors or robotic devices for part and runner separation. It is usually speed controlled and timed to grind the scrap at set rates. They may run continuously or when triggered as parts are fed into the feed section. They are usually dedicated machines, servicing at least two presses using the same resin. They are often computer controlled and tied into a closed-loop grind-dry and material-feedback system.

One of the most important features is the safety system. When accessing the system, safety interlocks must first open and shut down the system. Flyback doors at the hopper must always function to trap any plastic particles that might fly out of the feed hopper into the work area. Hand access into the feed hopper must be limited to prevent injuries that result from contact with the cutting system.

Granulator Problems and Maintenance

The granulators are the most abused and often the most neglected pieces of equipment on the production floor. Operators should read the manual and receive instruction in equipment operation and maintenance. Many quality problems can be traced, after the fact, to poor use or misuse of the granulator.

The five most common production problems associated with granulator performance are as follows:

1. Fines or irregular particles
2. Low output
3. Melted, fused, or overheated particles
4. Rotor stalling
5. High noise levels

Table 10.5 lists some of the most common granulator problems and their solutions. In addition to these problems, the knives must always be kept sharp and their material appropriate for the plastic to be ground. Granulator knives or blades come in two major configurations—"high shear" for easily cut uniform size parts and "landed" for hard plastics and large chunks. Each has a different cutting-edge blade design. The knife steel should be selected for the toughness and abrasiveness of the plastic to be granulated. The knife hold-down bolts and washers should be replaced every third or fourth time the blades are sharpened, because of thread stress fatigue and elongation.

The screens should always be inspected for wear, cracks, and plugged holes, and should be reoriented in the chamber if design allows. Throughput will suffer as the screen holes become elongated and trap oversize particles. In a more serious case, the metal can fail and fall into the regrind. In addition, replace worn, cracked, or elongated drive belts and recheck the tension on replaced belts after they have run for a week.

Maintenance, based on the machinery manufacturer's recommendations, should be followed periodically. When grinding harder and more abrasive materials, more frequent maintenance is required. Maintaining the condition of your granulator will ensure higher quality for your customer's finished parts.

PART REMOVAL, CONVEYOR SYSTEMS, AND ROBOTS

Control of part quality after ejection from the mold cavity is accomplished in three ways—operator removal, free fall to a conveyor system with part/runner separation, and robot removal with or without part/runner separation.

Operator removal is the least consistent. Larger cycle variations will and can occur due to operator inconsistencies. In time, these variations will alter the cycle times and result in varying part quality. Just a few seconds' variation from cycle to cycle will disrupt the temperature parameters on the mold and barrel. For critical parts, this may cause variation in dimensions and/or other factors that affect a part's quality and acceptance.

Some plants have installed sequenced, timed lights that provide a visual check on the accuracy of part removal. This helps operators develop a part removal rhythm, so that the cycle stays within its preset values. This system has improved part quality and enables relief workers to easily maintain the uniform cycles.

TABLE 10.5. Granulator Problem Solving.

Problem	Solution
I. Granulate condition	
1. Fines, dust	a. Knife worn, dull, or wrong edge angle
	b. Incorrect rotor-bed knife clearance (usually too small)
	c. Screen size too small
	d. Granulate take-away problem: Blower not running or belts loose; back pressure from blocked ducts or plugged filters; collection bin full (nonblower machine)
2. Ragged, frayed chips (variable bulk density bridging in machine feed hoppers)	a. Knife worn, dull, or wrong edge angle
3. Overheated material	a. Knife worn, dull, or wrong edge angle
	b. Granulate take-away problem: Blower not running or belts loose; back pressure from blocked ducts or plugged filters; collection bin full (nonblower machine)
	c. In-feed too fast
	d. Insufficient air flow (check for leaks from cutting chamber)
4. Melted material (usually film or fiber)	a. See 3a
	b. Incorrect rotor-bed knife clearance fiber (0.0005 to 0.001 in recommended for film)
	c. Too little air flow at rotor (open- or helical-type recommended)
	d. See 3b
II. Machine performance	
1. Low throughput	a. Loose rotor drive belts
	b. Granulate take-away problem: blower not running or belts loose; back pressure from blocked ducts or plugged filters; collection bin full (nonblower machine)
	c. Knife worn, dull, or wrong edge angle; knife clearance too wide/incorrect installation (usually accompanied by defects in granulate)
	d. Plugged screen (look for worn, funnel-shaped holes)
2. Rotor stalls	a. Overloading: too much material, too fast; parts too large or too thick for machine rating
	b. Loose rotor belts
	c. Overload switch defective or improperly set (should correspond to motor amp rating)
	d. Incorrect rotor direction (reverse motor polarity)
	e. Granulate take-away problem (see Ib above); plugged screen (see 1c above)
3. Machine vibration	a. Worn bearings (may also run noisy and hot)
	b. Loose flywheel
	c. Unbalanced rotor (check for adhering material)
	d. Bent rotor journals (caused by jamming from tool, loose bolt, heavy chunks)
	e. Loose belts ("galloping effect")

TABLE 10.5. (*Continued*)

Problem	Solution
4. Noisy operation (more than 85 dB)	a. Throat flaps taped open or missing b. Worn knives (sharpen; replace with quieter scissors blades, if not used) c. Knife motion (loose bolts) d. Insufficient or worn insulation (add, replace, install sound absorbing enclosure)
5. Metal-to-metal noise	a. Loose object (e.g., bolt) b. Knife-screen contact c. Knife-knife contact

Source: Adapted from Refs. [26] and [27].

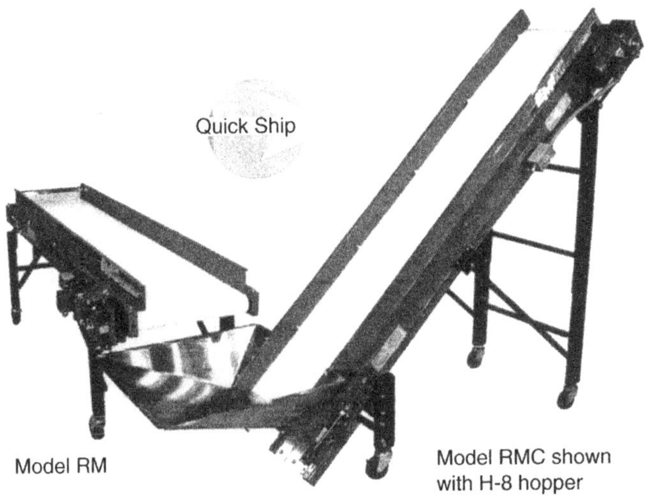

FIGURE 10.11. *Part conveyor. (Courtesy of EMI Corp.)*

Conveyor and Part Separator Systems

Conveyor systems move parts and runners to the next operation in the production cycle (see Figures 10.11). They transport the ejected part and runner to part separators (see Figure 10.12). The parts are then sent to their stations; the sprue and runner to the granulator. These are completely automatic operations, whose productivity is limited only by the auxiliary support equipment in the material flow system. Part conveyor systems can be located around, in, and under the molding machine.

The height of many conveyor systems can be adjusted to minimize the drop distance from the mold cavity to reduce part problems. Potential problems include scuffing, surface blemish, denting, or dimensional problems. These are usually associated with amorphous resins that are still hot and soft when

FIGURE 10.12. *Part/runner separator. (Courtesy of EMI Corp.)*

ejected from the mold cavity. The crystalline resins are usually rigid enough to prevent denting, but surface blemishes can occur from too high a drop weight.

Conveyors located under the mold to catch ejected parts can be continuous running or time-and-cycle indexed. Most molds are set to run automatically, with parts falling onto a conveyor to be sent to runner and part separators.

If the parts are not automatically degated by the mold operation, the part and runner system can be conveyed to an automatic separator (see Figure 10.12). Here the part and runner system have a series of pin-style separators. The detached parts go to another conveyor system, while the runner is sent to the granulator for feedback or to a collection station. Conveyor systems can also reduce operator handling by using part unscramblers. After separation, parts are moved downstream for different operations or packaging.

Controlled drop systems maintain correct part orientation onto the conveyor by chute, guide pins, or rails built into the mold. The conveyor system can also incorporate part quality control checks before conveying parts to the next station. Part quality can be greatly enhanced by the selection of the correct checking and conveying system. Three-plate molds, which have automatic part and runner separation in the tool, lend themselves to this system. Separate conveyors are used for the runner and parts to guide each to its destination.

Conveyor systems are also used for "weigh scale" box filling. This allows for precise bulk packaging of small parts into shipping or storage containers.

The system utilizes a conveyor tied into a scale and a photoelectric eye and control system (see Figure 10.13). This system can also operate on a cycle count method, if part weight is not involved.

For parts not sent directly to shipping, there are compact, beside-the-press, tote stackers (see Figure 10.14). They collect molded parts, using cycle count or part weight, for conveyance to the next operation. The conveyor system is

FIGURE 10.13. "Weigh scale" box filling. (Courtesy of EMI Corp.)

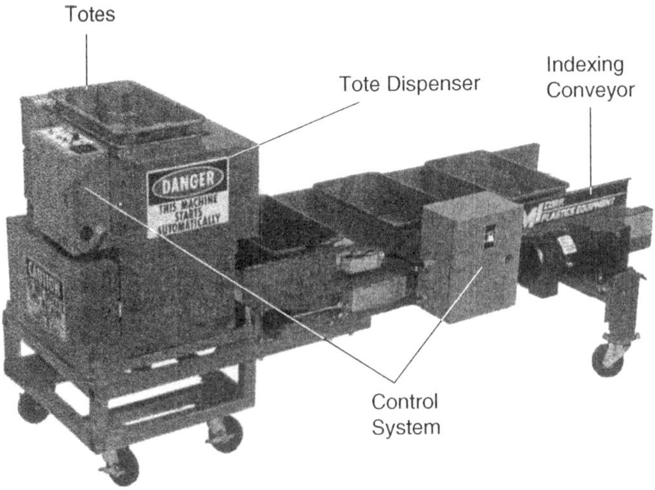

FIGURE 10.14. Tote stackers. (Courtesy of EMI Corp.)

cost and quality effective in moving parts downstream from the molding machine. Documentation shows that part quality can be increased if fewer operators are involved in its manufacture and the production equipment is tied into a process control system.

Robot Part Handling

Robots can improve part quality by maintaining consistent molding cycle times. Cycle consistency is important, as varying mold-open times will affect the resin's melt consistency. This may produce inconsistent and variable part quality. A typical platen-mounted robot is shown in Figure 10.15.

Using robots for part removal may slightly increase overall cycle time because programmed motions can be slower, but, the gain in quality and overall productivity offsets the extra time. If equipped with sensors, robots can also signal the controller if a sprue, runner, or part is still in the mold. They can be programmed to remove both parts and runners, placing each in its respective position.

Robots can be mounted beside the press or, in many cases, on the fixed half of the molding machine's platen. Using pneumatic and/or electric servodrives, they can delicately remove parts from the mold without marring their surfaces. Robots can also perform such secondary operations as degating, insert loading, part stacking, and assembly-line loading.

Companies are now employing multijointed selective compliance assembly robots (SCARA) in conjunction with part removal robots, conveyors, part

FIGURE 10.15. Platen/shelf-mounted robots. (Courtesy of K. Kuka, Augsburg, Germany.)

collectors, and/or packaging equipment for beside-the-press assembly, inspection, and packaging. These independent work cells complete all checks and packaging for product shipment.

Robots, if properly selected for the machine—not the job—can be reprogrammed. The easiest is the electricservo type, as programming is done with computer software, and not with pneumatic controls that require manual stops. The selection of robots must be well thought out; the supplier must be aware of all functions required. Some common problems associated with robot operation and their solutions are listed for future reference in Table 10.6.

Robots will not improve a bad mold design, but they can improve part quality by maintaining consistent cycle times, assuring part and runner removal, ensuring proper inspection, and performing part placement in box or on a conveyor for travel to the next station. Productivity and quality consistency are the main rewards of using robots.

QUALITY INSPECTION EQUIPMENT

By incorporating automatic quality-inspection equipment in the manufacturing loop, the molding machine's process control settings can be monitored and compared with the finished part's manufactured parameters. This can be accomplished by using computer checkpoint-control setups based on cavity pressure feedback sensors. If cavity pressure during fill and packout does not meet set pressures, the part will be switched to the granulator automatically by a swing-gate part separator on the drop chute or conveyor line below the mold. This is accomplished by selectively locating pressure transducers behind the cavity knockout pins to record cavity packing pressure. The necessary packing pressure is determined during production start-up. This is discussed in greater detail in Chapter 11, under "Cavity Melt Pressure Control."

Another quality check uses weight as the deciding factor. This check assumes that when parts are molded consistently to maximum part weight, based on cavity dimensions, they will meet customer requirements. This is verified during production startup to determine the exact part weight necessary to meet tolerances.

For very small molded parts, weight testing is done for each cycle. With larger parts, one is separated from the runner system and weighed immediately after ejection from the mold. Because weight variations from cycle to cycle in the runner system do not always affect part quality, parts are separated from the runner system and either fall onto the scale or are placed there by a robot to record their weight.

The new scales are accurate to within fractions of a gram and very fast, computing weight in approximately 1.5 to 3 seconds. The molded parts are then compared with the preset acceptable part-weight values. The scales

TABLE 10.6. Robot Problem Analysis.

Problem	How to Identify Problem	Solution
Loss of air	Molding machine stops automatically. Visual and audible alarms triggered.	Fix the air compressor.
Incorrect air pressure	Robot moves too fast or too slow. Air cushion not working.	Visually inspect the air pressure gauge.
Mold-close safety switch inoperative or bypassed (mold damage potential)	Check safety switch operation.	Replace switch.
Improperly adjusted speed controls	Robot slams into its shock absorbers in overspeed mode; cycle time increases in underspeed mode.	Adjust controls as required to achieve desired stroke.
Water in air lines	Air filter fills with water.	Replace filter. Add dual filter. Install an air dryer at the compressor.
Loose wire or bad contact	Robot stops moving. Alarms.	Check the assigned wire number listed in the manual. Track it down in the robot.
Proximity switch misalignment caused by impact. Position sensor switches poorly adjusted.	Robot stops moving at the same place during its operational sequence. Alarms.	Check the assigned wire number, track it down, and replace if necessary. Do not check using metal objects to trigger proximity switches. This can cause unexpected machine motion. Monitor switch status on programmable controller or make voltage checks. Check mechanically actuated switches in same manner
Fluctuating air supply	Intermittent motion. Difficulty in controlling speed.	Ensure adequate air supply. Check solenoid valves for proper operation. Check incoming air filters. Check pressure at all regulators. Inspect robot for kinked hoses, leaks.
Improperly sized air circuits	Operation is too slow.	Increase air pressure.
Poor cushion and cylinder selection	Sometimes cylinders slam together.	Adjust air speed and cushion controls.
Poor cylinder alignment	Cylinders bind.	Measure any alignment.
Insufficient air lubrication. Contaminants on self-lubricating cylinders	Resin buildup in valves, or sticking valves. Slow or choppy cylinder operation.	Wipe down cylinder with a cloth rag. Always maintain the same oil level in the reservoir. Check air lubrication to robot from oiler.
Excessive accumulation of oil on oil-feed cylinders	Low air pressure. Cylinder operates slower. Oil leaks.	Call the manufacturer if problem persists.

TABLE 10.6. (*Continued*)

Problem	How to Identify Problem	Solution
Shock absorber failure	Robot abruptly slams into a hard stop rather than decelerates.	Adjust shocks to proper position and damping, or replace shock absorber from robot spare parts inventory. Then adjust speed controls to avoid slamming into shocks.
Broken, kinked, leaking pneumatic lines or electric wires	Abrasion or kinks, erratic robot motions. Loose or broken wires or cables show up intermittently as system "bugs."	Visually inspect pneumatic lines. Keep spares handy. Check for loose connections on all wires. Visually check wire guide for cables riding on top of each other. Check for spiraling or twisted cable. Replace if condition exists. Check all valve cases, junction boxes, and terminal strips for proper connections. Adjust or replace if necessary. Use ohm meter to insepct continuity.
Worn seals	Robot exhibits fast motion when moving with gravity, and sluggish motion when moving against gravity. Air flows through the port when an air-line is removed from the underpressurized side of the cylinder.	Replace seals.
Solenoid valve failure	Robot axis of motion actuates manually by depressing the valve actuator but not electronically at the robot operator station.	Replace valves.

Source: Adapted from Ref. [10].

generate a response—accept/reject—prior to unloading the parts onto a conveyor for saving or scrap.

This system can also signal operators if molding machine parameters begin to drift or go out of process control, as uniform part weight is a very accurate method of determining part and cycle-to-cycle process control quality. Part weight variations can signal the beginning of processing, resin, and mold problems, so that an operator can check these problems immediately.

Part weight systems are also used as mold protection devices. If the calculated shot is not recorded every cycle, the machine's clamp operation will not close the mold. This ensures that all parts have been ejected from the cavity, so that the mold is clear and free to close for the next cycle.

QUICK MOLD CHANGE

With expanding just-in-time delivery requirements, quick mold change (QMC) is of great interest to captive and custom molders, who need to implement this fast mold change over while maintaining part quality. QMC can be accomplished in 15 to 30 minutes, as opposed to the 4 to 16 hours associated with traditional methods. With JIT manufacturing methods, more frequent and smaller orders, and shorter lead times, quick change overs are necessary. The emphasis is always on producing quality parts for shipment that will lower part inventory levels but increase the capacity and utilization of the molding equipment. Considerable planning is required for QMC. The tooling equipment—both molding machine and auxiliary, personnel, and planning—must all be adapted to QMC, if it is to be successful. Management must insure that the assets and quality checks are put in place.

The first item to consider for QMC is the mold, which must be designed for a quick change operation. The molding machine must be equipped to accept the mold in a horizontal or vertical change procedure. Adapting the mold, the molding machine, and auxiliary equipment for QMC involves standardization of interfaces and automation of the steps necessary for changing the mold quickly.

Then, determine which mold performs best in which machines and which auxiliary equipment is needed for the scheduled job. Third, determine which machine and auxiliary parameters must be set, which resin used, and how fast and accurately can the system produce quality parts. All operations must be planned and documented, so that everyone involved, from purchasing to production, is aware of their part in making QMC work.

QMC Requirements

For a quick mold change to work, the following steps must be followed:

1. Mold designed for quick change operation and startup. Mold must be in operating condition and cleaned after last run.
2. Molding machine adapted for QMC and able to mount and run the mold.
3. Scheduled change over should have written procedures that are understood by personnel involved.
4. Material ordered.
5. Material dried and available for feed to press.
6. Cooling equipment available to fit mold and molding conditions and cycle temperature control.
7. Mold transport/change equipment available and in good working order.
8. Personnel trained in QMC procedures.

Determining which molds are best suited for QMC is keyed to the run size and production schedule. The piece-part cost decreases as manufacturing lot size increases, because the cost of mold change over is amortized across an increasing number of parts. Therefore the frequency and time spent performing QMC has a definite effect on part cost. When the molding machine is idle, the molder bears the cost of this lost production time in the piece-part price. The production run time should also be long enough to cover change over costs, while making a profit on the parts produced. Evaluating which molds should be considered for QMC is very important. All molds may not justify QMC modification, for the following reasons.

For large lot sizes—100 or more hours of run time—manual mold change (4 to 16 hours) may be justified, as mold change occurs infrequently. With lot run times from 10 to 100 hours, involving intermediate lot sizes, a semiautomatic mold change system with a 30- to 60-minute change over may be appropriate. In this case, some manual operations are used in conjunction with some level of automation to speed the mold change process.

For smaller lot run sizes typical of JIT scheduling, involving 10 hours or less—a fully automated system with a mold change time of 5 to 15 minutes is cost justified. Labor rates are sometimes used to determine which molds and machines are set up for QMC. With more JIT schedule requirements and the ability to produce more quality parts in a shorter time, this should be the main determining factor.

Key Factors

The six key factors for a QMC operation are as follows:

1. Run size justified; mold and machine adapted for QMC operation.
2. Mold handling in and around machine, equipment, and space.
3. Mold locating in the machine—vertical or horizontal; rollers and guides with mold locators.
4. Mold clamping, automatic hydraulic quick release with positive locking.
5. Utility connections, prepositioned in tool as quick disconnects.
6. Preheat/purge machine; resin dry and fed to hopper. Mold preheated if required.
7. Auxiliary equipment available and preset to molding requirements.

Implementing QMC

QMC also offers safer handling of molds. This affects employee morale positively. Adopting the QMC philosophy also improves machinery maintenance, as the molding press and auxiliary equipment must always be in good operating condition. Any unscheduled downtime can seriously affect scheduled piece-part output.

When the decision is made to implement QMC, the mold must be adapted to fit the machine's QMC clamping system, which ranges from manual to hydraulic. For each machine or series of machines, consult a QMC supplier. The new mold requires oversized standardized base plates to adapt it to the machine's standard clamping position. Figure 10.16 shows both the mold and machine clamping and multicoupler attached to the movable platen. Figure 10.17 shows a backing or base plate (used interchangeably) with the mounted utility connection required for QMC operation. Existing tools can be modified for QMC, and the standardized base or bolster plates (used interchangeably) can be mounted to adapt the old tool for QMC.

If QMC is contemplated in the near future, it is less expensive to order a new tool with QMC base plates than to convert a tool after it is made. Base plates are standardized, but they may not fit between all machine tiebars and may not fit the clamping and positioning systems for all QMC machines. There are many different suppliers of QMC systems, and the mold and machine must be compatible. Verify the system to be used and talk with your molders or inplant personnel to see which system is required. Always verify the molder's system, equipment, and capability for performing QMC with the system you plan to use. If the tool has to be run in another press or in another molder's plant, will your QMC mold system fit and work?

The position of loading the mold in QMC is restricted to the system on the molding machine to be used. Most QMC systems mount obstructions on the platens (see Figure 10.18). These systems consist of mold clamps, automatic multicouplers for mold temperature control, and the mold ejector system coupler. The mold must be mounted vertically or horizontally. Most QMC systems are horizontal, as they are faster and handle heavier tools. The only drawback is the precise space requirements necessary to get the change table up to the press to load the tool. If the presses are too close, then vertical loading, which uses an overhead crane or A-frame, may be required. Evaluate each plant to determine which system is best.

FIGURE 10.16. *The components of QMC: (1) Clamps with capacities to 90,000 lbs. (2) Multicouplers for utility connections. (3) Ejector couplers for 6000- and 20,000-lb. models. (4) Mold-changing tables for molds to 16,000 lbs. (5) Offline preheating stations. (Courtesy of Clifford E. Drake, Manager, Quick Mold Change Systems, ENERPAC, a unit of Applied Power Inc.)*

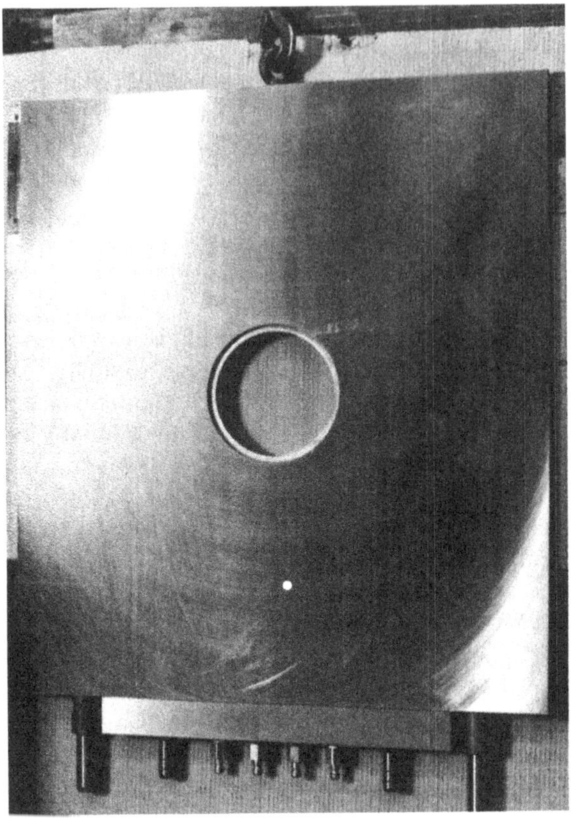

FIGURE 10.17. QMC back plate with "strip coupler" utility connection for air and water hookups. (Courtesy of Clifford E. Drake, Manager, Quick Mold Change Systems, ENERPAC, a unit of Applied Power Inc.)

Many molders have concluded that horizontal mold changing is easier, faster, and safer if a mold changing table is used. There are three basic types—one is manual and two are automatic. The manually operated table uses a scissors lift system to replace the forklift or overhead crane. The automatic tables come in two varieties. One is dedicated to a specific molding machine and operated from either the machine's controller or a discrete programmable controller that allows unmanned mold changes. The second table, or cart, has a more versatile design to service more than one machine. It requires operator assistance to change the mold. The automatic QMC systems also favor a pre-heated tool on the table, thereby saving valuable startup time.

Locating the mold on the machine's platens is accomplished off the standardized mold backing plate. Depending on the loading method, this positions the tool for correct alignment and clamping as shown in Figures 10.19A (horizontal) and 10.19B (vertical).

FIGURE 10.18. *Mold changes, automatic ejector coupler, and multiple utility connections on a movable platen. (Courtesy of Clifford E. Drake, Manager, Quick Mold Change Systems, ENERPAC, a unit of Applied Power Inc.)*

An alternative method is to locate the QMC mold using studs mounted on the mold. This requires modification of the mold's platens to match the stud plate. It eliminates the standard mold backing plate used for the other two systems. This method is shown in Figure 10.20. Mold clamping would be similar for all systems.

Once in position, the mold is manually or hydraulically clamped. This quick-clamping feature is a major time saver over the old manual method of clamps and bolts. The three clamping types are Belleville washer, toggle, and wedge clamps. The Belleville washer system has more parts and is more expensive. It is used mainly on small machines using hydraulic pressure to retract a series of Belleville washers as the mold is loaded. When the hydraulic pressure is removed, spring force clamps the mold in place.

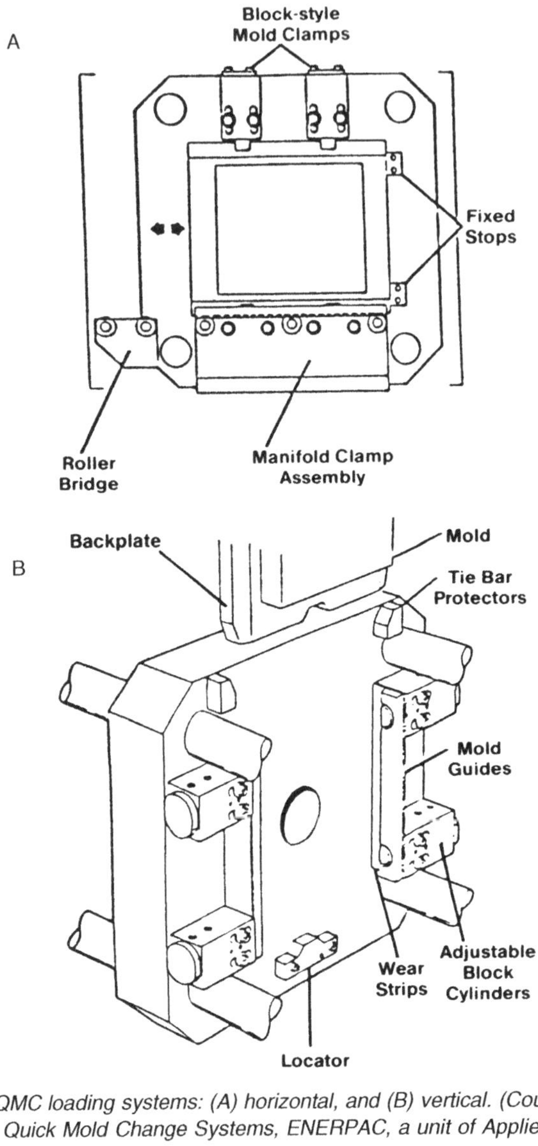

FIGURE 10.19. *QMC loading systems: (A) horizontal, and (B) vertical. (Courtesy of Clifford E. Drake, Manager, Quick Mold Change Systems, ENERPAC, a unit of Applied Power Inc.)*

The toggle clamp uses a hydraulic cylinder to advance a clamp linkage similar to the toggle action on the molding machine's mold clamp system. To unlock the mold, the clamp must be brought back over the center. The toggle-clamp system is sized so that the force of opening the machine cannot unlock the clamp.

Wedge clamps—the most popular—use a hydraulic cylinder to move wedge blocks to clamp and lock the mold in place. The wedge may be inside the cylinder or external to it and always requires a higher force to unlock than to

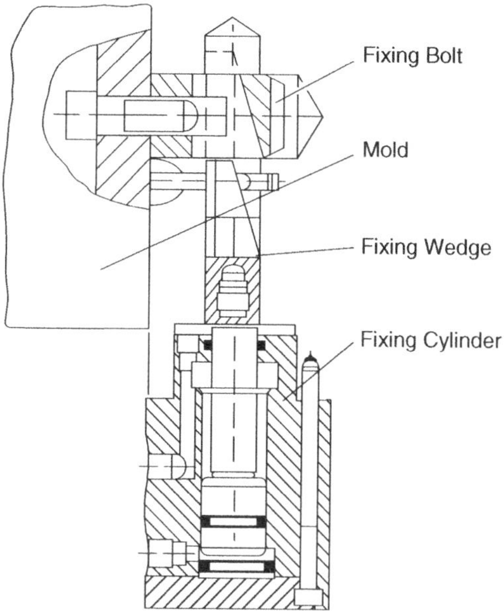

FIGURE 10.20. *Tool locating system. This tool location system uses studs mounted on the mold rather than the backing plate location method. (Courtesy of Clifford E. Drake, Manager, Quick Mold Change Systems, ENERPAC, a unit of Applied Power Inc.)*

clamp. These are all "positive locking" type clamps that stay closed even if hydraulic pressure is lost.

In smaller presses, less than 75 tons, the manual system for securing the mold to the machine's platens is often similar to the master unit die (MUD) system. This is the wedge-type clamp, using manual racket assemblies to replace the hydraulic system. Even with the smaller presses, however, the trend is away from the consuming manual system. The competitive edge can be lost over the hydraulic lockup.

For successful QMC, the utility connection to the mold must be handled safely. Connections for water, air, oil, high-pressure hydraulics, and mold operation must be considered. Back plates, as shown in Figure 10.17, with prewired and plumed standard utility connections are required for QMC to benefit from this system. When the mold is set or removed, automatic connections are made using "clean break" couplers, which release little or no oil or water on separation. Electrical connections can also be made through the back plate, but should be spaced away from the "clean break" couplers. Most molders prefer separate electrical hookups directly to the mold, using color-coded or different plug sizes for correct connector hookup. The QMC multicouplers are located to suit their horizontal or vertical installation configurations.

Precise alignment for these "clean break" connections is required, and a power assist unit may be required to overcome the high-coupling forces needed for these connections. If clean-break couplers are used, then oil, water, and electrical connections can be located in the same header box. For more tightly controlled mold temperature, you may also require additional cooling lines to the mold. These lines are often located in an additional, external, cooling header. The hookup plugs and connectors from this header to the mold should be sized differently to ensure that the in-and-out cooling circuits are hooked up correctly. These headers are external to the tiebars but are often permanently attached to the molding machine's frame. The cooling lines are then routed to the mold, away from other machine and mold operations.

The ejector plate connections can also be made either manually or hydraulically. Actuated quick-connect devices attach the ejector pins to the machine's ejector plate. These devices are the screw, wedge-lock, and ball-lock mechanisms. The screw engagement ejector, which is time consuming but simple, is made by a racket system installed on the machine. The wedge lock, or collet

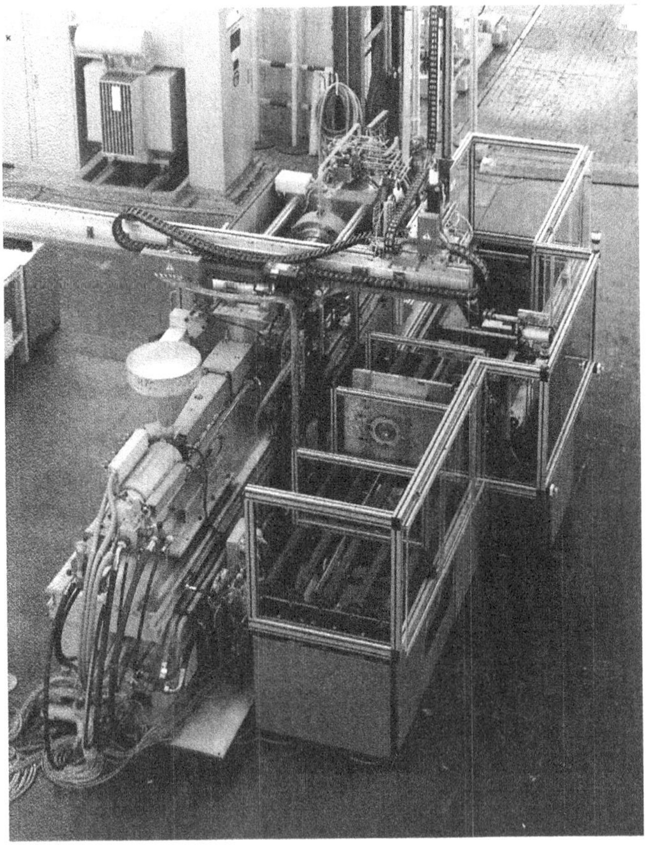

FIGURE 10.21. Quick mold change operation. (Courtesy of Clifford E. Drake, Manager, Quick Mold Change Systems, ENERPAC, a unit of Applied Power Inc.)

closure, uses hydraulic pressure to close the collet in place. The widely favored ball-lock mechanism uses a cluster of spring-loaded balls that grip a ring around the ejector pin.

Effective QMC requires advanced planning, scheduling, and personnel training. All tools and special equipment must be available with the mold preheated, if required. Auxiliary equipment must be in place and operating according to setup procedures. Resin should be loaded. If all of these items are not ready, the time saved in changing the mold will be lost. The full QMC system is shown in Figure 10.21.

Other key elements in the QMC equation are choosing the right molds, scheduling the machines, and selecting resins and their colors. The goal is to minimize the number of variables when the change is made.

Planning, training, and accurate scheduling are the keys. Running molds and materials with increasing temperature profiles and materials, from light to dark, can save considerable time and money. This may require running some jobs out of sequence and possibly storing some parts ahead of firm orders, but better use of your equipment and fewer variable changes may result in higher profits. This will also give you just-in-case inventory for your JIT customer orders.

Dedicating specific machines and molds for the same or similar materials may also be a solution. Examples are multiple hoppers, natural resin with color feeders easily interchanged, removable feed hoppers, and vacuum material evacuation units. With good planning, training, support equipment, and well-written procedures, all of these systems will make QMC work.

Good planning, scheduling, and personnel training are required to bring all the various components together for a successful production run. Procedures for setup and operation of the production cycle, coupled with a good maintenance program for your auxiliary equipment, will greatly assist the plant and production personnel to produce quality products.

11

Processing

Above all else, the consistent production of high-quality molded plastic parts is the most important function of your equipment. When production begins, the manufacturing equipment will be adjusted and set to control the molding cycle; it must remain in process control during the entire production cycle. But variations will occur in material and machine processing conditions that have to be monitored and accommodated. Production personnel must find solutions to any processing problem and make adjustments to meet part specifications. Successful companies have capable manufacturing equipment and trained personnel, who will can recognize process deviations and make adjustments.

PRODUCTION STARTUP FOR PROCESS CONTROL

Production startup procedures must be followed to incorporate the correct machine startup settings. If there are any known material or equipment deviations, they should be communicated in writing to the production personnel before startup. These should cover changes in material, machine settings, available equipment, or any other change that affects the production cycle. Once production begins, alarms should trigger any equipment deviations outside of set parameters. Once production starts, the operator should monitor and record equipment variables and molding cycle operations. The molded parts should be checked to be sure they follow specification and, if necessary, the cycle adjusted. Before final cycle adjustments are made, the manufacturing

Total Quality Process Control for Injection Molding, by M. Joseph Gordon

equipment must be in thermal equilibrium with cycle operations at steady-state conditions. This will, be determined by measuring temperature, time, pressure, process measurements, and the molded parts. Parts should not be saved, until all the above conditions are met.

Once quality parts are being produced, monitor the production cycle at regular intervals to catch any changes due to material or equipment variations. These changes must be corrected in "real time," so that the production cycle stays within its control limits.

Production control cannot wait until the manufacturing run is completed to verify the acceptability of the molded parts. It must supervise the manufacturing operation, know what cycle and part parameters to monitor to verify quality, and then monitor equipment settings. This is necessary to ensure control is maintained through the production run.

In the past, the operation ran from startup through complete part production with little, if any, monitoring of deviations or checks on part quality. This method used statistics to accept or reject a production run according to an acceptable quality level (AQL) sampling plan. The AQL method was first developed to grade mass-produced units during World War II. It was based on a statistically developed plan for a specified lot and sample size that accurately represented the whole lot. AQL acceptance tables and limits were originally established to eliminate complete inspection of parts and to establish an agreed on part-quality level. The customer and supplier used the Master Sampling Table (one of the many developed) shown in Table 11.1 to determine acceptable quality level. Then, based on these measurements, the lot was judged good or bad. If a lot was rejected, then the manufacturing equipment and operator procedures were checked. The rejected parts were segregated to be scrapped or thoroughly inspected to salvage the good parts.

ACCEPTABLE QUALITY LEVEL LIMITS

At first, the biggest problem with AQL was that defective parts, if not detected in the sample size, were accepted. Statistically, these were few, but always possible. This method is now out of date, too expensive, and unacceptable in meeting current quality requirements. Using AQL methods, total quality cannot be assured. This is a serious drawback, because customers are not allowing defective parts to be knowingly shipped to their plants for detection by their incoming quality inspection department. The goal is to prevent bad parts from reaching the assembly lines.

The tighter the AQL selected, the larger the sample size. As the AQL increased, the level of overall lot part quality decreased, as it allowed acceptance of a greater number of defective parts. For example, using an AQL of .25 and a lot size of 3199 or less, a minimum of 125 parts must be inspected for acceptance (see Table 11.1). If no out-of-specification parts were found in this sample, the lot was accepted. If two were found, it was rejected. But if

TABLE 11.1. Master Sampling Table.

		Acceptable Quality Level																													
	Sample	0.25		0.5		0.75		1		1.5		2		3		4		5		6		7		8		9		10		12	
Lot Size	Size	A	R	A	R	A	R	A	R	A	R	A	R	A	R	A	R	A	R	A	R	A	R	A	R	A	R	A	R	A	R
499 or less	40	*		*		*		0	2	0	2	0	3	1	4	1	4	1	6	2	6	2	7	3	7	3	8	4	9	4	9
	50	*		*		*		0	2	0	3	1	3	1	4	2	5	2	6	3	7	3	8	4	9	4	9	5	10	5	11
	60	*		*		*		0	3	1	3	1	3	2	5	2	6	3	7	4	8	4	9	5	10	5	11	7	12	7	13
	70	*		*		*		1	3	1	3	1	4	2	5	3	6	4	8	5	9	5	9	6	10	7	11	8	13	8	13
	80	*		*		*		2	3	3	4	3	4	4	5	5	6	7	8	8	9	8	9	9	10	10	11	12	13	12	13
500 to 799	40	*		*		*	1	*	2	0	3	0	3	0	4	1	5	1	5	1	6	1	7	2	8	2	8	2	9	4	10
	60	*		*		*	1	0	2	0	3	1	4	1	5	2	6	2	7	3	8	3	9	4	10	5	11	5	11	6	12
	80	*		*		0	2	1	3	1	4	1	5	1	6	3	7	3	8	5	10	5	11	6	12	7	13	8	14	9	15
	100	*		*		0	2	1	3	2	4	2	5	2	6	4	8	5	9	6	11	7	13	8	14	9	16	10	17	12	18
	120	*		*		1	2	2	3	3	4	4	5	5	6	7	8	8	9	10	11	12	13	13	14	15	16	16	17	18	19
800 to 1299	40	*		*	1	*	1	*	2	*	3	0	3	0	4	0	5	0	6	1	6	1	7	1	8	2	8	2	9	2	10
	60	*		*	1	0	2	0	2	0	3	0	4	1	5	1	6	2	7	2	8	3	9	3	10	4	11	4	12	5	12
	80	*		0	2	0	2	0	3	1	4	1	5	2	6	2	7	3	8	4	10	5	11	5	12	6	13	8	15	8	15
	100	*		0	2	0	2	0	3	1	4	1	5	2	6	3	8	5	10	5	11	7	13	7	14	9	15	10	17	10	18
	120	*		0	2	1	3	1	3	2	5	2	6	3	7	5	9	6	11	7	13	8	14	9	16	11	18	12	19	13	21
	160	*		1	2	2	3	3	4	4	5	5	6	7	8	9	10	10	11	13	14	15	16	16	17	18	19	19	20	22	23
1300 to 3199	50	*	1	*	1	*	2	*	3	*	3	0	4	0	4	0	5	0	6	1	7	1	8	2	9	2	10	3	10	3	11
	75	*	1	*	1	0	2	0	3	0	4	0	5	1	5	2	7	2	8	3	9	4	10	4	12	5	12	6	14	6	15
	100	*	1	0	2	0	2	1	4	1	4	1	5	2	6	3	8	4	9	5	11	6	12	6	14	8	15	9	17	10	18
	125	0	2	0	2	1	3	1	4	2	5	2	6	3	7	4	9	5	11	7	13	8	15	9	16	11	18	12	20	13	21
	150	0	2	0	2	1	3	2	5	2	5	3	7	4	8	6	10	7	13	9	15	10	17	11	19	14	21	15	23	16	25
	200	1	2	1	2	2	3	4	5	5	6	6	7	8	9	10	11	13	14	17	18	17	18	20	21	22	23	25	26	27	28

A—Acceptance number; R—Rejection number.
*No acceptance at this sample size.
Source: Adapted from Ref. [5].

one defective part was found, then 75 more parts had to be measured for a total sample size of 200 parts. If no additional defective parts were found, the lot was accepted; if one more was found, the lot was rejected.

The higher the AQL limits, the smaller the sample. Start with an AQL level of one and a lot size of 499 or less. If no defects were found in the first sample size of 40 parts, the lot was accepted. If two were found, the lot was rejected. But if one defective part was found in the 40 samples, 30 more samples had to be inspected. If no additional defective parts were found, the lot was accepted; if three defective parts were found, it was rejected. But if a total of two defective parts were found, than 10 more samples were inspected. This was the last point of accepting or rejecting a lot of this size for this AQL level.

This process, which took place after production, was very time consuming, open to human error, and expensive. It led to high scrap rates, with no provision for adjusting operating conditions. The only feedback production received was whether they were making good or bad parts. A company often had to overproduce to ensure sufficient good parts or, if the lot failed, perform a 100 percent inspection.

Because many parts may have one or more problem areas, a method was needed to classify and quantify defects. The statistical measuring and monitoring technique used to evaluate parts is the Pareto analysis technique, which looks at nonconformance of manufactured products for specific problem areas. This can include both the evaluation of purchased components and the manufactured product. With total quality process control (TQPC), this analysis after molding is not desired, as faults in the manufacturing cycle should be corrected during the real-time manufacturing process.

However, it is a good technique for judging and classifying the quality of parts supplied by a vendor, such as inserts, screws, or other components used in the final product. Incoming parts are inspected; defects are noted and ranked as to type, quantity, and severity. In some cases, the specification of the part may be too rigid for the vendor. The final outcome may be a relaxation of the specification, provided the final part quality is not affected. But in other situations, either select a new vendor or require improvements in the vendor's final part inspection. This is especially true of molded-in inserts or items encapsulated in a plastic part.

An example of a Pareto analysis for a female, knurled, threaded molded-in insert is shown below. The defects are ranked by degree of seriousness for later analysis and resolution with the supplier.

Pareto Distribution Analysis.

Specification	No. of Defects	% Defective	Accumulative %
Thread size	51	38	38
Insert length	42	31	69
Insert diameter	36	27	96
Knurling	5	4	100
	134		

As a result, the vendor can improve control on these key features and ensure that the customer receives parts to specification. Should defective parts pass through, a form of restitution can be negotiated. This technique focuses on detecting and quantifying defects, as well as possible resolutions of problems based on vender capability and cooperation.

Manufacturers can no longer afford the additional costs incurred when a product does not meet customer requirements. The supplier also cannot afford just-in-case inventory to cover production problems. Because AQL inspection knowingly includes some defective parts in the lot, the only way to eliminate defective parts totally is by a 100 percent inspection performed by either the molder or the customer's incoming inspection department. But, with parts going direct to the assembly line, this is not acceptable. Customers are eliminating incoming inspection and are requiring suppliers to guarantee the quality of incoming parts. With the emphasis and responsibility now being put on the producer of the molded parts, new methods of guaranteeing part quality had to be developed.

Quality must begin with the part's design and continue through the building of the mold and control of the material and production process. Management must provide the appropriate assets, direction, guidance, training, and working conditions. Manufacturing can only make quality parts if this support is provided and the process is capable.

Often, manufacturing is forced to make the parts with equipment, a mold, and material that is marginal at best. In most cases they succeed, but this requires extra time, work, and costs. Management is now recognizing that production is just one link in the chain of part quality. Officials are now buying better equipment with process controls, making molds that will be capable of holding part tolerances, providing operator training in process control techniques, and selecting resins and suppliers to meet quality requirements. This has led to the incorporation of the techniques of TQPC for the manufacture of plastic parts. With a good part design and capable manufacturing equipment, successful production can proceed.

NETWORKING PRODUCTION

Obtaining good production quality involves networking all of the manufacturing machine's control parameters. The microprocessors on the injection molding machine should be used with closed-loop, self-adjusting, and parameter-averaging feedback machine controls. This keeps the molding process within established limits. Some machines can also tie in, monitor, and control the process-line auxiliary equipment.

This can be successfully accomplished by using statistical process control (SPC) concepts, which require the continuous tracking of process parameters for changes in machine operations. By using statistical quality control (SQC) methods to monitor and track selected machine, process, and part variables (e.g., part weight, injection, pack pressures, and other process parameters), the system will indicate when a part is out of specification or if a change in one of the production control parameters has occurred. SPC monitors and controls the key-part and process-control limits by electronically networking them into real-time machine and part-output information.

This information is often tied into a central or local computer system, which monitors the process control limits for uniformity and compliance. If any problems occur in the manufacturing system, they can be identified and analyzed. The operator, machine controllers, or host computer systems can then implement corrections to continuously adjust process variables as required. For more serious problems that the system cannot handle automatically, an alarm will signal the need for additional attention to solve the problem. This electronic coupling and monitoring of machine process variables, along with part quality inputs, will instantly confirm part quality and that the processing parameters are within the manufacturing window.

THE INJECTION MOLDING PROCESS

To understand how process control is used to maintain part quality during the injection molding cycle, the methods and processing variables associated with the manufacture of molded parts must be investigated. The injection molding cycle involves a reciprocating screw-injection molding machine that processes a plastic resin from pellet or powder form into a molten state. The molten plastic is then injected under pressure into a mold cavity, which forms the part and, after cooling, ejects it from the mold. To accomplish this, the controls and settings of the molding machine's process parameters are based on the resin, mold, and support equipment's processing conditions. Auxiliary equipment is also an integral part of the process. Each resin has different processing parameters as well as settings that vary from machine to machine. When molding machines are changed, the parameters must be readjusted to suit the new machine's process variables. These new settings use the first machine's settings as a reference point. No two machines are ever the same, because wear varies. This is especially true of the 400-ton and smaller machine sizes, as they are the workhorses of the industry.

Each plastic resin will have supplier-recommended processing conditions, such as barrel-heater band-zone temperature settings, melting- or softening-point temperature, injection pressure, injection-speed profile, time needed to fill the mold, melt viscosity range, mold temperature settings, allowable regrind use, and resin setup time. The setup operation translates this information into molding cycle machine settings. To ensure successful molding and manufacturing, a machine setup and operating procedure is established and followed.

This is necessary so that at the start of a new job, or when rerun, the parameters are the same. It is a real-time document that is updated as new information is acquired or settings are revised.

Mold Startup Procedure

When a job is put on the production schedule, it is the responsibility of production control to coordinate all aspects of the process until parts are shipped or released to another department for secondary operations or assembly.

If the scheduling has been properly coordinated, purchasing has ordered the resin with sufficient lead time for incoming quality control to have tested the material and/or received a letter of certification from the supplier that the resin meets the required specifications and falls within processing limits. If this is a job that has been molded before, these material value limits should be on file. If it is a new program, then these limits will have to be developed with input from the resin supplier. This may be as simple as obtaining a supplier's letter of certification or it may involve establishing new material parameters during the initial startup and testing program.

Continuing the startup procedure, the production department must make sure that all auxiliary equipment is available and in good operating condition. The department must also verify that the molding machine will also be ready—purged and in good operating condition—to hang the mold and start production.

The setup person and subordinates will check the job setup procedure to learn what specific equipment is required. This includes material preparation, such as drying, as well as blending and feeding procedures. If the job setup procedures are written and followed correctly, change over and startup time can be minimized. Setup time varies from minutes to many hours, depending on the preparedness of the startup team and the availability of equipment. This is where good scheduling pays off for optimum use of a plant's equipment assets.

Startup procedures, which specify the equipment and tools required to perform the task, must be reviewed and equipment availability verified. Remove the tool or mold from storage. Verify its condition and the fact that all part revisions have been made. This should be indicated on the engineering sign-off sheet and mold maintenance records from the mold's last run.

After all necessary auxiliary equipment has been approved, the setup team must review the mold setup procedure. Each mold's setup procedure notes all requirements for correctly installing and removing the mold from an injection molding machine. If the mold has been installed in more than one machine, there should be a written procedure for each. These operations require a necessary skill level to protect and safeguard the personnel, machine, and mold from damage or injury.

Setup personnel are responsible for correctly installing the mold in the injection molding machine and setting up the auxiliary equipment. They are also responsible for setting all molding process parameters and verifying that all equipment functions correctly before turning the job over to a production operator. This includes getting the machine and mold on cycle.

Monitoring Mold Setup and Startup Procedures

A new technique used by some custom molders is to videotape the techniques for changing molds and starting production. This can be used to critique the startup procedures and as a way of improving and training new personnel in correct and safe startup procedures. The video can highlight problem areas that affect set-up time and can explain correct mold installation procedures, along with correct fittings and support equipment. It can also illustrate the procedures, material requirements, machine settings, and other factors required to produce quality parts to customer specifications. The video, if properly edited and updated, can be a useful teaching and troubleshooting guide for all major production jobs.

In most cases, change over and startup occur at the end of a scheduled production run. If a machine is idle, the mold may be installed hours or shifts before its scheduled startup time. In these cases, approval must be obtained from the production manager and shift foreman. The state of completion of the setup must be recorded, and if not done in one shift, passed on to the next shift. Therefore, when actual startup begins, the setup operator will know what has been completed. No mold should ever be left partially installed or removed during a shift change without the authorization of the molding room or production manager.

If this occurs at shift change, the mold must always be firmly clamped to the molding machine platens. The ending and beginning shift supervisors, along with each shift's setup supervisor, must be aware of the change over status and stage of completion. This ensures that the process technicians are kept up to date on the change over, especially if the mold is scheduled to start up on their shift.

Setup Operator Responsibilities

The setup operator has total control over setting and removing molds as well as starting up and shutting down production runs. Specifically, the operator should do the following:

1. Supervise and install molds in injection molding machines in a safe and correct manner to ensure proper functioning.
2. Ensure that all auxiliary equipment is connected and operating as required.

3. Startup the molding machine and establish a molding cycle that will produce parts that meet customer requirements before releasing the machine to production.
4. Record molding cycle conditions for startup and note any nonconforming conditions of the mold, cycle, and auxiliary equipment.
5. Ensure that all safety equipment and interlocks are functional and within tolerance.
6. Check out auxiliary equipment and hardware daily to be sure it is operational and in calibration, with clean filters and no unsafe or damaged items.
7. Make sure all equipment, machines, and molds in machines are in a safe condition when not running.
8. Maintain a clean and safe shop area, free froze resin, water, and hydraulic oil spills or purgings. Pick up and return all tools and equipment to their respective storage areas.
9. Assist in problem solving when parts go out of tolerance and provide operator training to monitor machine conditions.
10. Perform other duties as required and requested by the molding room production manager to ensure the safe operation of the molding department.

Each molding department has its own specific mold installation procedures. In general, following procedures can be followed and modified as required for each plant's operation:

1. Check the injection molding machine to ensure that the last setup conditions have been removed. Verify all safety equipment is in good operating condition and that no equipment problems exist from the last production run. Return any special or auxiliary equipment to its proper storage area if not required for this setup.
2. Review the specific mold's setup procedure for special and auxiliary equipment and tools, and verify that they are available and at the molding machine.
3. Stage the mold for transfer to the molding machine in a safe manner. Molds are very heavy and must be moved to the machine in a manner conducive to easy and safe installation. Be sure the mold transfer equipment is adequate and that lifting lugs, chains, and safety straps are securely in place.
4. Stage the mold at the press and check that all items required for installation are available. Review with the setup support personnel the procedure for installing the mold and their respective duties. For the setup to proceed, the following terms should be understood:

Platen(s): The portion of the molding machine to which the mold is attached. The fixed platen, or "A" plate, is in front of the barrel's nozzle, where the plastic is injected into the mold. The fixed platen holds the locating ring that the mold mates to during installation in order to center the mold to the nozzle. The movable platen, or "B" plate, is attached to the clamp ram that clamps the mold tightly during the injection cycle. The ejector ram that operates the mold's ejector system is also attached to the moving platen.

Locating Ring: Located on the fixed platen to align the nozzle of the injection molding machine's barrel with the sprue bushing on the mold.

Sprue Bushing: The mold's sprue bushing mates with the barrel's nozzle to form a pressure-tight seal through which the plastic is injected into the mold. Exact seal off is required to resist injection pressures as high as 20,000 PSI without leaking.

Clamp Plate: The portion of the mold used to clamp the mold halves to the platens. The top plate is clamped to the fixed platen and the bottom plate to the movable platen.

Mold Clamp: The metal device used to clamp the mold to the machine's platens.

Clamp Toe: The front portion of the clamp used to clamp the mold clamp plate to the machine's platen.

Clamp Heel: The heel portion of the clamp, where the adjusting bolt hole is located, plus the attachment of the clamp to the platen. The adjustment bolt hole is usually sized the same as the bolt used to attach the clamp to the platen.

Adjustment Bolt: The short bolt with a knurled head used to make the critical adjustment in the distance between the heel of the mold clamp and the platen to match the thickness of the clamp plate.

5. Preadjust the mold clamps to fit the thickness of the mold's clamp plate. The toe and heel should be parallel to the platen. If the heel is adjusted too far out, the mold could slip during production. Final adjustment is usually made when the tool is in the press before the final clamping adjustment.
6. Adjust the mold for its insertion between the machine platens and tiebars by either a vertical or horizontal method of installation. Be sure all personnel are safely away from the mold during this operation and that the mold is securely attached to the loading equipment.
7. Align and position the locating ring on the mold, with the ring on the fixed platen. Slowly close the machine's movable platen until contact is made with the movable half of the mold, but do not develop any clamp pressure. Adjust the mold's position until it square with the machine's platens.

8. Apply clamp pressure on the mold to hold it in place that equals the clamp tonnage required for production.
9. Install the mold clamp for the stationary, or "A" platen, side per setup procedures and torque each bolt to its specified level. The mold is now self-supporting, but leave installation supports in place until the movable half is clamped. Open the clamp and install the ejector rods if not done before mounting the mold. Be sure the ejector rods function properly before reclamping the mold to full-clamp pressure. Install the clamps on the movable half to the platen. Torque all bolts again and remove the mold installation equipment.
10. The mold's machine-clamping setup must be done correctly. With the center hydraulic piston clamp, the platens will close parallel to each other. With the toggle clamp, care must be taken that the platens are parallel and tiebar loadings are uniform. This is necessary for even clamp pressure. The mold should be sized and adjusted to the injection molding machine's clamp tonnage to provide adequate, but never excessive, pressure. Excessive clamp pressure on the mold can result in platen or mold-plate deflection coining the mold at its parting line. This is important for toggle-clamp machines; each tiebar takes its equal stress loading on lockup. If tiebar adjustments are not correct, unequal clamping occurs and flashing of the mold may result or the mold may be crushed locally.
11. Connect auxiliary equipment, mold chiller, core pulls, robots, or other special features as required. When connecting the mold's water lines, blow shop air through each circuit to be sure they are open and clear. Then carefully verify that the inlet and outlet connections are correct. Most mold-cooling problems occur at this point because of connections to the wrong lines.
12. Set operating settings for auxiliary equipment and verify that they are correct by taking cavity temperature readings no sooner than 15 minutes after equipment startup. This may take longer when heating the mold, which is a large heat sink. Verify that the correct nozzle is installed on the barrel. The nozzle tip radius must fit the sprue bushing correctly and the tip's flow channel—straight or tapered for amorphous or crystalline resins, respectively—must suit the type of resin to be molded for prevention of drool.
13. Enter the parameters for the molding cycle into the machine's controller. After heatup, begin slow material feed while cycling the machine to create the necessary melt temperature. Take air shots to verify melt temperature, button up the press, and begin molding initial parts in the manual mode.
14. Operate the machine until temperature equilibrium is achieved in mold and melt temperature, and parts are to tolerance. Verify the process

parameters and record their settings on the molding record sheet. Turn the press over to production for part's manufacture.

15. Police the area and return all tools to storage; clean up any spills and purge. Instruct the operator on any special conditions to be monitored and answer questions as required.

This is the general procedure for a cold startup. On a change over, this procedure is slightly modified. After the mold is hung and all auxiliary equipment is operating, the injection machine is purged of material from the previous run and the new material barrel temperature profile is set. A purge material may be needed to clean out the barrel fully. Be sure that the new and previously used materials do not decompose on contact, causing a gas formation. A neutral purge resin, such as polyethylene, may be needed to clean out the barrel before adding the new material. Once barrel heat-up is achieved, set the screw to turn at low rotations per minute (RPMs) and start material feed.

INJECTION MOLDING STARTUP

Setting the injection molding cycle consists of the following operations, which should be performed after the machine and mold are operating:

1. The molding machine's feed hopper gate is opened and predried resin is fed the slow moving screw.
2. With the nozzle backed away from mold's sprue bushing, set the screw turning at operational RPMs while the resin is conveyed forward through the feed, transition, and into the metering section. During this transition, the material absorbs heat from the barrel and begins to melt. The molding machine is now an extruder that melts the resin through conduction, heat from barrel, and barrel/screw shear heat. As the resin moves down the barrel, it is compressed, melted, and pumped forward in front of the screw tip by the shallowing screw flights. Material feed continues with the melt extruded from the nozzle and air shots taken until the molten resin is up to operating temperature. When the melt quality seems uniform, melt temperature is taken. If within molding parameters, the machine's nozzle is brought forward into contact with the mold's sprue bushing, the mold is closed, clamp pressure is applied, and the molding cycle begins. Now, the screw is forced back by the increasing volume of molten resin being built up ahead of its tip at the front of the barrel. The volume of melt produced in front of the screw is based on the number of ounces of material needed to fill the mold and runner system. A slight excess in volume is added to provide about

a ½ inch of melt cushion, or pad, in front of the screw for maintaining packing pressure after the injection stroke has been made. The volume of resin buildup is controlled by a limit switch that stops the screw's rearward travel and rotation when tripped. If melt temperature is off, machine adjustments are made until it is within the required zone.

3. Hydraulic screw-back pressure may be applied against the screw's rearward motion to create additional screw-shear heat if difficulty is encountered in producing a good melt. This may be required if rapid cycling of the press results in minimum travel to melt the resin, unmelt pellets occur, or the screw is inefficient in producing a uniform resin melt temperature. Heat-sensitive and glass-reinforced resins should use minimal pressure, if necessary, in the range of 25 to 50 PSI. This is because an increase in back pressure will result in higher glass-fiber breakage, which results in lower physical properties in the molded part.
4. The molding cycle continues with a shot of resin buildup in front of the screw. The screw is then stopped from rotating, and the injection cycle begins with the screw forced forward by hydraulic injection pressure. The screw is now acting as a ram that forces the resin into the mold cavity. The check ring at the screw tip is forced rearward, sealing off any material backflow over the screw. The resin is ejected into the mold at a selected first-stage high-injection pressure and speed. The injection-pressure and speed settings can be selected to suit the resin and mold filling characteristics desired. To determine the maximum injection pressure to use, calculate the number of square inches of part and runner surface area and the clamp tonnage of the molding machine. This must be computed so that the mold does not spring open during the injection stage, causing flashing. As resin is injected, the trapped air is vented through the cavity venting system. To calculate injection pressure, divide the pounds of clamp pressure, converted from tons, by the number of square inches of surface area of the part and runner:

$$\text{injection pressure} = \frac{\text{clamp pressure (tons)}(2000\,\text{lb/ton})}{\text{mold cavity and runner surface area}\left(\text{In}^2\right)}$$

For example, the maximum injection pressure for a mold with 100 square inches of surface area for a 350-ton clamp machine would be 7000 PSI of pressure. This would be 700 PSI gauge, as there is usually a factor of 10 of gauge pressure to actual injection pressure. The injection pressure varies based on surface area of material and machine clamp pressure. The time to fill the mold is based on the shear sensitivity of the material plus the cavity's gate size, and it can vary from slow to fast speeds with variable injection pressures. Today's machines have controllers that can be programmed to give the molder any fill pressure pattern and time desired.

5. After filling the mold from 90 to 95 percent during the first stage, high reduced injection is reduced by half to a lower second packing pressure stage to finish filling the cavity. If injection pressure was not reduced rapidly, the mold cavity would flash. This packing pressure is then held on the cavity until the gates freeze off and no more material can be forced into the cavity. During second-stage packing and before the cavity gate freezes off, the part cools and shrinks. As a result, pressure must be maintained on the melt to pack in as much resin as possible. This compensates for the resin's shrinkage during cooling.
6. After gate freeze-off, the cavity is sealed and the screw can start rotating to build up the volume of resin for the next shot. The part, which is now cooling in the cavity, continues to shrink according to its molecular structure. The part must now cool enough to become rigid, so that when the mold opens it can be ejected from the cavity without warping and the mold's ejector pins will not penetrate the surface to distort the part. Part cooling time is based on the type of resin, part thickness, crystallinity of the resin, mold temperature, and complexity of the part. This is normally the longest time period in the entire cycle.
7. The mold then opens and the part, sprue, and runner are ejected. The mold then closes with maximum clamp pressure applied, the screw injects more resin, and the cycle continues.

Figure 11.1 shows the different operations and simulated times for a typical injection molding cycle.

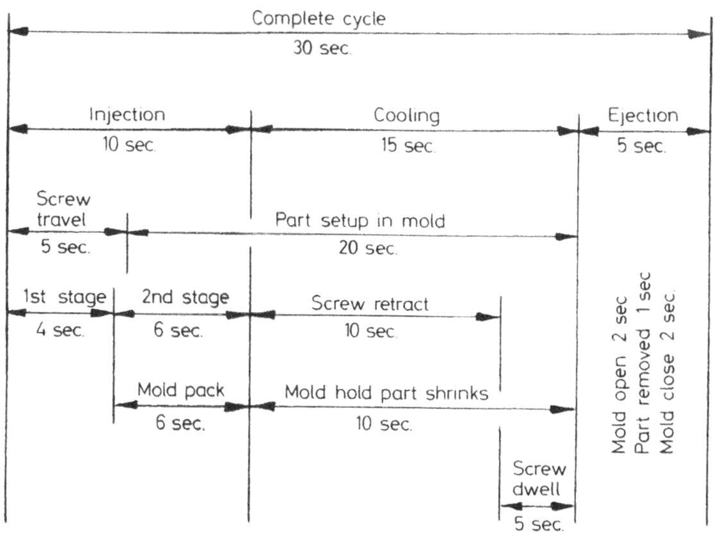

FIGURE 11.1. *The injection molding cycle.*

Setting the Cycle

The procedures are very similar for setting up new and existing molds. For a new mold, all cycle parameters for temperature, pressure, and time settings must be established. For existing molds, the information should be available from the last run. But if the existing mold is to be run in a different molding machine, the setup person can use the last production run's recorded process conditions for startup. Later, when cycle "equilibrium" is reached, the process variables can be fine tuned for optimum cycle and part quality.

The time required for the process to reach equilibrium varies based on the machinery used, molding machine, mold size, resin, and auxiliary equipment. Only after equilibrium is reached; temperature, pressure, and time profiles are holding; part dimensions and weight are consistent from cycle to cycle should parts be saved for the customer. This may require 15 to 30 minutes or more, especially for the mold, which does not reach equilibrium quickly. On startup, part measurements must be taken long enough to determine that the mold is in temperature balance. A pyrometer should be used to measure the steel temperature of the mold's cavity surfaces. Never rely on feed-water temperature as an indicator of mold cavity temperature. Heated molds must often be insulated from the machine's platens, which act as huge heat sinks that drain heat from the mold. Large molds are often preheated to reduce startup time.

A melt temperature can be taken just before saving parts by opening the mold on the regular cycle time and taking an air shot through the sprue into an insulated container at "reduced" injection pressure. The operator should be very careful, wearing long insulated gloves and eye protection. If the melt temperature is within limits, continue the cycle and begin saving parts. This method of obtaining a quick melt temperature causes only minor cycle interruptions. It is preferred to interrupting the molding cycle by backing the nozzle from the mold to take an air shot that results in melt and mold temperature variations.

Written procedures should provide each mold's processing requirements, including necessary equipment, preferred molding machine, and start-up and production control settings (e.g., temperature, cycle times, injection rate, and pressures). This assists plant personnel in verifying that all support equipment is available and the operator has the information to set up and run the job without delays or problems. Production must also know the required part quality checkpoints and the correct quality measurement tools to use.

Startup Procedure

Startup begins with a checklist of required equipment that is easily verified on a master list if bar codes are used. If bar codes are not used, then use the checklist to verify that all items are available, correctly hooked up, and properly set. Items to be verified are as follows:

1. Correct molding machine is available and ready to accept mold.
2. Molding machine is purged and cleaned from last run, with all systems calibrated and functioning correctly.
3. Machine has the correct screw for the resin and nozzle type (reverse taper or general purpose is on the barrel) to be used.
4. Resin type, product number, lot number. Resin is dried and in the hopper.
5. Mold is mounted to machine platens and adjusted for correct clamp pressure. Cavity dimension is verified to be within tolerance and with correct sprue bushing to mate with machine nozzle.
6. Heater or cooler is hooked up to the temperature control connections. Coolant flow and temperature settings are verified. Use a pyrometer to verify the coolant temperature feed to the mold is correct. Always check cavity surface temperatures to know when the correct temperature is reached and in equilibrium. Once molding begins, cavity temperatures may have to be readjusted.
7. Molding machine's barrel temperature profile is set. Once achieved, begin slow resin feed with the screw turning at low RPMs. Pressure settings are made, screw travel to build up shot volume is adjusted, and the screw's RPMs increased to produce a uniform melt at the required temperature.
8. Cycle times, screw injection pressures (first and second stage), and injection profile and time are set. Pack and hold pressure and times as well as mold open and close times are set, along with their pressure and speed settings.
9. Verify and operate the system. Take a melt temperature reading before bringing the barrel forward to mate with the mold's sprue bushing.
10. The first shot, injecting melt into the mold, is usually calculated to be short, especially for a new tool. The pressure settings are also set lower to prevent damage to the tool. Overfilling can result in flash, thereby locking the part in the mold and possibly damaging the mold. By shooting short, you can easily increase shot volume and pressure until a full cavity is obtained.

Once a full shot is obtained, the cycle can be fine tuned by adjusting only one variable at a time. Allow time for the variable changed to reach equilibrium and determine its effect on the part. Temperature changes always take longer to reach steady-state conditions. Pressure and time are quick response variables that are observed on the next or following cycle. Increasing back pressure to raise melt temperature may require half the barrel's melt and shot capacity before being noticeable. While these variables are being adjusted, the molding operation continues on the new cycle parameters.

Increasing injection speed will also create shear heat in the resin, as it flows into the cavity through the gate. This reduces the resin's melt viscosity and, if too fast, it causes burning, discoloration, and part or property problems. Each resin's flow is based on its molecular structure and recommended melt temperature and has a recommended injection rate, based on suggested cavity gate sizes to keep resin shear and melt flow within acceptable polymer limits. This avoids degradation and loss of properties. Typical processing parameters for plastic material melt and mold temperatures are shown in Table 11.2. For the ABS resins and their alloys, the processing ranges for a single generic polymer are listed in Table 11.3. Each material supplier will have a recommended set of molding conditions for its resins. These typically are as follows:

1. Barrel zone temperatures and nozzle and melt settings
2. Injection speeds and pressures
3. Drying times and temperature
4. Mold temperature settings
5. Gate size recommendations

TABLE 11.2. Processing Parameters.

Material	Temperature (°F) Melt	Mold
Nylon 6/6	540–580	130–220
Nylon 6	520–540	130–180
Nylon 6/12	520–540	130–180
Polycarbonate	580–625	160–190
Polypropylene	390–450	100–160
Polyethylene	380–450	50–110
SAN	450–500	160–180
ABS	380–525	120–190
Polystyrene	450–500	100–180
Acetal	380–420	180–250
Polyphenylene/Sulfide	550–650	150–275
Polysulfone	650–700	200–300
Polyethersulfone/Polyetherimide	625–750	200–300
Polyetheretherketone	680–780	300–400
PBT	450–480	100–200
PET	560–590	200–250

ABS, acrylonitrile butadiene styrene; PBT, polybutylene terephthalate; PET, polyethylene terephthalate; SAN, styrene acrylonitrile.

TABLE 11.3. Injection-Moldable ABS: Recommended Processing Factors[a].

Grade	Melt Temperature Range (°F)	Injection Speed	Mold Temperature (°F)	Maximum Regrind (%)	Gating (in) (Rectangular)	Shrinkage (in/in)
General purpose ABS	425–500	Slow to moderate	120–150	Up to 20	0.125–0.250 wide 0.094–0.125 deep	0.006–0.009
Flame-retardant[b] ABS	380–460	Slow to moderate	130–180	Up to 20	0.125–0.250 wide 0.094–0.125 deep	0.005–0.008
Flame-retardant ABS/PVC alloys[c]	380–410	Slow to moderate	120–140	Up to 20	0.125–0.250 wide 0.094–0.125 deep	0.005–0.007
High-heat resistant ABS	450–525	Slow to moderate	140–200	Up to 20	0.125–0.250 wide 0.094–0.125 deep	0.005–0.007
High-impact resistant ABS	450–500	Slow to moderate	120–150	Up to 20	0.125–0.250 wide 0.094–0.125 deep	0.006–0.008
Plating-grade ABS	450–530	Slow as possible	100–180	Up to 20	0.125–0.250 wide 0.094–0.125 deep	0.005–0.007
Plating-grade ABS/PC alloy	475–525	Moderate to fast	150–200	Up to 20	0.125–0.250 wide 0.094–0.125 deep	0.006–0.008

[a]Data based on $^1/_8$-inch wall thickness.
[b]At no time should the melt temperature of flame retardant ABS exceed 475°F.
[c]At no time should the melt temperature of ABS/PC alloys exceed 410°F.
PC, polycarbonate; PVC, polyvinyl chloride.

6. Allowable regrind
7. Shrinkage projections

Material suppliers also have molding manuals that note special recommendations and other information that can assist in establishing a good cycle to produce a quality part to specifications.

Each mold and resin combination will have varying molding cycle times. After establishing a satisfactory molding cycle, maintain complete records on all the processing variables. These records should include temperature and pressure settings; cycle times; and mold, melt, and equipment temperatures for the injection molding machine, as well as auxiliary equipment settings. These molding conditions, once established, will be used again if the job is rerun to reduce startup time and ensure the same quality of parts. Any fine tuning that needs to be made to optimize the molding cycle should also be recorded.

New machines have microprocessor control systems with built-in memory units that store all machine operational variables for later use. These systems can also control and store the mold and auxiliary equipment processing information. When coupled with host computers, printers, diagnostic units, and central Computer-Integrated Manufacture (CIM) systems, they also store, print out, and analyze the manufacturing cycle information.

The process control unit is usually machine mounted for easy set-up and monitoring of process settings and conditions. Once the process variables, machine, mold, and auxiliary equipment are inserted through the selection of different menu screens, processing begins. Process monitor alarm limits can be set to alert the operator/set-up person if the process drifts. Depending on the software, the process control unit can then output data of the process variables selected. Examples of these are shown in Figure 11.2.

A typical molding data record sheet for cycle and process information is shown in Figure 11.3. Most machine manufacturers will supply forms for recording their machine's variables in case storage problems occur with the microprocessor unit and the computer data are lost. When using the data record sheets, each significant process change or new lot of resin used should be recorded. This gives the production department a record should a quality problem occur later. An increasing number of customers are now requiring complete documentation of resins molded and process conditions or statistical process control charts for the manufacturing cycle.

It is important to understand how the many variables associated with injection molding affect the overall cycle and impact part quality. A change in one variable, such as increasing melt temperature to increase flow, may affect another variable in a positive manner, but it may cause part brittleness. A good way to see how these variables affect each other is to enter them in a matrix and evaluate them against each other, which will be discussed later in this chapter. This evaluation assumes that molding equipment and controls are working properly and displaying correct readings.

A

B

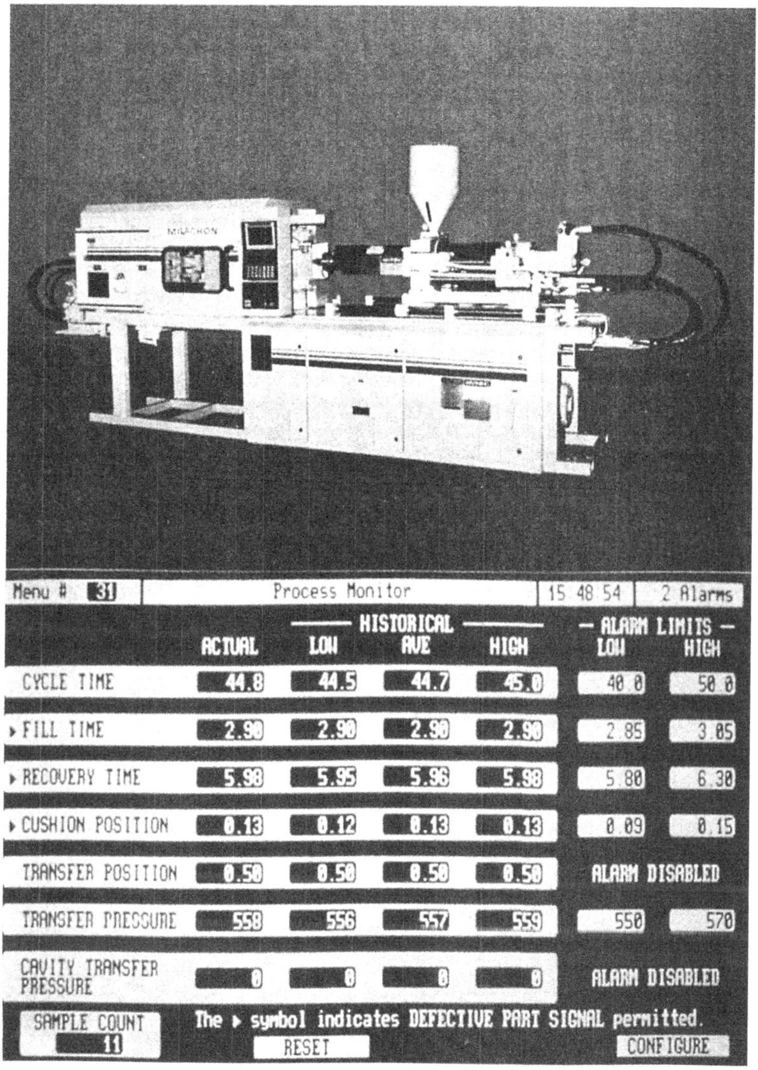

FIGURE 11.2. (A) Machine mounted process control unit, with three (B, C, and D) available menu screens. (Courtesy of Cincinnati Milacron.)

Shut-Down Procedure

At the end of the workday or production run, the molding machine will be shut down, and there is a correct procedure to be used. Many quality problems result from poor shut-down practices that cause difficult startups the following day or residual resin contamination when a new job is run.

When a change over or shut down is planned, the machine's nozzle should be backed away from the mold, the material shut-off gate at the

C

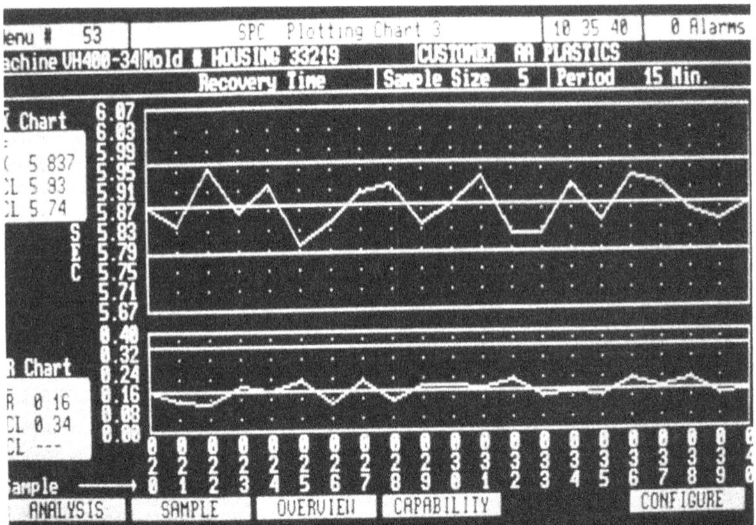

D

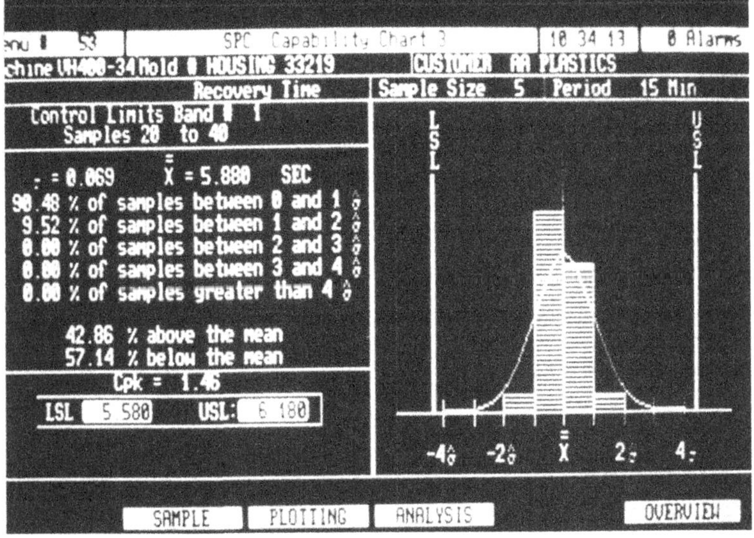

FIGURE 11.2. (continued)

hopper closed, and the material in the barrel pumped out. If material change over is also planned, the hopper should be emptied. The unused resin should be put into a sealed, air-tight container and identified with a label if its original container is not available. A good barrel and screw purging compound is then run through the machine, with the temperature settings adjusted for the purge material. This is usually a medium-to-high density polyethylene that cleans out the old resin. Once the purge material shows no traces of

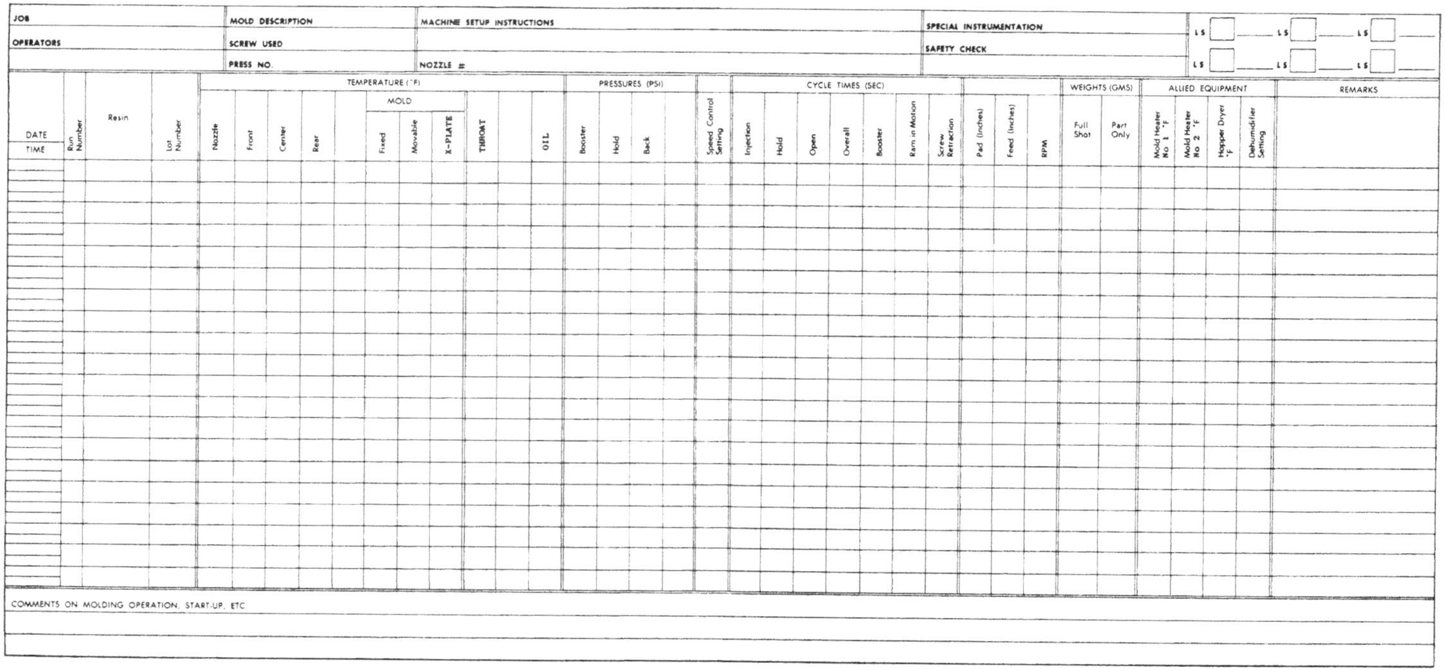

FIGURE 11.3. *Molding data record sheet.*

residual resin, the barrel heats are turned off and the barrel and screw are pumped dry.

Depending on the resin used, when overnight shut down is scheduled, the screw and barrel are pumped dry. A poly purge can be run through the barrel as described for the change over procedure. The poly purge should always be used if flame-retarded polyvinyl chloride (PVC) or another high outgassing resin was last run. This reduces the problems of corrosion, decomposition, and contamination of the resin after the barrel heats are turned off and the residual polymer continues to cook. It is always easier to purge the poly purge than the current resin, which may now have carborized in some sections of the barrel and screw. Such carborization causes startup and clean out problems when the machine starts up the next time. The poly purge may also save having to pull the screw if the residual resin badly decomposes during cooldown and contaminates the melt on startup.

The change-over and shut-down procedures, along with the recommended purging compound, should also be included in the job procedures. If in doubt as to what purge compound should be used, consult with the resin supplier.

OTHER MOLDING VARIABLES

Besides the normal molding machine, material, and auxiliary equipment variables, other molding variables are associated with the plant and its environment. Items that are often overlooked, but definitely can affect the overall cycle and quality of molded parts are as follows:

1. Molding room temperature, humidity, and air quality
2. Electrical power supplied to the molding equipment
3. Cooling tower water temperature
4. Air flow through the plant and around equipment
5. Cleanliness of plant production area
6. High-pressure air supply

These variables should be monitored to ensure that they are within appropriate limits.

Plant Environment

Maintaining uniform temperature and humidity in the molding area is important. With the heat generated by the machinery, during the summer months the plant environment, if not controlled, can lead to intolerable working conditions and overwork of the cooling system. The injection molding machine's barrel can be insulated to contain the heat for good temperature control and heat transfer to the resin, thereby saving energy. Failure to do this is costly in lost heat and can lead to worker fatigue and inattention.

A high humidity level can cause condensation to form in the mold and at the machine's feed throat. When humidity is high, condensation easily forms on cool surfaces. If the molding machine's feed throat is cooled, the condensation can get into the resin and cause molding problems. Moisture dripping into or forming on the mold cavity surfaces can also cause part problems. Good venting and air quality in the molding production area are necessary to remove fumes given off by the resins during the molding operation. Some of the additives used in molding resins—plasticizers, lubricants, flame retardants, and in general outgassing from acetal and PVC—must be vented from the plant work area. According to Occupational Safety and Health Administration (OSHA) guidelines, these fumes may have to be filtered and treated prior to expulsion into the outside atmosphere. These fumes and gases can also contaminate the cavity surface and provide an unhealthy work environment for personnel in the work area.

As discussed earlier, all materials, chemicals, and resins should have a Material safety data sheet on file. It can be used to answer questions on health and air quality, actions to protect workers, and OSHA requirements. These sheets should be filed in a binder and accessible to workers in the production area.

Electrical Power

An area often overlooked, but critical to part quality and equipment performance, is adequate and consistent electrical power feed. As plants expand and add more equipment, voltage fluctuations can occur. To avoid this, install special power transformers that ensure a constant line feed to equipment. Fluctuations in voltage cause equipment temperature variations and affect the efficiency and operation of electrical and hydraulic motors and pumps. Other quality problems have been traced to power fluctuations that caused equipment to run erratically. In addition, low voltage can cause motors to overheat and burn out. Your local power company can answer your power requirement questions.

Cooling Systems

The temperature of the plant's cooling water should be maintained as consistently as possible to control the temperature of your equipment and the molding operation. Tower cooling water is usually the primary cooling source and its temperature must be monitored. Adjustments must be made for daily variations in water temperature. Tower cooling water quality must also be treated and filtered to ensure that its heat removal capability is properly maintained. Many large plants use tower water to remove heat from their main refrigeration units and cool the hydraulic oil in molding machines. Cooling tower water systems are closed-loop systems that work most efficiently at water temperatures below 85°F. "City" water is not recommended to cool your equipment. It is expensive, may contain minerals that cause clogging, can reduce heat transfer in your condenser, and may vary in temperature year round.

Plant Airflow

Air flow through the plant should create a comfortable working environment and, if possible, be temperature controlled and humidity regulated. Hygroscopic resins, in particular, pick up moisture by absorption when exposed to humid air, but all resins can become contaminated by surface condensation on the pellets when going from a cold to warm environment. This can occur when the cold package is opened and exposed to the warm plant environment. Protecting the quality of the resin is the first priority. Resin stored in warehouses, with temperatures below the plant's temperature, should be brought into the plant at least 24 hours before processing to warm up and avoid condensation.

Cycle and part quality problems are often caused by portable fans used to cool off the work areas in nontemperature-controlled plants. When these fans blow on the manufacturing equipment, it causes variations in temperature. This can create process quality problems for the molded parts, as the injection molding machine and settings are affected by this air flow. When fans are used, the molding parameters must be monitored and process settings adjusted to keep part quality within specifications. If fans are used, direct their air flow away from the machine and mold.

Housekeeping

Good housekeeping in the plant from the floor to ceiling is important from a safety and part quality viewpoint. The feed hopper must always have a protective cover. Any open resin container must be covered to prevent contamination from overhead dirt, stray pellets, or trash. Vacuum systems should be used to clean spills and dirt around the molding machine and auxiliary equipment. Do not use high-pressure air to clean resin or equipment, as it only blows the contaminants around the plant. All oil and water leaks must be immediately repaired. Be sure that high-voltage electrical equipment is safely installed and monitored by trained personnel. Overhead areas should be cleaned at least once every 6 months and more often if needed. Some plants run their machinery water, air, and electrical lines in service trenches. These must always be clean and dry. The same applies if these lines are overhead. They should be routed in paths for easy serviceability and away from feed hoppers, cranes, and access to mold changing areas.

Many plants use high-pressure air systems to operate support equipment. These must have filter and cleanout systems to ensure that the air is always free of moisture and operating at required pressure settings. Otherwise, service problems may occur.

Pyrometers for Temperature Readings

Pyrometers are instruments for taking temperature readings of your equipment. They can be handheld or briefcase-sized units, with either digital or analog readout, a shielded cord, and two heat-sensor tips. The needle-shaped tip records melt temperature; the other—a circular-fiat tip sensor about ½ inch in diameter—measures surface temperatures. Each manufacturing area must use them during the manufacturing operation.

Taking temperature readings of melt, mold, and support equipment, and then comparing them with the equipment's reading is important for process control. These readings are used to verify that cycle conditions are met and staying in calibration. All too frequently, electrical gages fall out of calibration or fail. A well-maintained and calibrated pyrometer can quickly alert the operator to problems, such as heaterband burnouts, thermocouple reading errors, and incorrect mold cavity temperature settings.

Most plants do not use pyrometers as often as they should to verify the temperatures. The most widely used pyrometers are small units that can easily be held in one hand, while the operator uses the sensor tip to obtain the temperature with the other hand. Pyrometers are sensitive instruments and care must be taken that they are not dropped and damaged. Their accuracy is normally +/– 1°F depending on supplier and type.

The digital pyrometer has a slower temperature response time than the analog type, but they give a more accurate reading. Their slower response time, however, can cause longer interruptions when taking temperature readings during the molding cycle. To overcome this time lag, precondition the sensor tip before use by placing the probe against a surface whose temperature is close to that being measured. The digital pyrometer is more exact in that it indicates a specific number. It is used to verify machinery temperature outputs and thermocouple readings when response time is not critical and a specific number is desired.

The analog pyrometer has a very rapid response time and shows its readings with a needle pointer on a temperature scale. The finite markings on the scale are its degree of accuracy, which are close enough for most measurements. It is used to take mold and melt temperatures, as it records the temperature within a few seconds after the tip is placed on the cavity surface or in the melt. Preconditioning is never required.

Mold Temperature Balance

All the mold cavities must be at their required temperature setting during molding to obtain uniform part dimensions. This may require separate cooling circuits in the mold to balance cavity-to-cavity temperature. If a cooling circuit is too long, there could be a 20° to 50°F difference between inlet and outlet mold temperatures. For amorphous resins, mold cavities should be within +/– 10°F and for crystalline resins, +/– 5°F. If variances greater than these are recorded on the mold cavity surfaces, then part dimensions may vary. This is a major cause of part dimension problems; its cure involves adding more independent cooling circuits if possible.

All molds may not have the same temperature profile across their surfaces or even in each of their halves. Depending on the part's geometry, where it is gated, its critical dimensions, flow lengths, and how the mold cavity is built, some sections may require cooling, whereas others require heating to properly mold the part. Some molds may have cooling on one half and heat on another, if warping caused by part design is a problem. If differential cooling and heating cause one part surface to shrink more in one cavity than another, then warpage may still occur. If the mold's cooling circuit design is correct and temperature control problems still occur, then verify the controller unit's coolant temperature. Remember that cavity temperature is the key, not the temperature reading at the controller. The controller is often set higher or lower to obtain the correct cavity temperature once molding begins because the melt's temperature influences the mold cavity's temperature control.

The mold temperature controller also verifies the coolant flow rate. The temperature controller must provide a specified fluid flow rate to ensure mold cavity temperature control. Always verify that pump output pressure and flow are correct. If control of the cavity temperature still cannot be maintained, check the mold for plugged or corroded cooling lines. But before pulling the mold, verify that the cooling lines are correctly installed. Poor installation has been the major cause of many mold cooling problems. If part dimensions still vary or molding problems cannot be fixed by coolant flow, then a redesign or replumbing of the cooling system may be necessary to produce the correct cavity temperatures.

Resin Melt Temperature

To produce good parts consistently the resin's melt temperature must be within the supplier's recommended range. If resin processing temperatures are too low, unmelted particles may result. Too high a melt temperature can cause resin degradation or mold flashing. A melt temperature reading should be taken before molding begins and just before shutting down the machine at the end of the workday or production run. The most common method is to back the nozzle away from the sprue bushing and take an air shot into a container. However, this disrupts the cycle and may not always give an accurate reading. The second and most accurate method is to leave the mold open and take air shots into a container under reduced pressure during the cycle. Catch the melt in an insulated cup and read its temperature. This results in a more accurate measurement.

After collecting the melt in an insulated container, use the pyrometer's needle tip to stir the melt until the highest reading is obtained. With a digital pyrometer, preheat the tip to about 20°F above the anticipated melt temperature. This reduces measurement time while the melt is cooling down and the measurement is taken.

Proper calibration of the pyrometer is critical; the analog is more prone to drift and vary than the digital. Temperature measurements should be taken when the process is in equilibrium and on cycle. Measurements taken during process changes—for both melt and mold temperature—will not be accurate until equilibrium of the system is reached. Repeat temperature measurements should be taken after sufficient time has elapsed to ensure that equilibrium has been reestablished. It is interesting to note that the expected melt temperature based on barrel heaterband settings is often higher than these settings, because the screw adds considerable shear heat to the melt and back pressure amplifies this temperature.

The setting of the screw's RPMs, which pump the resin forward, are based on having the screw retracted and the next shot ready to inject within 75 to 85 percent of the mold hold-time after gate freeze-off. If melt temperature is too low or high, the screw's revolutions may have to be changed or the machine's back pressure or barrel temperature settings adjusted. These three variables are the only melt temperature control settings available to the molder.

Machine Pressure Settings

The injection pressure setting on the molding machine is usually a hydraulic ratio of 10:1 based on the area of the barrel's diameter divided by the area of the hydraulic cylinder's diameter. Injection and packing pressure can vary on a machine because the viscosity of the hydraulic oil is controlled by the cooling system. For consistent pressure output, the hydraulic oil's specified temperature must be maintained. This is important if adjustments need to be made. Selecting the correct injection and packing pressure based on cavity surface area and machine clamp tonnage were discussed earlier.

FINE TUNING THE CYCLE

Once machinery and mold process variable equilibrium is reached, the molding cycle should be fine tuned. Now is the time to begin recording new pressure, time, and temperature readings, so you will know what changes caused what variations. During start-up, one strives to obtain equilibrium for the molding cycle and steady-state conditions, while approaching quality part requirements.

Control by Part Weight

A good method for helping to establish process control is to mold parts to maximize part weight. This assumes that each cavity's dimensions were cut correctly and that each cavity is being fed melt through a runner and gate system that is sized to uniformly fill and pack out the part cavity to its maximum capacity. Then, when maximum and consistent part weight is obtained by adjusting the process variables, the operator will know the mold's capability to produce parts to this set of processing conditions.

This does not mean that stabilizing part weight always produces correct part dimensions. Part dimensions will only be correct if the cavity's dimensions were cut accurately and the material's shrinkage values were calculated correctly. If the gate is too small and processing conditions not right, part dimensions will vary and so will part weight. To analyze a molded part's dimensional problems during processing, see the section on "Determining the Missing Variable."

In a single-cavity or balanced-runner mold, establishing uniform part weight is fairly easy. Pressure drops are uniform throughout the mold system and part weight, cavity to cavity, should be uniform. With an unbalanced mold, this is not true and part weights vary slightly from cavity to cavity. This difference derives from nonuniform part packing pressures in the runner system and varying cavity gate freeze-off times. Unbalanced tools should only be used for the least demanding molded parts, nonmating or stand-alone parts, and parts where minor dimensional variances will have no effect on appearance or performance.

To determine if maximum cavity part weight has been achieved and if dimensions have been maximized and stabilized, check the screw forward or cavity packing time. For some parts, weight may appear to stabilize, but if dimensions are varying, the controlling factor is gate freeze-off time. Screw forward time and part weight are used to establish the correct gate freeze-off time and make sure the cavity is being packed to its maximum. To determine when maximum part weight is achieved, use an accurate digital scale and carry the reading to at least two decimal places. Weigh the part or parts, if very small, and record the results for each cycle. Each part must be uniformly degated to assure accurate measurements. Using the current cycle's screw forward time, take a part weight measurement. For the next and succeeding cycles, increase screw forward time one second until maximum part weight is obtained. Maximum part weight is reached when there is no weight gain on the next cycle. Thus the gates minimum freeze-off time. Any time less than this will cause the cavity to depressurize as the screw begins to retract. If the gate is not frozen off, molten resin can flow out of the cavity, and part weight and dimensions will vary. Then using the screw forward time, continue molding for at least five cycles to verify that the weight remains constant. Simultaneously, check other critical part dimensions.

If part weight will not stabilize, section the part with a saw at the thickest area and see if it is packed out. If there is any porosity or voids, the part is not packed out. This may be externally visible as sink marks. Usually if sink marks are visible on the surface, part weight will not stabilize. If sink is visible raising melt and mold temperature or varying injection speed, injection and packing pressure, and time may eliminate the problem. If this does not work, consider increasing the gate opening. This extends gate freeze-off time and allows more resin to be packed into the part, thereby reducing voids and sink marks. But before enlarging the gate, explore the other processing variables mentioned. Once the gate is enlarged, the cycle time will increase and it is expensive to bring the gate back to its original size. Another item to check before opening the gate is whether the venting at the end of the flow length is adequate. Lack of venting in the cavity can cause problems in long, thick parts. Use the masking tape technique earlier described to verify if venting is the problem.

Maximum part weight also requires that on the injection stroke, the melt pad or cushion is maintained on the melt injected into the mold. At the end of the injection stroke, the screw should never bottom in the barrel. A minimum of ¼ to ½ inch of melt pad must be maintained in front of the screw tip. This ensures that secondary packing pressure can be held on the mold cavity for part packout prior to gate freeze-off. Failure to maintain a consistent melt pad indicates a worn checkring with melt flowing back over the screw. The cavity pressure then varies as does part weight on each cycle. The only cure is to shutdown and replace the screw's checkring. When part dimensions are met, part weight will stabilize and can then be used as a control to verify that the molding process does not change from cycle to cycle. To do this, the weight scale is tied into the molding machine's process control unit and adjusts the cycle through closed-loop feedback if part weight varies. There are very accurate digital scales that can be placed close to or under the mold to accomplish this task. Degated free falling parts or robot placed parts are weighed each cycle to ensure that the process stays in control. Part weight systems are discussed in Chapter 13.

REGRIND EFFECTS ON PART QUALITY

There are many variables associated with molding plastic parts. If regrind is allowed, the cycle may have to be readjusted, particularly when molding critical parts. Regrind should always be dried and well mixed with the virgin resin in the hopper. Timing the virgin resin feed with the regrind feed into the hopper will aid in obtaining a better mixture. Regrind should be evenly dispersed over the surface of the feed hopper, so it will be well mixed as it moves down the hopper into the feed throat of the machine. Regrind causes a lower melt viscosity with limited effects on physical properties, if kept below 25 percent. The lower viscosity also makes the melt more fluid and thus faster flowing. This can increase injection-fill rates and gate shear heat effects, both of which can cause flashing of the mold cavity. If such problems occur, reevaluate the benefits of using regrind in terms of the part quality required.

DETERMINING THE MISSING VARIABLE

Injection molding is a complex method of manufacture because of the large number of variables and the complexity of their interactions. To better understand these interactions, one must understand all the variables.

Part tolerances are established by the molded part's end-use specifications. The ability of the molder to produce parts consistently to the specified tolerances is influenced by the following:

1. Processing conditions and their fluctuations
2. Mold design, including cavity dimensions, draft, gate and runner size, and the cooling system
3. Material lot-to-lot variability and shrinkage rate
4. Part design, uniform sections, flow in the part, and pack-out pressure

Physical properties and part appearance are affected by the following:

1. Processing conditions
2. Material properties
3. Material property changes during molding
4. Weld-line strength and voids/porosity in thick sections
5. Internal stresses induced during mold filling
6. Differential shrinkage in the mold cavity

Therefore, control of part quality in terms of tolerances and properties demands a solid understanding of the interactions of the following:

1. Processing equipment performance
2. Uniform processing conditions
3. Material properties and variables
4. Mold design
5. Part design

Assuming that the part and mold are designed to meet customer requirements and the resin is consistent, the processing team is responsible for manufacturing the part. More than 25 different processing variables can affect part quality, and it is often necessary to identify which variable(s) affects the part's dimensions the most. This happens when the molding process is in equilibrium, but one or more critical dimensions are not yet achieved.

In the past, if production tested one processing variable at a time, it might have taken hours or days before the problem was corrected. Therefore, in most cases, more than one variable was changed at a time. What often happened was that a part met the specification on one cycle but drifted off on the next or succeeding cycles as the processing variables continued to move to their new set points. Operators had to be taught to wait long enough for the molding process to reach equilibrium between each change. Otherwise, additional adjustments would prove necessary.

When this occurs, operators will solve the problem based on experience or the processing troubleshooting guide. In most cases, this is successful. If not, they hold all variables but one constant and through trial and error try to bring parts into tolerance. In some cases, more than one variable is causing the problem. This technique works, but it is expensive, takes time, and often may not detect the root cause variable(s). When this happens, the operator has to adjust the system constantly to make good parts without actually solving the problem.

As a result, a system was needed to identify and test the process variables that contribute the most to meeting part dimensional requirements successfully. Taguchi, in his Design of Experiments (DOE), has developed such a method. It involves the selection of key process variables that act on the part's dimensions. By their careful selection and testing, which requires a minimum amount of time at their maximum and minimum settings, this method determines which variable(s) directly affects the part dimension under study.

The Taguchi experiments are used once a process is in control—the process parameters are optimized, but the parts still do not meet or drift in and out of specifications. The process can be affected by the resin, machine settings, mold temperature, shop environment, or any number of items in the molding system. The molding process is in control but not totally capable.

The Taguchi method uses orthogonal arrays—rows of experiments (factors or variables) versus the trial (runs) to be performed. The variables for each factor and run are established using quality control problem analysis techniques. These factors are then tested with each run using a different combination of these variables.

Taguchi's orthogonal arrays capture the most significant variable combination levels for testing. For example, to test independently seven variables at two setting levels, high and low, the complete orthogonal array would be large: 2^7, or 128. This means to test each variable, at its high and low level, in combination with all the other variables, at each of their levels, would take 128 separate trial runs. There are many combinations of arrays and levels of testing. The tester determines which levels to test and the amount of time the plant can spend to find the elusive variable.

Taguchi Problem-Solving Techniques

The computer programs *Mold Fill*, which is limited to two-dimensional, simple parts, and *Mold Cool*, can assist the design and tool engineers in identifying problem areas. If analysis indicates a possible problem, then action can be taken to redesign the part and mold.

Once the molding process is capable, there are always going to be some minor equipment and resin deviations that require control changes. Total process quality control requires keeping all 25 plus variables within the processing conditions. Resins will vary from lot to lot, the mold cavity will wear, cooling temperatures will change, and machinery will drift and wear. Therefore, control limits for the manufacturing cycle need to be established to show whether the system is in control.

Molding machines with computer-controlled process control with closed-loop microprocessor feedback units will continually adjust the molding machine's cycle parameters to stay within preset conditions. But as parts wear and resins vary, they can only self-adjust within specific process and control windows. As changes in the system occur, there must be other visual indicators to show part quality is being maintained. Therefore, control charts are used. They are generated from machine processing conditions or from sampling parts at predetermined intervals to verify all systems are capable and in control.

PROCESS CONTROL CHARTING

During manufacture, controlling the part quality is the most important function of production. The methods used to verify control must be established. The statistical tool most frequently recommended is the process control chart, which was pioneered by Walter A. Shewhart. Control charting requires the understanding of only a few statistical fundamentals for practical application. It can be easily implemented and, once in place, will show production a real-time report card of progress. The data generated can be given to customers to show compliance with their requirements, along with an indication of problem areas that can be corrected in real time.

Many approaches are used to determine the limits or specifications needed to keep a process in control. Some are determined by tests, some by experience, and others are just "picked out of the air." Often, control procedures are written in the form of manufacturing procedures or exist in the heads of veteran operating personnel.

Experienced personnel separate manufactured part variations into usual and unusual. They make these distinctions within the limits of the process being discussed; if parts fall outside these limits, then something unusual has occurred that must be corrected.

The control chart is a visual method of plotting and evaluating whether a process is in a state of "statistical control." It factually separates production personnel's distinction of variations into usual and unusual components. In essence, it compares actual production variations with the control limits established to make high quality parts. When these limits have been computed and accepted for production, the control charts assume the visual role of aiding in the quality control of materials, parts, and assemblies.

The economics of manufacture are the primary driving force in determining if the usual variation is within the specification limits. This is based on how tight the tolerances are set and how much the customer is willing to pay to have parts produced to these specifications.

Manufacturing Limits

There are generally three sets of manufacturing limits as follows:

1. Control limits—limits established to produce acceptable parts.
2. Process limits—limits of process variables to produce acceptable parts.
3. Specification limits—limits the process must fall within to make acceptable parts.

Normally, the specification limits are the wider of the three, followed by process and then control limits. Their relationship is shown in Figure 11.4.

Because it is risky to depend on the process information acquired by production people from years of experience, accurate recordkeeping should be supplemented by control charts. This information is valuable when new employees are brought on board or new supervisors are appointed. The control chart greatly reduces the operator's learning curve for process and part control.

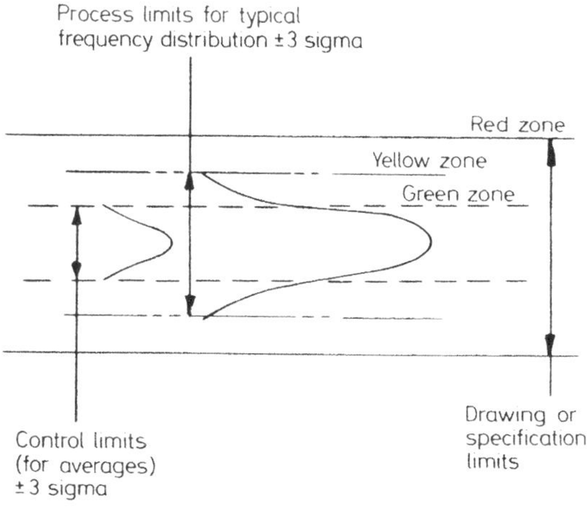

FIGURE 11.4. Manufacturing limits. (Adapted from Ref. [5].)

Besides product variances, control charts show whether the process is in control. This is usually indicated by a bunching of points along the central tendency. Often, it may show wider fluctuations that may indicate the process is having control problems. Shifts or swings may also show a tendency away from control because of machinery or material problems, or other factors on the production floor.

Control Charts

The two basic types of control charts are as follows:

1. Measurements or "variables charts," which are used when actual readings are recorded—the so-called X, R, and s charts.
2. Attribute charts, which use visual or go/no go data—the fraction or percentage defective charts called P charts.

The two different situations in which these charts are used are as follows:

1. When establishing a new process—either one that has undergone extensive changes or one that investigates ongoing control after a preliminary frequency-distribution analysis has demonstrated initial control. Data are collected on part quality characteristics and control limits, and central tendency values are then calculated. Hence, this condition is termed one of "no standard given."
2. When central tendency and spread values are initially established. This condition is known as "standard, give." It assumes that the process is in control based on whatever data were used to establish the limits, arbitrary or real. These are based on production or service specifications and their requirements or on a target value established between the customer and supplier for the parts.

The approach for calculating the control limits for these two charts is based on the laws of probability. The methods of calculations, which vary between the measurement and percentage charts, will be discussed in detail.

The process to follow in setting up control limit charts is as follows:

No Standard Given

1. Select the quality characteristics to be studied.
2. Record data on a required number of samples with an adequate number of units per sample. The minimum is five samples.
3. Determine control limits for the sample data.
4. Analyze the state of control in the process.
   A. Too much variation.

B. Parts move in and out of control.

C. Well-controlled process.

When establishing control limits, several samples will often be out of control. In this case, trace down the problem in the process and repeat steps 2 and 3 until the process is in control.

5. If the established control charts and limits are to be used for process control, go to steps 6 and 7 of the Standard Given condition that follows and proceed.

Standard Given

1. Select the quality characteristic to be studied.
2. Establish the central tendency value and the spread to be used. All available data must be used to show that control exists.
3. Determine the control limits from these values.
4. Establish that these control limits are economical, practical, and required.
5. Establish the control limit values and plot them on graph paper.
6. With the manufacturing process in control, begin recording results from production samples selected at periodic intervals.
7. Take corrective action if the characteristics of the production samples exceed the control limits.

An example of the graph paper used to plot the data is shown in Figure 11.5.

Measurement-Process Control-Chart Calculations

The basic principles for computing measurement-process control-chart limits are similar to those for calculating the frequency distribution of three-sigma limits. The data required is as follows:

The *mean* (or average) is the most commonly used measure for central tendency. It is the sum of all the readings (X) divided by the total number of readings (n) for a specific trial run. It is denoted by $\bar{X}$.

$$\frac{\bar{X} = X_1 + X_2 + X_3 + \ldots X_n}{n}$$

More frequently, it is written using the greek capital letter sigma (Σ) to denote the sum of the Xs as (ΣX)

$$\bar{X} = \frac{\Sigma X}{n}$$

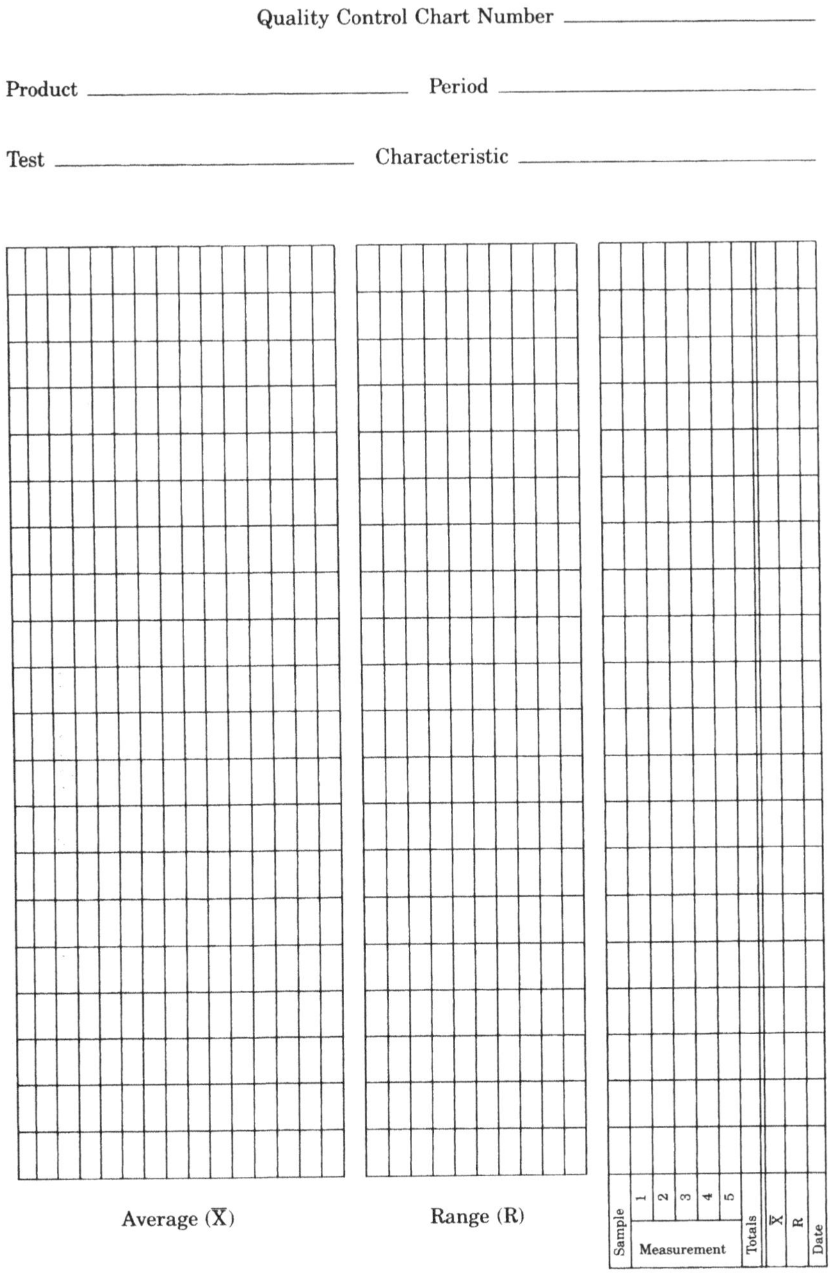

FIGURE 11.5. *Quality control chart. (Adapted from Ref. [5].)*

For example, if there were five readings—4, 5, 6, 7, 8,—the average or mean would be:

$$\bar{X} = \frac{4+5+6+7+8}{5} = \frac{30}{5} = 6$$

The *ranger* (R) is the difference between the high and low values recorded for a specific trial run. For the example above, you have

$$R = X(\text{high}) - X(\text{low}) = 8 - 4 = 4$$

The *standard deviation* is a measure of the spread of individual readings for a trial run. It is usually computed for samples drawn from larger lots and in these cases is known as the *sample standard deviation.* The sample standard deviation is the positive square root of the sum of the squared deviations of readings from their average, divided by one less than the number of readings:

$$S = \sqrt{\frac{(X_1 - \bar{X})^2 + (X_2 - \bar{X})^2 + \ldots + (X_n - \bar{X})^2}{n-1}}$$

For lots with multiple readings.

$$S = \sqrt{\frac{\Sigma(fx - n\bar{X})^2}{n-1}}$$

where, fx = frequency of readings within each lot.

Thus, in the similar series for X:

$$S = \sqrt{\frac{(4-6)^2 + (5-6)^2 + (6-6)^2 + (7-6)^2 + (8-6)^2}{5-1}}$$

$$S = \sqrt{\frac{4+1+0+1+4}{4}} = \sqrt{\frac{10}{4}} = 1.581$$

When comparing range versus standard deviation, it is readily observed that the range is very easily obtained—the difference between the high and the low values. But in samples larger than 15, it does not take into account the effect of multiple readings. With more samples, there is a greater probability of getting a wide-of-the-mark reading. With standard deviation, this effect is minimized by incorporating all the readings and balancing out the mavericks.

The primary difference is that a smaller number of readings is used to calculate and establish the central tendency and spread values. Because frequency distributions in industry tend toward normality, the three-sigma value for control charts has proven to be most useful and economical. The formulae and supporting chart data are shown in Figure 11.6 and Table 11.4.

As noted, two sets of calculations are given—one for range and the other for standard deviation. The choice of use is up to the individual. To calculate the data, some production personnel prefer to use sample sizes of more than ten and calculate their control limits based on standard deviation, but the five-unit sample has become the industry norm.

In computing the control limits for a "no standard given" condition, the following eight steps establish process control. The range will be used as the spread, and formulas IA, 1B, 2A, and 2B will be applied.

Control Chart Calculation Procedure

1. Select the quality characteristic to be controlled: length, thickness, warpage injection pressure, time, packing pressure, etc.
2. Collect data dy selecting an adequate number of lots or cycles based on a set time or frequency method. For this example, a part dimension will be used. For a process, use successive cycle data in increments of five successive cycles. Therefore, use 20 lots as a start and select five successive samples to be measured for the specific characteristic from each lot. The lots should be selected at set intervals every hour, half hour, or after 20 to 30 cycles, and the sample data should be recorded in successive order of selection.
3. Compute the average and range values for each of the 20 lots.
4. Compute the grand average $\bar{X}$ of the averages of the 20 lots as well as the average of the range R of the 20 lots.
5. Compute the control limits based on these lot averages and ranges.
6. Analyze the average and range values for each lot relative to these control limits. Determine whether any factors require corrective action before the control limits are approved.
7. Determine whether control limits are within economical limits for the molding cycle process.
8. Use the control limits in active production to be sure they produce parts within the limits established.

Control Limit Calculations

1. The characteristic to be measured is pin length in an eight-cavity balanced runner tool on a 30-second cycle.
2. Assuming 240 studs are made every 15 minutes, randomly select five samples from this lot and measure them every 15 minutes to establish the data for the control limits.

No Standard Given

When Range Is Used as Measure of Spread

Average:	Lower limit $= \bar{\bar{X}} - A_2\bar{R}$	(1A)
	Center line $= \bar{\bar{X}}$	
	Upper limit $= \bar{\bar{X}} + A_2\bar{R}$	(1B)
Range:	Lower limit $= D_3\bar{R}$	(2A)
	Center line $= \bar{R}$	
	Upper limit $= D_4\bar{R}$	(2B)

When Standard Deviation Is Used as a Measure of Spread

Average:	Lower limit $= X - A_3\bar{S}$	(3A)
	Center line $= \bar{\bar{X}}$	
	Upper limit $= \bar{\bar{X}} + A_3\bar{S}$	(3B)
Standard Deviation:		
	Lower limit $= B_3\bar{S}$	(4A)
	Center line $= \bar{S}$	
	Upper control limit $= B_4\bar{S}$	(4B)

where X = grand average
R = average range
S = average sample standard deviation

Standard Given

When Range Is Used as Measure of Spread

Average:	Lower limit $= \bar{X}_o - A_{ao}$	(5A)
	Center line $= \bar{X}$	
	Upper limit $= \bar{X} + A_{ao}$	(5B)
Range:	Lower limit $= D1ao$	(6A)
	Center line $= R_o$ (or d_{2ao})	
	Upper limit $= D_{2ao}$	(6B)

When Standard Deviation Is Used as a Measure of Spread

Average:	Lower limit $= \bar{X} - A_{ao}$	(7A)
	Center line $= \bar{X}_o$	
	Upper limit $= \bar{X}_o + A_{ao}$	(7B)
Standard Deviation:		
	Lower limit $= B_{5ao}$	(8A)
	Center line $= S_o$ (or C_{4ao})	
	Upper control limit $= B_{6ao}$	(8B)

where Xo = value of the average adopted for computing the center line and control chart limits.
Ro = value of the range adopted for computing the center line and control chart limits.
So = value of the sample standard deviation adopted for computing the center line and the control chart limits.
ao = value of the lot or population standard deviation adopted for computing the center line and control chart limits.

When computing the lot or population standard deviation, the lot size and number of measurements must be sufficient to categorize the lot with extreme values averaged out.

FIGURE 11.6. *Control chart formulae. (Adapted from Ref. [5].)*

TABLE 11.4. Factors for Computing Central Lines and Three-sigma Control Limits for $\bar{X}$, S, and R Charts.

Observations in Sample, n	Chart for Averages			Chart for Standard Deviations					Chart for Ranges				
	Factors for Control Limits			Factors for Central Line	Factors for Control Limits				Factors for Central Line	Factors for Control Limits			
	A	A_2	A_3	C_4	B_3	B_4	B_5	B_6	d_2	D_1	D_2	D_3	D_4
2	2.121	1.880	2.659	0.7979	0	3.267	0	2.606	1.128	0	3.686	0	3.267
3	1.732	1.023	1.954	0.8862	0	2.568	0	2.276	1.693	0	4.358	0	2.574
4	1.500	0.729	1.628	0.9213	0	2.266	0	2.088	2.059	0	4.698	0	2.282
5	1.342	0.577	1.427	0.9400	0	2.089	0	1.964	2.326	0	4.918	0	2.114
6	1.225	0.483	1.287	0.9515	0.030	1.970	0.029	1.874	2.534	0	5.078	0	2.004
7	1.134	0.419	1.182	0.9594	0.118	1.882	0.113	1.806	2.704	0.204	5.204	0.076	1.924
8	1.061	0.373	1.099	0.9650	0.135	1.815	0.179	1.751	2.847	0.388	5.306	0.136	1.864
9	1.000	0.337	1.032	0.9693	0.239	1.761	0.232	1.707	2.970	0.547	5.393	0.184	1.816
10	0.949	0.308	0.975	0.9727	0.234	1.716	0.276	1.669	3.078	0.687	5.469	0.223	1.777
11	0.905	0.285	0.927	0.9754	0.321	1.679	0.313	1.637	3.173	0.811	5.535	0.256	1.744
12	0.866	0.266	0.886	0.9776	0.354	1.646	0.346	1.610	3.258	0.922	5.594	0.283	1.717
13	0.832	0.249	0.850	0.9794	0.332	1.618	0.374	1.585	3.336	1.025	5.647	0.307	1.693
14	0.802	0.235	0.817	0.9810	0.406	1.594	0.399	1.563	3.407	1.118	5.696	0.328	1.672
15	0.775	0.223	0.789	0.9823	0.428	1.572	0.421	1.544	3.472	1.203	5.741	0.347	1.653
16	0.750	0.212	0.763	0.9835	0.448	1.552	0.440	1.526	3.532	1.282	5.782	0.363	1.637
17	0.728	0.203	0.739	0.9845	0.466	1.534	0.458	1.511	3.588	1.356	5.820	0.378	1.622
18	0.707	0.194	0.718	0.9854	0.482	1.518	0.475	1.496	3.640	1.424	5.856	0.391	1.608
19	0.608	0.187	0.698	0.9862	0.497	1.503	0.490	1.483	3.689	1.487	5.891	0.403	1.597
20	0.671	0.180	0.680	0.9869	0.510	1.490	0.504	1.470	3.735	1.549	5.921	0.415	1.585
21	0.655	0.173	0.663	0.9876	0.523	1.477	0.516	1.459	3.778	1.605	5.951	0.425	1.575
22	0.640	0.167	0.647	0.9882	0.534	1.466	0.528	1.448	3.819	1.659	5.979	0.434	1.566
23	0.626	0.162	0.633	0.9887	0.545	1.455	0.539	1.438	3.858	1.710	6.006	0.443	1.557
24	0.612	0.157	0.619	0.9892	0.555	1.445	0.549	1.429	3.895	1.759	6.031	0.451	1.548
25	0.600	0.135	0.606	0.9896	0.565	1.435	0.559	1.420	3.931	1.806	6.056	0.459	1.541

Adapted from Source: ASTM STP 15D from American Society for Testing and Materials.

3. Referring to sample lot No. 1 of Figure 11.7, the five readings to compute average are:

$$\begin{array}{r} 0.498 \\ 0.500 \\ 0.505 \\ 0.503 \\ \underline{0.503} \\ 2.509 \end{array}$$

The mean is:

$$\bar{X} = \frac{\Sigma X}{R} = \frac{2.509}{5} = 0.5018$$

The range is:

$$R = .505 - .498 = 0.007$$

Similar calculations are made for the other lots respectively.

4. Determine *grand average* or the mean and *average range*.

$$\bar{\bar{X}} = \frac{\bar{X}_1 + \bar{X}_2 + \ldots\ldots\ldots \bar{X}_n}{n}$$

$$\bar{\bar{X}} = \frac{\Sigma \bar{X}}{N} = \frac{10.0292}{20} = .50146$$

where N = number of lots measured.
For the average range:

$$\bar{R} = \frac{\text{sum of lot ranges}}{\text{number of lots}} = \frac{\Sigma R}{N}$$

$$\bar{R} = \frac{0.091}{20} = .00455$$

5. Computing the control limits.

Averages:
Lower Control Limit (LCL) = $\bar{\bar{X}} - A_2\bar{R}$
Control Center Line (CCL) = $\bar{\bar{X}}$
Upper Control Limit (UCL) = $\bar{\bar{X}} + A_2\bar{R}$

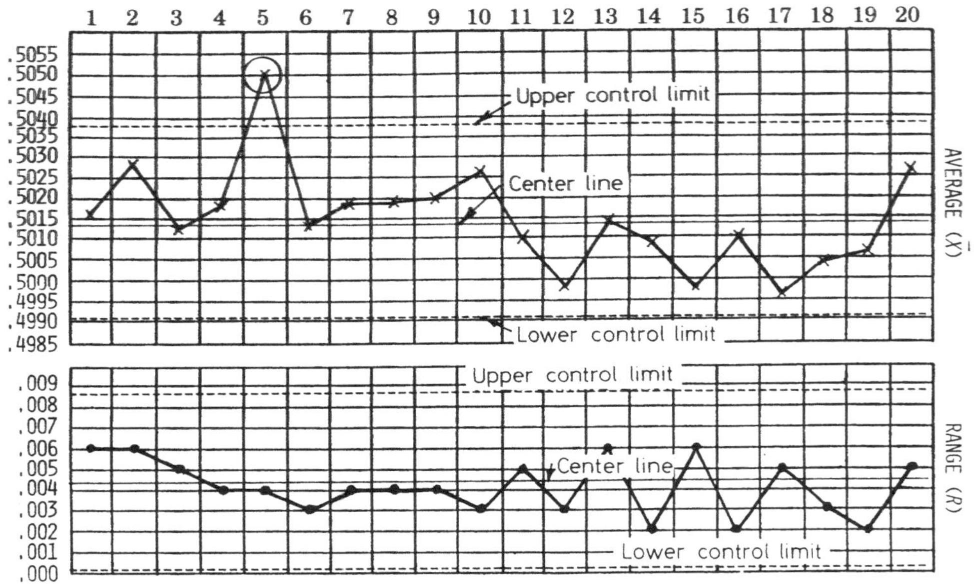

Sample		1	2	3	4	5	6	7	8	9	10	11	12	13	14	15	16	17	18	19	20
Measurement	1	.498	.504	.500	.499	.505	.503	.503	.502	.502	.504	.503	.500	.504	.501	.499	.502	.497	.499	.501	.505
	2	.500	.501	.498	.502	.505	.501	.500	.498	.501	.501	.497	.500	.502	.501	.502	.499	.498	.499	.499	.504
	3	.505	.506	.502	.503	.507	.501	.505	.503	.505	.502	.502	.500	.504	.501	.498	.502	.501	.503	.503	.501
	4	.503	.502	.503	.502	.503	.500	.502	.502	.503	.502	.502	.498	.499	.500	.501	.502	.502	.500	.500	.501
	5	.503	.500	.504	.502	.506	.501	.500	.503	.500	.503	.502	.501	.498	.501	.499	.500	.500	.501	.500	.502
Totals		2.509	2.513	2.507	2.508	2.526	2.506	2.510	2.508	2.511	2.512	2.506	2.499	2.507	2.504	2.499	2.505	2.498	2.502	2.503	2.513
$\bar{X}$		.5018	.5026	.5014	.5016	.5052	.5012	.5020	.5016	.5022	.5024	.5012	.4998	.5014	.5008	.4998	.5010	.4996	.5004	.5006	.5026
R		.007	.006	.006	.004	.004	.003	.005	.005	.005	.003	.006	.003	.006	.001	.004	.003	.005	.004	.004	.004
Date/time																					

FIGURE 11.7. *Quality control chart—control limits.*

Ranges:
Lower Control Limit (LCL) = $D_3\bar{R}$
Center Control Limit (CCL) = $\bar{R}$
Upper Control Limit (UCL) = $D_4\bar{R}$

For a sample size of five, the constants are taken from Table 11.4.

$A_2 = 0.577$
$D_3 = .0$
$D_4 = 2.114$

Substituting these data in the above formulas yields:

Average (LCL) = $\bar{\bar{X}} - A_2\bar{R} = 0.50146 - (0.577)(0.00455) = 0.4988$
Average (UCL) = $\bar{\bar{X}} + A_2\bar{R} = 0.50146 + (0.577)(0.00455) = 0.5041$
Range (LCL) = $D_3\bar{R} = (0)(0.00455) = 0$
Range (UCL) = $D_4\bar{R} = 2.114(0.00455) = 0.0096$

These control limits are then plotted on the graph in Figure 11.7. The center lines can also be plotted for visual clarity when reading the chart, although some leave these off.

When analyzing the data, be sure that out-of-control readings are not the results of human error. Such errors account for approximately three out of 1000 bad readings. Therefore readings for this sample lot should be repeated as a control measure. The same person should take all the readings with the same gauge. Until all questions are resolved, that specific lot should be set aside and properly identified. An example of using this technique for recording and charting part weight is shown in Figure 11.8. For an in depth analysis of control charting refer to Ref. [5] in the bibliography.

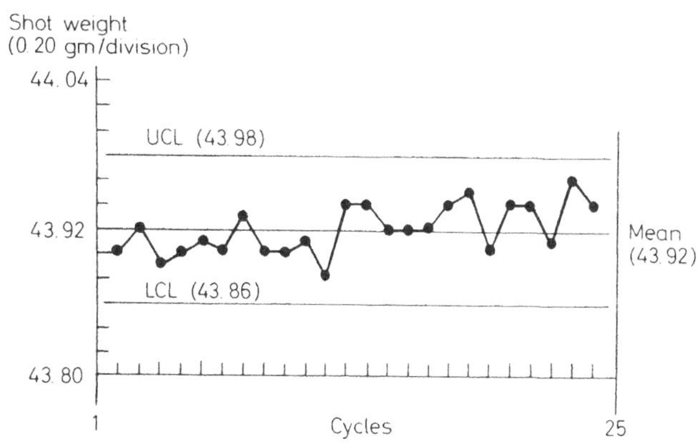

FIGURE 11.8. *Part weight control charting (Adapted from Ref. [8].)*

With the newer process controllers, the control chart data output may be directly available. This output may be generated by internal software or from signals pulled off the machine. These signals can be fed into a local or central process control charting software program for generating real-time output charts.

This technique can also be used to generate machine process control limits, if not part of the program for monitoring the molding machine's variables, such as the following:

1. Injection pressure and speed/time for
   a. first stage injection
   b. second stage—cut over to pack
2. Holding pressure on cavity
3. Cavity pressure against core pins or sensors in the tool
4. Mold cavity surface temperature
5. Cycle times for machine operations
6. Screw RPMs
7. Barrel/nozzle temperatures

These process control readings will give the operator real-time indicators of the manufacturing process and are very valuable in maintaining process control for the cycle and part manufacture.

Percent and Fraction Control Charts

These control charts can be used in two ways: to control part attribute quality to indicate process control capability during manufacture. In developing the attribute chart, a unit is classified according to whether it meets the specification limits. Frequently go/no go readings are represented by the percentage of units that do not meet requirements. This value is calculated by dividing the units not passing inspection by the total number of units inspected. Thus, if four units do not meet the requirements out of a series of 100, the fraction is $4/100$ and the percentage is 4.

The second use indicates the control maintained during manufacture and points out areas where improvement could be effected. It is not used for defect analysis, where the use of Pareto charting is preferred.

Percent Control Chart Formulae

When using percentage data, the average or mean, expressed as a percentage, is generally used as a measure of central tendency. The standard deviation is a measure of the spread in successive sample percentages.

The average for percent expressed as an integer is symbolized by $\bar{p}$. The symbol p (a percentage) represents the proportion of defective or nonconforming units found in a single sample of n units. If the sample size n is kept constant for a successive set of K lots, the average value of p is represented by $p = \Sigma p/K$. This p is plotted as a solid line across the p chart and represents the expected proportion of unacceptable units to be found in a random sample of n units.

$$P = \frac{\Sigma p}{K}$$

When sample sizes vary within the lots, p can be calculated by dividing the total number of defectives (c) by the total number of units (n) in the series.

$$\bar{p} = \frac{c \times 100}{\Sigma n}$$

With constant sample size, the standard deviation of $\bar{p}$ is symbolized by Sp. With constant sample size, it can be calculated as follows:

$$Sp = \sqrt{\frac{\bar{p}(100 - \bar{p})}{n}}$$

$\bar{p}$ = average value for percent defective

Control Limit Calculations for Measurement Data

P is the three-sigma value for *percent defective* or nonconformance.

Percent Defective

No Standard Given

$$\text{Control limits} = p +/- 3\sqrt{\frac{\bar{p}(100 - \bar{p})}{n}}$$

Standard Given

$$\text{Control limits} = p_o +/- 3\sqrt{\frac{p_o(100 - p_o)}{n}}$$

where, p_o is the adopted or selected standard level for average percent defective.

Fraction Defective

Standard Given

$$\text{Control limits} = \bar{p} +/- 3\sqrt{\frac{\bar{p}(1 - \bar{p})}{n}}$$

where, $\bar{p}$ is the decimal value for average fraction defective.

No Standard Given

$$\text{Control limits} = p_o +/- 3\sqrt{\frac{p_o(1-p_o)}{n}}$$

where, p_0 is the adopted or selected decimal value level for average fraction defective.

When the lower control limit value is negative, zero is used for this value in the above formulas.

To calculate the $\bar{p}$ or p_0 charts, use the following procedures:

1. Determine the quality characteristic to be controlled. It may be a single characteristic or the total of all defective units found during examination.
2. Select an adequate number of samples from each lot (for example, twenty, as in the previous problem). These are taken again in successive order and recorded during specified intervals (cycles, hours, shifts, etc.), preferably by the same people.
3. Compute the average sample size. As is the case where the total number of parts inspected in an hour or day is the sample size, it should equal average hourly or daily production.
4. Compute the average number of defectives found hourly/daily.
5. Compute control limits based on calculation of steps 3 and 4.
6. Analyze percent-defective values for each sample with relation to these control limits. Determine whether any factor needs corrective action before adoption.
7. Determine whether control limits fit economic considerations.
8. Use control chart during active production as a quality guide.

Example of Control Chart Calculations. There are two basic methods for calculating the control limits. The first uses the formula for standard/no standard given, and the second, the simplest, uses the charts in Figure 11.9, which have established curves that show the average number of defectives on the horizontal scale and control limits on the vertical scale. The curves represent minimum, average, and maximum control limits based on an average number of defectives. The data are recorded and plotted on a chart shown in Figure 11.10, with a completed percent defective chart shown in Figure 11.11.

MAINTAINING PROCESS CONTROL

Process control of the molding cycle can change for many reasons, such as equipment failures, material variations, control function irregularity, unau-

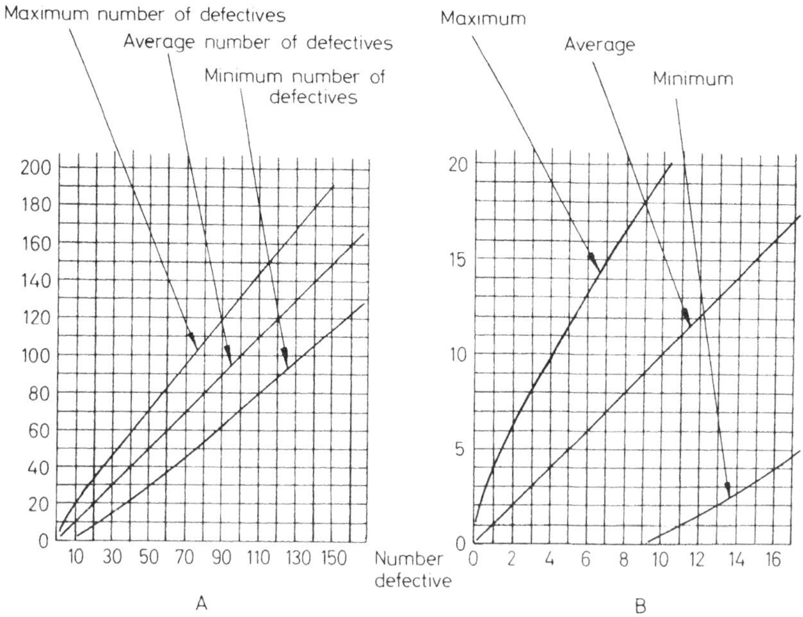

FIGURE 11.9. *Average number of defectives. (Adapted from Ref. [5].)*

thorized operator adjustments, and plant environment variations. To know that a problem is occurring, production must monitor the process either automatically with machine or auxiliary quality control equipment or manually by measuring key part parameters using a real-time sampling plan that measures parts as they are produced.

Precontrol

The critical nature of a job will determine the frequency of sampling and the number of samples drawn. As noted in the examples, a minimum of five successive samples are taken and measured at set intervals. The tools required should be at the machine or on a quality control test stand that is wheeled to the station. After measurements are made, the data are calculated and plotted on the control chart. Trends are noted from the last data plot. If corrective action is required, the proper people are summoned and adjustments are made and noted on the molding record. Once the process has equilibrated and the new points are plotted, rerun the tests to see if the adjustment has brought the part and process into specification.

This is known as "precontrol" and can be performed by the operator on one part at a periodic time or every so many cycles. One-piece checking is done as long as the process is running smoothly with no cycle interruptions. As long as the parts fall within process limits called the green zone, a single

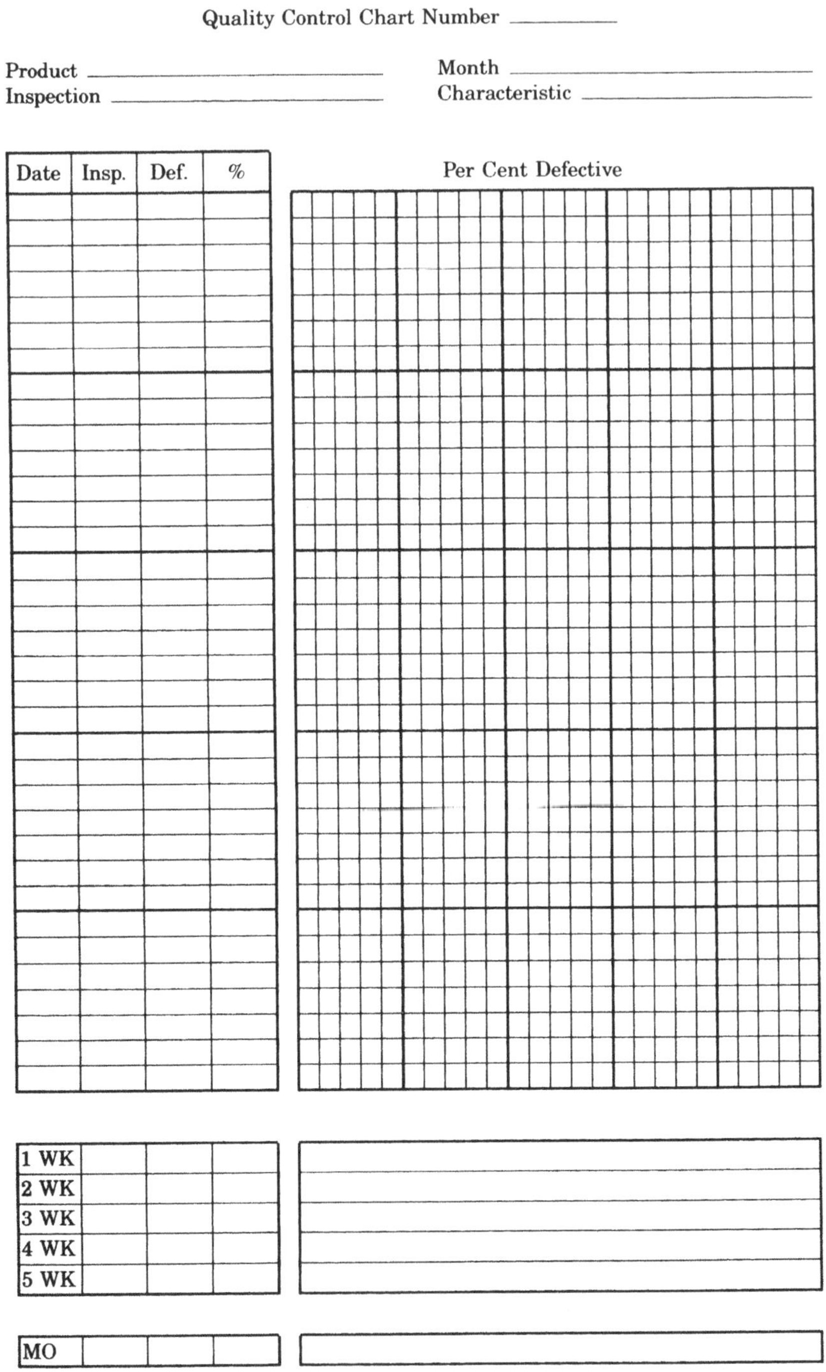

FIGURE 11.10. *Percent defective quality control chart. (Adapted from Ref. [5].)*

PRODUCT CLAMP RING

INSPECTION FINAL

MONTH JULY

CHARACTERISTIC ALL DEFECTS

DATE	INSP.	DEF.	%
1	1003	41	4.0
2	1001	46	4.5
3	1017	89	8.9
WE			
WE			
4	997	71	7.0
5	1000	30	3.0
6	986	21	2.0
7	991	41	4.0
8	1009	49	5.0
WE			
WE			
9	1000	40	4.0
10	1004	74	7.5
11	991	81	8.1
12	996	26	2.5
13	1010	50	4.0
WE			
WE			
14	1000	31	3.0
15	1000	20	2.0
16	1005	76	7.5
17	1000	54	5.4
18	997	60	6.0
WE			
WE			
19	992	72	7.1
20	1010	45	4.5
21	1000	70	7.0
22	1004	44	4.5
23	1015	60	5.9
WE			
WE			
24	991	36	3.5
25	992	39	4.0

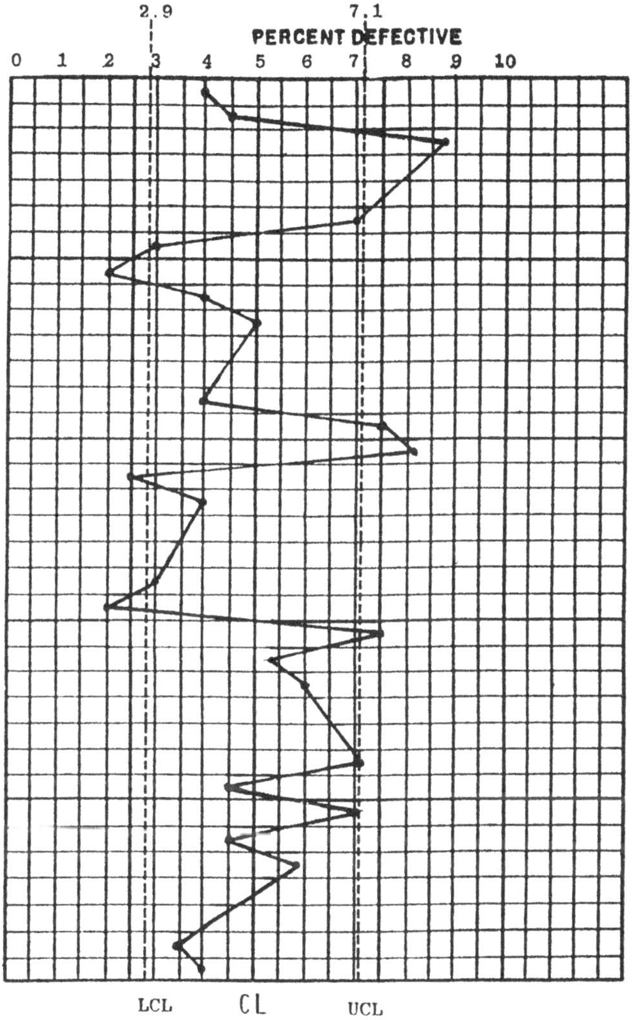

1 WK	3021	176	5.8
2 WK	4983	212	4.2
3 WK	5001	271	5.2
4 WK	5002	241	4.8
5 WK	5021	291	5.7
MO	23028	1191	5.17

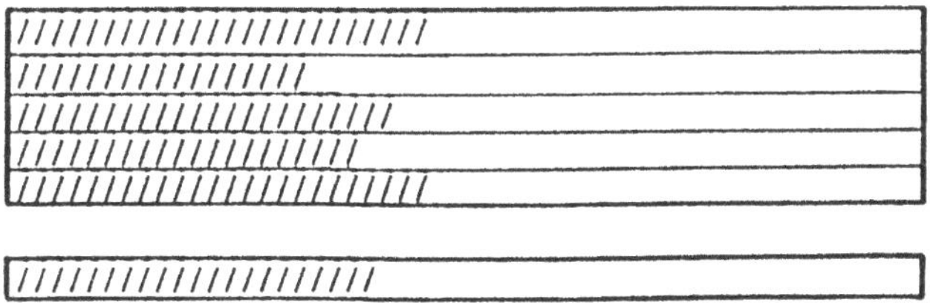

FIGURE 11.11. *Example of a completed percent defective quality control chart. (Adapted from Ref. [5].)*

sample measurement is adequate. Any measurements from outside of the green zone up to the specification limits fall in the yellow zone. Above this limit is the out-of-specification red zone—or scrap zone—as illustrated in Figure 11.4.

With the molding cycle in process control, quality checking begins by measuring an initial five samples. If measurements are within the green zone, then single samples can be measured at periodic intervals. As long as the measurements are green, the process is in control. But should a yellow measurement appear, the next cycle's parts must be checked. If green, the cycle can continue. But should the next fall in the yellow zone, an out-of-control trend may be occurring and assistance is required to bring the process and parts into control. If red occurs, parts must be segregated until green control is attained.

Whenever a new lot of resin or regrind is introduced into the manufacturing process, these tests must be run to verify that the process control settings are still valid.

Taking Measurements

To ensure test accuracy, only a trained operator should take the measurements and perform the calculations. The test equipment must always be in calibration and used correctly. If a holding fixture can assist in more accurately taking measurements on critical parts, it is highly recommended as a way to reduce human error.

When taking these measurements, the operator should also record the molding machine's process conditions, temperature, pressure, cycle times, as well as auxiliary equipment and mold conditions. The use of the pyrometer to measure heater band and cooling-line temperatures validates the equipment's controls and accuracy. The pyrometer readings should he within +/– 5 degrees of the equipment's settings. There will be a slight variance because the measurement is being taken outside the temperature sensing region of the equipment measured. Melt temperature should only be taken if the process is going out of control and all other measurements appear to be within tolerance for the cycle and equipment. This may indicate a bad thermocouple, controller, or gauge, or a heater band problem. If a hydraulic valve sticks, it may or may not be indicated by the pressure control gauge. These problems are harder to discover, but the resin's temperature may be an indicator, along with changes in the injection-pressure profile or the physical movements of the machine's operation.

To be of any value, control charting must be performed at set intervals by trained personnel. The information should be plotted in real time to be of use in controlling the process and the quality of the molded parts. The control charts should also be visibly displayed in the production area and at the molding machine. Keeping the molding process in control and always producing quality parts should be recognized as a mark of excellence.

QUALITY MAINTENANCE

Change in the manufacturing cycle is only welcomed when it improves the process, part quality, or economics of manufacture. When part quality moves out of specification because of equipment or material variations, it must be immediately caught and corrected. When parts are in control and then begin to drift, the machine operator must discover the process change, determine what has caused it, and make necessary corrections.

During any manufacturing operation, there will always be normal part and process variations about the central control point. But when out-of-control trends occur, answers must be found quickly to reduce scrap or a possible shut down. The experienced operator learns over the years how to recognize and solve the more than 25 possible molding and part problems. Too often, the solutions are not documented or passed on to fellow workers. When problems occur and are solved, they should be documented and discussed in operator training sessions. The problem and solution should also be filed in troubleshooting and problem-solving books kept on the production floor under specific section headings for easy reference. With the high turnover in experienced production operators, this is necessary to guarantee a minimum of lost production time and to assure the customer you have the capability and knowledge to produce parts to meet delivery schedules and quality standards.

SOLUTIONS TO TYPICAL MOLDING PROBLEMS

In the injection molding industry, there are typical processing problems that occur. Material suppliers furnish information on how to solve these problems with their materials. In most cases, the solutions are specific to the respective material but can often apply to other materials. In addition, many variables in machinery, materials, personnel, plant environment, and housekeeping create molding and part-quality problems. When more than one of these problems occur, the solution becomes more complex. Presented here are a few of the more common problems and their likely solutions.

Shot-to-Shot Variations

There are many likely causes for this problem and, if part weight is a quality check, it will show up immediately. Likely causes can be any of the following and should be included in your analysis of the problem. Try and begin with the most obvious and easiest to correct.

1. Inconsistent resin feed, wet resin, and resin variability
2. Nonuniform melt density in front of the screw
3. Poor barrel heat profile that causes unmelt or inconsistent viscosity

4. Varying or malfunctioning limit switch or timer for inject to hold pressure setting
5. Worn nonreturn valve, screw tip, or broken valve stud
6. Bad seating or broken ring on nonreturn valve
7. Nonuniform mix of virgin and regrind resin
8. Worn barrel causing material to back flow over the outer diameter (OD) of the nonreturn valve
9. Hydraulic system problems
10. Mold venting

Resin Feed. Inconsistent material feed can be caused by fines, longs, skins, and marriages collecting in the screw flights. These are mainly supplier-created problems. When the resin is poorly screened, longs and marriages can collect at the molding machine's feed throat and inhibit uniform material feed. Longs are fused or poorly cut pellets that are cut too hot during manufacture and stick together forming pencil-like sticks. Skins are formed inside the transfer piping when conveyed in the material supplier's or plant's material transfer piping and are thin skins of resin that flake off the inside of piping. Marriages are similar to longs, except that they are clumps of fused pellets. When feed seems erratic, the machine's feed throat must be exposed and these items removed. On most machines, the feed hopper can be shut off and rotated to expose the throat.

Poor material feed can also occur if the feed section of the barrel gets too hot. The resin melts prematurely and sticks to the screw flights. The screw flights become blocked or too slippery to convey the pellets down the screw. This condition is known as bridging. By observing the feed section and hand-feeding resin into the feed throat, you can identify this problem because the resin will pulsate in the feed section and not convey. The solution is to reduce the heat at the rear barrel zone or add cooling at the machine's feed throat. Cooling the machine's feed throat should be controlled so that condensation does not form. If this does not correct the problem, the screw may have to be pulled and the screw flights cleaned.

Poorly dried material may result in blemishes on the surface of molded parts, cause gassing and a frothy melt, or vary part weight, as moisture creates porosity or voids in parts. Excessive moisture also reduces melt viscosity, which creates a more fluid melt that can result in mold flashing or backflow over the nonreturn valve. Use a calibrated moisture meter, inserted into the dryer's primary air feed line to the hopper, to determine whether the problem is related to the dryer.

Resin inconsistency is usually a supplier problem and should be noted in melt-flow testing when the material is received. But it may not always show up. If this problem continues, change to another lot of material or try a new package. Whatever the solution, you should notify your resin supplier of the problem. The problem material should be removed from the hopper, with the lot and package number noted for later problem solving by your supplier. Segregate this material so it will not be reused until the problem is resolved.

Melt Density Variations. Nonuniform melt density is caused by a variation in resin density being fed by the screw. If a large percent of regrind is being used, differences in resin and regrind particle size can cause a variation in bulk density. This results in a varying melt density in front of the screw. On injection, this affects how the resin fills and packs out the cavity.

The solution may be backpressure to increase the density of the melt, improve the blending of regrind in the hopper to reduce the layering effect, or use of a smaller screen size on the pelletizer to bring regrind size closer to pellet size. Fines from resin and regrind can also contribute to this problem. The screw's compression ratio is another possible cause if a crystalline material, which requires higher compression ratios to promote better melting characteristics, is being molded.

Heat Profiles. Each material has a barrel temperature heating profile recommended by the supplier. It is amazing how often these profiles are set wrong, the machine's temperature controllers are out of calibration, or the thermocouples are not accurately registering the temperature. These cause many problems from unmelt to degradation and burning of the material. A low barrel-heat profile causes poor melting of the resin that can result in unmelt reaching the nonreturn valve. If the particle is caught between the ring or ball seat, it could hold them open and create a pressure leak at the valve during injection. This problem is spasmodic and may be visible in the molded part by varying part weight or by the screw turning counterclockwise during injection as a result of the backflow of resin from the checkring. This problem is easily solved by adjusting the barrel-heat profile. But a heater band may also be burned out or a controller failing. Check the heater band temperature with a pyrometer or take a melt temperature reading to prove this and then take corrective action.

Should the problem be cyclic in nature, with the shot-control problem occurring every third or fourth shot, the problem may be improper adjustment or operation of the barrel heater controls. In addition, a thermocouple may not be functioning correctly, properly seated, or positioned in the barrel to detect a wave variation in the heat level of the material. A heat wave causes a slight viscosity variation resulting in a cyclic variation of shot control.

If material contamination happens when a high-melt temperature resin is mixed with a lower-melt resin, this may show up at the nonreturn valve or the unmelt may pass through and block the part's gate opening or be visible in the molded part. If only a small amount of contamination is present, this problem is harder to identify. It is often intermittent and may disappear as the contamination passes through the machine. Parts produced during this period should be discarded. If the problem persists, inspect them very carefully for an unmelted pellet, particularly if the part is thin.

Inspection of the part and runner may show this when they are bent, as a break will most likely occur near the unmelted pellet. The only solution for this, because the degree of contamination is hard to determine, is to dump the hopper and reload with known good material. The resin feed system should also be inspected. If regrind is being used, check the granulator for foreign resin caused by improper clean out from its previous use. In any case affecting part quality, dump the hopper and segregate all suspected material, clean up and purge the system, and start again with resin of known quality.

If the barrel temperature profile is too high, the material may degrade and burn. Resin degradation can also be caused by too much back pressure or if the screw RPMs are too high. Overheating the polymer causes a lowering of melt viscosity, which can result in flashing of the mold, burning and color changes of the resin, poor part packing, and varying part dimensions. The solution is to check heater band temperatures, screw back pressure settings and screw PRMs, as well as to reduce the melt temperature to its recommended setting. Taking a melt temperature reading will verify this need and should be used to determine when the machine settings produce the correct melt temperature.

In some cases, the molder may have selected a machine with limited melt capacity and is trying to increase its capacity by raising barrel temperatures, back pressure, and screw RPMs. This causes the above problems. In these cases, you should select a machine with a higher screw length/diameter ratio (L/D) and compression ratio. Melt quality is too important to overwork the molding machine and resin. It only results in poor part quality and questionable part performance.

Cycle Times. Improperly set or malfunctioning limit switches or timer settings for controlling screw travel, injection, and hold pressures are other problems that affect part quality. If the first stage injection pressure does not cut out on time, the filling and packing of the mold cavity can result in flashed or underfilled parts. Replacement of the switch or timer is required if they cannot be repaired or recalibrated. The screw's return and plasticating motion must be regulated to build up the correct amount of melt for the next cycle. There is usually a limit switch that controls the screw's rearward travel and stops it at the correct position. If improperly set, the correct amount of melt will not be produced.

Precise switch over from high-injection pressure to low-pack pressure is required in the injection cycle. If cutoff is too soon, poor part filling and packout may result. If too late, the mold will flash and possibly be damaged as a result of high-injection pressure on the mold's internal cavity sections. This could be caused by timer settings or a problem in the hydraulic system that operates off these controls. If the timers are accurate, a sticking hydraulic valve may be the problem. Replacement of the problem valve is the only solution in this case. If the problem reoccurs, the hydraulic oil may be contaminated and have to be filtered or replaced.

Nonreturn Valve. Several types of problems can occur at the nonreturn valve, and the operator should be trained to recognize them. A broken, poor-sealing, or worn nonreturn valve can usually be identified by observing the coupling area of the screw, where it attaches to the hydraulic piston and screw-rotation drive mechanism. If leakage over or back through the nonreturn valve is occurring during injection, the backflow will put pressure on the helix angle of the screw flights and cause the screw to counterrotate. Also the screw will not be able to hold the set pad distance and bottom in the front of the barrel.

Nonreturn valves wear mainly on the outer diameter of the seal ring. Because the ring is made with a lower metal hardness than the barrel, it will wear faster. As wear progresses, the clearance between the OD of the ring and the inner diameter (ID) of the barrel will increase, providing a path for melt to backflow over the ring during the injection stroke. Over many cycles, this wear will increase until packing pressure cannot be maintained on the melt and the pad cannot be held. The only way to verify ring wear is to shut off the material feed, run the screw dry, and take the head off the machine. Then pull the screw far enough out of the barrel to inspect and measure the ring OD and barrel ID for clearance.

If the ring shows excessive wear, it must be replaced to maintain a tight seal with the barrel and packing pressure on the melt in the cavity. Maintaining a good seal with the barrel ID will also reduce barrel wear, which is a more costly and time-consuming problem to correct. Using a larger diameter check-ring is not recommended as a temporary fix. Barrel wear also indicates that the screw has worn. As the clearance between screw flight OD and barrel ID are worn, generating a quality melt also becomes a problem. Always maintain the recommended machinery supplier's clearances for the barrel and screw.

Cold Start Up. A problem that rarely occurs is a broken screw-tip stud. A screw-tip stud secures the nonreturn valve to the screw and can only break during a "cold startup." This occurs when the machine is shut down with the screw retracted and full of resin ahead of the screw tip and nonreturn valve. On startup, if barrel/resin heat soak is insufficient, the operator may rotate the screw before this section of material is sufficiently melted. This shears the screw tip's stud. After removing the broken stud section from the screw, total replacement of the tip is required. It is always a practice on shutdown to cut off resin feed to the screw and pump the barrel and screw dry. In most cases, if shutdown is overnight, a low-melt point purge resin such as polyethylene is run through the barrel to clean out any resin residue that may degrade or attack the metal. On startup, the polyethylene is easily melted and purged from the machine by the primary resin. This ensures a clean barrel free from degraded polymer residue.

Some molders purge with clear polycarbonate and, after running the screw dry, turn off the heat. On cooling, the polycarbonate shrinks and pulls all prior barrel and screw resin contamination with it. On startup, this residue is easily melted and purged from the machine leaving a clean screw and barrel for the next resin.

Decomposed resin particles can contaminate parts for many cycles. If any residue remains in the barrel, it may require pulling the adaptor and screw for cleaning before molding can be restarted.

Bad Valve Seating. Bad seating of the nonreturn or ball valve can cause indentations in the ring and ball by unmelted pellets when compressed during the injection cycle. This is usually evident in the ring or ball, as they are made of a softer steel. If the angle of the nonreturn valve's check ring seal surfaces were incorrectly ground, this could be the problem. As a result, shutoff takes place at the ID of the ring and not on the full face of the seal. A piece of thin shim stock should fit between the seal surface if the angles were ground correctly. As the check ring wears, it may also cause the resin flow path at the flutes or slots of the screw tip to decrease, leading to erratic shot weights and excessive screw/cycle retraction times.

For a ball-check valve, the ball may have worn and become out of round, causing erratic shutoff. If replacement is necessary, a new ball is used and the seat resurfaced to suit the new ball radius. Both of these conditions may be observed if the melt pad in front of the screw cannot be maintained during the injection and packing cycle. Loss or pad reduction plus screw rotation indicate nonreturn valve problems.

Regrind. The higher the percentage of regrind used, the greater its effect on melt viscosity. Most resins and parts can tolerate 20 to 25 percent of regrind without any loss in physical property strength or increase in difficulty in processing.

However when using regrind, some variations will occur with the melt quality of the material. If the regrind is not properly dried, screened for fines, particle-size controlled, or uniformly metered back into the feed hopper, problems can occur. Regrind of hygroscopic resins has a high affinity for moisture, especially when warm, and must be returned as soon as possible to the hopper dryer to avoid this problem.

Worn Barrel and Screw. All barrels and screws will wear. The degree of wear is consistent with the type of resin processed. Reinforced and filled resins will cause more wear because of their abrasive qualities. But some general grade resins also cause wear as a result of their chemical makeup and additive packages. This is one reason why the correct barrel temperature settings are important to correctly melt the polymers without excessive abrasive or chemical action occurring on the steels. After running these corrosive grades, purging with neutral resins is recommended to eliminate any chemical action.

As the barrel and screw wear, the operator may be forced to adopt new ways to create a good quality melt, such as increasing back pressure, raising barrel temperature settings, or increased screw RPMs. When these are required, the screw should be pulled and the barrel and screw measured for correct size and wear. If worn beyond machine tolerances, they should be repaired or replaced. All the above ways of increasing melt will likely cause polymer degradation, loss of properties in the molded part, poor processing, and varying part quality.

As the barrel wears, this can also cause other problems with the nonreturn valve. During the injection cycle, the nonreturn valve checkring expands outward against the wall of the barrel as a result of the pressure of the plastic on the ID of the ring. If excessive barrel wear has occurred, the expansion of the ring exceeds the elastic limit of the steel and fracture occurs. This will occur more frequently with small diameter screws with high L/D ratios. With time, screws may also begin to sag because of metal fatigue and cause increased barrel wear to occur by metal-to-metal contact with the OD of the checkring.

To avoid this, the molding machine's screw should be pulled periodically, and the barrel, screw, and checkring measured for wear. The performance of the injection molding machine is directly related to the tolerances on the barrel and screw. To perform these measurements, pull the screw, clean the barrel, and measure the barrel ID at room temperature. The nonreturn valve checkring clearance should also be gaged. If the difference between the barrel ID and the check ring OD is from 0.012 to 0.014 inches or greater, this will cause the ring to fail. This amount of barrel wear will also allow material to backflow over the ring, even if the nonreturn valve is within tolerance. Worn barrels can be resleeved or rebored and hardened with larger screws inserted to match the new barrel diameter. This is a cost and judgment call; a new barrel and screw may be a better decision. The barrel, screw, and nonreturn valve are the center of the molding operation and must always be within tolerance.

Hydraulic System Problems. All injection-molding machines must have preventive maintenance performed periodically. Too often, this is ignored until a major problem occurs. Then, the machine is lost to production until it is brought back up to performance specification. The hydraulic machine must have its piston seal rings in good condition to develop the pressure and cycle response times of the molding cycle. Leaking piston seal rings cause pressure and cycle time problems that result in a loss of process parameters and an inability to produce quality parts.

The machine's hydraulic oil must also be maintained at the temperature recommended by the machine supplier. The quality of the oil must also be pure, not degraded or contaminated. There are filter systems to screen water and ultrafine particles from the oil to maintain the correct operating viscosity and eliminate any wear particle that will affect the piston ring seals and valve operations. Follow the supplier's maintenance guidelines so that the injection molding machine is always operating within specifications.

Venting the Mold. The vents in a mold must be sized specifically for the plastic to be molded and the anticipated cycle fill time. The trapped air must escape at the same rate as the polymer is being injected into the mold cavity. If the vents in the mold are inadequate or become plugged, excessive cavity pressure can build up as the trapped air is not vented fast enough. This forces the mold halves apart and flashes the mold. Vents must be inspected and cleaned regularly based on the resin molded (some resins outgas badly) and depending on the material's additive package, which may plate out a residue on the cavity surfaces and eventually plug up the vents. Molds running 24-hours straight should be wiped down daily. In addition, molds should always be cleaned each day prior to shutdown, as the deposits will be easier to remove when warm than at cold startup.

When starting up a new mold, the venting system may not be sufficient to produce good parts. Similarly, as the cavity seat-off area wears, the vents may have to be enlarged. If this is a problem, check whether there is enough venting by placing a thin strip of masking tape on the mold steel adjacent to the suspected problem area. The tape will provide just enough additional localized clearance for any trapped air to escape during injection without damaging the mold surfaces.

If on the next few cycles the problem is eliminated or improved, this is a positive sign that the venting must be increased at this area. To locate the problem area on the parting line, use the following technique. With the tape off, shoot the next few cycles just short of filling the cavity and observe where the melt's front stops. Then place a small piece of tape at this area and pack out the mold. If good fill and no burning results, this is the area where additional venting should be placed. Be careful not to open up the surface area too much, as localized material flashing may result. Also, be sure that the vent extends beyond the shut-off line of the cavity seal.

When a new tool flashes and the tape technique does not solve the problem, it may indicate a gate-and-runner design problem that must be evaluated before the tool is released for production. Uneven flash on a part often indicates a core/cavity shift that can be verified by shooting short and measuring opposing walls. If a thickness variation exists, the cores have shifted and realignment will be required. When calculating the mold's required clamp pressure, use 1 to 3 tons/square inch of exposed resin surface area for amorphous resins and 3 to 5 tons/square inch for crystalline resins. The mold's parting-line edges must also be absolutely sharp and flush, or material can work into these crevices and wedge the mold halves apart during the injection stage. This mismatch can be the result of core/cavity slippage or platen/mold rigidity problems.

To verify this problem and check out parting-line integrity, apply die maker's blue on one mold surface. The mold is then closed, clamped, and then opened. If the blueing is not transferred to the opposing half, the problem of nonparallel parting lines is located and a solution developed to fix the misalignment. This may be caused by mold distortion, poor clamp alignment, or nonparallel mold surfaces.

To identify trace deposit buildup on the clamping surface, use the extremely sensitive low-pressure clamp control to sense mold deposit buildup and use it to signal the operator that cleaning is required before damage is done to the mold or flashing occurs.

Cavity Melt Pressure Control

Cavity melt pressure can be integrated into the process control system to monitor and control shot-to-shot uniformity. With improved pressure-sensor hardware, plus modern machine control processors, cavity pressure is a sensitive barometer of overall process control and a valuable supplement to fine tuning the critical points in the molding cycle.

The dynamic pressure of the melt moving into the cavity is a by-product of many process variables and machine settings that affect melt temperature and viscosity. This includes screw RPM, resin shear rate, injection speed, barrel and mold temperatures, resin composition, and regrind levels, as well as temperature and viscosity of the oil in the hydraulic system.

When cavity melt pressure is used as a control device, with a pressure sensor in the mold, it can switch the machine's hydraulics from first-stage high-injection pressure to second-stage lower-packing pressure at the proper instant in the cycle. Using the proper pressure sensor, it can time the change over to coincide with peak pressure in the cavity as it completes filling to avoid flashing the mold, but ensure maximum cavity packout pressure for each cycle. This mold-mounted sensor can also compensate for variable melt density when filling the cavity, as full pressure will only be sensed when the cavity is full and the melt compressed to the required density to trigger the sensor. The cavity pressure measurements can be used to monitor the cycle for process quality assurance. When interfaced with a recorder, they provide quality control documentation of shot-to-shot packout pressure measurements and their timing.

If cavity pressure is used as a control for part quality, the sensor must be interfaced with the machine's microprocessor controller to control hydraulic switching from injection to pack pressure. If an older machine is retrofitted, then a separate controller is required. The most common hook-up is to bypass the timer with the pressure sensor signal to trigger the pressure transfer. Since timing is critical, these hookups, of which there are many depending on the age of the equipment, should be coordinated with a specialist in the machine control system.

Figure 11.12 shows pressure sensor locations to monitor melt, hydraulic, and cavity pressures. The most effective site to be monitored is the cavity or runner, depending on mold layout. Current technology uses timers and screw position to trigger the pack-to-hold pressure. Precision molding requires

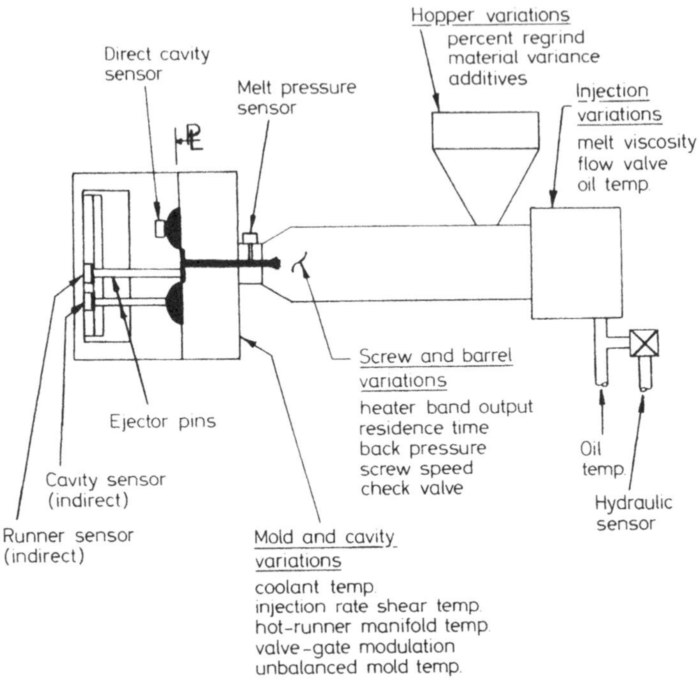

FIGURE 11.12. *Pressure sensor layouts. (Adapted from Ref. [18].)*

higher repeatability. Preset timers cannot compensate for the effect of drift in process settings and changes in resin density when filling and packing the cavity. As the molding machine reaches equilibrium operating conditions, the viscosity of the hydraulic oil drops. This results in faster ram travel and cavity fill rates in machines with inadequate control compensation. This can lead to mold flashing, overpacking, and/or resin degradation caused by an increased shear during injection and filling the cavity through a small gate.

During the injection stroke, the screw position is usually sensed by a limit switch or linear transducer that signals the pressure switch-over point. This guarantees that a constant volume of resin is injected into the mold cavity but not a constant weight. Resin variation in density caused by temperature variations, regrind, and additive level and leakage around the nonreturn valve, will affect the final part weight and packing pressure. These are ignored by the screw-position pressure transfer.

Injection profiling during mold filling controls the injection rate at preselected screw positions. This aids in solving such part-fill problems as fill pattern and fiber orientation, inhibiting shear-sensitive resins from degradation, and preventing cores from shifting. But it does not address weight and dimension problems. Injection profiling does not correct any process variable that affects the amount of melt injected into the mold. It operates on preselected time, position, and pressure settings.

Other methods that have been suggested used pressure measurements to monitor process control. Sensors were also located in the machine's hydraulic system or melt stream at the nozzle but did little to compensate for machine and resin process variable changes What was needed were sensors as far forward as possible in the molding system to sense changes in melt density. This implied that the pressure sensors should be located in the mold cavity to sense cavity pack out pressure accurately, which only occurs when the cavity is full.

Different types of mold cavity sensors can be used—either removable or fixed. The removable sensors are more versatile and can be used in more than one mold for economical utilization of equipment. These slide sensors are positioned behind the ejector pins and are installed from the plate edge on slots in the plate. Melt cavity pressure is transmitted either by an active stroking or dummy ejector pin positioned in the cavity surface. They are indirect pressure sensors, because the melt does not directly contact the sensor.

The fixed, or button sensors, are usually permanently mounted in the cavity surface. The benefits of using direct sensors are greater accuracy, less mounting space, no ejector pin, and smaller tip sizes, down to 0.10 inch in diameter. They can also be machined to match the contour of the cavity. Some suppliers can furnish the sensor to monitor both pressure and temperature by incorporating a thermocouple into the sensor unit.

Placement of the sensors is critical for accurate melt pressure readings. Optimum location depends on the specific mold and part. For single cavity molds, place the sensor midway between the gate and the end of the flow in the cavity. It should be in an area free of turbulence and jetting so it can sense premature gate freeze off. If it is too far from the gate, the signal will come too late for proper control.

For monitoring large, long flow parts, an end-of-flow sensor will confirm correct filling and packing Quality control for large parts may require one or several sensors near the end of flow for fill confirmation, with the pressure control sensor near the gate for correct timing to switch the hydraulics to pack pressure.

For multicavity molds making smaller parts, a balanced runner system is preferable. The sensor is located in the runner system midway between the gate and the last straight runner section away from the sprue and any runner intersections.

Many problems are self-evident and easily diagnosed, while others fail to respond to the obvious solutions. See Appendix E; Table E-1 list common injection molding problems and solutions. To assist in troubleshooting, Table E-2 illustrates a troubleshooting guide of process parameters, and Table E-3 lists common injection molding problems and solutions. The latter can inform the operator which way to adjust a molding parameter to obtain the desired result to solve a problem.

Each material supplier also publishes a molding guide or at least processing conditions for their resins that list recommended ranges of barrel settings, drying temperatures, injection pressures, and mold temperatures. They may also include information on machine size, L/D ratio, and screw type. These guidelines are used to establish the initial cycle conditions. Adjustments are usually required to produce quality parts within the cycle parameters given. In the back of these molding guidelines brochures, there is usually a troubleshooting guide.

The important item to remember when problems occur is to make a single machine or process adjustment at a time and, depending on the adjustment, wait a sufficient length of time or number of cycles to be sure the process is in equilibrium before determining if the adjustment cured the problem. Often, inexperienced operators make several changes on one cycle, but because they were not in equilibrium, the parts or problem reappeared or another problem occurred on the next cycle. If making single adjustments will not solve the problem, the system mold, machine, and process may not be capable of maintaining process control. If so, reevaluate the total system to determine the cause of the problem. The use of the Taguchi methods may be required to bring the process under control.

Solving molding problems requires experience, good equipment, and training in equipment operations and troubleshooting procedures. Solutions should be documented and become a part of the process procedure for the mold and machine. This ensures that if the problem occurs again, this solution can be tried first. This information should also be given to the production supervisor, as it may indicate a change in preventive maintenance scheduling or procedures. These solutions can also be filed in the department's troubleshooting binder and can be referenced if other molds and machines experience similar problems. They should be filed by the type of problem (e.g., flash, burn marks, sink marks, and warpage).

CONTROLLING AND MONITORING PROCESS VARIABLES

Total quality process control involves networking all injection molding equipment into a machine processing network. A local control or central computer system should be tied into the machine's microprocessor to control and monitor variables. The method employed depends on the equipment, its output capability, and resources used to control the manufacturing cycle. This is known as process line integration (PLI). It ties together the molding machine and its auxiliary equipment, as shown in Figure 11.13.

PROCESS LINE INTEGRATION

To guarantee continuous part quality, molders are now controlling all the equipment from a central or local controller that can also output data in the

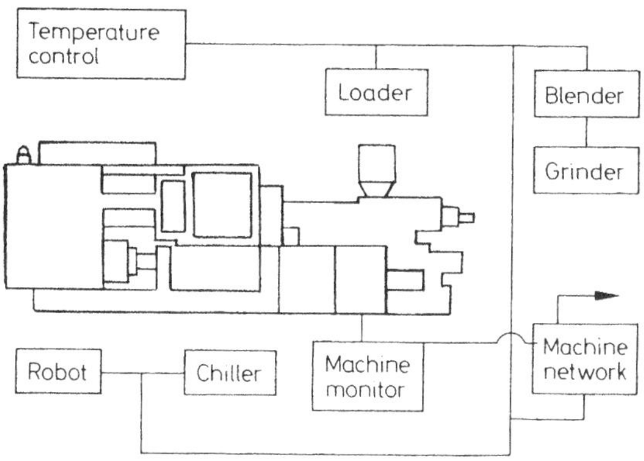

FIGURE 11.13. Process line integration. (Adapted from Ref. [12].)

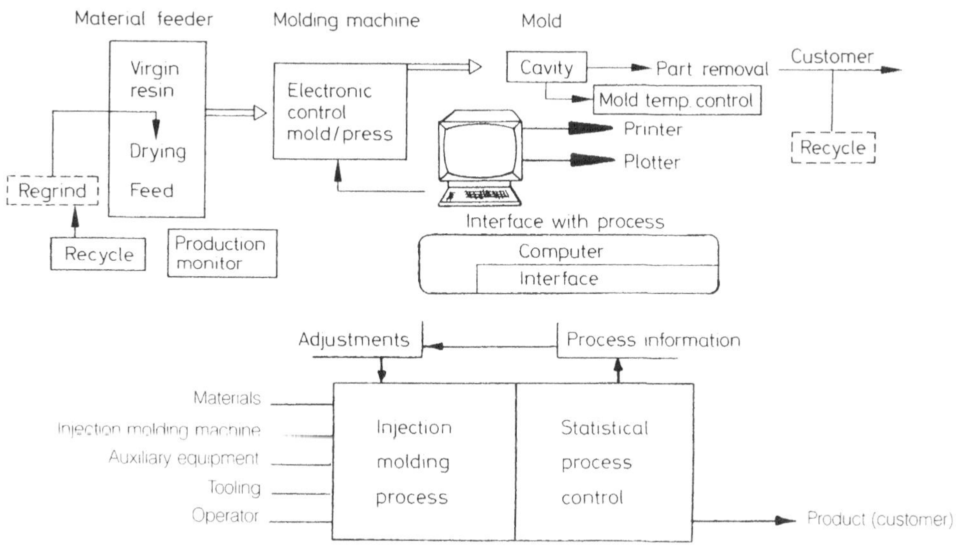

FIGURE 11.14. Statistical process control looping. (Adapted from Refs. [10] and [11].)

form of process control charts. With just-in-time (JIT) manufacturing methods and parts arriving at assembly lines without incoming inspection, customers require assurance that the parts meet specifications. To meet this challenge, molders are adopting PLI.

PLI connects all auxiliary equipment with the molding machine, as shown in Figure 11.14, through a central or individually dedicated "cell" (shared data acquisition system). It collects data on temperature, time, pressures, and other variables that can be evaluated in real time. The output can then be compared with preprogrammed limits. PLI uses closed-loop feedback to keep the network's equipment continuously within preselected limits. Should a variable move outside of the control limits, PLI alerts an operator to fix the problem before the parts go out of specification limits.

The benefits of PLI are twofold—automated setups and diagnostics. Automated setups ensure that the correct processing parameters are put into the machine and auxiliary equipment every time. It also reduces setup time and ensures part quality and consistency. Diagnostics assist the operator in identifying problems, thereby reducing maintenance and freeing them to analyze and improve the process.

Most auxiliary equipment has some form of standardized microprocessor unit or can be retrofitted at a relatively low add-on cost. The software for the control and networking of the system's equipment has become increasingly sophisticated, with the emphasis on ease of use and low maintenance. The Society of the Plastics Industry has completed a procedure known as "Communications Protocol." It is an example of the steps that machine builders are taking to make integration easier and more cost effective.

Process line Integration Benefits

The benefits of using PLI are as follows:

1. Storage of all final machinery settings for auxiliary molding, and support equipment. These settings are from the last run before shut down and include a printout.
2. Automatic downloading of all production parameters to the machinery at startup.
3. Time savings at start of production run (JIT or QMC) by ensuring that all set points are correct. Tweaking of equipment is minimal to unnecessary. No operator error.
4. Improved part quality through improved consistency by eliminating any guesswork in repeating the quality of a prior production run.
5. Analysis of molding problems is faster and easier, as all information from auxiliary equipment and the molding machine is centrally located and accessible.
6. Audible alarm systems alert operators when control set points are exceeded. This allows operators to react quickly to solve the problem.
7. Equipment can operate automatically with one operator serving more machines and monitoring their performances.
8. Consistent part quality shot-to-shot as all systems are in control and monitored consistently.
9. Ability to produce SPC charts for internal use and to provide evidence of part control to customers.
10. Ability to predict future failures because of machinery wear by statistically tracking variations in machine performance in the form of trends in deviations from set points.

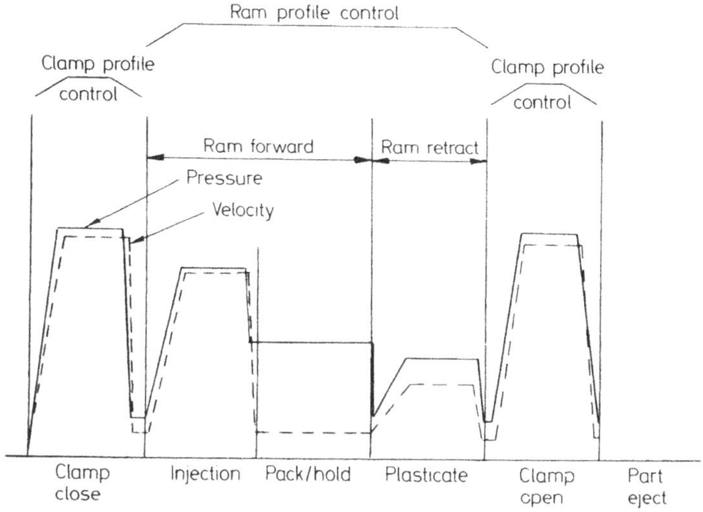

FIGURE 11.15. Molding cycle pressure/velocity profiles. (Adapted from Ref. [12].)

Controlling and monitoring the production process not only yields higher and more consistent part quality, it also increases productivity and profits. It improves cycle times, tracks down time and production output with round-the-clock reporting, and reduces scrap.

The systems are real-time tools. When molding cells are tied into a central system, they become even better management tools. Color screen monitors can transform numbers into a graphic display of plant operating conditions. Figure 11.15 illustrates a typical molding cycle, with pressure and velocity outputs for the molding machine's hydraulic system. By continuous analysis and comparison, it can visually show that process control is being maintained by the unit's cycle-to-cycle operations.

Process Line Integration Scheduling

The PLI system can also assist in scheduling, provide job status information, and offer management summary data for each shift, day, or week of production time. System software can translate the data into many common reports, such as follows:

1. Job status: percent complete, time to completion
2. Material requirements: current supply, projected need
3. Material usage: per job, per machine
4. Mold usage
5. Downtime: by machine, by job, by cause
6. Maintenance reports: by machine, by cause

7. Time and attendance reports
8. Production by department
9. Machine performance: temperature, pressure, output
10. Quality control: scrap rate by machine, scrap rate compared with processing parameters
11. Production output

Selecting a System

With all this real-time power and output capability, the main problem faced by processors is how to select a production monitoring system. Assess your requirements by a systematic review of operations and what you feel is necessary to meet quality goals. You should address the following questions of capacity, available information, flexibility, and choice of systems.

Capacity

1. What is the maximum number of machines to be monitored?
2. What is the amount of permanent storage space (disk drive) for historical database?
3. Does the monitor store data while the computer is active?
4. What is the maximum number of computer remote terminals CRTs?
5. What is the color or monochrome CRTs?
6. What is the maximum number of keyboards?
7. Can reports be customized?
8. Is software code available to you for modification?
9. Can several users access the system simultaneously?

Information Availability

1. Will data, as cycle time, be shown as a running average for a given number of cycles, or only for the last cycle?
2. What processing conditions can be monitored—temperature, pressure, times? And how many parameters will the system handle?
3. What is the machine power consumption?
4. Are out-of-limit alarms available?
5. Is there a predetermined list of downtime causes and can it be expanded?

Flexibility

1. How easy is the system to upgrade? Is additional hardware or software required if you further automate equipment with process controllers or a CIM system?
2. Can monitoring system be integrated with other management information functions, such as accounting and costing?

3. How compatible is the system with other system and equipment controllers?

Which Supplier System

1. Does the supplier furnish and install a complete system?
2. Is equipment compatible with you current and future equipment?
3. Is online telephone service available?
4. What is the type, length, and quality of operator training?
5. Is system uses proprietary hardware or software, is the vendor a reliable source for service and replacements?
6. Is there an opportunity to receive software updates?
7. What types of updates are planned to keep the system current with technology?
8. What is the cost of the system compared with its benefits?

Quality will be judged at this point in the manufacturing process. Did we produce parts within the cost and quality projections quoted and agreed to by the supplier and customer? At this point the final question is answered—did we succeed?

12

Part Testing at the Machine

Confirming the quality of the molded part at the press, with minimum reliance on operator judgment, is desirable but not always possible. With the use of total quality process control, the molder can be assured that the manufacturing process consistently produced parts in tolerance at a cycle fast enough to earn profits. After the cycle is optimized and in control, periodic checks on the molded parts must be performed to be sure that another variable that could contribute to defective parts has not entered the system.

How many times have good parts been produced one lay and then, for no apparent reason, a quality problem occurs on the same cycle conditions? The only way to detect subtle process or material changes is by monitoring the finished article with a dimensional aesthetic and/or a physical monitoring system.

SELECTING THE TEST

The customer may require the molder to perform tests of a destructive or nondestructive nature on the parts to verify that the resin and processing conditions will meet end-use requirements. Some resins are very sensitive to moisture and processing conditions. A problem that cannot be detected visually may show up in other tests. This implies that some sensitive resins may not always exhibit the physical property strengths required after molding.

Total Quality Process Control for Injection Molding, by M. Joseph Gordon

Processing variables, such as melt temperature, injection rate, pressure, and mold temperature, as well as resin variables, may affect the part's end-use properties even if the molding process is thought to be in control.

Many molders use molded part weight as one check to verify that the molding cycle is producing quality parts consistently. But part weight can remain constant, while other problems that may cause rejection of the part, may be occurring. Therefore, end-use part testing or visual inspection may be required, either online or offline, when the parts have cooled.

Part testing at the machine may consist of measuring or testing a set number of parts—usually a minimum of five—at a set cycle interval to verify conformity with customer requirements. Tests may occur immediately after the part is ejected, or after a predetermined cool-down period. The closer the tests are performed to actual production time, the more useful the results are for effecting the necessary corrections.

When performing the initial checks on the mold, process, resin, and part, determine whether any more tests or checks will be necessary to verify part and process quality. Dimensions and molded-in stresses, which can result in part warpage, are the most difficult to verify while the part is still warm. Based on their material makeup and processing conditions, all plastics will experience a varying amount of post-mold shrinkage on cooling. In addition, post-mold shrinkage and relaxation of molded-in stresses will continue over a period of time. Any measurements taken right after molding must be adjusted once the part has stabilized. The nonreinforced crystalline resins have very high mold shrinkages. On cooling, they continue to shrink over longer periods of time than amorphous resins, depending on processing conditions.

After molding, the dimensions of some resins, such as nylon, are affected by moisture. Dimensions will increase when the part is exposed to high humidity. Moisture also has an annealing effect that causes post-mold stress relaxation, dimensional changes and warpage, and some additional post-mold shrinkage. Some customers require that parts be placed in humidity chambers to pick up a specified percent by weight of moisture to increase toughness. This is often required if an end-use test is necessary during or right after assembly. However, moisture conditioning may not be necessary, particularly if the part had been designed correctly or produced with a different resin. Nylon 6 zippers are always moisture conditioned to increase their toughness.

If thick-sectioned parts are ejected from the mold while still semimolten, they may have to be placed in a holding fixture to resist warping. Because these parts are highly stressed, impact tests are often performed to be sure that molded-in stresses will not affect end-use function.

Another form of testing exposes the parts to end-use temperatures. This verifies that part dimensions will remain within end-use operating parameters. Molded parts that will be exposed to higher operating temperatures than the mold-cavity temperature will experience post-mold shrinkage. Parts exposed to elevated temperatures should therefore be molded under conditions where maximum shrinkage is obtained in the mold. The goal is to minimize shrinkage when the part is in service. It is mainly the crystalline resins that exhibit higher in-mold and post-mold shrinkage. Crystalline resins used in cold molds of 80°F and lower will experience higher post-mold shrinkage when exposed to temperatures above this point.

Before implementing an inspection system for the molded part, determine the critical elements of the test, the benefits to be realized, and the test method and frequency. The different types of testing that can be conducted to verify part and process quality are as follows:

1. Verify molding parameters are in control.
2. Test end-use performance of molded parts.
3. Perform dimensional checks.
4. Conduct assembly requirement checks on critical areas.
5. Execute aesthetic surface checks.

These tests are not the normal American Society for Testing and Materials (ASTM) tests, where raw materials are evaluated for physical properties, but they are specific tests to evaluate particular requirements of the molded part.

VERIFYING MOLDING CONDITIONS

To verify that molding parameters have not changed, measure the part or check its weight. Other changes that could affect part quality include use of an incorrect amount of regrind, variation in material filler content or additives, molded-in stresses caused by subtle changes not noted in the process variables, viscosity variance affecting part toughness, moisture in the resin, and use of the wrong material.

DESTRUCTIVE TESTS

The test usually employed to verify molding conditions is an impact, or bending test, that is controlled and regulated. It is usually a destructive test that measures the energy absorption ability of the material and molded part. This test is often associated with end-use testing of the part. Testing can be performed at the press after a preset cooling time. In some resins, testing warm parts can give misleading results. These are go/no go type tests. They can also be used to check part weld-line strength at critical areas.

Among the most common impact tests are the falling dart and pendulum methods. These tests are easy to perform. The part is positioned in a fixture and the impact force directed to a specific area each time. The results are repeatable and reproducible.

During the initial molding trials, determine the test method and the load absorbing values that the part must withstand. Tests are run to determine the critical points on the part and the maximum load values that can be used to verify that the molded part is acceptable. Parts molded using these conditions are then tested. If the part fails, it indicates that a change has occurred. Process and material values must then be checked to discover why part failure occurred. It is important to support the part in a holding fixture so that the part's critical point is impacted each time. This test, which only takes a few seconds, is easily run on temperature-conditioned or normally molded parts. The part is removed from the conditioning chamber and placed in the holding fixture. The ball impacts the part, and the results are recorded with minimal, if any, change in part temperature.

Gardner "Ball Drop" Impact Test

The most frequently used test is the Gardner Impact, or ball drop, test. The simple apparatus can easily be set up at the press, with an operator or technician performing the test and recording the results (see Figure 12.1). Variations can be adapted to suit particular cases, but the principle is the same. This test uses a known weight that falls a set distance and impacts the part at a

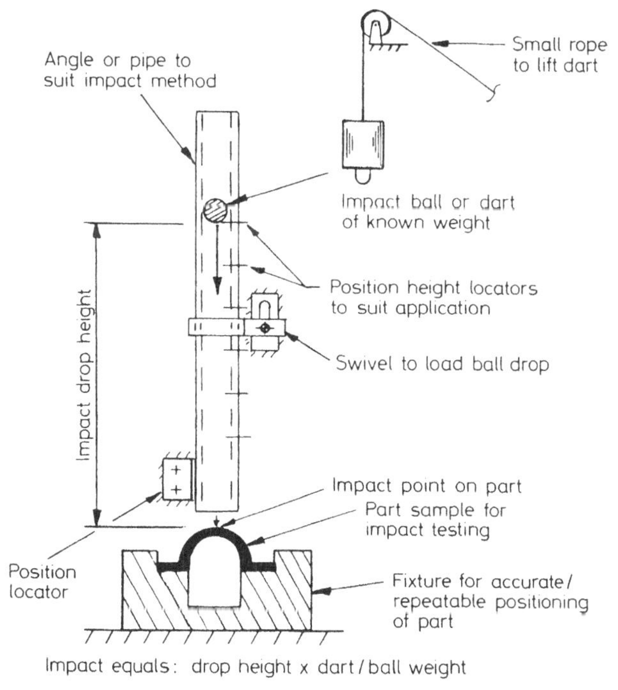

FIGURE 12.1. Gardner impact test apparatus.

preselected spot. When using the ball drop test, the radius of the ball, or impact dart, should be based on the material supplier's recommendations. Too sharp a radius on the ball or dart may, for some resins, concentrate too much energy at the impact point and lead to erroneous results.

It is necessary to define what constitutes a successful test and what constitutes failure. The test conditions must also be selected so that a failure indicates a significant process or material variation. If the impact force is selected so that a minor variation initiates a failure, but does not affect end-use performance, the test is too rigid. Once testing establishes the amount of force needed to cause failure, reduce the pass value by 5 to 10 percent to allow for normal variations. Parts that pass the test may be retained as examples but should not be included with finished parts for customer use.

NONDESTRUCTIVE TESTS

Nondestructive tests are visual inspections of the finished part. These may include dimensional, form and fit, stress/strain, aesthetic, and color checks. Dimensional checks may incorporate go/no-go gages or the part may be put in a measuring fixture and measurements made for critical dimensions. Using a fixture and fixed gages or calipers reduces operator errors and leads to uniform measurements each time. Check that gages and calipers are in calibration before measurements are taken. Operators must be trained in the use of the measurement tool, as the results will be used to verify the quality of the molding and parts. Measurements should always be taken after a predetermined cooling period to avoid shrinkage errors and to have uniformity of measurement data.

Optical Comparators

Optical part comparators are often used by the quality control (QC) department away from the molding area. Parts are compared with an outline of the feature to be verified. It can be a go/no-go type of evaluation or actual measurement checks.

More sophisticated vision-testing systems use stored computer images of the part. During testing, the parts pass under or through a specified light source, from infrared to visible to ultraviolet wavelength ranges. Incandescent and strobe lights are used to monitor part contours, features, and dimensions.

Lasers are quickly becoming the state-of-the-art medium for measuring part dimensions because of their straight-line orientation. It is superior to other light or wavelength sources, which are affected by light refraction. Lasers are even used to measure blind-holes and contours. Their readings are compared with stored data from a master part or pattern in the computer's memory. The analysis is fast, and parts can be conveyed and shunted to good/

bad points on the line as determined by the system. When using lasers, the parts must be in a fixture and properly oriented when passing through the beam. This can be automated to occur as parts are removed from the mold.

Stress/Strain Part Evaluation

The quality of the molded shapes of clear materials can be determined by the absence or presence of residual strains. This is a good quality check for a variable that has previously been hard to detect and analyze. The presence of "lock-in" strains can seriously affect the performance of plastic parts. These strains are introduced by differential shrinkage, nonuniform flow, overpacking the cavity, uneven cooling from thick/thin sections, stretching, and other strain-inducing procedures during manufacture. On occasion, residual tensile stress combines with service stress can cause premature part failure in snap-in joints, molded hinges, and areas of high-service stress near gate or at weld lines. Corrosion-cracking is another form of failure that usually occurs at lower-strain levels when parts are exposed to chemicals and solvents.

All residual part stresses are not harmful—for example, in oriented fibers and film. In addition when parts in service experience tensile stress, they can be made with molded-in compressive stress that neutralizes the effects.

Polarized Light

With polarized light, strains in clear parts can be seen but not evaluated. This is overcome by using a linear-wedge comparator, which is an artificial crystal cut to produce a linearly variable retardation. The strain and resultant stresses in the part can be calculated from a scale attached to the body of a compensator. These procedures are illustrated in Figure 12.2, with a strain pattern shown in a plastic acrylic lens (see Figure 12.3). This manual procedure requires a trained technician, but an electrooptic analysis system eliminates human error and permits on-line inspection.

The spectral contents analysis (SCA) uses a personal computer and special software to produce a spectral signature. When digitized and compared with a standard photoelastic response, the strain level can be calculated (see Figure 12.4). The equipment setup for these measurements is shown in Figure 12.5.

Opaque parts can be evaluated to a certain extent if a clear material (e.g., acrylic, polycarbonate, or styrene) is molded in the tool. By varying molding conditions, conclusions can be reached on appropriate packing and injection pressures, as well as other process variables. Residual molded-in strains are similar to detect. Any end-use loads on the clear part will show the effects of stresses in simulated service. Because the analysis of stresses in opaque parts is more difficult, finite part analysis may be necessary to determine the actual stresses. Because this is only an analytical technique, end-use testing may also be required.

Aesthetic Part Checking

Aesthetic part checking is more subjective in nature. As a result, this testing requires customer approved samples for the operator to compare with the molded parts. These samples must be stored in a humidity- and temperature-controlled location. Visible parts usually require aesthetic checks. This may consist of noting the surface finish on a part, the right color, or the absence of contamination. During processing, some resins develop mold deposit residues on the cavity surface. Over time, these can cloud or destroy the surface appearance of the part. If residues accumulate, machine conditions should be adjusted, specifically melt temperature, vent clearances checked, and the cavity surface cleaned every so many cycles. Certain mold release by-products can also build up on the cavity surface and create problems. To solve this, there are special mold releases for specific materials and finishing operations. If mold-release products are used to a large extent, the cavity should be reworked. Other problems, including orange peel, jetting, glassy surface, splay, and gate blush, must be caught by the operator or QC team during routine machine/part inspections. These problems must then be corrected by adjusting resin and processing conditions. Specific solutions are noted in the trouble-shooting guide.

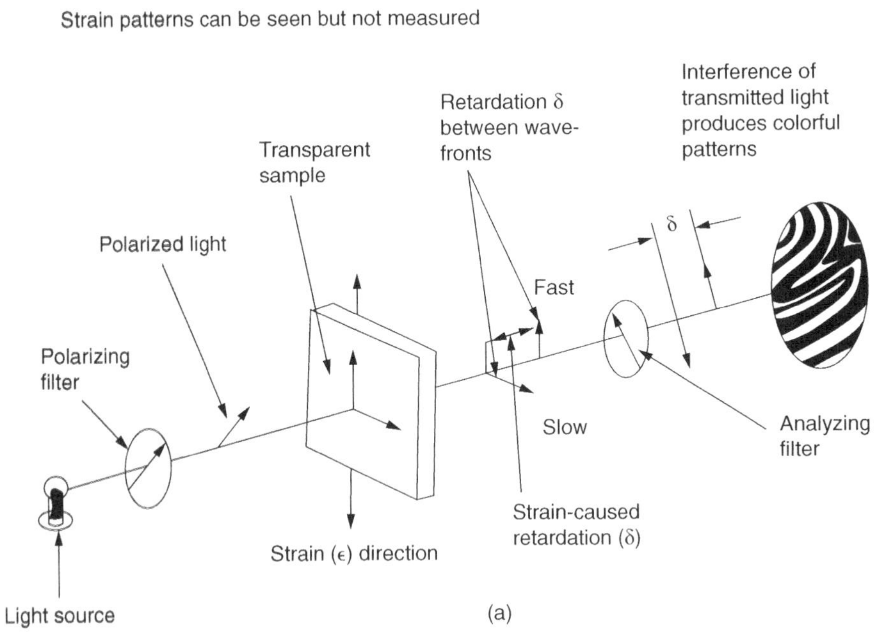

FIGURE 12.2. (a) Polarized light shows strain patterns in transparent parts. (b) Strainoptic™ Polariscope Polarimeter System PS-100. (Courtesy of Strainoptic Technologies, Inc., North Wales, PA)

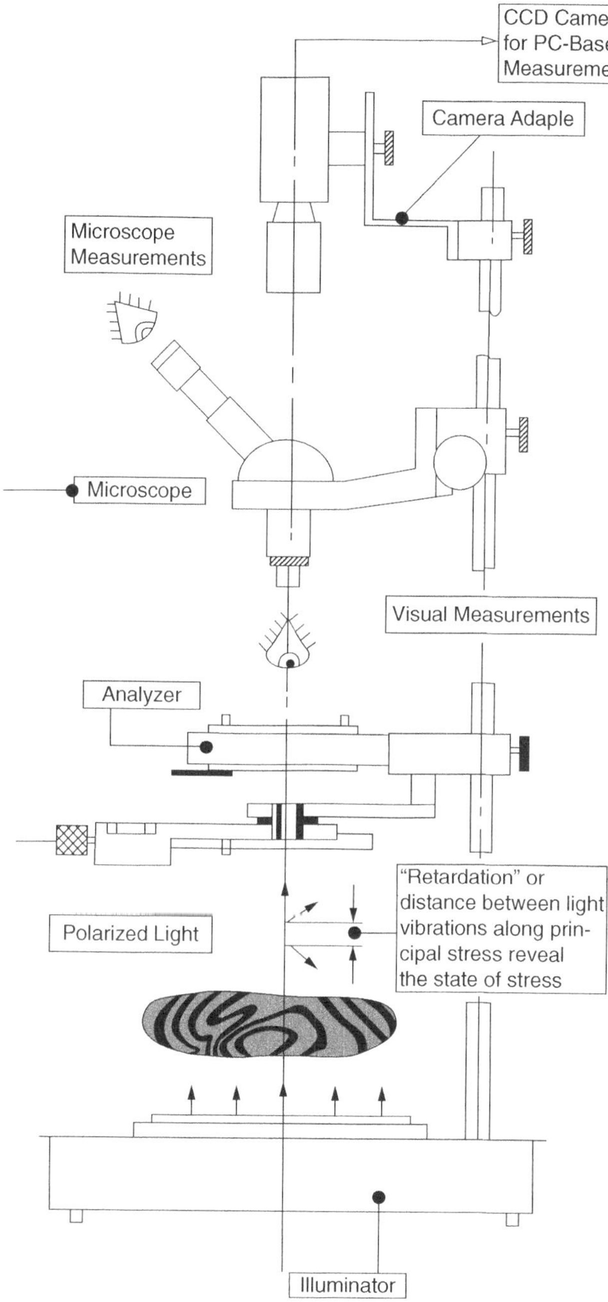

(b)

FIGURE 12.2. *(Continued)*

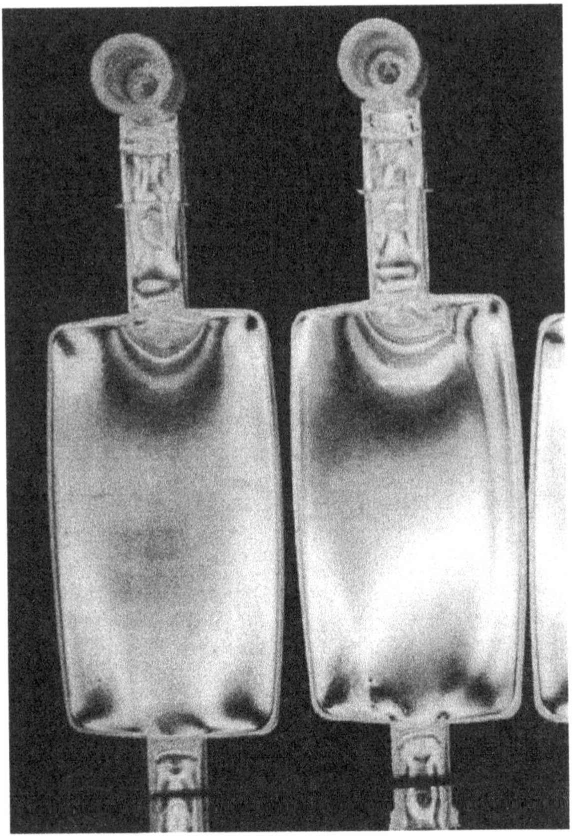

FIGURE 12.3. *QC of strain pattern in a compact disc. (Courtesy of Strainoptic Technologies, Inc., North Wales, PA)*

Part contamination is usually considered a surface problem. It can occur from degraded resin in the barrel or from resin contamination that was not caught at incoming inspection. Although this happens infrequently, the operator should visually check parts that are exposed in the final products. Contaminated parts should be thrown away and never put into the granulator for regrind, where it would only create additional part quality problems and result in lost production time and material. For clear parts, a photoelectric cell with a light source performs this checking automatically. If the light beam is broken, an alarm summons an operator. All contamination problems must be solved immediately to reduce production loss and protect part quality.

Color Checks

The control of color is very important and subjective. It is also becoming more difficult, as suppliers are moving away from cadmium and other heavy metal

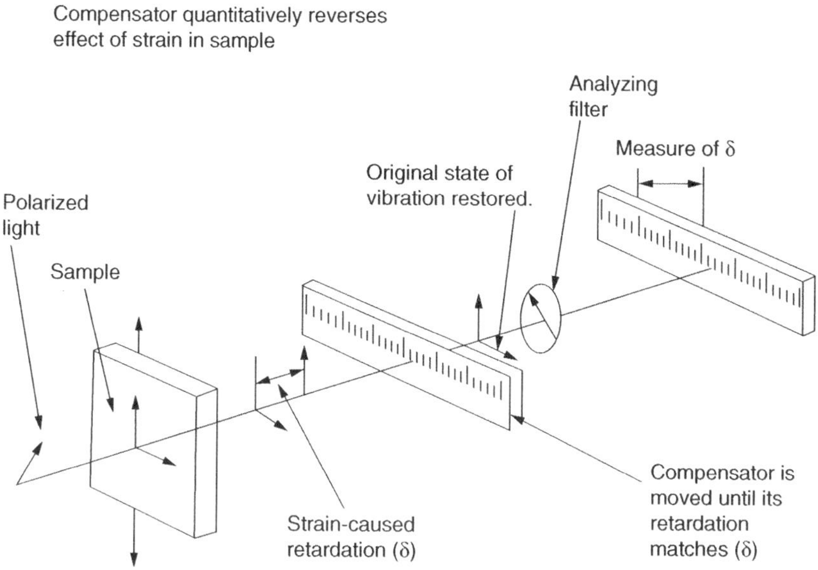

FIGURE 12.4. *Measurement of strains in transparent parts. (Courtesy of Strainoptic Technologies, Inc., North Wales, PA)*

pigments to organic pigments with lower melt-heat stability. This requires tighter control of processing conditions to avoid overheating the material. This burns out the pigment and causes a color change. Check with your material supplier for the recommended heat settings and the allowable machine resonance time before degradation occurs. Most suppliers run these tests and the data is available to the processor.

Color concentrates do not normally require drying, because the resin is a low-melt polymer of ethyl vinyl acetate (EVA), polyethylene (PE), polypropylene (PP) or the same base resin used for the part. These concentrates are very low blend ratios and most are nonhygroscopic. Therefore, their moisture level is so low that it does not affect the molded part's processing characteristics. Should they require drying, you must know the base resin and the pigment system. Then, use their minimum times and temperature drying conditions to avoid degrading the concentrate.

Processing compounded colors and some salt-and-pepper blends requires greater care. For these materials, the hopper drying temperature may have to be reduced to avoid degradation of the pigment colorant. Each pigment's heat stability varies. Check with your supplier to be sure that the color is not affected by too high a dryer temperature or too long a residence time. Pigment processing temperature ranges are listed in Chapter 16.

The visual testing of molded parts therefore becomes especially important for matching the customer's color. Knowing when to verify the color to a standard becomes more critical, as color matching is generally a judgment call.

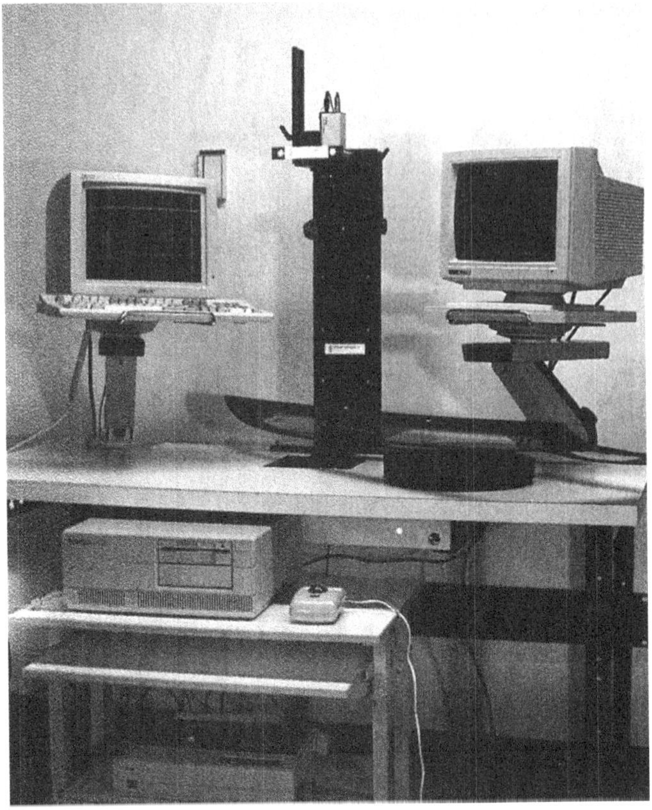

FIGURE 12.5. Digital image analysis system for birefringence inspection in molded lenses. (Courtesy of Strainoptic Technologies, Inc., North Wales, PA)

It is also advisable when molding individual parts that fit together in an assembly to schedule them in the same production period. This allows a comparison of the matching colors of mating parts at the same time.

In evaluating color, remember that the color of the part will change after cooling. For some parts, the color may not stabilize for several days. Another variable is moisture, which can affect the color of parts stored in open containers over a long period. Therefore, if parts that are susceptible to moisture are not being used right away, store them in moisture-proof containers. After being boxed, store these parts out of the sunlight in a temperature-regulated warehouse. This ensures that when they are finally used, the color will be as close as possible to the standard.

When judging part color at the molding machine, a color standard should be available for reference. Verify the color after the part cools. With customer approval, a standard color plaque can be obtained from the material supplier. However, it should be replaced every six months because of fading. When comparing colors, select a person with good color perception. The part and standard should be compared under a known and predetermined light source. The standard's surface should replicate the part's surface finish or a section of the part, as color tone is affected by light refraction off of the part's surface.

More frequent color checks need to be made if color compounded material is not used. When using salt-and-pepper color blends or metering pigments at the feed hopper, color control is more difficult.

TESTING OF PLATED PARTS

Molded parts that are to be plated often require tests to determine the point at which molded-in stresses could relax and cause cracking or poor adhesion of the plating. To determine the appropriate test for each material, consult with the plater or material supplier. For platable nylons, the red dye test indicates the amount of molded-in stress. If stress is too high, the material should not be used for plating. In addition, machine conditions should be modified to lower the molded-in stress level. These tests should be conducted as soon as possible after molding to make any necessary processing adjustments. This is one of the main reasons that companies using metal plating prefer to do the molding themselves.

POST-MOLD SHRINKAGE TESTING

An area often overlooked during part design is the effect of the environment on part dimensions. The molding cycle produces parts based on the cavity's dimensions and selected processing conditions. These factors must take into account the product's end-use environmental conditions. The dimensions of the molded part must remain stable through any future conditioning or environmental changes.

Crystalline polymers experience the greatest dimensional changes, particularly if not processed in a hot mold using high-injection pressure to pack out the part. The mold cavity and processing conditions should be selected to achieve the correct mold shrinkage for the part's dimensions. The dimensions of moisture-sensitive materials, such as nylons, will grow and shift after molding. The part's processing conditions must be selected to compensate for these later changes.

CONDITIONED PARTS

Some parts are oven, oil, air, and moisture conditioned before shipping and final assembly to make sure they perform as intended. This added expense should only be incurred if absolutely necessary. The QC department and the operator checking the parts as they come out of the mold must know that the part dimensions before conditioning will produce the required final dimensions. This is particularly important in the automotive, power tool, and electronic industries, where plastic parts are exposed to nonambient operating conditions. Failure to recognize these problems could result in misalignment of a motor's bushings and shaft positioning, where shifting of dimensions can cause premature failure of the part. This is where design, tooling, manufacturing, and the resin supplier must coordinate all the available information to produce a part that will perform in all environments.

Part and dimensional testing at the machine ensures that the process produces good parts, the manufacturing process was in control, and that part dimensions and surface aesthetics will meet end-use requirements.

13

Part Handling and Packaging

The degree to which parts are handled after molding has a definite bearing on the final quality of the finished article and, therefore, on profits. The cost of labor—its efficiency and ability to control part quality after molding—is an ongoing concern of management. Management is always striving to upgrade its labor pool for more efficient utilization of available talents. One method is to modify and automate handling of the finished part.

This can be accomplished by conveyor systems. These systems, which range from simple to sophisticated, interact with the injection molding machine's microprocessor control unit. Conveyors transport parts from the molding machine to the end-of-belt separators. The parts pass through cooling or cleaning stations to secondary assembly or decorating stations, and then through manual or automatic inspection points to automatic packing systems. These conveyors are available in all sizes and configurations to handle the most delicate parts. There are also low-profile designs that fit under the mold and transport parts to work stations or other conveying stations for further operations.

PLANNING

The planning for part handling begins at the design and manufacturing planning stage. Items to be considered in part handling are as follows:

Total Quality Process Control for Injection Molding, by M. Joseph Gordon

1. Part size, weight, and complexity
2. Volume of parts
3. Type of tool: two plate, three plate, or hot runner
4. Molding machine microprocessor control and integration capabilities
5. Part inspection method and requirements
6. Production process for part removal—operator or automatic
7. Part/runner separation: operator, end-of-belt—robotic
8. Surface cleanliness of part, secondary operations and packing
9. Productivity requirement
10. Automation of part handling/packing system
11. Quality requirements of parts
12. Dunnage type for moving and/or part storage (cardboard-high dust factor)

All the above factors need to be evaluated at the start of a new manufacturing program. How the molded parts are handled is key to the productivity and profitability of the operation. It also has a decided effect on the quality of the finished part.

Once these decisions are made, manufacturing and part-handling procedures can be written. These procedures are used to inform and control part removal by the operator, automatic free-drop, or robots.

Large parts often require operator assistance for removal from the mold. When this is required, the operator should be trained to remove the part correctly and efficiently, as delays affect the molding cycle. In these labor-taxing situations, operators should also be spelled more frequently. A light or audible tone system timed to the molding cycle operation assists operators in developing a rhythm for removing parts.

When parts free fall from the mold, they must not be damaged. There are chutes and slides to prevent this, along with padded low-profile conveyors to limit the drop height. Parts are then collected by an operator or a conveyor system and sent to the next station.

The use of robots to remove parts is becoming more common, and they can be programmed to perform more than one function. For example, robots can remove parts and the runner system as a single unit or separately, placing each in its proper location, and signal the mold for closing. When robots are used, the cycle may be slightly longer, but it will be uniform. Robotic handling of parts is efficient and the initial investment is recouped faster than most people realize.

Just how part removal from the mold is to be handled depends on the type of mold, its operation, the number of cavities, and what is required of the parts after molding.

PART REMOVAL

Most two-plate molds eject the part and runner as a unit. Depending on the parts' gating, they may be separated when the mold opens. The parts and their runners free fall and are then degated by an operator or conveyed to an automatic part separator, as shown in Figure 13.1.

Three-plate and hot-runner molds automatically degate parts from the runner when the mold opens. The part and runner are dropped independently and can be segregated by a baffle positioned under the mold. They can then be directed to their respective locations using another conveyor system. Hot-runner molds only produce finished parts; the runner system is eliminated. This is more efficient, and parts can be removed and directed to their next work station by various methods.

The injection molding machine's control unit is a major factor in controlling part handling. Any auxiliary part-handling equipment, such as robots, can interface with the machine's processing unit for timed removal of parts. This is particularly for use with programmable microprocessor control units.

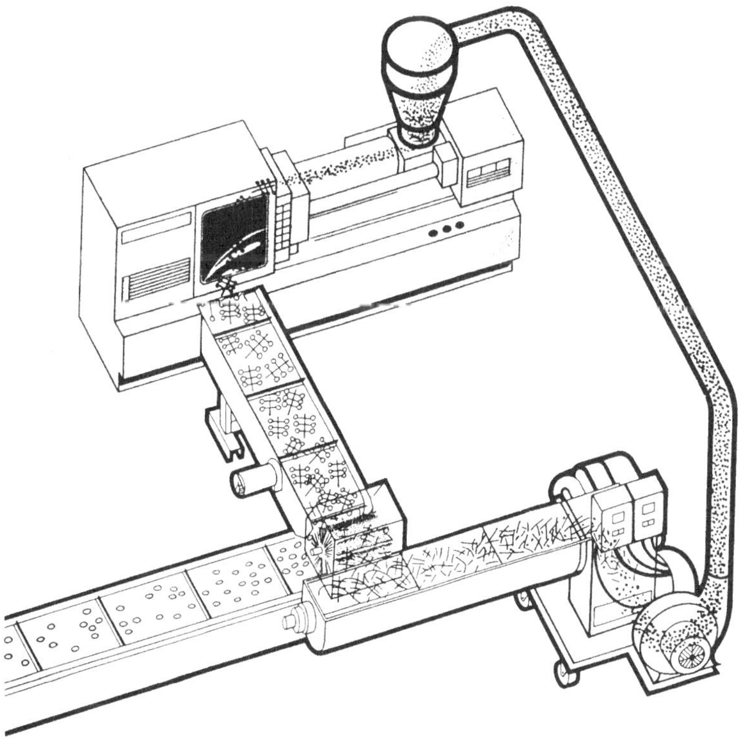

FIGURE 13.1. Part separator—manual or automatic. (Courtesy of Ball and Jewell, Sterling Inc.)

At the press, operators should only be used if downstream automation is not available, and they can perform a required secondary operation to justify their cost. In some cases, assembly operations have been moved press-side. This results in greater productivity, better secondary operation performance, and higher part quality. This may include pressing inserts into parts, welding operations, or other assembly operations. The machine operator performs these operations while waiting for production of the next cycle of parts. If inserts are pressed while the part is still warm, the holding power can be increased because the material shrinks on the insert and insertion force will therefore be less. For welding operations, the parts are very dry; there is no moisture contamination at the weld joint. This results in a better weld, fewer rejects, and improved part performance. Many secondary assembly problems can be eliminated, by press-side operators, if they have extra part handling and storage capabilities. This is where smart scheduling pays off.

For automatic molding, the parts must be rigid enough to withstand the drop from the mold without out distortion or damage. To avoid these problems, the parts may require a longer cooling time or a different method of part removal. In addition, special part catching and cushioning systems are available.

There are also water-submersible conveyor systems for cooling thick-walled heavy section parts (see Figure 13.2). These systems are designed with sealed bearings and are enclosed in rust-inhibiting materials. Temperature and water flowing through the enclosure must be regulated to control cooling and flush away contaminants. The conveyor can run continuously or be timed by molding cycle to cool down the parts before submersion or to maximize cooling time in the water.

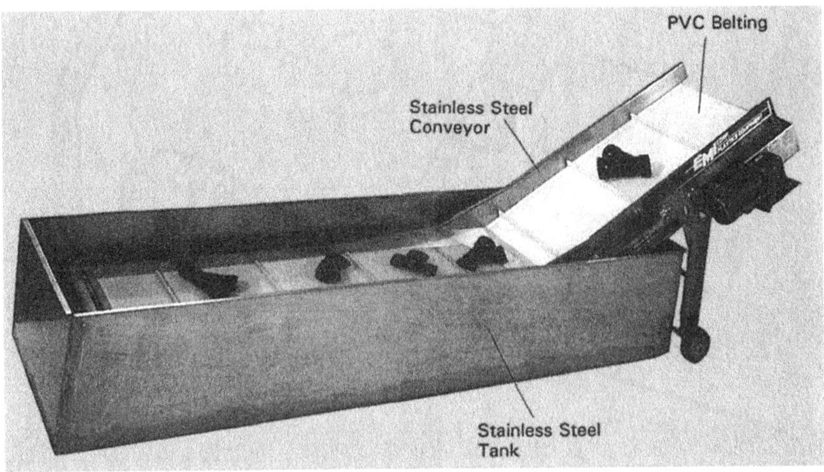

FIGURE 13.2. *Water submersible conveyor. (Courtesy of EMI Corp.)*

Large, complex, critical, or fragile parts that could be damaged or marred during removal must be protected. This may require operator assistance and special handling. If robots are not employed, special chutes are often used under a mold to cushion and direct parts from the machine. This added expense often means the difference between accepted or rejected parts.

PART HANDLING AND PACKAGING

For visible parts, the exposed surface must be protected. Operators wear soft cotton white gloves and the parts are conveyed in special containers. Some customers provide reusable shipping containers to provide this protection and to reduce the cost of paper containers. These reusable containers give additional part and surface protection, as well as aid inventory control and ease of delivery of parts to the customer. They are usually more expensive, but reuse justifies the cost for long production runs. Often referred to as dunnage-type containers, they are used exclusively in the automotive industry to ship parts to the assembly plants.

Special containers may also be used to package the parts for secondary operations either in-house or at the customer's plant. Parts that will be decorated later are often packaged in dust- and moisture-proof containers to protect the surface and ensure good surface aesthetics and adhesion of the coating. Part protection and packaging should be discussed during the development of the program, as extra costs will be involved. The part supplier may be able to offer other options on packaging to reduce expense and obtain better quality.

It is important to understand requirements of each step in the manufacturing operation. Your extra efforts in part handling and packaging may yield a higher return and better part quality.

AUTOMATIC PART PACKAGING

For a manufacturing operation that uses a multicavity mold with lightweight mass-produced parts, an automatic part and runner handling system is recommended. After the parts are ejected from the mold, they are collected and conveyed to a part separator station then sent to an automatic packer.

There are now "smart box" packaging systems that communicate with the molding machine's microprocessor. This system tracks the number of parts produced and controls the packing of parts into shipping boxes. These boxes are automatically filled by part weight or piece count, automatically sealed, and entered into inventory using bar-coded labels on the cartons.

Small parts are normally packaged by part weight, which is more accurate and less expensive than piece count. Boxes are filled by conveyors timed to the scale's accuracy, once the box's weight is zeroed out, down to 0.001

pounds. When a new box is put on the scale, it is automatically filled to a present box weight that corresponds to an accurate piece count. This can be done in two ways—by feeding parts into a weigh hopper or shipping parts directly to the box. Either way when full, the box's net weight is printed on a bar-coded label that is adhered to the box. The carton is then sealed and directed to inventory or shipping. Part weight measurement is becoming an acceptable method for packaging and billing.

The labels also include additional information from the production cycle that aids in lot and product traceability. This information, which is also logged into the production and customer record file, includes the following:

1. Weight of box
2. Number of parts
3. Time, date, and shift information
4. Customer identification and part numbers
5. Package number

This packaging system can automatically tie the parts produced in with the process control charts generated by the process controller. These can be given to the customer as a production control record of product quality. The system selected for part handling must fit the productivity and quality requirements of the customer. Each system and piece of equipment must be judged based on its merits in the part-handling system.

ROBOTS

Robots provide versatility, repeatability, and a tireless programmable machine to fit almost all part-handling operations. Figure 13.3 shows a transverse robot. Robots interfacing with automated conveyer systems will also offer productivity and quality improvements, such as follows:

1. Better use of the workforce by the elimination of low-skill repetitious functions. Workers can be retrained for higher-level positions.
2. Improvement of part quality, with more consistent processing and handling of components.
3. Increased productivity through accuracy, consistency, and accelerated production integrated into downstream operations.
4. Flexibility to meet changing manufacturing requirements.

Robots are versatile and can usually be used for more operations than originally thought appropriate, especially if planning is done ahead of time. How many machines does the robot fit or will it be dedicated to one machine? How many degrees of motion must it have? Can it interface with the molding machine's process controller for operational control, timing, and feedback? If

FIGURE 13.3. Three-axis servo traversing robot. Traverse robots are designed for high-speed part removal from 50- to 700-ton injection-molding machines. They are fast, stable, and precise. The design of these robots incorporates the following features: high-speed vertical takeout, achieved with dual-action pneumatic cylinder combined with a belt-drive mechanism; main arm supported and guided by high-accuracy linear rail supports; new three-phase induction motor with a pulse modulated drive provides variable ramping and multiple-stop capabilities accurate to 0.039 inches. (Courtesy of AEC Inc., AEC/Application Automation, 801 AEC Dr., Wood Dale, IL 60191)

so, the robot can aid defect detection by segregating defective parts told to do so by the controller.

Robots can also perform secondary operations, including degating parts and placing them on to conveyors or in fixtures for other operations. During part removal, the robot can inject mold spray, if required, into the cavity.

Robots are very efficient for loading inserts in a mold, which can greatly increase productivity. They can also pass parts to other robots for secondary operations such as adding inserts, assembling components, decorating, or completing other finishing operations, without the possibility of contamination.

Quality is the main reason for using robots. They aid in obtaining more uniform and consistent cycle times and reduce part-handling problems. The repeatability of functions helps ensure a stabilized molding cycle, which eliminates variability in part quality. Robots do not tire, and studies have concluded that the more "hands-off" an operation is, the better the quality. The molding machine's microprocessor or an integratable computer process control system must be used to regulate the entire system by monitoring the process, checking machine variables, and verifying product quality.

Automation is part of the key to quality molding. All the techniques described are steps that can be used to produce high-quality molded parts.

14

Part Design Influence

The quality of a part begins with the designer's knowledge of the correct use of plastic to manufacture a part. The design is a result of this knowledge and experience, along with input from material suppliers, tool builders, and production personnel. This help is necessary to ensure that the part performs properly when manufactured in the chosen material, that it can be made in the shape desired, and that it will function as required. The part design, mold capability, and manufacturing control principles are all tied together, as shown in Figure 14.1, to achieve Total Quality Process Control.

SELECTING THE CORRECT DESIGN PARAMETERS

All plastic designs should begin with a checklist that assists in analyzing the part and understanding its end-use requirements. By answering these questions, most material selection questions will be answered. Once the material is selected, the tooling and processing questions can then be addressed.

Each plastic part has specific design requirements. How the designer calculates the effects of end-use physical force and the selection of the resin has a definite bearing on how well the part performs and how it should be produced. The drawings and specifications must be specific in critical areas to avoid misinterpretations in the tool-building stage that could yield a faulty or hard-to-produce part. The part is only as good as its design, the

Total Quality Process Control for Injection Molding, by M. Joseph Gordon

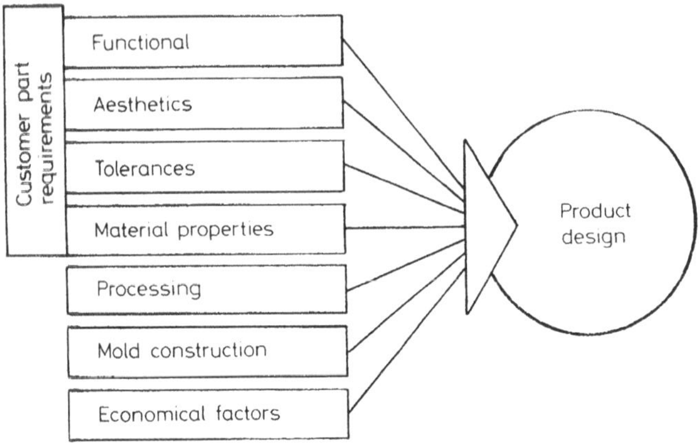

FIGURE 14.1. *Factors affecting product design. (Adapted from Ref. [5].)*

mold used to make it, and the material from which it is made. Because all plastic materials have key features, such as high strength, toughness, dimensional stability, good wear, and molding properties, the plastic must be selected based on end-use requirements, manufacturing capability, and reproducibility. See also the author took from Wiley on "Industrial Design of Plastic Products," 2003.

MATERIAL SELECTION

If the wrong resin is selected and the design is poor, the part is destined for failure. Even if the best resin is chosen, a poor design, inadequate manufacturing equipment or tooling, and varying molding parameters can doom the part.

Plastics must flow uniformly into a mold cavity, be packed out with minimum stress points, have good weld-line strength when opposing melt fronts meet, and have uniform wall thickness to enhance melt flow to eliminate sink marks and reduce warpage. After the part is molded, it must be removed from the cavity without distortion, without surface appearance problems, and according to the tolerances specified.

Plastics are very versatile building materials. In the hands of a knowledgeable designer, they can, in many cases, perform at a level equal to or better than metals. The basics of good part design apply to both materials, but for plastic parts, the designer needs to know stress, deflection, creep, temperature and humidity curves, and chemical and electrical data. The resin's product datasheet information is never used for design purposes. It is only a preview of the resin's physical properties in a dry-as-molded condition. The design information for a plastic material is available from the material supplier along with recommendations concerning part design. Such information includes temperature/humidity and stress/strain/creep curve data charts, and dimensional, shrinkage, and tooling suggestions for molding the part to meet customer requirements.

A question often overlooked by the designer is whether the selected resin can be molded into the part as designed. Are there flow length problems and will the resin's shrinkage be uniform for the dimensions and tolerances required for the part? Another question, how the resin will process, is often overlooked until the mold is in the press and parts are being made for the first time. Because a material's flow and shrinkage rates vary according to the type of polymer used and processing conditions, this is often too late a stage for changing the resin, because the mold cavity dimensions were sized for a particular material. Similarly, the flow of the resin will greatly affect the moldability of the part and the extent of molded-in stresses during fill and packout.

Therefore, resins cannot be easily changed after the mold is built. A new resin may require costly and time-consuming mold changes. It is the designer's responsibility to be sure at the start that the resin of choice will fulfill all, or most, of the requirements for the part in terms of physical properties, moldability, environment, code, and agency needs. This should be done before finalizing the cavity's finished dimensions, especially if several materials are being evaluated and their shrinkage rates vary considerably.

The most common design problems and often overlooked factors will now be discussed. Total quality process control begins here, but it carries on through the entire manufacturing process.

PART DESIGN FOR END-USE APPLICATIONS

Plastic parts must be designed for end-use application. When tested right out of the mold—in their dry-as-molded slate—parts should pass all requirements. This is usually the toughest criterion for a part. Parts in their dry-as-molded condition are more prone to failure, because they have not had any time for relieving molded-in stresses and conditioning to the environment. Once a part has become conditioned, it absorbs more energy and relieves molded-in-stresses that can negatively affect part performance. The conditioning of parts is often recommended using hot oil, water, or moist air, along with oven annealing. Most experienced designers believe that these secondary conditioning operations and expenses can be eliminated with a properly prepared design. In addition, there are many new resins and plastic alloys available to solve the most demanding requirements. This choice is the designer's. If knowledgeably exercised, it can result in a real savings in manufacturing time and costs, plus the elimination of molding and end-use problems.

The most common and easily solved problems associated with plastic part design are the ones most often forgotten during the design of the part:

1. Lack of or too small radii at corners
2. Nonuniform part thickness

3. Flow/fill problems—voids, sinks, or weld lines
4. Drag and scuff marks
5. Molded in stresses and warpage
6. Surface aesthetics and part color

Radii

The leading contributors to plastic part failure are sharp internal corners or molded-in notches. Designers invariably fail to recognize no internal or sufficient radius on a part is the main reason for premature failure. If radii are specified on drawings, they are usually too small and never considered in high-stress areas as critical for the functioning of the part. All internal corners must have radii specified to reduce the stress concentrations at these critical sections. A sharp corner at a stress point can magnify the force factors three to four times at that point on the part. A fillet or radius of 0.5 to 0.6 times the wall thickness is recommended in reducing stress concentration at these points. Figure 14.2 best illustrates this optimum reduction of stress at an internal corner, as the stresses at a corner in a clear part show under polarized light. Impact testing and molding clear parts with and without a radius have visually illustrated, under polarized light, the stress reduction at radiused corners versus sharp nonradiused sections. Materials with low elongation qualities are very susceptable to stress cracking four external and internal forces, such as a boss, when a screen expands the outer diameter as during assembly.

Radii are also required on the exterior corners of such parts as gears and pulleys (see Figure 14.3), when a square-cornered shaft must fit into a plastic part. Radii on external corners also aid material flow, sink reduction, and stress reduction to improve part impact resistance (see Figure 14.4). The greater the radius, the higher the impact load the section can absorb.

For example, a sharp 90 degree corner is required for design requirements. Figure 14.5 shows several methods of reducing stress at a corner with a radius, while maintaining the right angle. When a part is subjected to impact, radii become even more critical in assisting the absorption of energy at corners and reinforcing ribs. But there is a point at which the radius may affect part aesthetics or packaging. When this point is reached, the designer may have to consider an impact-modified material to solve the problem.

The effects of radii on molded parts of polysulfone when subjected to falling-dart impact testing are shown in Figure 14.6. Where the radii were increased, the part absorbed more energy without or before failing. Impact tests on the external corners showed a fivefold increase in energy absorption, with more generously rounded corners. At the top, the increase in energy absorption was tenfold, even though a failure was recorded. When ribs are added to increase section stiffness, they can reduce the part's ability to absorb energy by preventing the walls from deflecting. Ribs are used for strength.

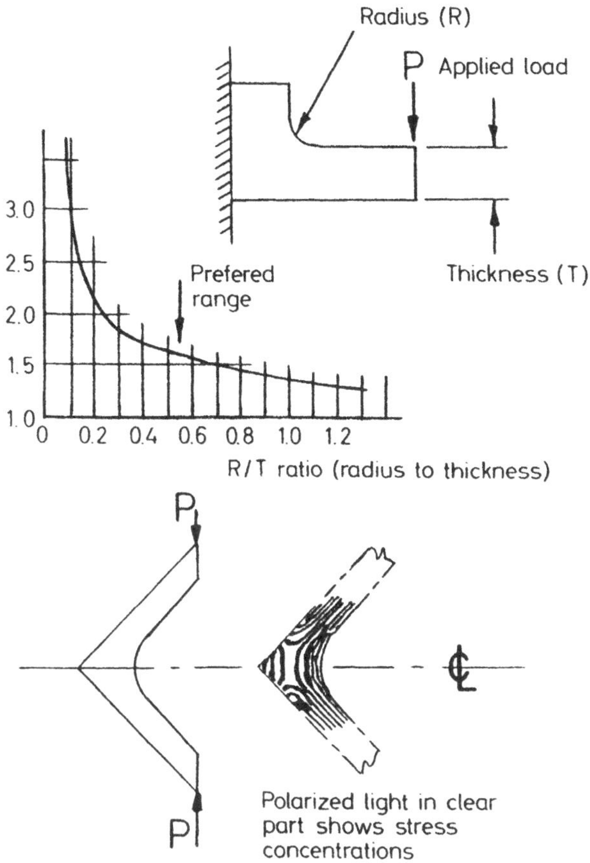

FIGURE 14.2. *Stress concentration factors. (Adapted from Ref. [1].)*

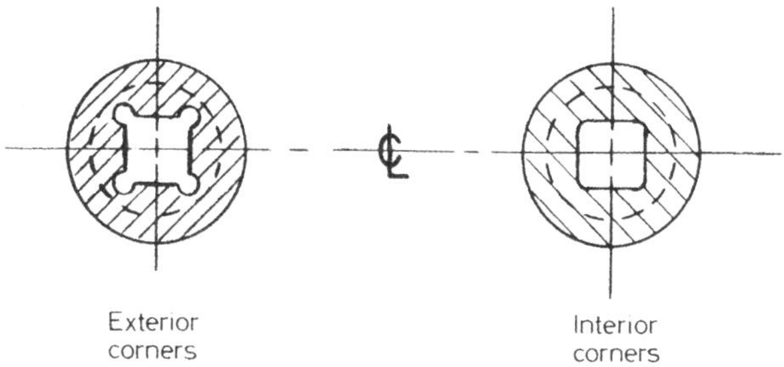

FIGURE 14.3. *Exterior/interior radii. (Adapted from Ref. [1].)*

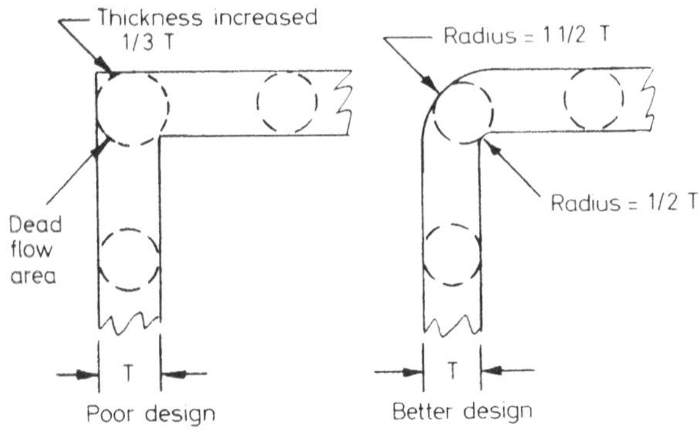

FIGURE 14.4. Examples of corner radii. A good design improves flow and increases part toughness.

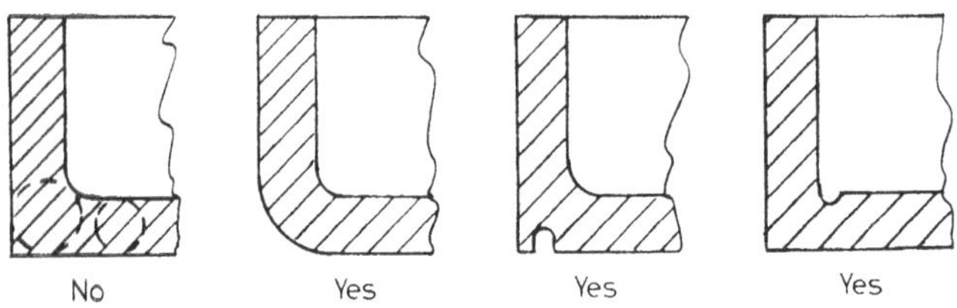

FIGURE 14.5. Methods of reducing corner stress with a radius.

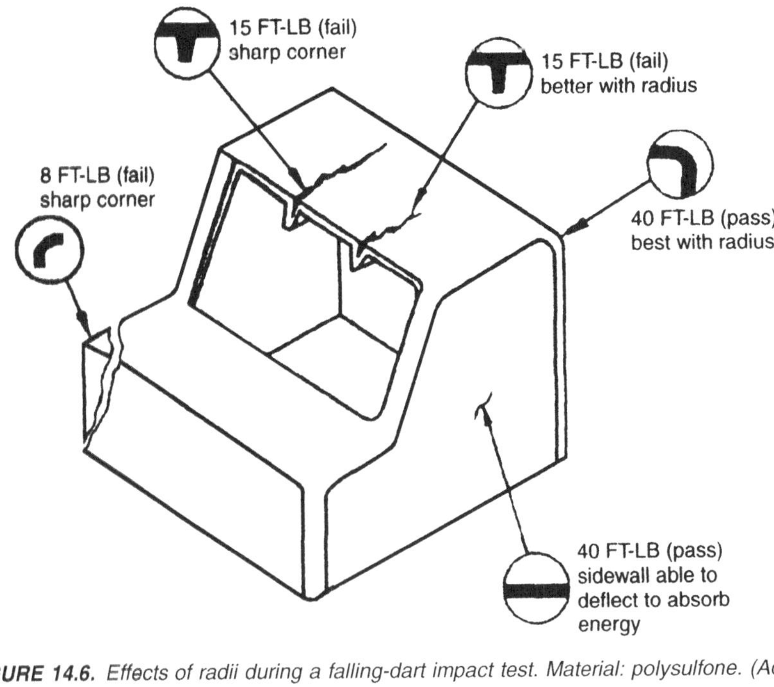

FIGURE 14.6. Effects of radii during a falling-dart impact test. Material: polysulfone. (Adapted from Ref. [3].)

With a properly designed rib section and the use of an impact-modified resin, an otherwise unacceptable part may perform satisfactorily in service. Figure 14.7 shows the effect of radius size for polycarbonate, an excellent energy absorbing resin, in an impact test. The larger the radius, the more energy the part can absorb before failure.

To better understand the behavior of the thermoplastic materials and why some are stronger than others, one must know how they behave under standard American Society for Testing and Materials (ASTM) testing of physical properties, as recorded on data sheets.

All plastics exhibit a stress/strain curve typical to the one shown in Figure 14.8. The curve shown is for an unreinforced polymer with a typical dry-as-molded (DAM) elongation of 50 percent. With a reinforced or filled polymer,

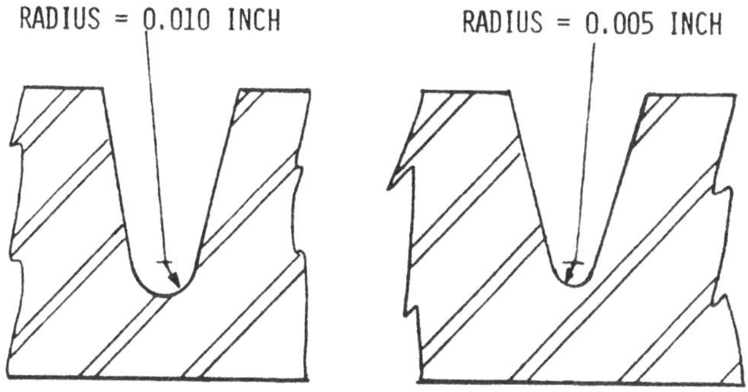

FIGURE 14.7. Effect of radius size for polycarbonate during an impact test. (Adapted from Ref. [9].)

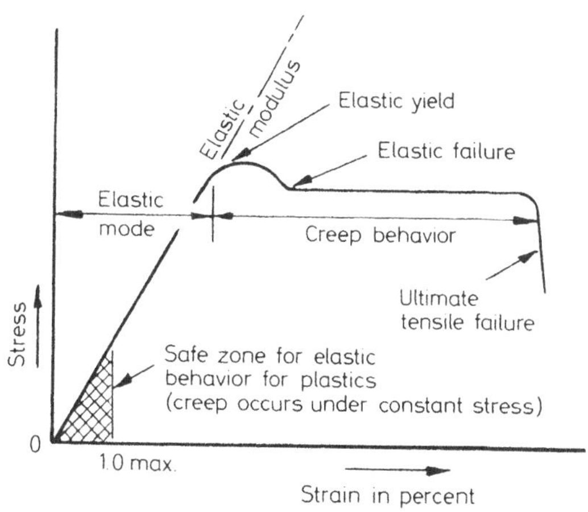

FIGURE 14.8. Stress versus strain at fixed temperatures for plastics. (Adapted from Ref. [15].)

failure will occur at or near the top of the curved section, depending on the amount of filler or reinforcement. Reinforced or filled resins have lower elongations—10 percent or less—with a lower ability to absorb energy and deflect in impact situations. Reinforced or filled materials will require more generous radii to lower the stress concentrations when subject to loads or impacts. With short or long glass-reinforced resins, this is often disputed when notched izod, data sheet values are reviewed. The data often indicate higher impact values for more highly reinforced resins. This is just the effect of the glass fiber/resin matrix strength at the notch; it is not the true energy absorbing value of the base resin. The higher the glass loading, the more generous the radius necessary to increase part toughness.

Nonuniform Part Thickness

Nonuniform wall thickness is the second leading contributor to plastic part failure. The problems, which are illustrated in Figure 14.9, can result in wider tolerances, voids and sinks, warpage, poor fill as a result of pressure drops in the mold cavity, and molded-in stresses. Examples of these problems are illustrated in Figures 14.10, 14.11, and 14.12: The part shown has a flange with a part thickness of ½ inch in the solid section, with a full ring-gate thickness of ⅛ inch. The material is a glass-reinforced resin that is molded and then machined to its final dimensions. These problems result from poor design that prevents the resin from being fully packed out during molding. Had a more uniform section been used, all three problems would have been eliminated.

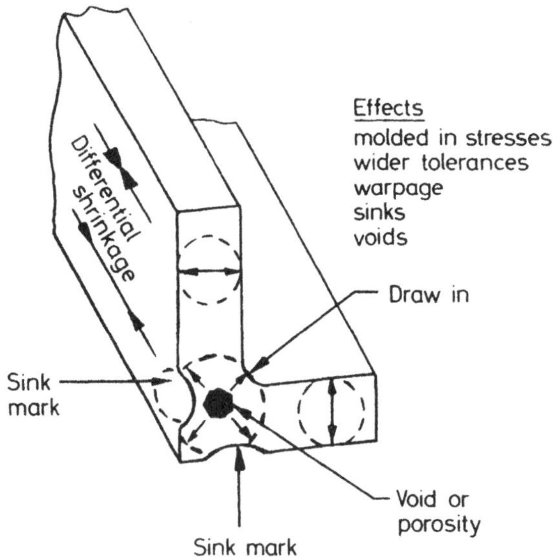

FIGURE 14.9. Effects of nonuniform wall thickness on molded parts. (Adapted from Ref. [6].)

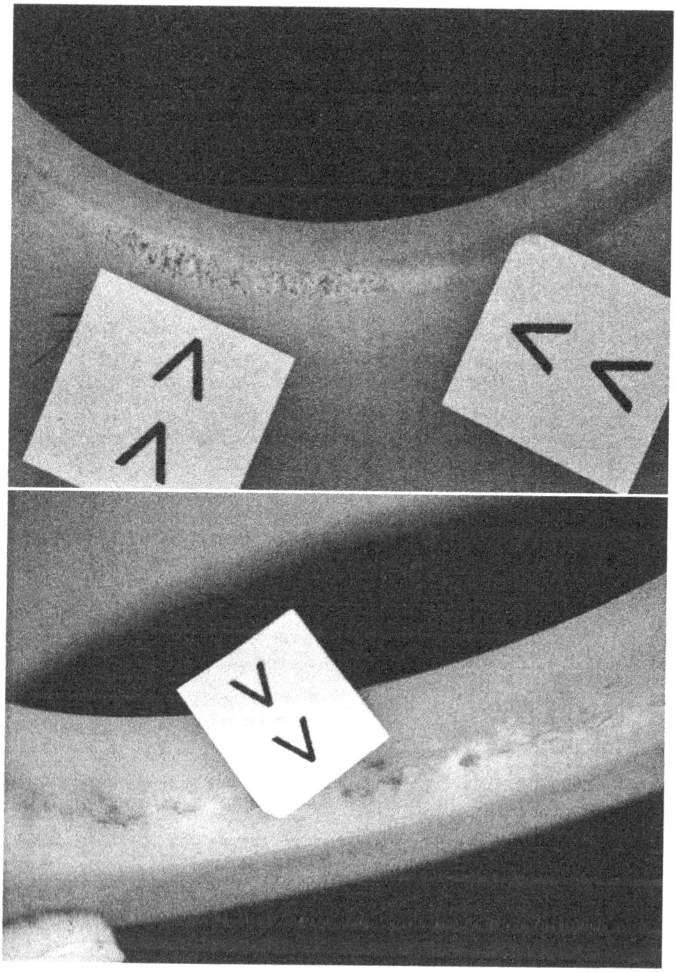

FIGURE 14.10. *Effects of nonuniform wall thickness on porosity.*

One cause of molded-in part stresses is thick/thin sections, which create differential shrinkages when the part is cooled. The hotter, thicker section shrinks more and this contributes to wider dimensional tolerance variations, warpage, sinks, and voids. It also causes flow problems around corners, as a result of pressure drops, and dead pressure flow areas. When nonuniform section thickness are used and a thin section must feed a thicker section, pressure drops occur in the thicker section. This may result in incomplete filling of the part which leads to voids, sinks, and porosity. The thin section could also be over-packed in an attempt to fill the thicker section. This problem, illustrated in Figure 14.13, is to be avoided at all cost. The designer must transition and core-out thick sections to obtain as uniform a part thickness as

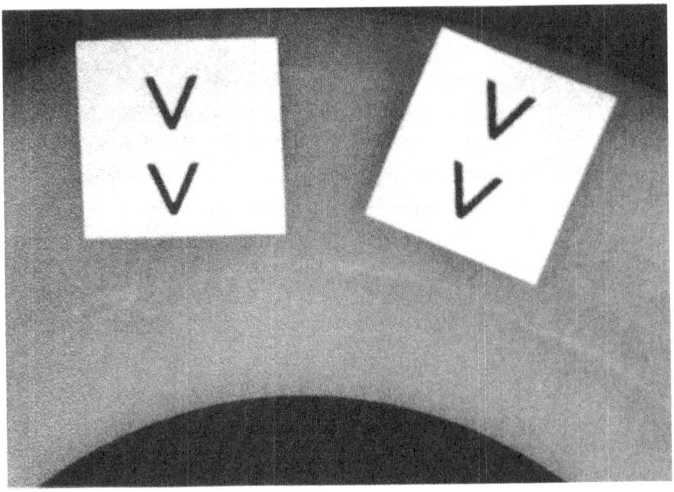

FIGURE 14.11. Radial cracking resulting from nonuniform wall thickness.

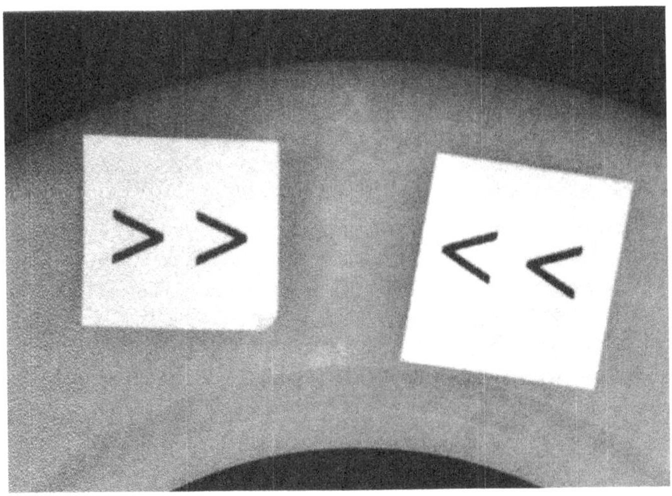

FIGURE 14.12. Radial weld line and voids resulting from nonuniform wall thickness.

possible (see Figure 14.14). This improves part strength, reduces surface and internal problems and sinks and voids, and saves material.

Therefore, all parts should be designed for uniform wall thickness and should always be gated into the part's thickest section to ensure continued melt flow for good packout. Recommended material wall thicknesses are listed in Table 14.1.

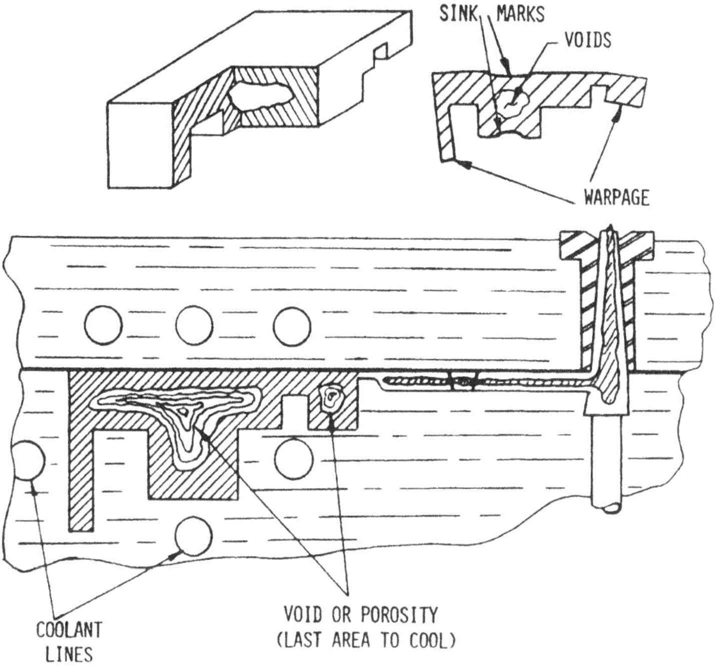

FIGURE 14.13. *Thick/thin sections create part problems in fill and pack out.*

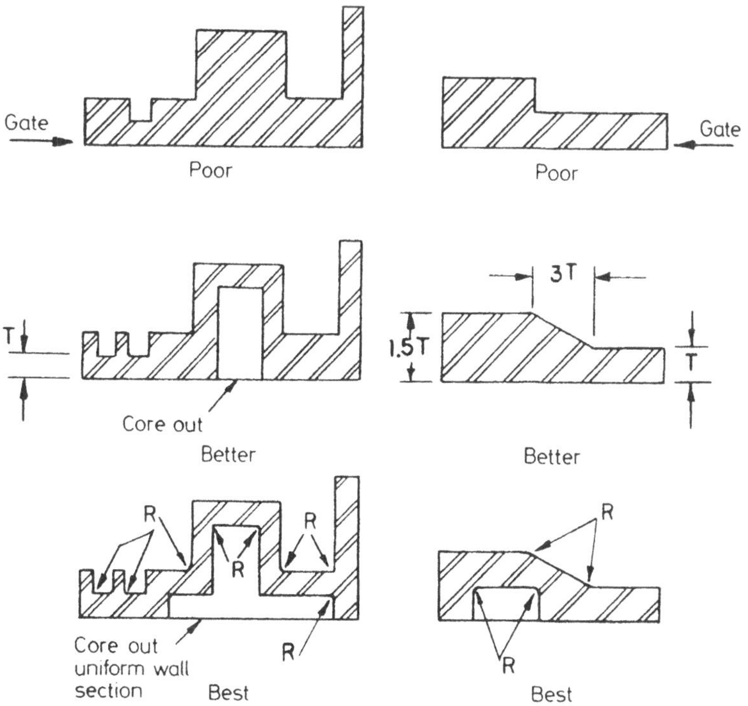

FIGURE 14.14. *Uniform wall thickness improves part strength and minimizes problems.*

TABLE 14.1. Recommended Wall Thicknesses for Thermoplastic Molding Materials Thermoplastic.

Materials	Minimum (in.)	Maximum (in.)
Acetal	.015	.125
ABS	.045	.140
Acrylic	.025	.150
Cellulosics	.025	.187
FEP fluoroplastic	.010	.500
Long-strand reinforced resin	.075	1.000
Liquid crystal polymer	.008	.120
Nylon	.010	.125
Polyarylate	.045	.160
Polycarbonate	.040	.375
Polyester	.025	.125
Polyethylene (LD)	.020	.250
Polyethylene (HD)	.035	.250
Ethylene vinyl acetate	.020	.125
Polypropylene	.025	.300
Polysulfone	.040	.375
Noryl (modified PPO)	.030	.375
Polystyrene	.030	.250
SAN	.030	.250
PVC (rigid)	.040	.375
Polyurethane	.025	1.500
Surlyn (ionomer)	.025	.750

ABS, acrylonitrile butadiene styrene; FEP; HD, high density; LD, low density; PPO; PVC, polyvinyl chloride; SAN, styrene acrylonitrile.

Source: Adapted from Ref. [2].

Examples of uniform wall thickness are shown in a series of right and wrong illustrations in Figure 14.15, which indicates the correct ways to use plastics in the design of a part. The materials are too valuable to waste and nonuniform walls only cause processing and quality problems later on.

All plastic parts should be cored out as much as possible for better utilization of the material, reduction of molded-in stress and part warpage, and minimizing of the molding cycle. If additional strength and/or rigidity are required, the use of ribs or changes in part geometry are employed (see Figure 14.16):

Ribs for Strength and Quality

When additional part strength and/or rigidity are required, a thinner rib section is the best design. Two or more smaller ribs, or one larger but thinner rib, are better than a single, large, thick rib. This is proven by calculating the stiffening effects of the smaller ribs versus the large rib. When thinner rib sections are used, molding and part problems are reduced and the impact

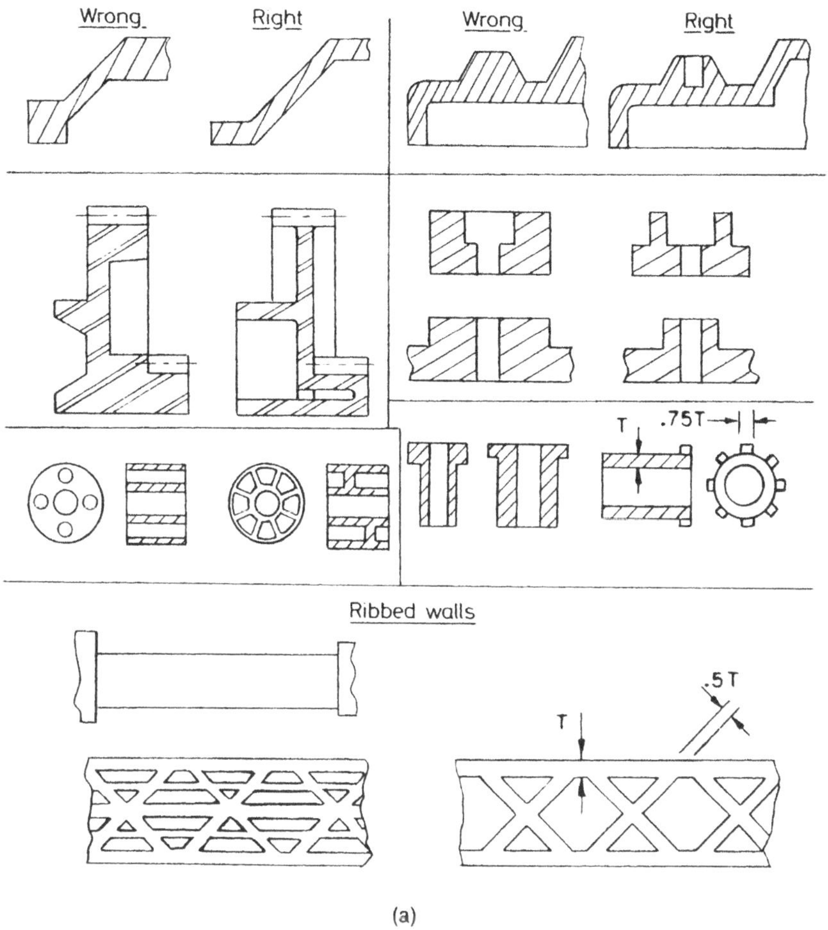

FIGURE 14.15. Examples of designs for wall thickness. (Adapted from Ref. [1].)

resistance of the part is increased by lowering molded-in stresses at these areas. Designers can use the moment of inertia calculations to obtain a rib design that reduces material and part problems while creating a stronger and tougher part. Good and poor rib designs are shown in Figure 14.17. Good design utilizes the plastic material in the correct proportions to obtain maximum results. Good proportional rib design also avoids molding problems, thereby resulting in faster molding cycles with fewer dimensional and quality control problems.

One area often overlooked is color variation in thick/thin sections. The more dense or thick the section is, the more intense the color appears at this section. This is more noticeable in translucent or semiopaque colors, and depending on their section thickness, with certain types of resins and designs.

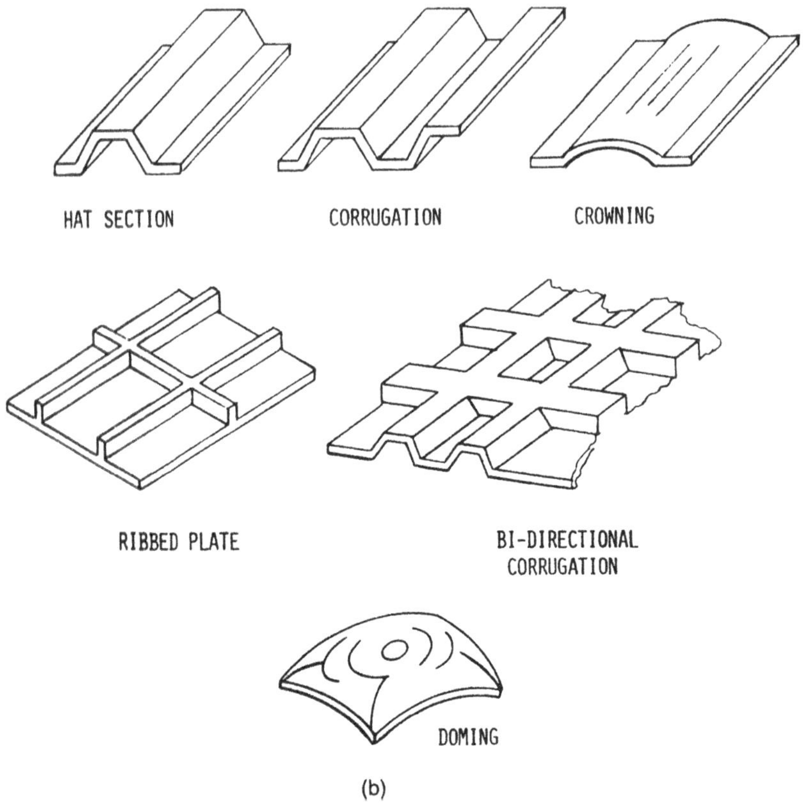

FIGURE 14.16. *The use of ribbing and part geometry to enhance strength and rigidity. (Adapted from Ref. [5].)*

Weld-Line Considerations

Whenever the material flows around an obstruction, hole, cutout, or boss, a weld-line occurs where the melt flows meet on the opposite side. This can become a critical part problem, if the weld-line forms at a high stress point. The designer must note these high-stress areas on the part design drawings as an aid to the mold designer, who selects the part's gate location. When the mold designer determines he part's cavity layout and gate location, this potential problem can be resolved. In many cases, this is not noted until part failures occur. If known in advance, there are many ways to minimize weld-line effects through changes in part design and gating. The solution may be to change the gate location to move or minimize the weld-line effect at these obstructions. Multiple gating can also shift the point at which converging melt fronts meet. This may also impart a mixing action at the melt front that improves weld-line strength. The convergence of melt fronts around obstructions is shown in Figure 14.18, with the addition of flow tabs.

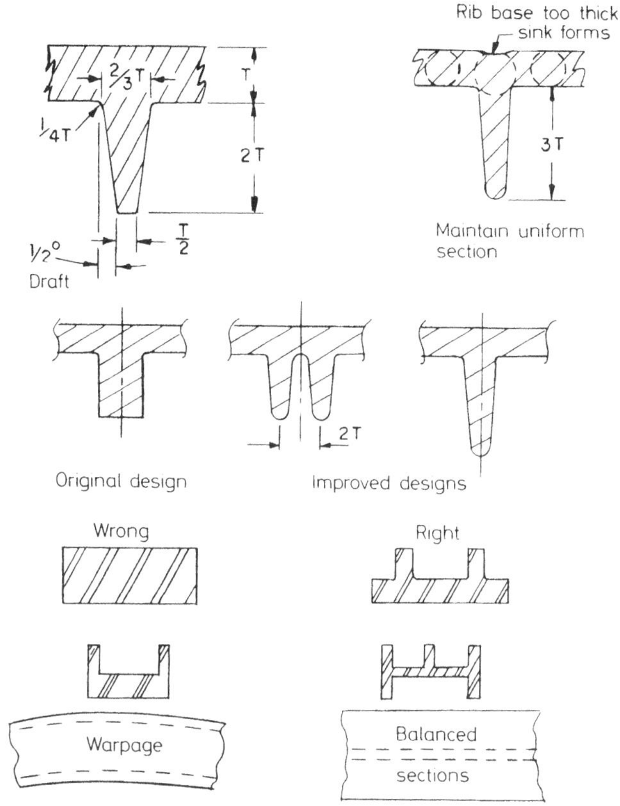

FIGURE 14.17. *Examples of rib designs.*

A method for improving weld-line strengths at the edge of a part is to use flow tabs that are later removed. Flow tabs are often used at the point where weld lines form to allow the melt fronts to meet and flow together. The mold cavity at the weld-line areas must be adequately vented. Otherwise, the air in the cavity will not allow a smooth meeting of the melt fronts. In fact, the melt fronts will be slowed down by the slowly escaping air and may vibrate into each other as the air is pulsated out of the cavity. If escaping air is not vented quickly enough, it may also cool the melt fronts enough so that a poor bond is formed. Additional localized heating, along with adequate venting, is often used at weld-line areas to ensure good unions of the melt fronts.

To avoid weld lines in parts requiring holes—blind or through the section—various methods can be used particularly if they occur at high stress points. Holes can be marked or spotted and later drilled out in a secondary operation. Examples of this technique are shown in Figure 14.19, with spotted or partially molded holes for later drilling. If countersunk blind holes are used, retain at least ⅓ of the section thickness so that material flow will minimize the

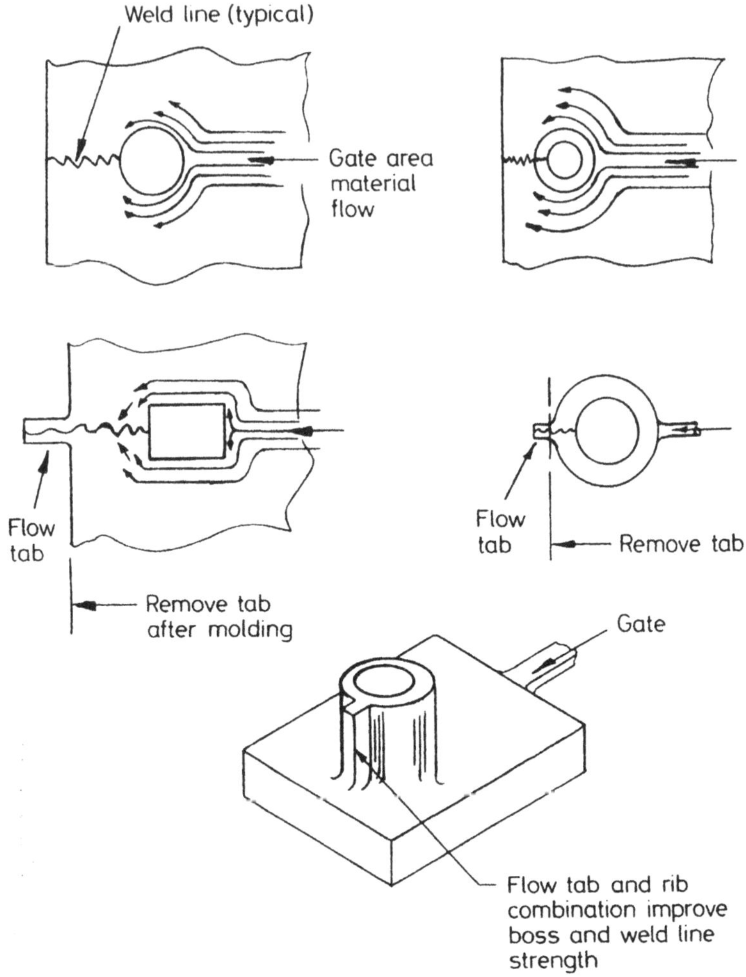

FIGURE 14.18. *Material flow patterns around obstructions. (Adapted from Ref. [2].)*

weld-line formation opposite the hole. The preferred method for hole designs is shown in Figure 14.20, utilizing the above methods to aid in eliminating part cracking and tear out. For blind holes the minimum thickness of the bottom should be no less than $\frac{1}{6}$ diameter. This eliminates bulging of the mold surface, as shown in Figure 14.21A. Figure 14.21B shows a better design with rounded corners that reduce stress concentration.

When reinforced materials are used, the part's weld-line strength should be calculated based on the parent resin's strength. This is because the reinforcements (e.g., glass fibers) do not flow across the weld line and knit with the opposing melt front. Even if an external flow tab is used, the possibility of reinforcement interflow is not probable until the fronts are a good distance

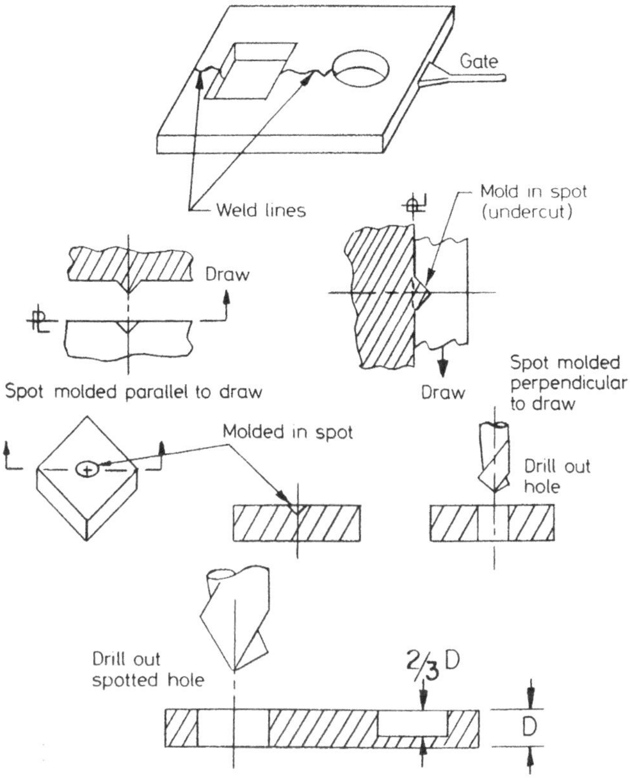

FIGURE 14.19. *Marking or spotting holes to be drilled in a secondary operation. (Adapted from Ref. [6].)*

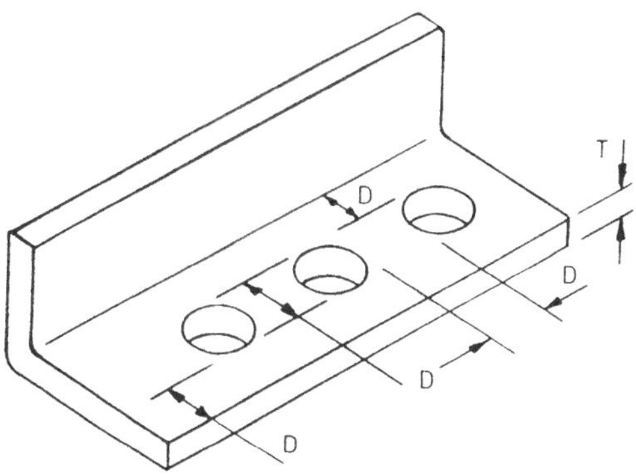

FIGURE 14.20. *Preferred method for hole design. (Adapted from Ref. [6].)*

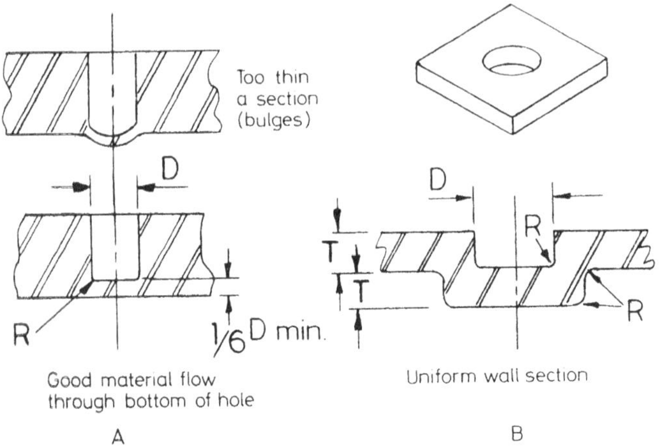

FIGURE 14.21. *Examples of designs for blind holes. (Adapted from Ref. [2].)*

from the obstruction. With some part designs, the weld lines may be very visible. For example, the use of colors or reinforced materials results in noticeable surface variances at the knit line. The melt temperature of the resin can be raised and the mold can be heated to improve weld-line strength. Amorphous resins usually have more problems with weld-line strength because of their lower softening temperatures and higher melt viscosities. The crystalline resins, with their sharper melting points, can cool and freeze off quicker and form a poor weld-line joint. Flow tabs with good venting, increased melt temperature, localized cavity heating, and hotter mold temperatures can reduce weld-line effects and improve joint strength.

Surface Appearance Problems

It is important not to damage a part's surface during ejection from the mold. Drag and scuff marks result from improperly polished cores or cavities with very little taper or draft. Scuff marks occur mainly on texturized surfaces dragged against the cavity as the mold opens or the part is ejected. To eliminate this problem, increase the draft angle by a minimum of 1 degree for every 0.001 inch of texture depth.

When a textured surface is used, apply the finish uniformly to avoid color variations over the visible surface of the part. Translucent resins and thick/thin parts that are backlighted can use a textured surface to help mask these problems.

The designer can use texturizing to hide or minimize surface defects caused by rib or boss sink marks. If this is not possible, redesign the ribs or bosses to reduce surface effects as earlier discussed. If texturizing is not possible, other means must be explored to reduce the surface effects. The designer can use several methods, but they must be decided at the beginning of the project. If ribs are involved, then thinner more abundant ribs may be required, along with uniform wall thickness. For bosses, the designer may consider an alter native assembly method or have the boss ultrasonically attached to supporting ribs after the molding operation. In severe cases, the designer could request that the mold builder add section thickness locally on the opposite surface of the expected sink mark. As the resin shrinks at that point, the extra material would create a smooth, flat surface appearance. It is an art to be able to estimate how much to increase the section at these points. This is usually not done until all other measures have been unsuccessful in eliminating the problem.

Designers must also be aware that some reinforced resins can offer a reflective, nonglassy surface finish. This appearance can often be obtained using a highly polished cavity surface and running the mold cavity temperature at more than 150°F. The ability to do this depends on the type of resin selected and how fast it solidifies on the hot cavity surface. The designer should ask the resin supplier if this is possible before specifying a reflective or glossy surface finish on the part. The crystalline resins, with their sharper melting temperatures, have this problem. Most experienced molders know which resins, when used with high-mold temperatures, can obtain a rich resin surface of only 1 to 2 mils thick. The hot cavity temperature will bury the reinforcing medium by allowing the resin melt to come to the surface before solidifying. This usually requires a resin with a slightly longer setup time or one that is modified to obtain the desired surface effect.

Bosses

Bosses are used as assembly points for mating parts. They should be designed for strength of attachment without creating surface or appearance problems. To eliminate quality problems, consider locating the bosses behind a part's nonvisible surface. If bosses and ribs are incorrectly designed, they will create voids, sink, and weld-line problems. Both open and blind boss designs must have radii and be of uniform wall thickness. To avoid nonuniformly thick sections, the boss can be supported and strengthened with ribs. Examples of these are shown in Figure 14.22, with recommended dimensions for design. The type of fastener used in the boss will determine its final dimensions, and each material supplier can recommend boss designs for their respective materials. This is based on the screw type or insert required for the attachment, plus its required holding strength.

There are other ways to design bosses that require through holes to reduce surface effects and increase boss and weld-line strength. The technique is similar to that used for blind holes (see Figure 14.21A), where material is allowed to flow both around and through the bottom of the boss. The base of the boss can be drilled out in a secondary operation to complete the part's requirements. Part designers need to think about assembly function, molding, and appearance when designing parts.

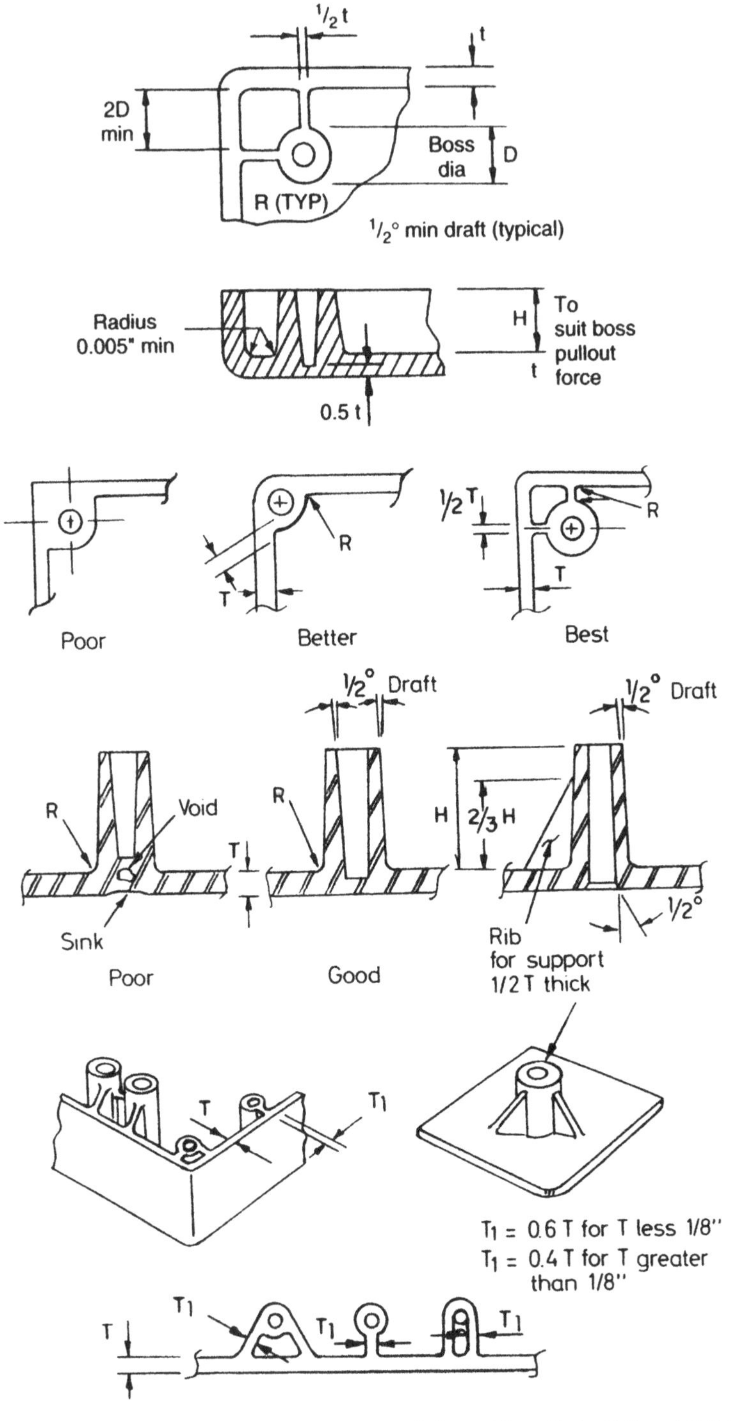

FIGURE 14.22. Examples of boss designs. (Adapted from Refs. [11] and [12].)

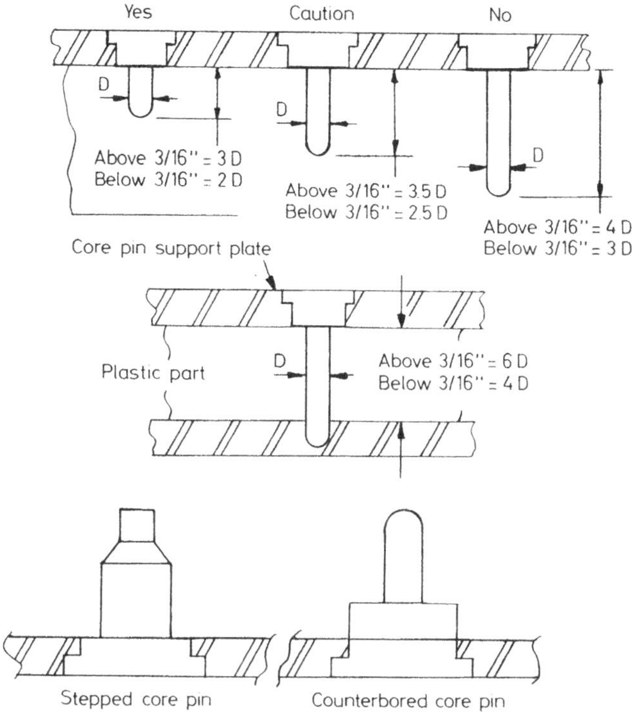

FIGURE 14.23. *Examples of core pin design. (Adapted from Ref. [6].)*

When blind holes are molded in a part, the core pins that form these holes must be rigidly supported. This is especially true of core pins whose length-to-diameter (L/D) ratio is greater than five. The core pins must be stiff and strong enough to avoid shifting and bending under high injection pressure and speed when the polymer is injected into the cavity. Examples of recommended and unacceptable designs for core pin length as compared with pin diameter are shown in Figure 14.23. If these pin length-to-diameter guidelines are exceeded, a stepped core-pin design may have to be used. Remember to consider cooling these core pins, as they become extremely hot during molding. For holes in a side wall, use retractable core pins—either cam operated or a split tool if the side wall can be tapered or sloped to alow this design (see Figure 14.24).

THREADS

Threaded plastic parts can be produced with both external and internal configurations. The external thread is usually created by locating the parting line on the center line of the thread (see Figure 14.25). If this is not possible, or

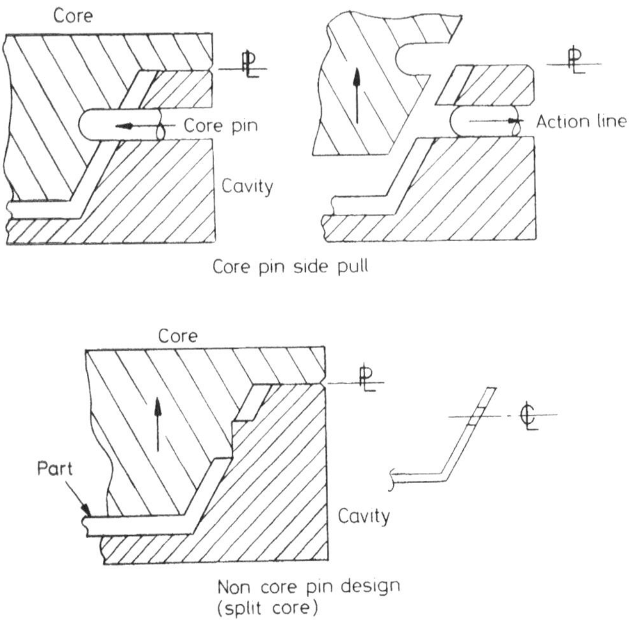

FIGURE 14.24. *Noncore pin design (split core).*

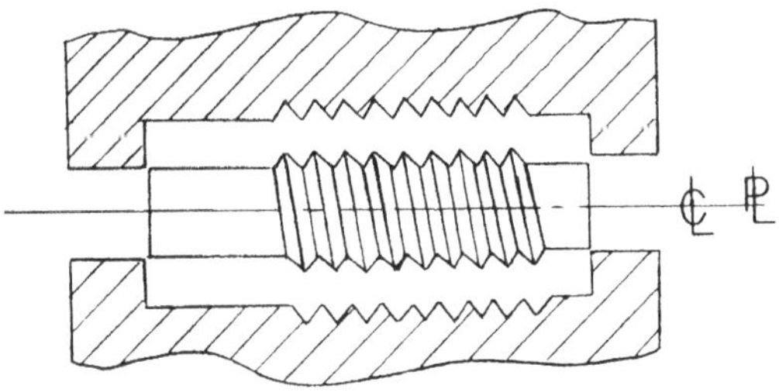

FIGURE 14.25. *External thread molding.*

the threaded part is in the direction of mold operation, an unscrewing thread-forming device or cam-operated side cores are used in the mold.

For parts requiring internal threads, an unscrewing device or collapsing core can be used or stripped from the mold if the threads are well rounded and their depth and number of undercuts are minimized. Examples of each are shown in Figure 14.26, with both external and internal strippable thread forms and an optional unscrewing core. If unscrewing is used, an individual part should be indexed in the mold so it will not rotate in the cavity during core removal.

Stripping threaded parts requires thread forms with proper radii. If sharp thread forms are used, they may spit on stripping or become so locked on the core they cannot be removed. For parts to be stripped, as in an internal

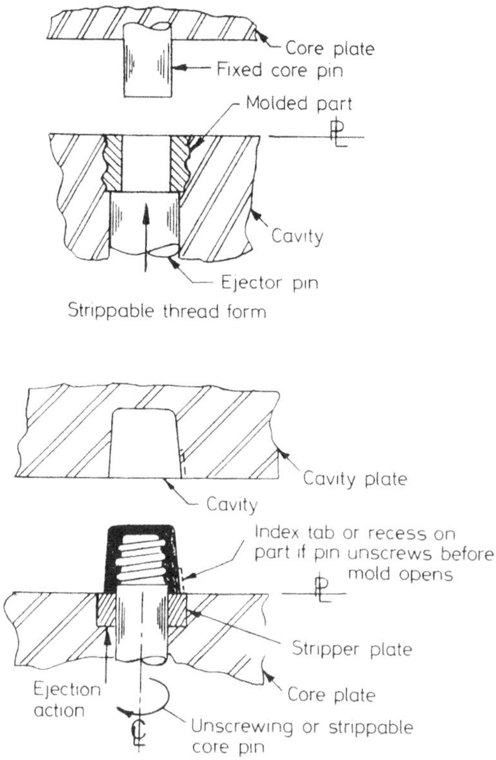

FIGURE 14.26. Unscrewing or strippable thread forms. (Adapted from Ref. [6].)

threaded boss, cavity steel must be removed from the outside diameter of the boss. This allows the boss to expand when the core is stripped. The part and boss must be well supported during the stripping action, so that the part will not warp. The boss should be strong and stiff enough not to collapse or fracture. Threaded parts with a diameter-to-wall thickness ratio of 20:1 or greater should be ejectable if the resin has the required elongation to be ejected off the core pin.

The allowed strippability of a resin, expressed as percent strain at a material temperature of 150°F, is shown in Table 14.2. This is the minimum anticipated part temperature when ejected from even a cold mold, unless the part was extremely thin. These values should not be exceeded when determining a part's strippability. The calculation for percent strain and an example are shown as reference.

Major diameter: 1.250 in Minor diameter: 1.157 in

$$\frac{1.250-1.157}{1.157}\times 100 = 8\%\ \text{Strain}$$

TABLE 14.2. Strippability of a Resin.

Material	Strain at 150°F (%)
ABS	8
SAN	N.R.*
Polystyrene	N.R.*
Acetal	5
Nylon	9
Acrylic	4
Polyethylene (LD)	21
Polyethylene (HD)	6
Polypropylene	5
Polyallomer	15
Polycarbonate	N.R.*
Noryl**	N.R.*
Surlyn***	10

*Not recommender.
**General Electric Trademark (PPO/Styrene blend).
***DuPont Trademark (Ionomer).

$$\text{Percent strain} = \frac{\text{Major thread diameter} - \text{Minor thread diameter}}{\text{Minor thread diameter}} \times 100$$

Some glass reinforced resins can also be stripped from a mold as long as the part temperature is high enough and the percent strain is not exceeded. The hotter the part is during the ejection cycle, the greater the amount of allowable material strain. For example, a 33 percent glass-reinforced $\frac{6}{6}$ nylon can be stripped from a 100°F mold if the strain does not exceed 1 percent, or from a 200°F mold if a strain of 2 percent is not exceeded. To ensure proper part ejection, consult with the material supplier for design and resin recommendations.

Molded-in threads should be terminated at a minimum of $\frac{1}{32}$ inch from their ends. This reduces material/part fretting from repeated assembly and disassembly operations and eliminates compound sharp corners at the end of the threads (see Figure 14.27).

UNDERCUTS

Undercuts are an important feature of plastic parts. They are used primarily to form attachment and assembly features that minimize assembly hardware time and part cost. Parts designed with external and internal undercuts are formed in three basic ways—by split cavity molds, collapsible and pulling cores, and stripped from the mold in a manner similar to screw threads. Depending on the degree of undercut and the return angle forming the under-

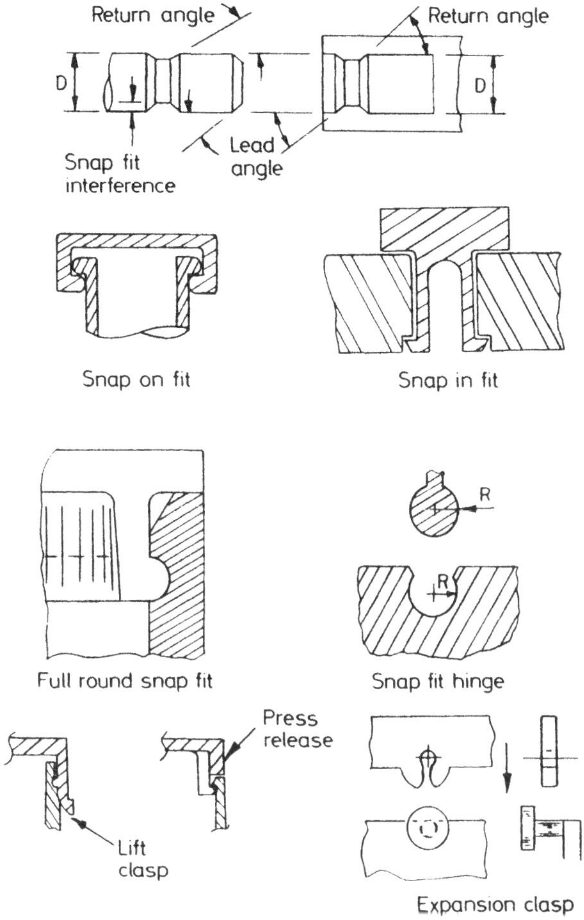

FIGURE 14.27. *Examples of molded-in threads. (Adapted from Ref. [6].)*

cut, most parts—including some of the reinforced materials—can be stripped from a mold as shown in Figure 14.28. Only when the return angle approaches 0° is the undercut deemed nonstrippable. This type of internal undercut can be formed by using two separate core pins, as shown in Figure 14.29. Depending on the core pin's L/D ratio, one end can nest inside the mating pin for stability (see Figure 14.29B), or they can fit tightly together (see Figure 14.29C) to eliminate flashing where the pins meet.

Other nonstrippable undercuts use through-the-wall cores, as shown in Figure 14.29A, or an offset ejector pin system, as shown in Figure 14.30. Because of the steel's bending strength, the undercut depth is restricted to the thickness of the pin or plate used to form the undercut and must be assisted

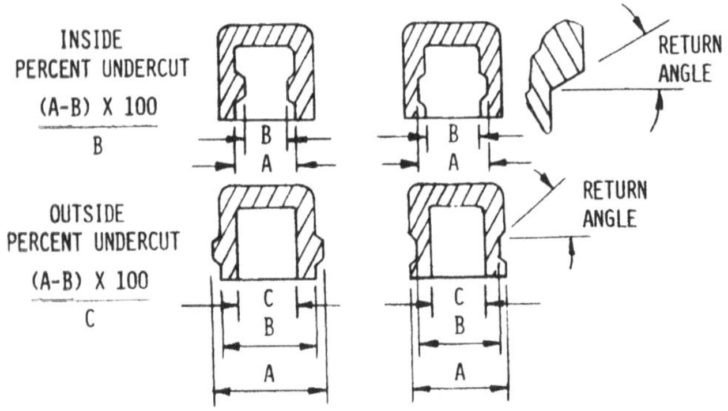

FIGURE 14.28. *Allowable undercuts. (Adapted from Ref. [6].)*

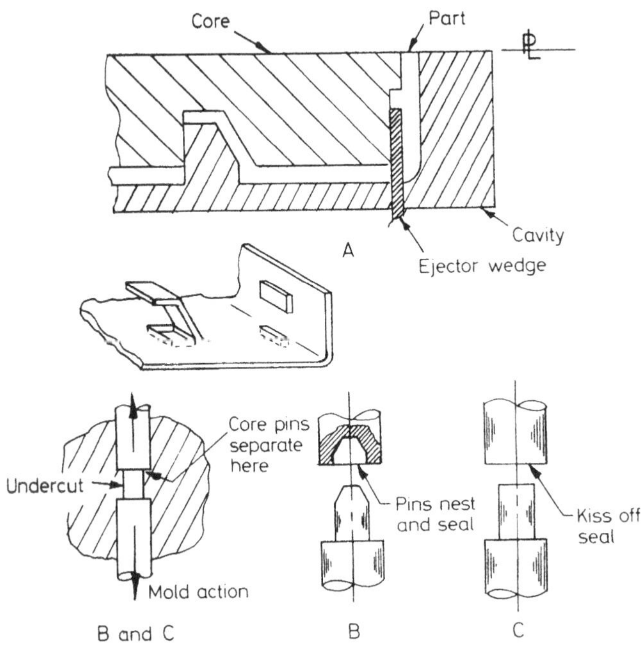

FIGURE 14.29. *Internal undercuts with two separate core pins.*

by other knock-out, ejector systems. A minimum taper of 2 degrees must be on the undercut to ensure correct release during operation, especially when molding high-shrinkage resins. This reduces the bending stress on the sliding core and the wear as it slides against mating metal parts.

Undercuts can also be stripped in glass-reinforced resins if the design does not exceed the material's elongation. The percentage of allowable undercut will be in the 1 to 2 percent range for most materials. Generous radii and

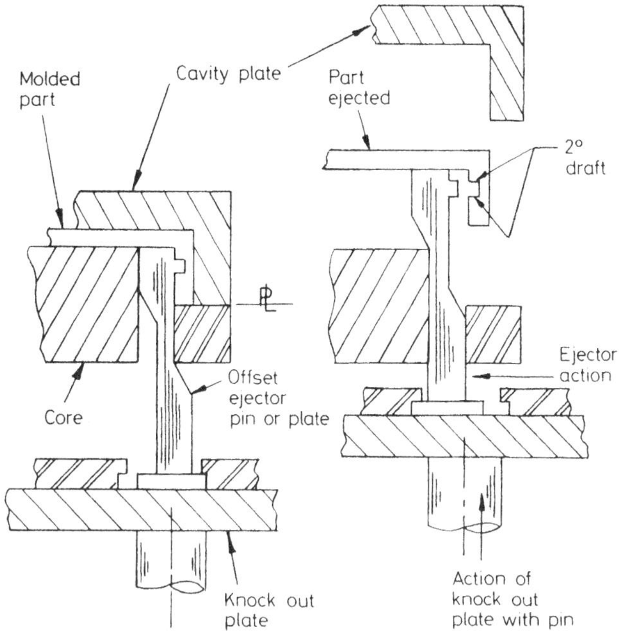

FIGURE 14.30. Offset ejector pin undercut. (Adapted from Ref. [6].)

release angles are used to ensure smooth release and reduce stress concentrations during ejection.

Parts used when stripping undercuts are usually round. Rectangular or closed-wall shapes, like a box with a full circumference undercut, are not strippable. If a full-lipped undercut was used, the container would bow and lock up the part at the corners. This would destroy the part or it would stick in the cavity or on the core. Undercuts placed at the center wall sections would be strippable if kept short and if supporting cavity steel is first removed. This allows the part's side wall to deflect on ejection. The undercut section of the stripped part must be capable of deflection to be ejectable.

INSERTS

Threaded inserts are used for assembling parts. They are either molded in or attached in a secondary operation as press fit or can be inserted into the part ultrasonically. They are used primarily to increase the assembly holding force of the parts, reduce material creep at the assembly points, and access areas requiring frequent service.

Inserts are expensive and can be either a help or hindrance depending on the requirements of the part and how they are installed. Inserts, like bosses and ribs, should only be used when required. Alternative methods of part assembly, using molded-in snap or press fits, screws, or ultrasonic welding, should be considered first. The four main reasons for using inserts are as follows:

1. Threads are under continuous load or stress, or parts require frequent disassembly or access for repair.
2. To meet close tolerance on male or female threads.
3. To permanently attach two highly loaded parts, such as a pulley to a gear shaft.
4. To provide electrical conductance between two parts.

If inserts are necessary, the question of whether to use molded-in inserts or having a secondary operation must be answered. In parts made with high-shrinkage material, where exact dimensional location is the critical factor or there is a concern with boss stress fracture, ultrasonic insertion should be considered. With a proper fixture, the insert location can be exact. When using ultrasonics, the stress on the boss is significantly lowered two to three times versus a molded-in insert.

The factors to be considered for using molded-in inserts are more complex. Although molded-in inserts are less expensive and permanent, there are up-front considerations to be evaluated before their use is selected. Problems that may occur with molded-in inserts are as follows:

1. Loading of inserts by hand will disrupt the molding cycle; robots may prolong the cycle, but will even the cycle out.
2. Inserts may float or become dislodged, thereby damaging the mold.
3. Inserts will require degreasing and possibly preheating to reduce boss stress.
4. Rejected parts with inserts are expensive to salvage.
5. Tight shutoff is required at the insert mold face to avoid flash in the threaded area.

Weld-line strength around inserts is also a factor. Boss insert design for molding should follow established criteria for minimizing weld-line formation and avoiding surface aesthetic problems. With reinforced resins, weld-line strength may be only 60 percent of the unreinforced resin's physical properties because of poor bonding at the knit line. But boss strength at the mold line can be increased by locating a rib at the weld-line junction. Multiple ribs are also used for additional strength and rigidity. Suggested minimum wall thicknesses for inserts of different diameters for various plastic resins are given in Table 14.3.

Other areas to be considered when using inserts are as follows:

TABLE 14.3. Minimum Wall Thickness for Inserts.

	Diameter of Inserts (in.)					
Plastic Resins	.125	.250	.375	.500	.750	1.00
ABS	.125	.250	.375	.500	.750	1.00
Acetal	.062	.125	.187	.250	375	.500
Acrylics	.093	.125	.187	.250	.375	.500
Cellulosics	.125	.250	.375	.500	.750	1.00
Ethylene vinyl acetate	.040	.085	N.R.*	N.R.*	N.R.*	N.R.*
FEP (fluorocarbon)	.025	.060	N.R.*	N.R.*	N.R.*	N.R.*
Nylon	.125	.250	.375	.500	.750	1.00
Noryl (modified PPO)	.062	.125	.187	.250	.375	.500
Polyallomers	.125	.250	.375	.500	.750	1.00
Polycarbonate	.062	.125	.187	.250	.375	.500
Polyethylene (HD)	.125	.250	.375	.500	.750	1.00
Polypropylene	.125	.250	.375	.500	.750	1.00
Polystyrene	N.R.*	N.R.*	N.R.*	N.R.*	N.R.*	N.R.*
Polysulfone	N.R.*	N.R.*	N.R.*	N.R.*	N.R.*	N.R.*
Surlyn (ionomer)	.062	.093	.125	.187	.250	.312

*Not recommended.

Source: Adapted from. Ref. [2].

1. No sharp corners—well-rounded knurling. Undercut provided for increased pullout strength.
2. Insert should protrude more than $\frac{1}{16}$ inch into the mold cavity to provide good sealoff, as shown by the heavy line, and protect threads from resin contamination as shown in Figure 14.31A.
3. Thickness below the insert should be a minimum of $\frac{1}{6}$ of the insert diameter to minimize sink and increase weld line strength (see Figure 14.31B).
4. Consider toughened grades of materials with higher elongations to reduce boss stress cracking.
5. No oil or grease on inserts, which must be clean.
6. Preheat inserts for high-shrinkage resins, i.e., greater than 0.010 inches per inch, to pre-expand the insert and minimize postmold shrinkage. It also improves weld-line strength.
7. Conduct a thorough end-use test program that cycles the part through temperature ranges anticipated in service as well as stress and vibration loading. This is done in the prototype stage to detect any problems with the chosen assembly method.

INSERT LOADING

An item often overlooked that can greatly affect cycling consistency and part quality is how the inserts are loaded into the cavity. Robots will give uniform

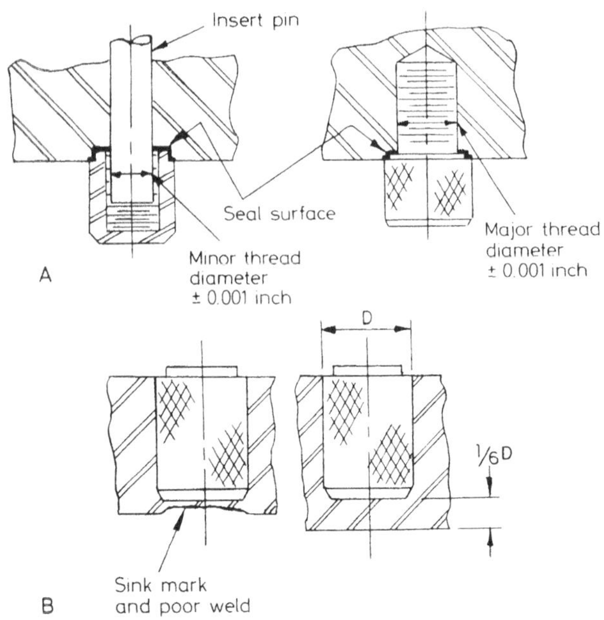

FIGURE 14.31. *Molded-in inserts. (Adapted from Ref. [2].)*

insert loading time as compared with an operator. But if robots are unavailable, an insert loading fixture can be used. Fixtures solve many problems, including the following:

1. Ensuring threads are to specification; this can be a major problem after molding is completed.
2. Loading of all inserts at one time and correct sealing to mold surface.
3. Fast and easy loading, with no cycle disruption.

When inserts are used, the method of insertion and how they are to be retained in the mold must be considered.

Ferrous metal inserts can be held in place by a pin (female) or hole (male), plus a magnetic insert retainer block. This ensures that they will not jar loose during mold closing. For nonferrous inserts, the pin or hole must be within +/− 0.001 inch to hold it tightly in place and reduce flash from forming in the thread area. Inserts with shoulders are preferred to obtain flash-free shutoff using both a vertical and horizontal cavity seal surface.

To reduce molded-in insert problems for parts with multiple inserts, design the mold for vertical or shuttle-press operation. By using a shuttle, or rotary, multiple station mold, inserts are easier to load. Parts are molded and ejected in separate sequential operations to maintain a uniform cycle.

All of these factors must be considered for inserts before a design is finalized. Otherwise, serious problems can occur, leading to delay and poor part quality. Part quality is the key and although the cost of insert molding is almost equal to that of postmold insertion, the tooling and insertion methods must be considered to obtain the desired results.

Alternative methods of part assembly and the use of inserts will be discussed in the next chapter.

INTEGRAL HINGES

With the right design techniques, multiple-piece functional assemblies can be produced in the mold simultaneously. This can be done with molded-in integral hinges. This method keeps all parts of the assembly together and increases the part's worth and functionability. The only prerequisite is that all parts must be made from the same material and can perform the desired function. Molded integral hinges are often used for parts, such as housings, with internal components that can be serviced or to hold all components together until final assembly. For example, electrical cord plugs enclosing electrical contacts are molded in halves with a living hinge and snapfitted or screwed together after the internal contacts and wires are installed. Resins with high dry-as-molded elongation, such as polyethylene, polypropylene, nylon, acetal, and EVA, are the primary materials used for these parts. Molded-in living hinges, capable of hundreds of flexes, are designed using these materials.

Typical living hinge designs are shown in Figure 14.32 for polypropylene and nylon. To increase hinge life, the part is flexed several times after molding—while still hot—to orient the web and give it additional flex and tear strength. This is similar to cold heading or coining a hinge section by compression in a die after molding, which is another technique.

Parts with living hinges may require multiple gates, as the hinge section is often too thin to fill an adjacent thicker section. As a result, multiple gating of the part is often required. When using multiple gates to feed the part, the gating must be located so that the fill of each section is timed to avoid the formation of the resulting weld line at the hinge area. Because high frictional heat will occur at the hinge area, cooling lines should be near this point to prevent lamination of the melt and hinge weakness. Where the material flows meet beyond the hinge, the cavity should be hotter to ensure a good union of the two opposing melt fronts when they meet. Vents must also be more generous at the parting line to vent the air from the meeting melt fronts. Select high-flow resins with high DAM elongations.

By following these guidelines and by discussing part and tooling limitations and requirements, the designer should provide the production group with a part design and tool capable of producing parts to meet the quality expectations of the customer.

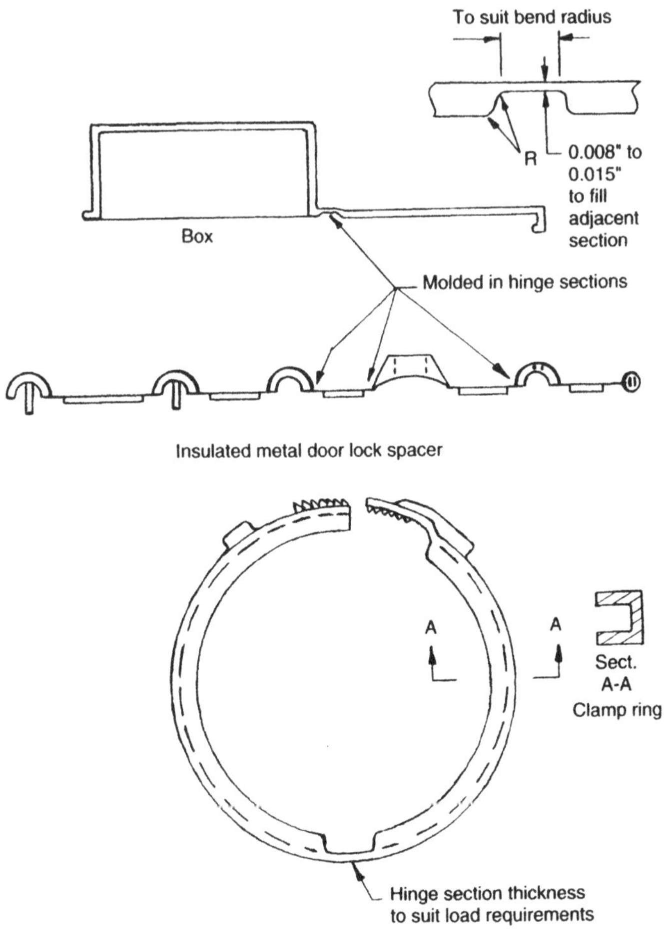

FIGURE 14.32. *Living hinge designs.*

These last few ideas illustrate bow a designer must think and use available assets as well as customer input to design a better part with lower tooling and production costs. These ideas may also provide other part features the customer may include to make a product of better quality that is less expensive and better able to compete in the marketplace.

These factors are even more important when plastic is considered for the redesign of a metal part. The versatility of plastic may allow for combining many separate metal parts. These can be cams, springs, bearings, snap fits, and so on. In this way, a good product design and manufactured package can be developed to meet the customer's requirements for function and part quality.

15

Assembly Techniques

Plastics provide more versatility for the assembly operation than any other material. Their properties lend them to mechanical, thermal, chemical, and electrical methods of joining to similar or dissimilar materials. Before decisions are made on assembly or joining, many questions need to be answered. These involve the part's function, its materials, type of assembly, method of assembly, assembly equipment, and required quality control. The assembly method also depends on the volume of parts, the human and machine element for control of the process, and the need to keep costs under control and earn a profit. Accomplishing all of this begins with part design and experience.

PLAN FOR ASSEMBLY

Part design establishes the assembly operation. Once the assembly method is selected, cost must be addressed. This involves such items as: Can the operation be automated or must some operations be manual? Is assembly equipment standard or must special equipment and/or fixtures be developed? If special equipment is required, will plant personnel need training? Finally, what quality requirements need to be established for the part and process?

All these questions need to be answered, along with any requirements for decorating or printing information on the part. These operations can

---

Total Quality Process Control for Injection Molding, by M. Joseph Gordon

sometimes be integrated into the design or performed on the assembly line to reduce cost, speed up the operation, and control quality. The method to be used must be in line with the part's end-use requirements for reliability and cost.

Examples of successful assembly operations are household trigger spray pumps that are comprised of 12 to 15 separately molded parts automatically assembled by the hundreds of millions annually. Other examples are electrical, multipin, printed circuit board connectors with automatic insertion of metal contacts, with reject rates in the range of 0.05 percent or less.

Molded parts now enter assembly operations to be automatically arranged, stacked, fastened, pressed, bonded, riveted, snapped, screwed, welded, decorated, and inspected to produce the final product. These operations are integrated with the injection molding line, where molded parts are fed, either directly to the assembly line or placed in "work-in-process" bins awaiting delivery to the assembly line. From there, the parts are sent to a feeder setup to feed, track, orient, and disperse parts to the assembly machines. Before each operation, robots install parts on fixtures or, if not used, orient and load parts for delivery to the next assembly station. This continues until assembly of the part is completed and the product is packaged for shipment.

AUTOMATED ASSEMBLY

Automated assembly machines are tireless; they perform operations repeatedly and consistently. If well maintained, they always yield a high-quality product. This does not mean that hand assembly yields lower efficiency or lower quality. The use of automation must be justified by volume. Each program must be evaluated on its own merits.

Many plants still use manual assembly if the volume of products, assembly line adaptability, or the price of automated equipment does not justify automation. Most automatic assembly equipment is very versatile. Many machines have programmable computers that can modify the machine's functions to perform similar operations on a variety of parts. These machines can also store the operating instructions for recall if the part is run again. This versatility reduces capital expenditure, saves setup time, and reduces the chance of human error. The equipment's process controllers can also be integrated in local control stations for data storage or networked into a central control system. The information from these assembly stations can include cycle times, production rates, and downtime. They can also generate product control charts for production control, scheduling, and quality control, for both internal and customer use.

There are a wide variety of techniques for assembling plastic parts. As with any item, the holes, pins, and mating surfaces must line up and be within the design parameters. Quality control for finished assembly of mating parts begins with understanding the assembly process and the required tolerances.

The designer and tooling engineer must understand what is required of the assembly to prevent such errors as stress cracking of bosses or the use of the wrong type or size of screw. When assembling spin and ultrasonically welded parts, the joint must be molded to tolerance. If not done correctly, excessive flash or low joint strength may result. This could lead to scrapping the part or, at worst, a field failure. Such variables as the type and number of gates and their location, tolerance control in the tooling, and maintaining uniform molding conditions affect these part control factors.

AUTOMATED INSPECTION

Automated inspection stations can be integrated within the assembly operation to check any critical part or step to ensure that defects do not reach the next station. This is particularly important for a critical part housed within the assembly. By detecting the problem in real time, during the assembly process, the cost of repairs is minimized.

Automated inspection equipment is expensive, but with careful attention to its versatile features, including its process controller, it can be reprogrammed and adapted for future operations. Just as with robot selection, consult with the machinery suppliers to develop a clear understanding of a machine's features for current and future uses. The quality of your customer's product may depend on and demand this type of inspection equipment.

The designer, along with the tooling and manufacturing team, needs to understand what is required for the assembly of plastic parts and how to control the quality of manufacturing variables. As an aid, the various assembly techniques used for plastic parts will now be described.

ASSEMBLY TECHNIQUES

The basic assembly techniques are as follows:

1. Press fits
2. Snap fits
3. Spin welding
4. Ultrasonic welding
5. Angular or vibration welding
6. Induction magnetic bonding
7. Hot plate welding
8. Cold or hot heading
9. Mechanical fasteners, screws, and inserts
10. Adhesive or solvent bonding.

The design calculations for these various techniques will not be discussed in any detail, as each resin has its own recommended method of assembly. What will be addressed are the limitations and recommendations for each assembly technique that will assist the designer, tool maker, and production personnel to manufacture the part.

Press Fits

Press-fit assemblies are the easiest and most economical for plastic parts that require a minimum holding force. Press fitting involves the assembly of two or more parts using an interference fit with friction. This technique provides a simple and fast method of assembling plastic-to-plastic or plastic-to-metal parts. Press fits can eliminate screws, inserts, and adhesives, while producing a joint with aesthetically pleasing results. The assembly and disassembly forces can be calculated for similar and dissimilar materials. If the parts are exposed to thermal cycling after assembly, dissimilar materials may loosen. This is because differences in each material's rate of linear thermal expansion causes the materials to shrink or expand away from each other, thereby lowering part interference. Because all plastics creep under load, the designer must anticipate a loss in joint strength for a potential loosening of the press fit over time. The press-fit joint may never relieve itself to the point the parts fall apart, but retention forces at the joint will gradually decrease. Repeated thermal cycling of the part will increase the rate of fall off in joint strength.

The type of resin and its molding conditions will also affect press-fit joint strength. Crystalline polymers, molded in a cold mold and subjected to thermal cycling, will experience greater postmold shrinkage than an amorphous resin under the same conditions. Crystalline resins molded in cold molds (100°F or lower) do not achieve maximum shrinkage rates. Therefore, these materials should be molded in a hot mold (above 150°F) to avoid this problem. Postmold shrinkage of a part, which is a factor of time and temperature rates, may happen in days or weeks depending on the part's exposure to elevated temperatures.

Using a cold mold to reduce cycle time is fine, unless the parts will be exposed to higher temperatures in their end-use application. Therefore, testing parts and their assembly under anticipated end-use thermal cycles is obviously indicated. Problems with press-fit assembled parts have even occurred during shipping or storage, if the part's temperature exceeded the molding cavity temperature. If this becomes a problem, a positive assembly method, such as a snap fit, should be investigated. Press fits are typically a pin and socket attachment but may include a full-part circumference joint, as with a box.

To aid in press-fit part assembly, tapered lead-in angles on mating surfaces are recommended. They assist in lining up assembly points as shown in Figure 15.1. Control of warpage is also important, not only for part appearance and function but also for lining up assembly points. Draft on pins and part contours avoids hanging up the part in the mold cavity and warping during

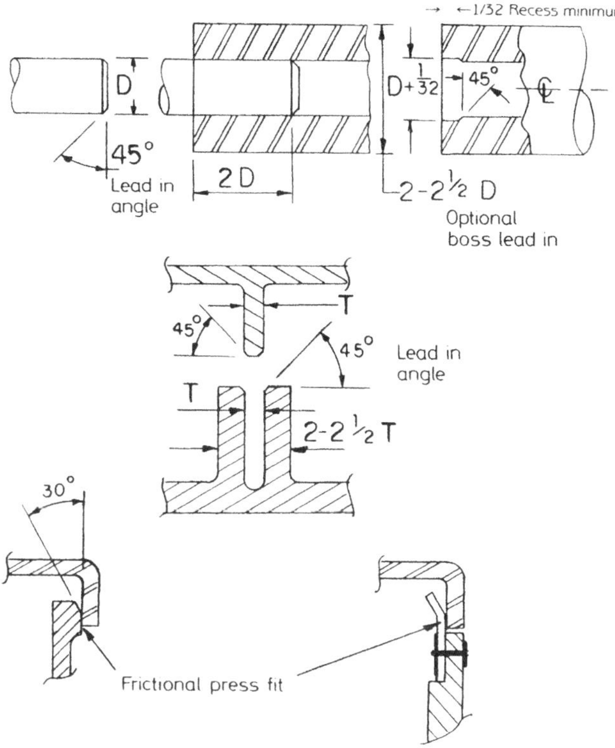

FIGURE 15.1. Press fits with tapered lead-in angles on mating surfaces. (Adapted from Ref. [1].)

ejection. The draft on mating parts should match to insure uniform press-fit contact.

Surface finish on the press-fit surfaces will also influence the assembly and holding force, as will the coefficient of friction of the materials. Most press-fit finishes are smooth. But to increase the holding power of similar plastics, a very light honing of the cavity's press-fit surfaces increases the holding force. The honed finish will allow the surfaces to flow or creep into one another, thereby increasing the retention force. The press-fit assembly should be tested to determine whether this is required for keeping the parts together or whether a snap fit would be preferred.

Snap Fit

Snap fits, which are a popular assembly technique similar to press fits, can provide either a permanent attachment or one that can be disassembled. They come in two basic designs: cantilever beam style, with a locking tab that deflects and engages a mating tab or undercut, and a pin style with a full cylindrical-raised lip or undercut whose circumference contracts or expands when engaging a mating undercut or lip which locks the parts together. Examples of these are shown in Figure 15.2. They are limited only by part size and properties of the material. Snap fits for positive-locking assemblies cannot be disassembled, whereas semipositive snap fits can, with extra force or effort, be taken apart. The design of the snap fit dictates this feature.

Plastics are well suited for this type of assembly because of their flexibility, elongation, and snap-back resiliency. An important requirement of the snap fit is that the locking tab be of sufficient length and flexibility to deflect without breaking (see Figure 15.3). Snap-fit assemblies can eliminate mechanical fasteners and bosses to obtain tight-locking features, with the assembly method molded into each part. One example would be automotive wire-harness connectors. Assembly and retention forces, which can be calculated for the snap-fit design, are based on the material selected. With most snap-fit designs, once assembly is completed the parts are under low-tensile or flexural stress. This greatly reduces the effects of creep. Usually, only tensile forces are present in

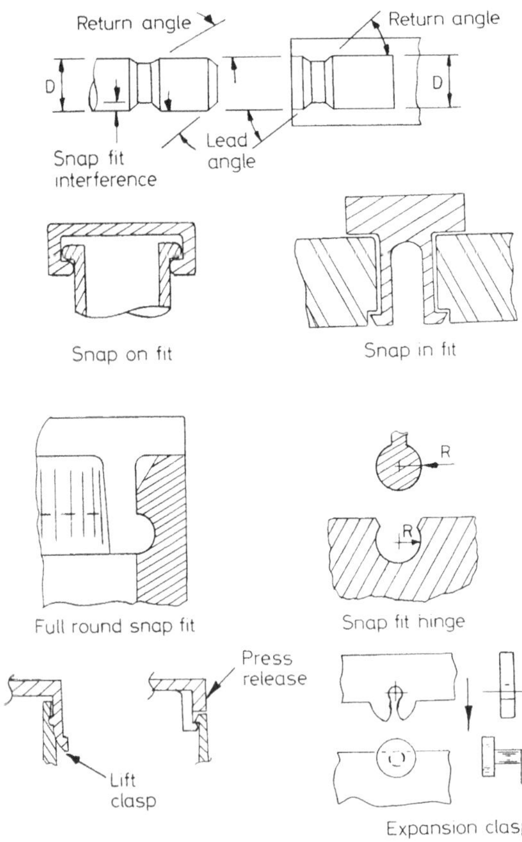

FIGURE 15.2. *Snap fits: the two basic designs. (Adapted from Refs. [1] and [4].)*

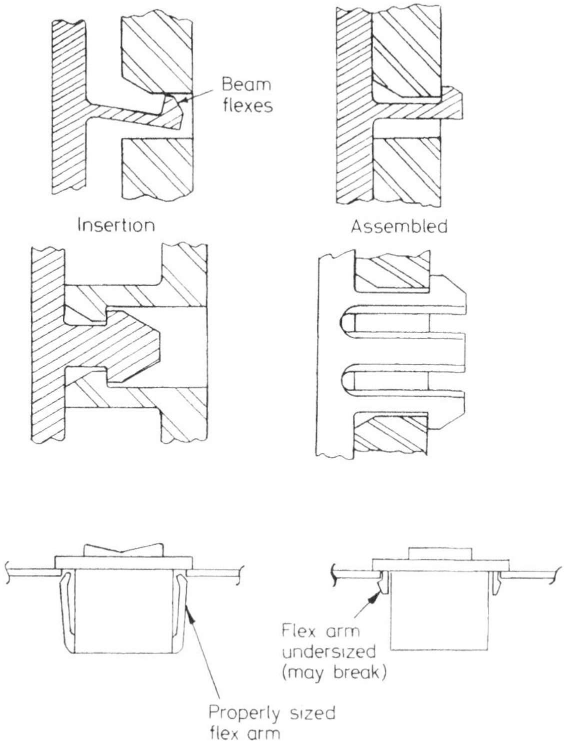

FIGURE 15.3. Cantilever snap fits. (Adapted from Ref. [4].)

some very tight-fitting cantilever snap fits, but tensile creep can still occur if the joint is kept under load. The ideal design is to have the assembly tight, but not under any great tensile or circumferential loading. By using a 90 degree locking tab, the joint will stay tight and the effects of material creep under load can be minimized.

One factor often overlooked during the design and prototype stages is whether the parts are to be assembled by hand or machine. The joint assembly forces can be calculated and, in most cases, the parts in the fixture orient the mating surfaces for uniform snap-fit assembly. The speed of assembly, often a critical factor, must then be considered. With low elongation materials, such as reinforced resins or material in the dry-as-molded state, a slow, steady assembly force of one-to-two seconds may be required for the material to flex or yield during assembly. In comparison, a rapid, fraction-of-a-second, assembly time with an automatic assembly machine may cause the joint to fail. This can occur when the material does not have enough time to yield and absorb the energy of the rapid assembly force and the cantilevered arms break off or the boss splits.

If the mating parts are not well aligned during assembly, excessive forces and deflections may cause some snaps to fail. For parts with multiple snap fits a fixture is required during assembly so that flexing parts are not overstressed and deflect uniformly. Therefore, the assembly force, time, and fixture of mating parts are important to maintain quality control.

The amount of interference or deflection of the plastic snap-fit assembly is critical, as each resin's elongation or amount of allowable strain must be considered. When designing the snap fit, do not overstress the joint or cause permanent deformation by overstretching the material. This often appears on the part as stress whitening and causes a color change at the high-stress point. If the resin's elastic limits are exceeded, the part could break or produce a weak snap-fit joint. Always obtain the material supplier's recommendations for the resin and joint design. Tapered lead-in angles on snap-fit parts are recommended for ease of assembly.

All cantilever snap fits should be well radiused to reduce stress concentration at the base of the beam. If cantilever snap fits are located on a part's sidewall, the wall may also deflect during assembly. This may allow a shorter, stiffer, and often lower elongation resin to be used. Each snap-fit design must be evaluated on a case-by-case basis for the length and size of the beam's section, amount of deflection, material used, and stresses involved.

Coupled with design is the control of part dimensions and tolerances. With snap fits, as compared with press fits, the tolerances can be more relaxed as long as the lip and undercut contact provides sufficient retention force for the joint.

After assembly, snap fits should be tight but not under any great amount of tensile of flexural (hoop) stress. Otherwise, creep may reduce joint strength. Part design will control this as long as the joint dimensions are in tolerance.

Welding Assemblies

Welding plastic parts for assembly can produce a fast, strong, and hermetically sealed joint. This natural assembly method uses the thermal properties of the polymer to bond the separate pieces. There are five basic welding techniques plus a new type, known as Focused Infrared Melt Fusion, which is gaining recognition. The five basic techniques are:

1. Spin welding
2. Ultrasonic welding
3. Angular or vibration welding
4. Induction or magnetic bonding
5. Hot-plate welding

All welding methods involve softening or melting mating plastic surfaces that are pressed together, melted, and allowed to cool and fuse. This forms a strong joint equivalent to the physical properties of the parent plastics. With reinforced materials, joint strength is restricted to the virgin resin's strength, because fiber reinforcement does not migrate across the weld line. Joint strength can be increased by enlarging the contact area to achieve equivalent or greater material property strength. The welding process can be easily automated and the bonding operation takes only a few seconds.

To obtain strong joints, the melting or softening points of each resin are important. With spin, ultrasonic, and vibration welding, frictional heat is created at the joint interface as the mating parts rapidly rub or vibrate against each other. Material with a lower melting or softening point will melt first. If melting points of the two materials are not fairly close, one will melt before the other. If this occurs, the melt will smear and fail to bond. Each half of the joint must melt within a specific temperature and time range to obtain a strong joint. Therefore the melting or softening points of each material, crystalline or amorphous, are important when using these assembly techniques.

Most welded parts are of the same material, which presents no melting point problems. Only when dissimilar plastics are joined does this become a factor. Amorphous resins permit a wider melting point temperature range than crystalline resins. The welding compatibility of plastic resins and some of their alloys are shown in Table 15.1.

If amorphous resins have softening points within 40°F of each other, they can be welded successfully. With the crystalline resins, the melt points must be within 5°F to obtain a strong weld. Alloys of these resins can also be welded, but the designer must verify compatibility with material and welding equipment suppliers before specifying this type of assembly.

The design and molded quality of the weld joint is important for obtaining a strong bond. The designer must know the weld joint's design and material's molding (shrinkage) specifications to be sure that part tolerances will meet assembly requirements. Too often, the wrong joint design or material shrinkage is used, resulting in a poor weld and low joint strength. The part's joint design, tolerances, and bond strength should be verified by building an inexpensive prototype or by using machined prototype parts. Suppliers of welding equipment can assist by offering experience and knowledge in joining similar parts. They also know the design specifications for each type of joint and can help design the mold cavity's gating and layout to achieve required part tolerances. They also have the equipment to assemble the prototype and can recommend the appropriate equipment, holding fixtures, and welding cycle for production.

The following general rules are used for welding plastic parts:

1. The melt point temperature range for amorphous resins is within 30° to 40°F, with 25°F preferred. For crystalline resins, it is within 5°F.
2. For best results, both parts should be made of the same material.
3. As for moisture content, parts made from hygroscopic resins must be kept dry as molded. They should be welded immediately after molding or stored in sealed, moisture-proof polyethylene bags. Moisture at the weld site will create steam, resulting in poor joint strength and aesthetic

TABLE 15.1A. Ultrasonic Weld Compatibility.

	1	2	3	4	5	6	7	8	9	10	11	12	13	14	15	16	17	18	19	20	21	22	23	24
1 ABS	O	O	O		O	O	X	X												X				
2 ABS/PO		O			X	X	X	X					O											
3 ABS/PVC			O		X	X	X	X											X					
4 Acetal				O																				
5 Acrylic					O		X						O							X	X			
6 Acrylic Multipolymer						O	X	X												X				
7 Acrylic/PVC							O	X																
8 ASA								O																
9 Cellulosics (CA, CAB, CAP)									O															
10 PRS										O														
11 PPO											O						O							
12 Nylon (H)												O												
13 Polycarbonate (H)													O	X										
14 PC Polyester (H)														O							X			
15 Polythylene															O									
16 Polypropylene																O								
17 Polystyrene (GP)																	O			X				
18 Polysulfone (H)																		O						
19 PVC (Rigid)																			O					
20 SAN/NAS																				O				
21 Polyester																					O			
22 Structural Foams																						O		
23 Polymide																							O	
24 Polyimide-Imide (H)																								O

Source: Adapted from Ref. [15].

ABS, acrylonitrite butadiene styrene; ASA; CA, cellulose acetate; CAB, cellulose acetate butyrate; CAP, cellulose acetate phthalate; GP, general purpose; NAS; PC, polycarbonate; PO; PPO; PRS; SAN, styrene acrytonitrile.

Weldability—read across Compatibility—read across and up.

O = Good compatibility X = Compatibile at times based on material blend.

H = Hygroscopic, should be dry before welding.

TABLE 15.1B. Ultrasonic Weldability Chart.

	Resin Type	Welding	Inserting	Staking	Swaging	Degating	Spot Welding	Near Field	Far Field
ABS	A	G	G	G	G	G	G	E	G
ABS/PO	A	G	G	G	G	G	G	E	G
ABS/PVC	A	G	G	G	G	G	G	F	P
Acetal	C	G	G	F	F	F	F	E	P
Acrylic	A	G	G	G	G	G	G	F	F
Acrylic Multipolymer	A	G	G	G	G	G	G	G	F
Acrylic/PVC	A	G	G	G	G	G	G	P	P
ASA	A	G	G	G	G	G	G	E	P
Cellulosics (CA, CAB, CAP)	A	F	G	G	G	F	G	F	P
PRS	C	G	G	G	G	G	G	G	F
PPO	A	G	G	G	G	G	G	G	G
Nylon	C	G	G	G	G	F	G	E	P
Polycarbonate	A	G	G	G	G	G	G	G	G
PC/Polyester	C	G	G	G	G	G	G	G	F
Polyethylene	C	G	G	G	G	P	G	F	P
Polypropylene	C	G	G	G	G	P	G	F	P
Polystyrene (GP)	A	G	G	G	G	G	G	E	F
Polysulfone	A	G	G	G	G	G	G	G	P
BVC (Rigid)	A	G	G	G	G	F	G	G	P
SAN/NAS	A	G	G	G	G	G	G	G	P
Polyester	C	G	G	G	G	F	G	G	P
Structural Foams	A	G	G	F	F	P	G	F	P
Polymide	C	G	G	G	G	F	G	G	P
Polyimide-Imide	C	G	G	G	G	F	G	G	P

Source: Adapted from Ref. [15].

A = Amorpous resins C = Crystalline resins.

Weld characteristics: E = Excellent G = Good F = Fair F = Poor.

problems. Do not touch joints with your hand as oil from your body will contaminate the joint.

4. For fillers and reinforcements, the stiffer the material the easier the welding. Filler content should be limited to 40 percent to avoid poor joint strength. For design purposes, use only the parent resin's physical property strength to calculate joint strength. Reinforcements do not cross the bond line.
5. Mold release agents should be removed from the joint area and, if at all possible, not used. All parts must be degreased.
6. Lubricants and some material modifiers in the resin will weaken a joint by contamination. This could reduce contact friction between parts. A material's normal internal processing aids for flow will not affect joint strength unless excessive amounts are added by the molder. Waxy processing aids should be avoided, as they migrate to the surface on molded parts.
7. Pigments, if oil based, can affect joint weld strength.
8. Gate location and type will vary for different joint designs and materials. Select gates to suit part and joint tolerances for dimensional control and to eliminate warpage, surface defects, and molded-in stresses.
9. Use fixtures to support the parts and to ensure consistent pressure during welding. Parts must be held securely and in correct alignment to prevent shifting.

Spin Welding. Spin welding is only suitable for circular parts. It is a rapid, inexpensive, and efficient assembly method for producing strong, permanent, and leak-free circular joints in 1 to 5 seconds. Parts up to 10 inches in diameter have been successfully assembled and the system can be automated.

Spin welds are made by holding one part stationary and rotating the mating part 10-to-40 feet/sec under approximately 700 PSI joint pressure for 1 to 2 seconds. The resulting friction heat at the joint interface causes the surfaces to melt. The rotating motion is then stopped, in ¼-to-½ second, and clamping pressure is applied until the joint solidifies. Parts are held in the fixture by vacuum pressure or keyed to molded-in holding points, such as ribs or blind depressions. Parts must be held securely so that the joint does not flex or distort during spinning and clamping.

Types of Spin Welding. There are basically three types of spin-welding techniques: the pivot, vacuum, and inertia methods. The pivot and vacuum methods use a driving motor to spin the top part against the stationary lower part to create the melt, as shown in Figure 15.4. The pivot tool shown in Figure 15.5 holds the top of the assembly in place, centering it over the lower stationary part, to be welded with a preadjusted compression spring tip that engages a small recess in the center of the part. In the welding sequence shown in Figure 15.6, the welding head descends and the pivot pin contacts and centers the lid

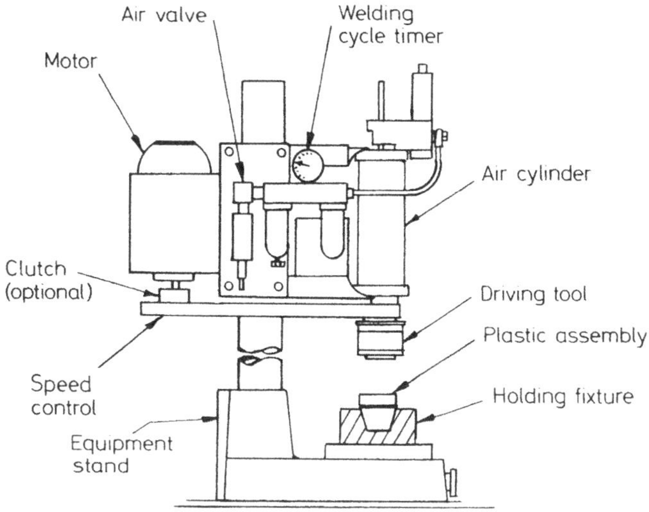

FIGURE 15.4. *Basic equipment for spin welding. (Adapted from Ref. [4].)*

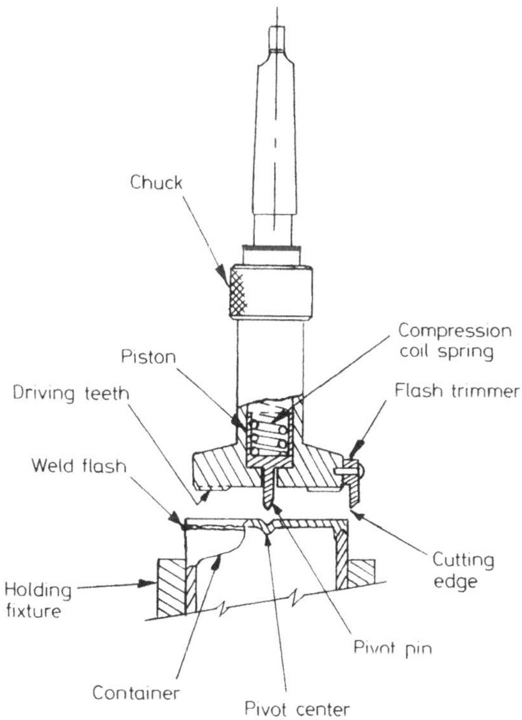

FIGURE 15.5. *Spin welding sequence. (Adapted from ref. [4].)*

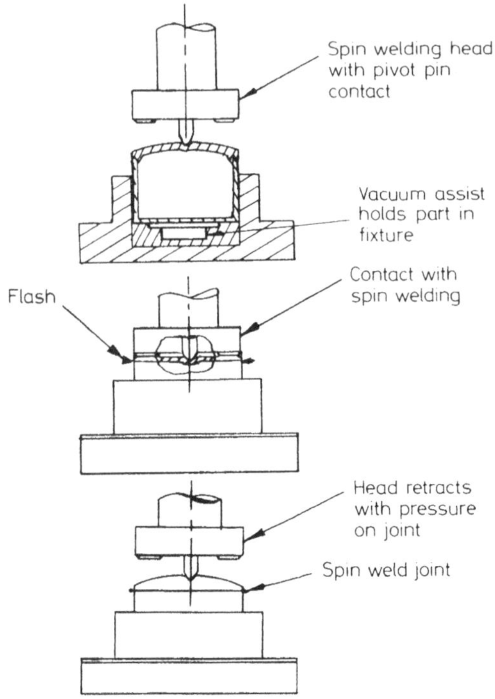

FIGURE 15.6. *Spin welding sequence. (Adapted from ref. [4].)*

on the part. The toothed-drive crown then contacts the lid and spins the top, while the pivot pin retracts. After melt is formed, rotation stops. The action is synchronized so that the tip descends as the toothed-drive crown stops and then rapidly retracts as the tip applies pressure and the joint solidifies.

The vacuum method is similar, except that the tooth-driving crown is attached to the end of a spinning air cylinder. When rotation stops at the end of the spin cycle, the air cylinder clamps the parts together. The downward travel and holding pressure for each method is preselected. It is based on the joint design and material selected to form a good joint without creating excess weld flash.

The inertia spin-welding method is more variable. It uses the principles of acceleration for a known mass to obtain enough kinetic energy to produce a good weld. A mounted drill-press unit is used for short runs. For larger production runs, use a commercial unit with an air or hydraulically operated clamping cylinder. The inertia method uses a part-holding spinning fixture of known mass that is accelerated to a specific speed. The spinning fixture is then disengaged from the motor and moved downward by an air cylinder that contacts the mating part. The kinetic energy of the spinning part is converted into heat at the joint interface to produce a melt. After a specific spinning time, molding pressure is applied to stop the rotation in a fraction of a second

(0.1 to 0.3 seconds) and to provide clamping pressure until the joint solidifies. This is shown in Figure 15.7, with trimming of flash occurring during welding.

The variables associated with spin-welding equipment are regulated by the machine's controls and timers. These variables change depending on the resins to be welded and part size. It is recommended that the material and spin-welding equipment suppliers be consulted on joint design and machine settings.

Spin-welded parts must be designed with sufficient wall stiffness to support the clamping forces. In addition, the top must be rigid enough not to bow during clamping. The top must also have indexing, centering, and driving points for spinning.

Production personnel are responsible for setting up, adjusting, and monitoring the welding cycle to produce good, strong joints every time. Among the variables they will encounter are the following:

1. Rotation speed
2. Clamp pressure and movement of clamp
3. Weld time
4. Pressure hold time
5. Vertical tool speed
6. Total cycle time (load-weld-unload)

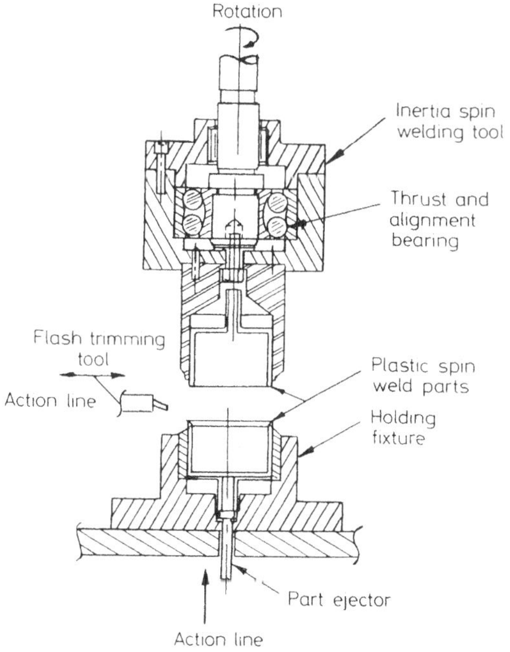

FIGURE 15.7. *Inertia tool for spin welding. (Adapted from Ref. [4].)*

Speed, timing, and holding pressure must be controlled within fractions of a second to produce good welds. If any part or equipment variable changes, sufficient heat may not be generated at the weld joint. If timing is inexact in stopping the head's rotation, melt smear can occur. This causes the melt to shear during solidification, resulting in a weak or no weld situation. The downward travel of the top to contact and create the melt, and then to clamp the parts together, must also be controlled precisely. Holding pressure must be exerted long and high enough for solidification to occur before pressure on the joint is released. The top and bottom holding fixtures must also support the side walls of the part to inhibit flexing at the joint during the welding cycle and to prevent the parts from moving when rotation stops and clamping begins.

Recommended spin-weld joint designs are shown in Figure 15.8, with flash traps to catch the excess material formed during welding. Flash traps contain the melt that flows out of the joint area during welding and eliminate flash trimming after welding. If flash formation does not affect functional or aesthetic features, a flash trap may not be required. Flash traps should be used

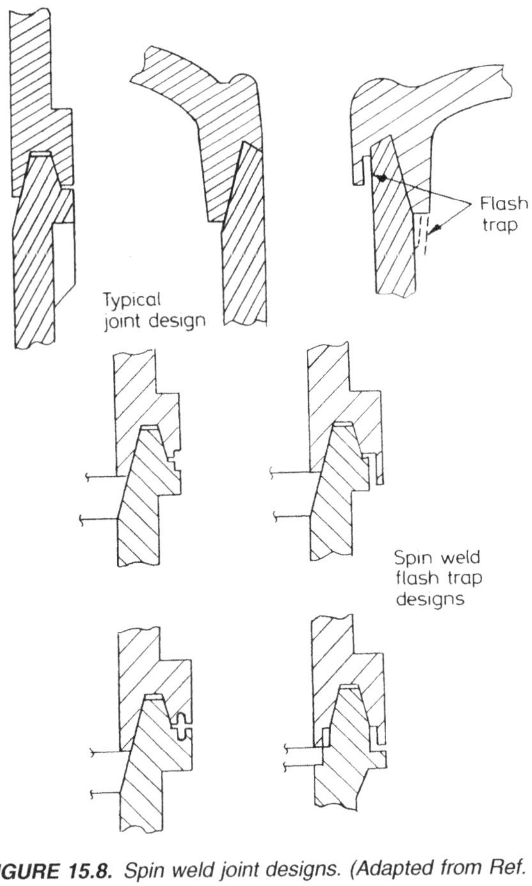

FIGURE 15.8. Spin weld joint designs. (Adapted from Ref. [4].)

for trapping flash on the inside of a joint which if it breaks off could interfere with the end-use operation of the part.

The part designer must ensure that the mold is built to obtain concentric parts, which is critical for good welding. This depends on how the parts are gated, number of cavities, cooling layout, and type of mold selected. Single-edge gating will produce an elliptical rather than a round joint (see Figure 15.9). Consider a three-plate or hot-runner tool that is center gated to produce a uniform concentric part and weld joint. By specifying on the drawing the degree of concentricity required, the mold designer will know the required quality of the weld joint.

Ultrasonic Welding. Because ultrasonic welding is not restricted to round parts, it offers more latitude to the part designer. Ultrasonic welding converts electrical energy into high-frequency mechanical vibrations. There are transmitted through a tuned welding horn to the joint interface to create frictional heat. This heating creates a melt at the joint interface, which, after the vibrating stops, is clamped tightly together and allowed to solidify. Part size and shape depend on the size and shape of the welding horn, because of size limitations on welding horn design. Large joint areas can be welded, but they require special horns or multihorn layouts. As with any assembly operation, contact with the equipment and material suppliers is recommended to reduce problems.

The basic equipment required for ultrasonic welding is shown in Figure 15.10; modifications are, of course, available from different suppliers. The

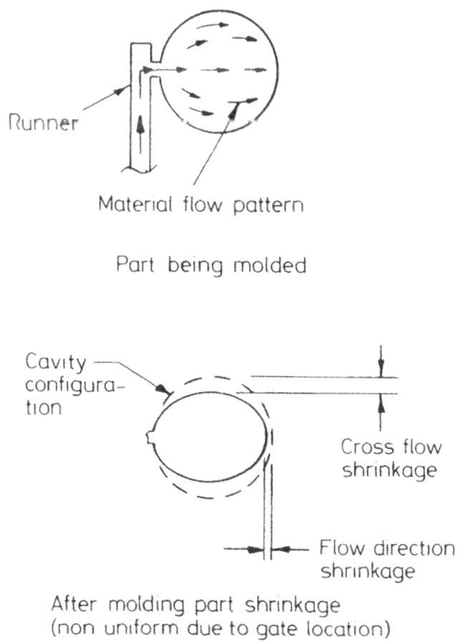

FIGURE 15.9. *Single-edge gating produces an elliptical joint. (Adapted from Ref. [2].)*

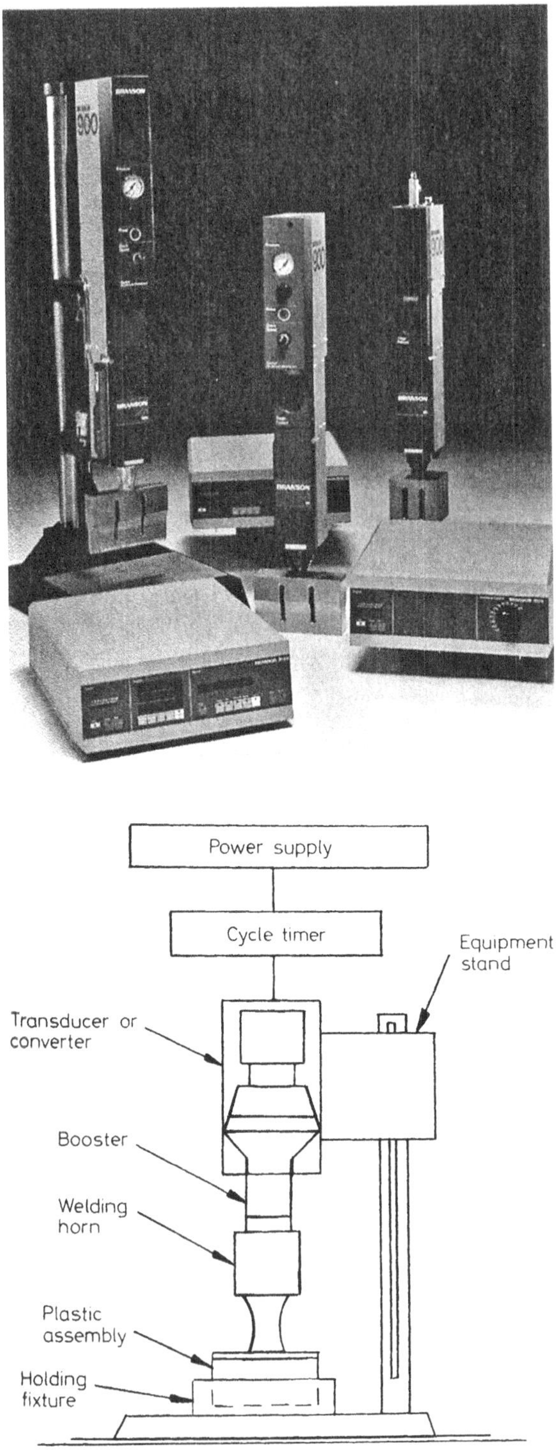

FIGURE 15.10. *Basic equipment for ultrasonic welding. (Courtesy of Branson Ultrasonic Corp.)*

FIGURE 15.10. Continued

machine regulates the welding cycle by controlling weld time, energy, contact, hold pressure, and position. Most welders use microprocessors to control the cycle and compensate for minor part variability.

Ultrasonic welding requires the following three major joint design considerations:

1. Initially, the contact area should be small and focused to concentrate the energy and reduce the time needed for melting.
2. Mating parts must be fixtured for alignment.
3. Mating surfaces around the entire joint area must be uniform and in contact with each other.

Joint and dimensional requirements will vary with amorphous and crystalline resins. Therefore, each must be considered separately. Amorphous resins are easier to weld. On the one hand, the vibrational energy is more effectively transmitted through the part and the process requires less energy to create the melt. Crystalline resins, on the other hand, require special joint designs, close distance between the horn and the joint, and higher amplitudes of energy because of the material's dampening characteristics and sharper melting temperature. This is shown in Figure 15.11A with the vibration intensity scale for these resins.

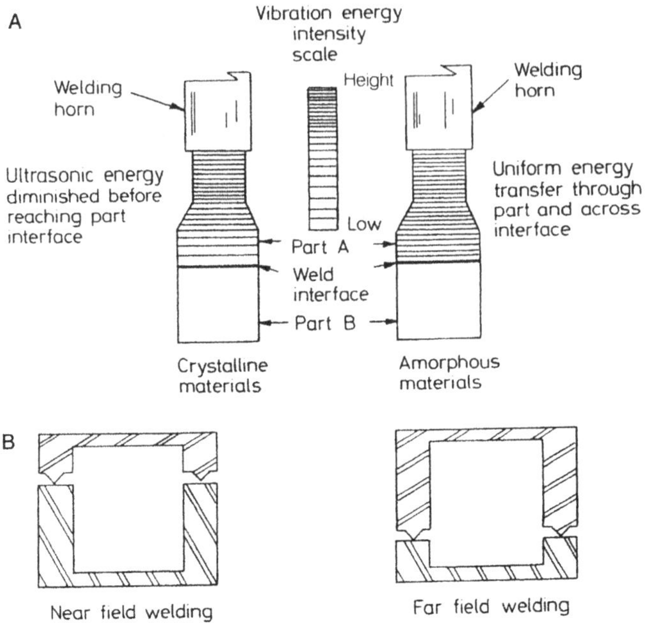

FIGURE 15.11. *Ultrasonic joint design for amorphous and crystalline resins. (Adapted from Ref. [4].)*

As with other tasks, consult the ultrasonic welding equipment or resin supplier for specific information about welding horn size and shape and the proper amplitude or frequency to use. They can also advise you about resin compatibility and whether near or far field welding will be acceptable with your part design. Both of these items must be considered in the early design stage, as shown in Table 15.1. Near field welding means that the welding horn is placed in close proximity to the joint. Far field means "across the part face" or a "sidewall distance" from the joint (see Figure 15.11B). Crystalline resins require near field welding because of the loss of energy, which is quickly dissipated in these resins to obtain a strong joint.

More examples of joint designs for amorphous and crystalline resins are shown in Figure 15.12. Amorphous joint designs require a triangular (energy director) section at or near the center of the joint on one part surface. For best results, crystalline resins require a shear-joint design that focuses the energy at the joint interface. When energy directors are used for crystalline resins, insufficient melting of the mating surfaces minimizes strength. The rapid solidification of the resin also causes spotty welding and marginal joint strength. With all ultrasonic weld designs, the part's sidewalls must be well supported to avoid possible deflection from the welding motion's downward pressure.

The part-holding fixture used for ultrasonic welding is often overlooked in planning, but it must be considered as it contributes to weld quality. When

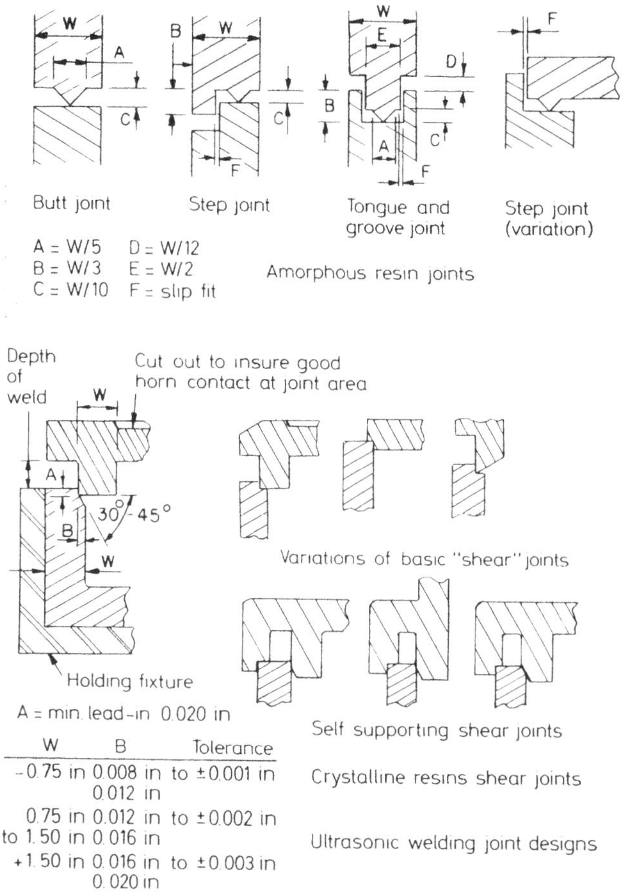

FIGURE 15.12. Additional joint designs for amorphous and crystalline resins. (Adapted from Refs. [4] and [7].)

designing the holding fixture, note that hard plastics need a soft nest and soft plastics need a hard nest. These nests tend to retain or focus the energy from the welding horn into the part at the joint area. If everything seems right but a poor weld occurs, the holding fixture may need additional support, either under or on the sides supporting the part. Nesting materials, such as pads of silicone, rubber, cork, or leather, work best and retain their resiliency over many welding cycles.

Flash traps can also be designed for ultrasonic weld joints. They help avoid trimming, ensure that flash will not interfere with internal operations of the part if it becomes dislodged, and with aesthetics (see Figure 15.13) of both energy director and shear joints. The depth of the welding has a bearing on the size of the trap area. If joint appearance is critical, a blind trap that captures the flash inside the joint area is preferred.

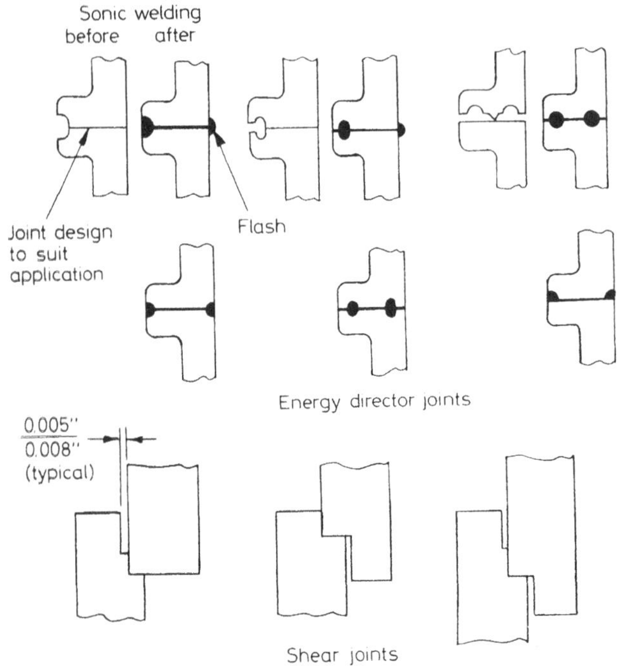

FIGURE 15.13. Basic flash trap designs. (Adapted from Ref. [4].)

To produce good ultrasonic welds, the welding horn must make solid and uniform contact with the molded part; the joint must uniformly contact the mating part. Even a slight amount of part warpage will produce poor joint strength because of nonuniform melting at the bond line. The variables for ultrasonic welding include the following:

1. Clean and dry joint surface.
2. A welding horn that is tuned and sized for the part and material.
3. Correct weld time, pressure, weld-horn travel, and energy amplitude applied to the weld-joint surface.

Ultrasonic welding can be automated with overall cycle times of 5 to 10 seconds or less, with in-line or rotary feed and positioning tables.

Vibration Welding. This is another form of ultrasonic welding, sometimes called angular welding, performed with the equipment shown in Figure 15.14. Vibration welding is performed on a single plane joint surface as shown in Figure 15.15. It is used to join irregularly shaped and large parts. In vibration welding, one part is held stationary while the mating part vibrates against it, under pressure, through a small relative displacement in the plane of the joint. This motion creates frictional heat that causes melting at the joint interface. The motion is then stopped, the parts are aligned, and increased holding pres-

FIGURE 15.14. Minisized vibration welder. (Courtesy of Branson Ultrasonic Corp.)

sure is exerted until the joint solidifies. Welding can be accomplished with either linear or angular part motion, depending on part geometry (see Figure 15.15). Angular motion must be used if an interference or shaft extends through both parts. The total relative motion at the joint surface is a 0.10 inch minimum, with melting occurring within 1 to 3 seconds time.

The amount of angular or linear motion between mating parts is dictated by part geometry and weld contact area. The pressure required to generate frictional heat at the joint is governed by part design and resin properties. Joint strength is close to parent material strength. Burst tests have shown that weld-line strength can exceed part strength based on the contact area. Welded joints in transparent resins, such as PC and acrylic, are clear. This indicates a totally homogeneous bonding of the surfaces.

Typical vibration weld joints for long, unsupported wall designs are shown in Figure 15.16. Joint-flange width should be 2 to 2.5 times the wall thickness for maximum strength. Traps can also be designed to catch any flash that might interfere with the part's end-use function and aesthetics, along with gripping tabs to position, vibrate, and clamp the part during welding.

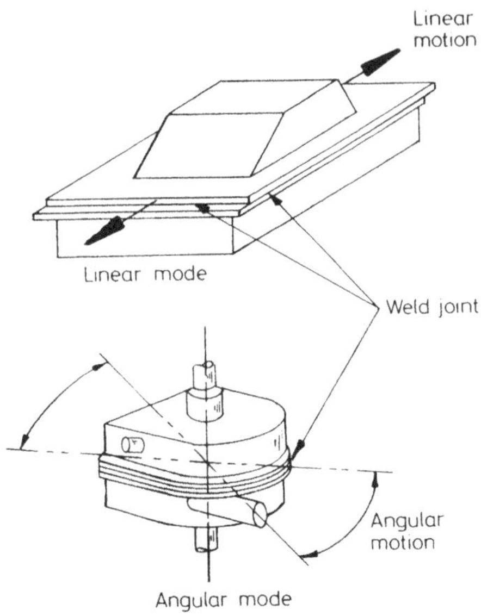

FIGURE 15.15. *Linear and angular part motion in welding. (Adapted from Ref. [4].)*

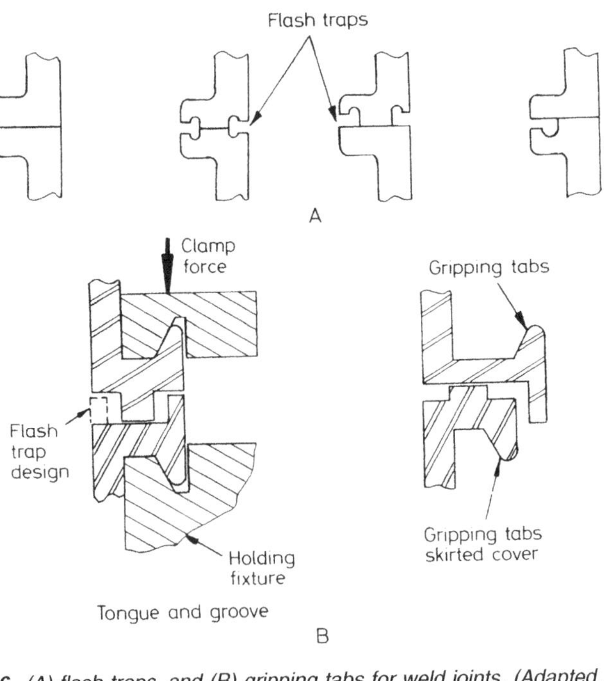

FIGURE 15.16. *(A) flash traps, and (B) gripping tabs for weld joints. (Adapted from Ref. [9].)*

If external appendages (e.g., ribs and tubes) project from the part, the fixture must be supportive to avoid their distortion or failure during vibrating. Parts should fit securely in their fixtures. Large parts should be keyed from a boss or center recess for correct placement in the fixture. Try to avoid keying parts off of their extreme edges. The tolerances for vibration welding assemblies are as follows:

1. Meltdown at the joint surfaces should be an average of from 0.016 to 0.032 inches, depending on the flatness of the joint and the resin used. The finished weld height should be within 0.004 inches or less of the weld depth.
2. After welding, part alignment accuracy should be 0.010 inches or less in the direction of the vibration.
3. In the direction perpendicular to motion, an accuracy of 0.005 inches is desired.

The variables to be considered in vibration welding are as follows:

1. Fixtures holding parts with projections must be supported during welding.
2. Part must be warp free and flat; the weld joint must be strong and stiff enough to withstand high-frequency vibration without buckling.
3. Frequency, time, and pressure must be accurately controlled to obtain a good weld in the same position every cycle.
4. The joint surface should be clean and dry, and resins within melting temperature range.
5. External ribs on the joint line can be welded but must be in the direction of the vibration.

Vibration welding can be automated for high-volume part fabrication using an in-line or rotary system. Typical welding cycles from load to unload range from 5 to 15 seconds.

Additional Ultrasonic Operations. Ultrasonics is very versatile for secondary operations. Its use can eliminate mechanical fasteners, permanently mount components, and reduce labor and assembly costs. The quality of the welds is controllable to tight tolerances. Ultrasonics is also used to install threaded inserts and head studs, as well as to rivet metal or plastic parts together. It is also used for spot welding, swaging, and forming operations.

Ultrasonic Inserting. The advantages of ultrasonic threaded inserts as compared with molded-in inserts are as follows:

1. Elimination of wear and damage to the mold.
2. Elimination of preheating and hand loading of inserts.

3. No molding cycle interruptions.
4. Dimensional tolerances of inserts can be adjusted to suit mating parts.
5. Greatly reduced boss stress and equivalent or better pullout strength.
6. No shoulder interference problems—inserts are flush with part surface—no flash at threads.

Inserts can be installed using near- or far-field welding, as shown in Figure 15.17, and they are locked into the part or boss when melt flows into the undercuts, flats, or knurling. This prevents pullout and rotation under torsional loads. A shear joint on the head of the insert captures and forces the melt into the undercut.

Ultrasonic Heading, Riveting, and Staking. Ultrasonic and heading or riveting uses a molded-in or a separate pin rivet of the same material to join plastic and metal assemblies (see Figure 15.18). The shear joint technique is used for both amorphous and crystalline resins to develop maximum strength. When a blind hole is staked, the air must flow through the stud or the hole. Stud welding requires precise center-to-center placement for good joint strength. If stack-up tolerances might be a problem, consider placing molding pins on the part and use staking with wider tolerances for the holes on the mating part. Flush or headed staking achieves almost the same joint strength without the tolerance problems.

Spot welding using ultrasonics (see Figure 15.19) is not recommended for crystalline polymers. There are no energy directors on the part to ensure that melting and bonding will occur. Amorphous resins, with their lower melting points, are better suited. Mating parts must be free of surface contaminants for this bond to be effective. The joint strength is marginal at best, particularly when compared with other ultrasonic joints.

Swaging or forming is similar to staking. It is used to capture or retain another component in place on the plastic part. If swaging or staking is used,

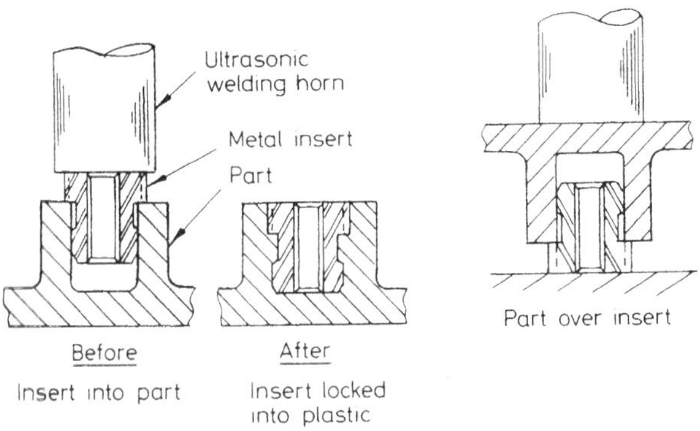

FIGURE 15.17. Welding of ultrasonic inserts. (Adapted from Ref. [4].)

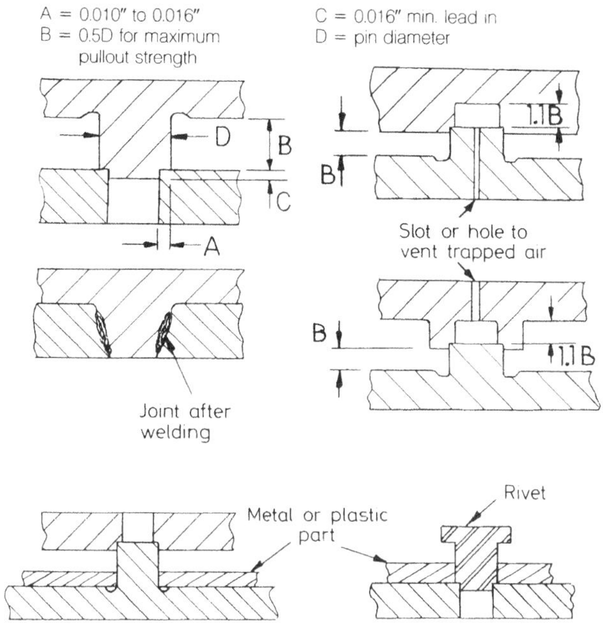

FIGURE 15.18. Ultrasonic staking and riveting. (Adapted from Ref. [4].)

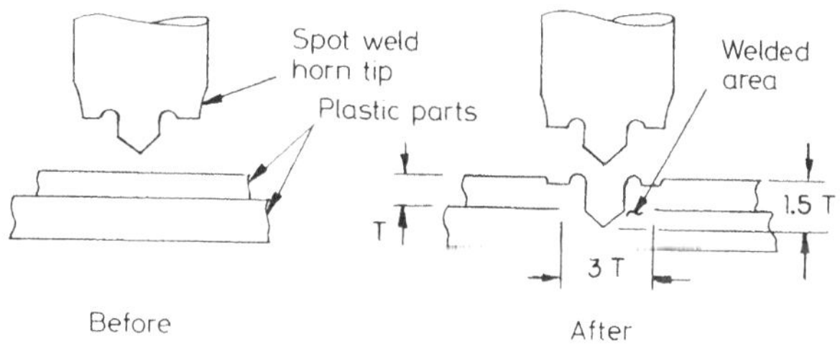

FIGURE 15.19. Spot welding using ultrasonics. (Adapted from Ref. [14].)

the designer must be sure that the encapsulated part will not be damaged if it comes in contact with the welding horn (see Figure 15.20).

Staking yields a strong, stress-free attachment created by the melting action of the horn. There are other stake profiles to choose from, including spherical, hollow, knurled, and flush. Radii should be used at the stud base to reduce stress concentrations at the joint. The staked part should also have a radiused hole on the surface to reduce the notch effect.

Induction/Magnetic Bonding. This technology involves induction and/or magnetic bonding of mating parts through the use of a foreign medium. The

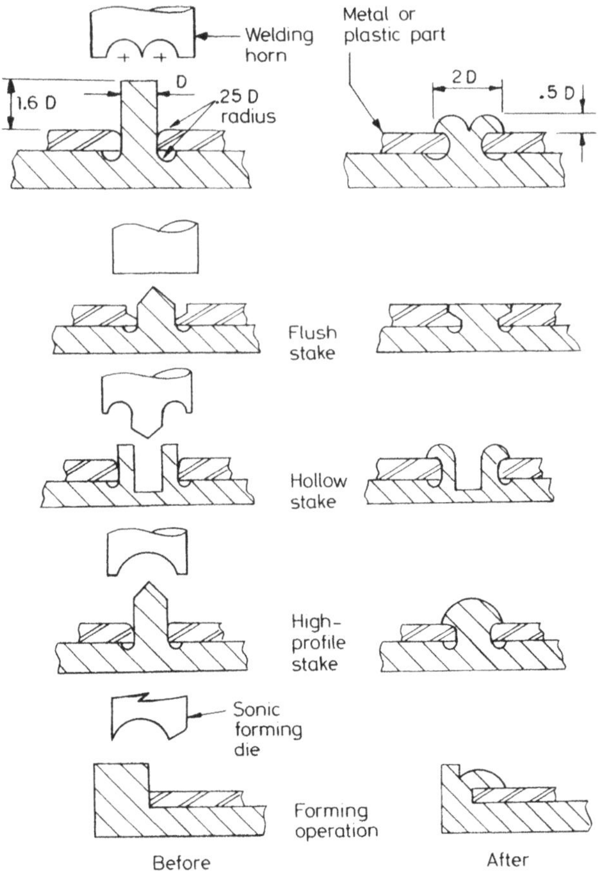

FIGURE 15.20. Ultrasonic staking and swaging. (Adapted from Ref. [14].)

medium melts or causes melting and fusing of the contact surfaces, as shown in Figures 15.21 and 15.22. Material compatibility for this assembly technique is shown in Table 15.2. Induction bonding involves inserting either a wire, mesh, or a compound in liquid, paste, or solid form filled with micron sized metal particles at the weld joint. When subjected to an electromagnetic field or direct current, the insert heats and causes the surrounding resin to melt and flow together at the contact surface. Simultaneously, the joint is subjected to a load of 100 PSI or greater to create a good bond. Because the heat is localized, cool down takes only 2 to 4 seconds.

Commercial compounds are available in extruded profiles, tapes, strips, or custom injection-molded shapes for creating good induction bonds. The sealant is placed at the interface. When exposed to an oscillating electromagnetic field under pressure, it melts, flows, and fuses to the adjoining surfaces.

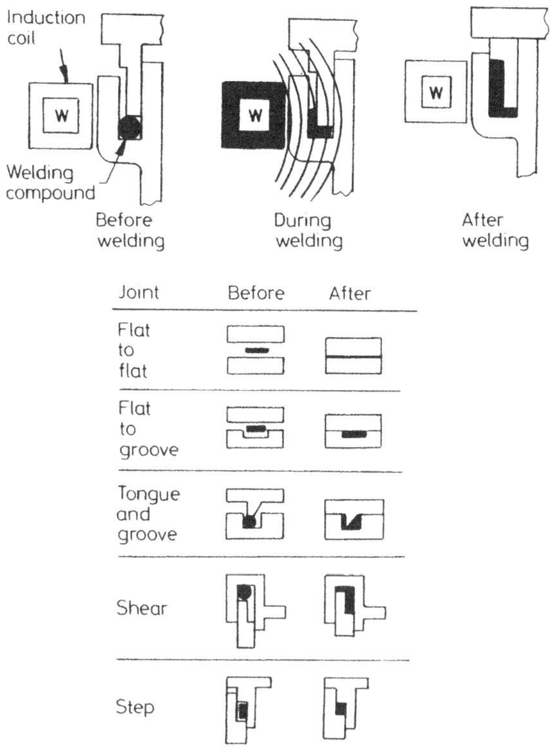

FIGURE 15.21. *Induction welding and joint design. (Adapted from Ref. [2].)*

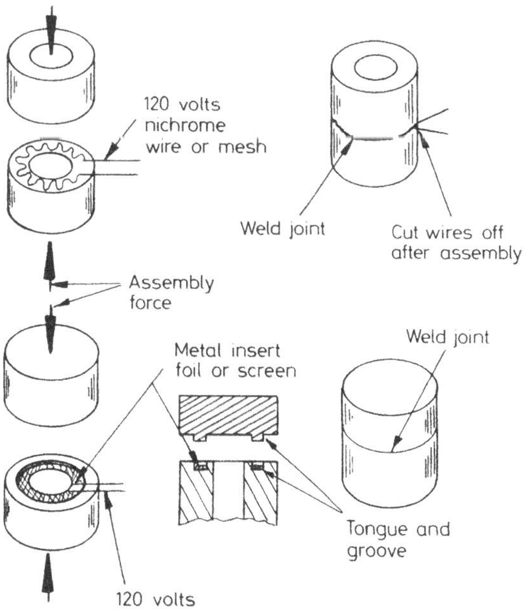

FIGURE 15.22. *Wire or mesh welding technique. (Adapted from Ref. [1].)*

Table 15.2. Bonding Plastics to Plastics with Electromagnetic Materials.

	A	B	C	D	E	F	G	H	I	J	K	L	M	N	O	P	Q	R	S	T	U
A ABS	X	X		X				X						X		X					X
B ABS/PC Alloy	X	X		X				X					X								X
C Acetal			X																		X
D Acrylic	X	X		X				X								X					X
E Cellulosics					X																X
F Modified PPO						X						X									X
G Polybutylene							X														X
H Polycarbonate	X	X		X				X					X	X							X
I Polyethylene									X									X			X
JUHMW-PE										X											X
K Polypropylene											X								X		X
L Polystyrene						X						X				X					X
M Polysulfone		X						X					X								X
N Polyurethane	X							X						X							X
O PVC															X						X
P SAN	X			X								X				X					X
Q TPE																	X				X
R TPE									X									X			X
S TPE											X								X		X
T TP Polyester																				X	X
U Paper	X	X	X	X	X	X	X	X	X	X	X	X	X	X	X	X	X	X	X	X	X

Source: Adapted from Ref. [12].

Note: X indicates compatible material bonding combinations.

Structural, hermetic, and pressure-tight welds are possible on parts that have bond-line lengths up to 20 feet long. A feature of this system is the bond line can be reactivated, with separation or disassembly possible for repair of components within the sealed area.

With the wire or mesh technique (see Figure 15.22), the weld is localized. For the best results, the entire surface area should melt and fuse. This technique is only recommended for parts requiring minimum bonding strength.

Hot-Plate Welding

Hot-plate welding is one of the oldest assembly methods for joining like plastics. It is losing its place as a viable and economical method because it is slower and not as easily automated as other techniques. One major drawback is the exposure of the joint to oxygen degradation during assembly. The amorphous polymers are better suited than crystalline polymers for hot-plate welding.

In hot-plate welding, the joint area is heated on a Teflon-coated surface for a specific time period and under minimal pressure until softening occurs. The two parts are immediately pressed together until solidification occurs, as shown in the welding sequence in Figure 15.23. The temperature of the heating surface and clamping pressure must be precise to be sure that joint tolerances are achieved and a weld bead is formed. The process can be automated. Some machines hold the parts in their heating and positioning fixtures by vacuum pressure. Irregularly shaped parts can be assembled with a single joint plane. Tolerances for the molded parts are critical and any warpage will create nonuniform welds and variable joint strength. Welding variables are heating surface temperature and time, clamp pressure, compression travel stroke, and clamp time. A joint is as strong as its parent material; reinforced resins are not recommended for this form of assembly.

Focused Infrared Melt Fusion

Focused infrared melt fusion is a new welding technique that uses a reciprocating, focused, and temperature-controlled beam of infrared thermal radiation over the joint surface to create melt without contact with the joint surface. When the predetermined melt temperature of the joint surface is reached, the heating source is removed and the parts are clamped together. This technique, which is shown in Figure 15.24, can be automated for an assembly line.

This technique can control the melting temperature of each half of the joint surface to allow welding of thick/thin and dissimilar materials. Each half is brought to its respective melting temperature simultaneously through the use of the infrared beam's temperature sensing control.

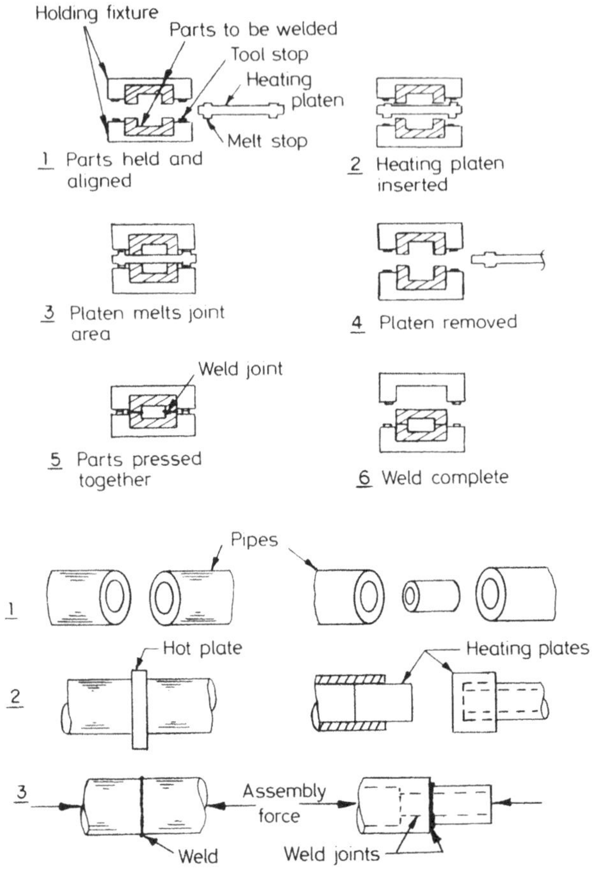

FIGURE 15.23. *Hot-plate welding sequence. (Adapted from Ref. [12].)*

Cycle time, including heating and cooling, depend on the plastic being welded and the degree of automation employed. Weld joints from as narrow as 1/16 to as wide as 48 inches and up to 8 feet long are possible.

Joint strength depends on the design chosen—usually a lap joint. Even if the plastics are not the same or chemically compatible, some bonding can be achieved. This process can weld a wide range of materials, from styrene to advanced composites, including materials of varying stiffness.

The key variables for this technique are as follows:

1. The surface temperature for each material selected must be set on each temperature controller.
2. Press cycle time, robotic stroke distance, and speed must be set.
3. Press-close distance, clamp pressure, and clamp time must be set.

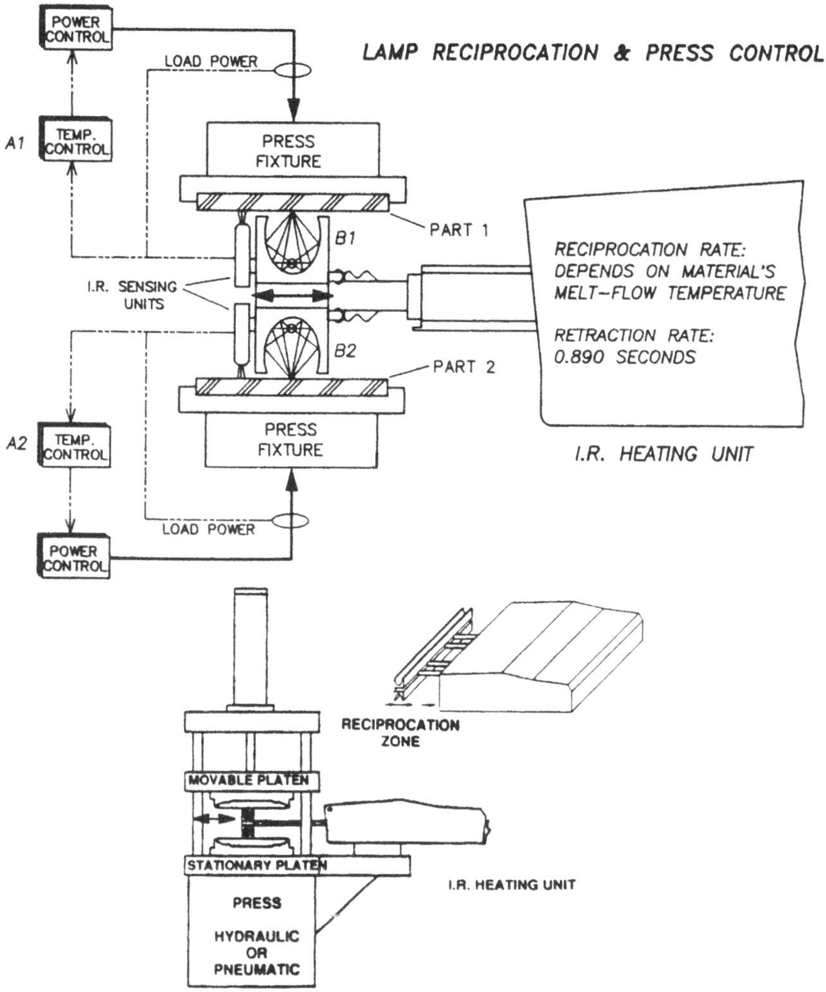

FIGURE 15.24. *Focused infrared melt fusion. (Courtesy of Branson Ultrasonics Corp.)*

This technology is now available from Branson Ultrasonics Corporation, Danbury, CT. It has exclusive worldwide rights to the focused infrared welding technology from the Entwistle Company and the inventor, Henry Swartz.

Cold or Hot Heading

Cold or hot heading is a fast, economical method for permanently joining plastic or plastic/metal parts. Similar to ultrasonic methods, it involves permanent deformation of a rivet or stud. This creates a locking head similar to

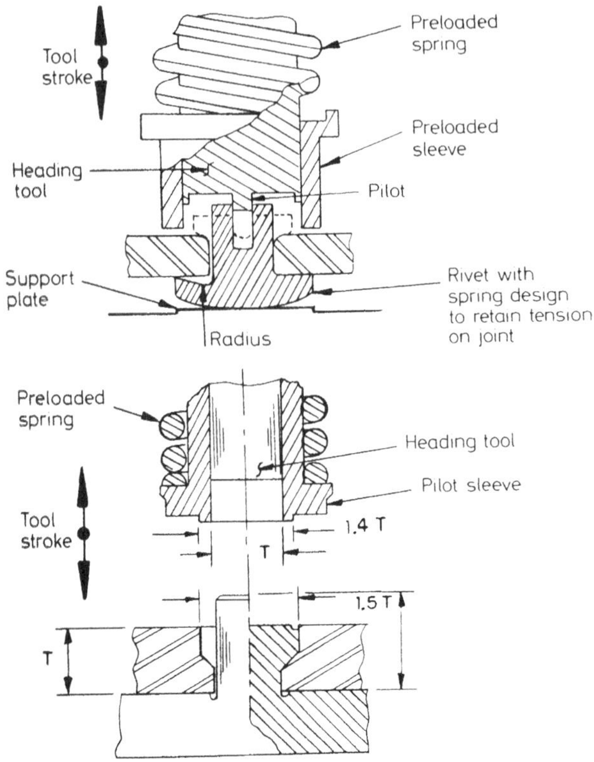

FIGURE 15.25. Heading operation and tool. (Adapted from Ref. [4].)

staking or swaging at either room or elevated temperatures by exceeding the elastic limit of the material. The heading temperature should reflect the temperature of the part's end-use environment, as some degree of recovery will occur if the part sees elevated service temperatures. Amorphous resins are more easily headed than crystalline resins. Examples of heading tools are shown in Figure 15.25.

The recommended length-to-diameter ratio for a good head is 2:1, as this reduces the column buckling effect with longer studs. In cold heading, the loading rate is critical because the resin must flow without fracture. When hot heading is used, the loading rate can be increased. A new method employs hot-air heating of the stud, whereas a cold tool forms the head. This reduces stress cracking, polymer adhesion to the die, and material stringing.

A radius at the base of the stud and on the edge of the circumference of the joined part is recommended. This reduces the notch effects and high stresses that develop at these points. If problems occur with cold or hot heading, the designer can fall back on ultrasonic heading to obtain a tight stressfree headed connection.

Mechanical Fasteners

Mechanical fasteners are composed of metallic screws and inserts. They are used to provide permanent or multiple assembly/disassembly of the plastic or plastic/metal or other material assembly. They are expensive but are used when the design requires a stronger attachment, repeated accessibility to internal components, or part design or use will not permit other forms of assembly.

Screws. Screws are the most common fasteners for assembling plastic parts. They are used for permanent assemblies and when access is required for repair. Selection of the wrong type of screw often leads to assembly, service, or quality problems. Screws are usually secured in a boss on the part or a mating assembly. Screws affect how threads are formed and the resulting holding force. Mechanical fasteners that use self-tapping screws are either thread-forming or thread-cutting. These screws are economical and the joint can be strong, problem free, and tight over the life of the part.

Thread-Forming/Thread-Cutting Screws. Thread-forming screws deform the plastic into which they are driven and form threads in the part. Thread-cutting screws physically remove material, like a machine tap, forming the screw thread. To select the right type of screw, the designer considers the resin's modulus of elasticity, which is a direct function of the material's elongation. Table 15.3 provides guidelines for selecting the correct screw. As noted, only the lower modulus materials are selected for thread-forming screws, as they impart considerable hoop stress into the boss area.

The different type of screw thread designs are shown in Figure 15.26. Each type and thread provides a unique assembly or holding benefit. The quality of the screws used must also be consistently maintained by the supplier. The plastic part must not have to absorb variances in screw quality.

The selection of thread configuration depends on the material and ultimate part function. The U-type screw is for permanent assembly. Other types, AB, B, and BP, are removable, but thread damage may occur. Thread cutting screws are not recommended for removal. If on reinsertion the threads are not lined up, accurately they may recut the threads and ruin the part. When

TABLE 15.3. Guidelines for Selecting Screw Types.

Resin Modulus of Elasticity	Screw Type	Thread Configuration
Up to 200,000	thread forming	AB-B-BP-U
200,000 to 400,000	thread forming	Trilobe-Hi-Low
	thread cutting	L-BF-BT
400,000 to 1,000,000	thread cutting	Hi-Low B-F-G-BF-BT
1,000,000 up	thread cutting	Type T

Source: Adapted from Ref. [4].

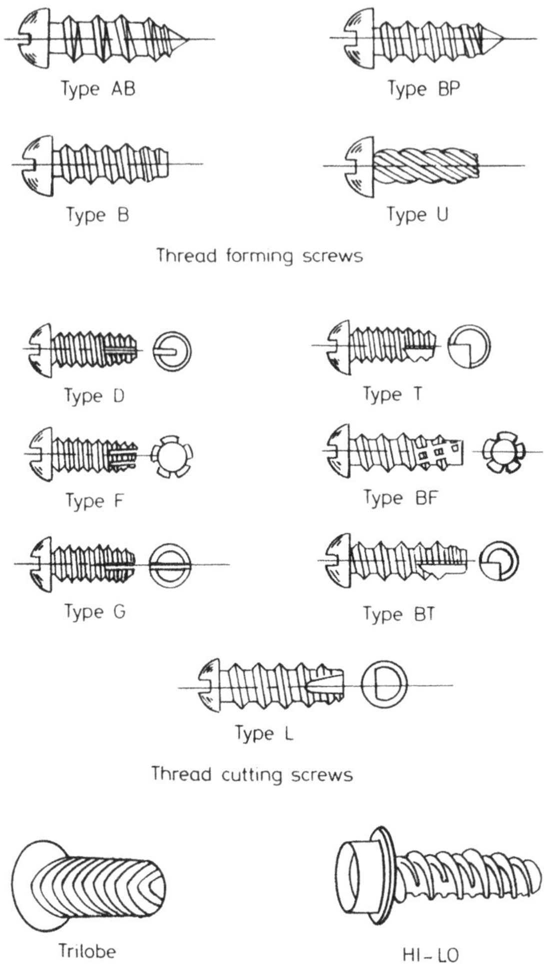

FIGURE 15.26. *Screw thread configurations. (Adapted from Refs. [1] and [4].)*

the type T screw is used and removed, the threads will always be destroyed. Reassembly is only possible by using the next larger screw size. If screw removal is anticipated, the boss diameter must be increased to handle the next larger size. If repeated assembly and disassembly are contemplated, always use a threaded insert to permit repeated removal.

The guidelines for self-threading screws, as shown in Figure 15.27, are as follows:

1. Thread engagement length should be 2.5 times the screw diameter.
2. Hole diameter should be based on 50 to 70 percent thread engagement. This can vary with type of fastener and resin.

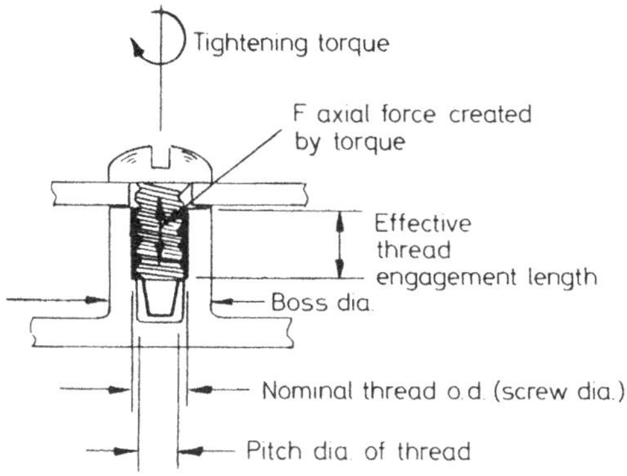

FIGURE 15.27. *Self-threading screw. (Adapted from Ref. [2].)*

3. Holes should be counterbored (preferred) or chamfered to aid alignment and reduce cracking.
4. Boss diameter should be 2.5 times the screw's diameter for best performance. If disassembly is anticipated, use 3 times the screw's diameter. Allow space in the bottom of the boss for thread-cutting screws to deposit debris.
5. Strip-to-drive torque should be at least 3 to 1 for hand assembly. With power tool assembly, 5 to 1 is preferred to reduce stripping the threads if torque cutoff settings go out of calibration.
6. Special screws for plastics, such as Trilobe® and Hi-Lo Plus®, cost more but give improved performance.
7. Screw manufacturers and material suppliers can make recommendations for your particular application.

The strip-to-drive torque calculation is important. During assembly—either by hand or power tool—the screw is tightened to the correct torque without stripping.

When a self-tapping screw is tightened, it produces a torque-to-engagement curve (see Figure 15.28). As the screw penetrates the plastic up to point A, the driving torque slowly increases. At point A, the head of the screw seats. Any further tightening to point B is used to torque the threads into the plastic for a strong attachment. If torquing continues to point C, the plastic yields and the threads begin to shear. From points C to D, the threads strip and the fastener fails. It is important to reach, but not exceed, point B. Workers need to be properly instructed not to exceed this torque level, and power tools must be calibrated to cut out when the drive-torque setting is reached. This can be calculated and verified by running prototype tests on bosses or flat plaques molded in the plastic selected for assembly.

Because a screw's strip torque is so important, it can be calculated at point C by using the following equations.

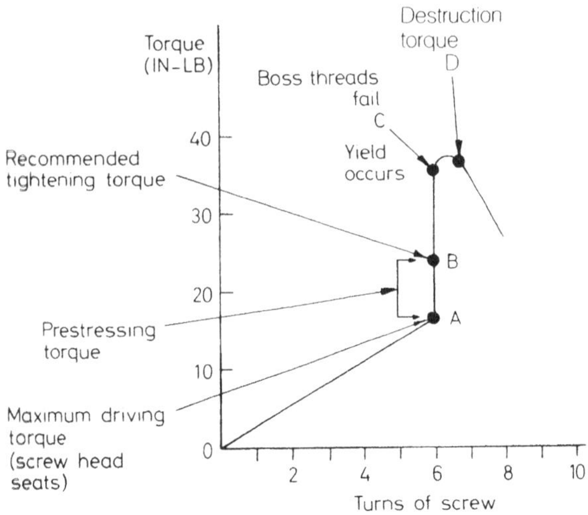

FIGURE 15.28. *Screw torque-turn plot. (Adapted from Ref. [4].)*

Stripping Torque:

$$T = F_r \frac{(p+2f_r)}{(2_r + f_p)}$$

where;

T = torque to develop pull out force
F = pullout force
r = pitch radius of screw
p = reciprocal of threads per unit length
f = coefficient of friction

Pullout force:

$$F = SsA = Ss(3.14)DpL$$

where:

F = pull out force
Ss = shear stress (S_t/1.732)
St = tensile yield stress of resin
A = shear area = (3.14) × Dp × L
Dp = pitch diameter
L = axial length of full thread engagement

As with other assembly techniques, elevated temperatures can cause creep, and testing will verify whether the threaded fastener will provide the required holding strength. Should testing prove the self-threading screws are not adequate, then use expansion, molded-in, or ultrasonic inserts. Threaded inserts should be used if frequent disassembly of the part is anticipated.

Inserts. Threaded inserts provide additional holding force with metallic machine threads and can be molded-in, ultrasonically inserted, or pushed and expanded into a plastic part. Inserts provide one-side mechanical joining and resist creep better than screws. They are permanent attachments to a plastic part. Different types of metal inserts are shown in Figure 15.29, with knurled or fluted outer surfaces that resist pullout and torque forces from the mounting screw.

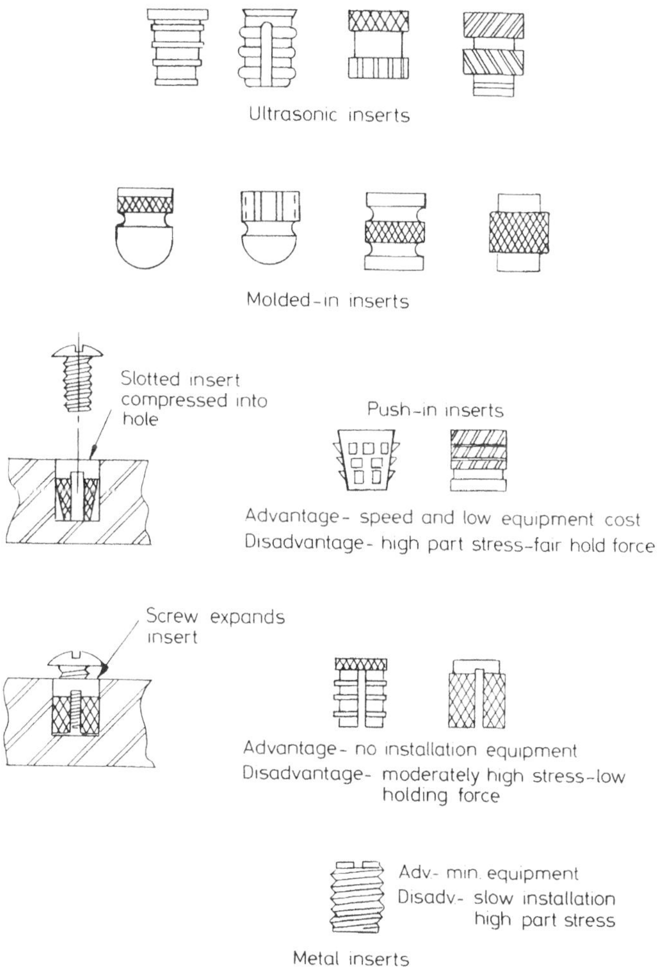

FIGURE 15.29. *Types of inserts. (Adapted from Refs. [1] and [4].)*

The molded-in and ultrasonic type uses the melted plastic, as it flows into the undercuts, as the retention device. The mechanical push-in type expands as the screw is driven in, thereby exerting holding pressure against the side wall of the part. The pressed-in mechanical inserts provide only moderate holding force but are easy to assemble with minimum installation equipment. The major drawback is the moderate-to-high stresses they apply to the plastic part. This can cause cracking of the part. The boss diameter for mechanically installed inserts must be 2.5 times the insert's diameter, and the plastic must be able to withstand the high-hoop stresses induced when the insert is expanded. Your material and insert suppliers can guide in the selection of the correct insert for your particular application.

Adhesive and Solvent Bonding

Adhesive systems are now more acceptable for bonding and assembling plastics both to themselves and other substrates. By eliminating the mechanical fasteners, part designs can allow more uniform wall sections, eliminate bosses, and reduce warpage. The elimination of holes and the resulting weld lines will also strengthen the parts. Adhesives allow stress and impact to be more evenly distributed throughout the part and withstand fatigue stress better. They can bond to dissimilar substrates and usually do not distort the surface area.

Adhesives lend themselves to robotic assembly better than mechanical fasteners, and creep is a minor factor. The biggest drawback is that the joints are permanent and not accessible for maintenance.

Improved hot-melt and water-base systems are now available. They overcome the problems with solvent systems that required special handling of air and control systems to protect workers and the environment.

The eight basic adhesive systems are listed in Table 15.4. Within each family, there are different suppliers for specific and generic material bonding. Some are designed for specific materials and temperature ranges, different cure times, and varying methods of application, from single to multicomponent systems.

A multitude of simple joint designs are possible when using adhesives (see Figure 15.30). The design depends on part geometry, bond/joint strength requirements, and aesthetics. Joint stress occurs in four ways: tensile, shear, cleavage, and peel (see Figure 15.31). In service, the joint should be designed to experience all four.

Tensile stress joints (Figure 15.31A) seldom experience all the force perpendicular to the joint. Shear (B) and cleavage (C) stresses are often associated with the joint design. Shear stress is parallel to the joint plane and concentrates the load at the ends of the lap joint. The bending effects can be relieved by reducing the stiffness of the adhesive with flexible doublers or rubber modifiers, and by limiting part motion at the joint interface. Cleavage stress concentrates the total tensile load at one edge, leaving the other edge unstressed. The joint design should be modified to transfer the stress to a more

TABLE 15.4. Adhesive Systems.

		Physical Properties		User Factors	
Adhesive Key No.	Type	Tensile Strength (PSI)	Peel Strength (PSI)*	Cure Time to Handle	Price/Liter
1	Acrylic, methyl methacrylate	3000–5000	20	5 min.	E
2	Acrylic, other monomer	3000–5000	25	30 sec.	E
3	Epoxies, room temp. cure	3000–5000	3	5 min.	L
4	Epoxies, heat cure	3000–5000	3–25	5 min.–72 hours	M
5	Cyanoacrylates	3000–5000	2–3	5–10 sec.	VE
6	Urethanes, with solvent	1500–2500	25–40	4–24 hours	M
7	Urethanes 100% solids	1500–2500	25–40	5 min. 24 hours	M
8	Solvent	1500–4500	5–20	2–5 min.	L

*lb/linear inch.

Cost factor: VE = Very expensive, E = Expensive, M = Moderate, L = Low cost.

Selection Guide to Adhesive.

Substrate, Plastic	Recommended Adhesives*
Acrylic	1, 6, 8, 5
Glass reinforced TPs	1, 2, 4, 6, 7, 3, 1, 5
Nylons	1, 6, 7, 5
Polycarbonates	2, 1, 5
Polyesters, T/S	1, 4, 6, 7, 5
Polyolefins	6, 1, 5
Polyurethanes	6, 7, 5
Styrenics, incl. ABS	1, 6, 4, 8, 5
Vinyls, flexible	1, 8, 5
Vinyls, rigid	1, 6, 8, 5
Substrate, Nonplastic	
Ferrous metals	2, 1, 7, 6, 4, 3, 5
Nonferrous metals (except copper)	2, 4, 7, 6, 1, 3, 5
Copper and alloys	4, 7, 6, 2, 3, 5
Glass	2, 4, 3
Wood	1, 2, 6, 7, 3, 5

Source: Adapted from Refs. [8] and [11].

*Listed in order of preference, toughness being the most preferred property. To pick adhesives for dissimilar substrates, select only from the matching key numbers listed for both materials (For styrenics to steel, for example, select 1, 4, or 5.)

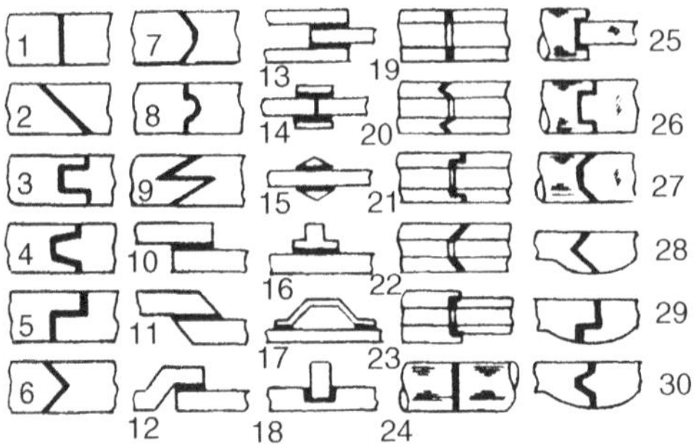

1. Butt joint
2. Scart joint
3. Square tongue-and-groove joint
4. Angled tongue-and-groove joint
5. Half-lap joint
6. "V" joint
7. "V" type joint with a flat
8. Round tongue-and-groove joint
9. Double-scarf lap joint
10. Simple lap joint
11. Tapered simple lap joint
12. Offset lap joint
13. Double-lap joint
14. Double-strap joint
15. Beveled double-strap joint
16. "T" section joint
17. Hat section joint
18. Recessed right angle joint
19. Tubular butt joint
20. Tubular "V" joint
21. Tubular half-lap joint
22. Angled, tubular, half-lap joint
23. Tubular lap joint
24. Rod-butt joint
25. Tongue-and-groove joint in a solid rod
26. Landed tongue-and-groove joint in a solid rod
27. Scarf tongue-and-groove joint in a solid rod
28. "V" type joint with increased bonding area for additional strength
29. Half-lap joint with increased bonding area for additional strength
30. Angled tongue-and-groove joint with increased bonding area for additional strength

FIGURE 15.30. *Adhesive joint designs. (Adapted from Ref. [1].)*

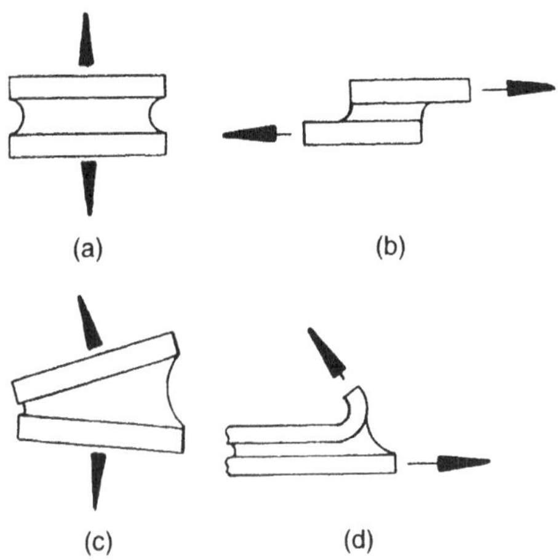

FIGURE 15.31. *Stresses on adhesive joints. (Adapted from Ref. [11].)*

favorable orientation in line with the joint interface. Peel (D) stress applies the joint forces to a thin line between the adhesive and part. To counter this, widen, stiffen or recess the peelable end member.

Adhesives come in different base resins and viscosities. Cure times may allow final adjustments and positioning of the part before full adhesion/bond strength is affected. Using adhesive bonding for prototype part assembly, to prove out a joint design, allows the designer some latitude before the primary assembly technique is selected.

Adhesives can also solve difficult assembly problems, particularly if screws, snap fit, ultrasonic welding, or other assembly techniques are not possible and a strong hermetic seal is required. Adhesion bonding can allow complicated multipiece parts, some even with internal seals, to be assembled easily, quickly, and even automatically.

Joint designs similar to spin and sonic welding can be used with flash traps to contain any excess adhesive within the joint area and not spoil the aesthetics of a visible part.

The part tolerances for adhesion bonding are the same as for the other assembly techniques. Keying features can be molded-in to orient or position internal parts correctly. The parts may be placed in a fixture until final curing of the adhesive occurs but with some adhesive systems the parts can be handled before joint strength is attained.

Some adhesives require part surface preparation or primers, but many new systems do not require this step. Urethane and epoxy adhesives tolerate surfaces that are contaminated with oil, but as a rule joints should be kept clean and dry for best results. Most adhesives exhibit good-to-excellent chemical resistance and can survive steam and hot water emersion.

Each system has its own cure medium—be it air, oven, ultraviolet (UV), infrared, thermal heating—and the system selected should be based on the following requirements:

1. Application method of adhesive
2. Secondary material (nonplastic)
3. Joint strength (tensile, shear, etc.)
4. Environment (hot-cold chemicals)
5. Cure time (shelf life of system)
6. Fixture required
7. Cure method
8. Automated or hand assembly
9. Primary plastic material
10. Worker/environment consideration

Some adhesives come in preformed shapes, and some are single or dual systems that mix at the dispensing head. All have varying shelf life after mixing and this must be considered when selecting the adhesive systems.

As with any assembly technique, consult with adhesive and material suppliers to be sure that the design, material, and adhesive system will accomplish the end-use requirement.

These 10 methods are the principle techniques for assembling plastic parts. The quality of the assembly begins with the designer's needs and ends with meeting customer specifications. The design must be adaptable to the mold being built for production. All must go together to produce a quality part.

16

Decorating Considerations

Plastics lend themselves to decoration for visual appeal, information transfer, and part protection. It enhances and adds value to the finished part. By using internal or external agents and decorating media, the part can be protected from the natural elements and chemicals. Some systems can also mask or reduce abrasion and wear effects, so the part will retain its value longer in the marketplace.

A part's appearance is a function of the plastic used, the surface finish of the mold cavity, and the processing conditions under which it was produced. If the processing conditions are tightly controlled, most plastics have a surface finish that duplicates the mold cavity finish. These conditions are primarily melt and mold temperature plus injection and packing pressure.

CONTROL OF THE PROCESS

Controlling the finish is easier with plastics that have not been modified with fillers, flame retardants, and plasticizers. These additives and modifiers can migrate to the surface during molding and produce an undesirable appearance. If processing conditions and internal colorants do not mask these surface defects, then a secondary finishing operation may be necessary.

The base resin also has a pronounced effect on the part's appearance. Some resins will always reproduce the finish of the mold cavity; others have to be

Total Quality Process Control for Injection Molding, by M. Joseph Gordon

processed under rigid conditions to obtain the desired effects. The designer must know these conditions and the base resin's natural color, if color matching is desired. Some resins have a neutral base color (e.g., crystal, white, cream, tan), while others are darker (brown to dark gray). Darker colors make color matching difficult to impossible for some systems.

The mold builder must also know the manufacturing conditions, so that shrinkage can be accurately calculated in determining the cavity temperature. This temperature has a definite effect on the surface finish of some resins. Generally, for crystalline resins, the hotter the cavity surface, the better the finish. The resin is drawn to the surface and buries the effects of fillers in the material. This also results in higher material shrinkage in the mold. Shrinkage rates are calculated under standard molding and mold temperature conditions for the base material. Most amorphous resins use cold molds for good surface appearance. Each resin system has its own natural colors and surface effects. Because all may not produce the surface appearance desired, a primary or secondary decorating operation may be required.

DECORATING TECHNIQUES

Primary decorating occurs in the mold cavity with pigmented resins, surface finish techniques, and information molded into the plastic part. Secondary decorating is adding value to the semifinished part for information enhancement, aesthetic value, and part protection. It includes painting, printing, labels, and plating. Plastics adapt to most decorating techniques as long as the technique is anticipated in the design stage and proper materials are selected to achieve the desired effects. Tight quality control must be maintained during production and final decorating. An error at this point usually results in scrapping the finished part.

Primary decorating is preferred, as it is reproduced accurately in the mold during each cycle. Secondary decorating involves more variables and equipment monitoring that can affect part quality. Because all decorating cannot be done in the mold, these decorating conditions must be tightly controlled. The designer must also be aware of these decorating requirements and specify them on the part drawings. To gain insight into the quality control conditions required for decorating, the manufacturing and decorating variables must be understood.

Color is the primary decorating medium for plastic parts. It offers part identification and a visual effect. Colors add value to a product, hide surface effects, provide protection, and reduce signs of wear. Most parts use colors compounded into the plastic resin.

The second most frequently used decorating technique is molding information on to the plastic part in the form of printed information, company logos, symbols, and user instructions. This information may he raised or depressed on the part surface. Later, in a secondary decorating operation, it can be highlighted with inks or foils of contrasting color to make it more noticeable and readable.

Other secondary decorating techniques add information to parts with painting, printing, silk screening, labeling, and metalizing. For these methods, the part's surface is more critical and the quality depends on the decorating method employed.

Information and product identification can be enhanced by different colors and surface textures that draw the customer's eyes to this area. Information in a potential wear area can also be protected by recessing it on the part. This is often not considered, as it is more expensive.

The part designer should consult the customer to find out what decorating or information may be required on the part. This information should then be specified on the part drawing. At this time, any special color matches or unique decorating techniques can be explored to be sure they are compatible with the part design and materials specified. Not all decorating techniques are compatible with every plastic. The base color of some resins may not permit the use of all colors, and some materials, such as the polyolefins, may require special surface treatment for the decoration's adhesion. All these questions must be answered before the drawings are released for tooling.

SURFACE PREPARATION

An area often overlooked is protection of the part surface before decorating. Plated and painted plastic parts must be produced as stress free as possible, so that the decorative secondary finish will not crack or craze.

With secondary decorating methods, the part surface must be dry and free of contamination, oils, natural skin greases, dust, mold release agents, and in some cases, resin processing aids. Their presence can severely affect the adhesion of the decorating medium.

If parts cannot go directly to the decorating line, they should be protected. Operators handing parts should wear white cotton gloves ,which are changed if contaminated. The mold cavity should also be kept free of mold release and oils, although there are some mold releases suitable for use on decorated parts. If mold releases are necessary, it is better to correct the mold cavity problem that requires their use. If in doubt about surface cleanliness, the parts should be degreased before reaching the decorating line.

Some plastics, such as the polyolefins, require surface preparation for good adhesion of the decorating medium. The decorating medium—ink, paints, and adhesives—must have what is known as "wetting" action on the part's surface. For good adhesion, the surface tension of the decorated part must have a surface energy at least 10 dynes/cm greater than the decorating medium. If this difference is not attained, the adhesives and water-based inks, which have a surface energy higher than polyethylene's surface energy, cannot "wet" the surface and develop good adhesion. Therefore, the surface must be specially prepared or changed to develop good coverage or adhesion. If this is not done, the ink will form spherical droplets on the surface and the adhesives will not stick. The surface energy numbers of resins are available from suppliers and must be investigated before a decorating technique is specified.

Several pretreatment methods can be used to develop high-quality decorating results. These are based on the material to be decorated, decorating medium, part shape and configuration, volume of parts to decorate, and pretreatment method selection. All pretreatment processes create a chemical change in the surface structure of the polyolefin part. This results in an increase in the surface energy of the treated part and promotes the necessary bonding sites for the decorating medium.

There are basically six pretreatment surface preparation techniques that can be employed. All but mechanical abrasion cause a chemical change on the part's surface. The techniques are:

Mechanical abrasion
Flame treatment
Chemical etching
Corona treatment
Plasma treatment
Electrical surface treatment

Mechanical abrasion is difficult to control, not recommended for printing surface preparation, and used mainly for small batch runs of parts. Techniques used are sand blasting and emery-cloth sanding.

Flame treatment is used mainly on low value parts because of its "deglossing" of the part's surface where increased printability or label adhesion is required. The parts are passed through a gas flame (e.g., natural or butane) at a relatively rapid rate. This can be automated at an initial low cost. It suffers from a low repeatability, high safety and operating costs, and is not considered for high value appearance parts.

Chemical etching is not an attractive choice because it uses hazardous and toxic chemicals. Parts are batch dipped on a timed basis. While preparing the surface, this pretreatment also microcleans the part for other operations. Drawbacks include disposal of the toxic chemicals, personnel hazards, and high operating costs.

Corona treatment is only good for two-dimensional objects but has excellent repeatability, high in-line speed, and good treatment results. It has replaced flame treating in many applications. The setup involves a high-voltage generator and two electrodes. An air flow blows the generated corona "plume" from the electrodes on to the parts. With the correct setup and masks, small areas of a part can be treated. The major drawback is that ozone and other by-products are produced. An ozone treatment system should be part of the equipment package.

Plasma treatment uses a batch method for surface preparation. The system prepares all surfaces to a high surface energy, is highly controllable, and microcleans all surfaces. It is very expensive, requires highly skilled operators, and cannot treat parts online.

Plasma treatment is done in a vacuum chamber with a radio frequency generator, an automatic impedance-matching network, a gas delivery system supplying various gases (nitrogen, argon, tetrafluoromethane, and others), and a microprocessor control unit. The parts are loaded into the chamber, a vacuum, is developed at high temperature, and the gases are injected under controlled conditions to create the gaseous plasma for surface treatment. The plasma system is used for high-value medical and electronic parts for better wetability for printing and adhesion.

Electrical surface treatment combines the best of all previously discussed preparations. It has three-dimensional and online capabilities with good repeatability. It is used for high-value parts or those needing online, multiple-surface, and repeatable surface pretreatment. The system consists of electrodes that conform to the treatment area, a high-voltage transformer, and a high-frequency generator/impedance matching set of electronics. It produces ozone and an air cleaning system is required. The electrodes must also be custom shaped for each application, similarly to electronic discharge machining (EDM) electrodes.

Generally, all polyolefins can be treated to obtain good adhesion for assembly, sealing, or decorating. The treatment level is time and the method sensitive, as internal additives can leach to the surface and cause the treatment to become ineffective. Any secondary decorating operation should be performed right after surface treatment to ensure part quality.

Structural foam parts made with internal blowing agents must be allowed to degas for a sufficient period of time before decorating. The material supplier can give you the time required based on the part thickness and blowing agent.

Every type of plastic can be decorated by some method. The designer must plan and know the system to be used. The decorating technique must be compatible with the resin. If in doubt, ask the supplier. Also have the material suppliers' assurance that they will send notification about any formulation changes in the resin, no matter how small, to determine whether affects the system. The same is true for the decorating equipment supplier.

MOLDED COLORS

The most common decorating technique is molded-in color. If the application is large enough in pounds of material, most suppliers will compound a special color-matched product for a part. If this is not possible, the customer can select from standard or prior-matched colors. Certain regulatory agencies have established standardized colors to identify parts. The National Electrical

Manufacturers Association (NEMA) uses specific colors to identify outlets, plugs, cords, light switches, and connectors. Suppliers who have resins that meet these requirements have color matched resins for these color standards.

There are three basic ways to make or obtain colored resins for molding parts. They are compounded colors, blends made with natural resin and color concentrates, known as "salt and pepper" (S & P) blends, and color additives blended with natural materials at the injection molding machine's feed throat by a color metering device. Major resin suppliers, who produce compounded colors in high volumes for major customers, carefully control color quality. This is the tightest obtainable control of color and resin quality. The suppliers also monitor the chemical and physical properties of these materials to ensure that each lot meets customer requirements. This is also done for S & P blends, but color may vary slightly because of blending variance. The material supplier verifiers the color before shipment by molding test plaques and comparing them to color standards under specified light requirements.

The designer must be aware that some pigment or color systems can have dramatic effects on a base resin's physical properties. Some systems can lower a resin's physical properties as much as 20 to 25 percent. This must be determined by a part testing program. If color concentrates are added at the feed throat only, a list of approved and tested color concentrates should be on the part drawing. The molder must be told by the color concentrate material supplier if any pigment or formula changes are made. The customer must then be notified and tests performed to verify that the physical properties of the molded part are not affected. This requires sufficient warning and testing time in case the part also needs specific agency approval. This is extremely important when consumer and personal safety items are involved.

Color concentrates come with a specified pigment loaded in the pellet. It should not vary from lot to lot but it may from supplier to supplier. The pigment loading, the lot number of the concentrate, and the let-down ratio for correct color and property control should be listed on the process setup and control sheets, along with the specified resin for the job.

For lower volume colors, many material suppliers are now using S & P color blends. The color match is just as critical as with compounded colors, and production personnel need to take additional steps to ensure that the part's color is always correct. Therefore, production personnel should reblend them before release into the machine's feed hopper. This is the only way to ensure that the right mix is fed to the machine. There is always the possibility that the color concentrate settled during shipment because of differences in pellet size and density.

Some molders use color concentrates that are blended at the molding machine's feed throat. This can be done by using automatic concentrate feeders by mixing the concentrate with virgin resin. The control of the color quality is then regulated by how thoroughly and accurately the concentrate is mixed and fed into the system.

Most major resin companies have developed specific color matches for their preferred customers. Depending on the color matching agreement, these formulations may be available to other customers. Many molders use color matching companies and color concentrate suppliers to provide color matches and supply compounded colored resin or concentrates for small and medium volume runs.

The use of heavy metal or inorganic pigments because of toxicity and environmental concerns has almost eliminated their use in plastic. Their use is restricted in many countries and for many applications. Cadmium-based pigments offer the best heat stability of the major inorganic pigment systems. But because of environmental concerns, they—along with other heavy metal pigments—are being replaced by less thermally stable and more expensive organic pigments. Depending on where the product is sold and agency regulations, select the right pigment system for your resin. Figure 16.1 lists typical organic pigments and resin processing temperatures. Note how close to the thermal border they are and how processing conditions must be controlled to not burn out the colors. The organic pigments have less heat stability. If the processing group is aware of this condition, it can size the molding machine and set process conditions for temperature, pressure, and residence time that minimize negative effects on the color of the plastic parts.

If regrind is used for colored parts, it should be fed back at low percentages of 5 or less. Repeated heating of organic pigments will cause them to burn out and degrade very quickly. Some pigments degrade in less than 15 minutes at melt temperatures, which causes part color shifts. Regrind testing should always be conducted in the early stages of a program to check out the allowable percentage and to determine whether color or physical properties are affected. If the color is critical, and must match adjoining plastic or painted parts, regrind should not be permitted.

The surface finish of the mold cavity also affects the shade of the color. Most color matches use parts or standards that are glossy or shiny. Tighter color matching can be obtained with these finishes than with dull or texturized surfaces. A texturized part surface will be different. Most color houses have standard texturized plaque molds to check whether the color is correct. Color matching is both an art and a science. An information checklist for color matching is given in Figure 16.2. Information and part samples for mating parts should also accompany the color request.

When buying pigments off the shelf ,the molder must know what additives are in the base resin. Some additives can affect the pigment and cause a color change during processing. Additives, such as copper-based heat stabilizers, may cause color changes in the pigmented resin. The pigment used must be formulated around these additives for the color to remain the same during processing. Always check that the pigment system is compatible with the base resin' additive and processing conditions.

Variable part thickness can also affect part color, especially with lighter colors and if the part is exposed to direct or back lighting. Color definitely

Resin	Process	150	200	250	300	350
Polyethylene	Molding		210	290		
Polypropylene	Molding		230	290		
Polystyrene	Most			260		
PVC	Most		230			
Engineering resins	Most				300	350

Chart shows processing temperatures typically encounted for widely used plastics. Chart below shows thermal stability of acceptable substitute pigments. Adequate colors cannot be achieved for high temperature polymers, particularly for yellows, oranges, and reds. PVC, polyvinyl Chloride.

Pigment Temperature Limits

Color	Pigment	$/LB	150	200	250	300	350
Red	Perylene	50			280		
	Quinacridone	30				300+	
Yellow/ orange	Diarylide	5		230			
	Disazo	20+			280		
	Pyrazolone	20+				300	
	Isiondoline	20+				300	
	Anthrones	20+				300	
	Hansas	20+				300	
	Iron oxide	1		230	280		
Green	Phthalic	10				300	
	Chrome oxide	2					380+
Blue	Indanthrone	35			280		
	Phthalo	10				300	
	Iron	2		230			
	Ultramarine	2				300	
Violet	Carbozole	35		220			
	Quinacridone	30				300+	
	Ultramarine	5				300	
Brown	Iron oxide	1			280		
	Ferrite	1				300+	
Black	Carbon black	0.5				300+	
	Iron oxide	1		230			
White	Titanium dioxide	1				300	
	Barium sulfate	0.5				300	

FIGURE 16.1. *Organic pigments and processing temperatures. (Adapted from Ref. [6].)*

Customer: Contact:
Name and address Name, Position, Phone#
Application: Replacing what material?
Part description ____________ Min. Thickness: ______ Weight: ______
Est. lbs/yr: Est. lots/yr: Could start by:
Date desired: Must have by:
Color name: Base resin:
Color Target Standard Enclosed or Part to Match:
Sample Requirements: Grained Plaques ( ) or other (specify):
Is molding sample required?
( ) Molding sample of pounds will be required. ( ) S&P blend OK.

Color Match Requirements

An exact match requires a flat, glossy target standard. For critical colors use a Master Target Standard. If plastic masters are not available will supplier provide them? If possible, use actual parts, adjacent parts or an assembled tool, and a pellet sample of one to ten pounds of current material.

Color Tolerances—Match ( ) Exact ( ) Close, or ( ) Wide
Production: Exact ( ) Close, or ( ) Wide
Part have to match paint ( ) Other plastics ( ) Polymer
Type ____ If so, are parts adjacent ( ), separated ( )

Light sources: Specify lights the customer will use in judging the color match.
Mark (1) for primary light source, (2) for secondary light source.
( ) Macbeth Daylight ( ) Macbeth Horizon ( ) Tungsten
( ) Cool White Fluorescent ( ) Other: ____________

Other Requirements

Mechanical Properties:	Desired	Required
Lightfastness:	Desired	Required
Pigment restriction:	Desired	Required
UV Required:		
Other comments:		

Saleman:	Date:
Designer:	Date:
Production Manager:	Date:
Purchasing:	Date:

FIGURE 16.2. *Color match request.*

varies under different light sources. A key person should be selected to approve the color match under the selected light source(s), listed in Figure 16.2 (Light Sources), with the customer. Color samples can then be sent to the molder to be verified using the same light source. Color standards should be replaced every 6 months at the molder's plant; they fade with time. The color sample should be protected and not exposed to direct sunlight but stored in a controlled environment until needed for quality checking.

SURFACE FINISH

To be decorated successfully, the surface finish of molded parts must be produced with low levels of molded-in stress, good surface finish with tight core shutoff, and kept clean before and during decorating. Reinforced plastic parts often exhibit a dull or glassy surface even in highly polished molds. With some resin systems, a hot cavity surface will draw resin to the surface, burying the reinforcement and producing a very high luster. If this is not possible, the designer may consider a lightly honed or texturized surface that will not be detrimental to the aesthetic appearance of the part. Some suppliers have textured plaque molds or molded samples that show the standard finishes and resulting colors that are attainable. Texturizing of exposed surfaces can also hide potential design, molding, and tooling problems, as well as weld lines, steel shutoffs, and other part surface defects. A good designer by planning ahead can ensure that parts will have good appearance and reduced molding and material problems.

Plastic parts must be decorated in a clean area. Depending on the technique, a precleaning or static electricity discharging using ionized air to repel dust particles may be necessary. Special equipment is available if these treatments are required. At this stage, ensure that, the decorating will not cause rejection of the nearly finished part. Tight quality control procedures must be followed. Before selecting any procedure, you should consult with the company supplying the decorating medium to ensure that the correct equipment, procedures, and training are available.

PAINTING

Plastic parts are painted for three reasons: to protect the substrate, to enhance appearance, and to give the product a uniform color. Paint coatings provide chemical and ultraviolet (UV) barriers that retard degradation of the base polymer, extend service life, prevent fading, cause surface crazing, and out roughness.

Paint also hides many surface defects by flowing into surface imperfections and presenting a smooth surface when dry. Paints can also reproduce different surface finishes and textures by the addition of fillers and modifiers. Textured paint surfaces give three to five times better surface wear than two coats of flat paint. The wear factor is obtained from the peak and valley effect. Abrasion must wear the paint peaks down before affecting the valley base coat.

Trying to match the color of plastic parts to painted metal parts can be very frustrating, and processing must be exact to control the color after the match is made. Many customers have chosen paint, thereby reducing the material and processing quality problem. For added appearance and protection for parts that receive rough handling, some customers also color the painted part's base resin. Then, if the paint is damaged in service, the visual effects are minimized on the part.

Paint Systems

The two basic paint systems for decorating plastic parts are solvent and water based. They have four basic components: resin, pigment, solvent or carrier, and additives.

The resins are plastics or blends of synthetic and natural materials that are dissolved by the solvent and hold the pigment. The molecular chain length of the resin controls its properties, as follows:

1. Epoxy paints are quite hard.
2. Vinyl paints are close to rubbery.
3. Urethane paints are very durable but can be softened by some oils.
4. Acrylic paints are very oil resistant but brittle.

The pigments are the color of the paint and can come in metal oxides, minerals, and organic dyes. They are used to provide color, UV protection, X-ray shielding, electrical conductivity, and impede some specific environmental processes.

The solvent is a compound, or mixture of compounds, that ensures the paint flows evenly onto a substrate by a selected application method, spray, dip, or flow process, and then it evaporates as the paint dries or cures.

Additives are used as leveling agents, viscosity controllers, wetting agents, anticaking agents, crosslinking agents, and plasticizers. Without these additives, a rough or uneven surface and poor adhesion would occur. Slow curing, poor strength, and coating brittleness could also result. Each paint pair supplier can recommend the correct paint system for the part's plastic resin and end-use application. Paints are really true adhesives; they adhere to the plastic substrate by mechanical and/or chemical adhesion.

There are three basic methods for painting plastic parts—spray painting and flow and dip coating. The coating methods are less preferred because of the availability of exact color-matched plastic materials. Flow and dip coating drench or dip the parts in a paint bath and allow the parts to dry while slowly rotating them in an oven for even coverage. These processes are economical and have low equipment costs, but are slow.

Most parts are spray painted by hand or with an automatic spray gun. Almost all paint lines are automated—the parts flow from loading to the spray booth to the oven and unloading station. There are five basic spray gun techniques for applying an even, uniform coat, as shown in Figure 16.3. (A): A stationary gun or guns with moving workpiece or large parts painted only on side. (B): Reciprocating gun traversing back and forth in a controlled manner with stationary work piece. (C): Spray guns oscillating back and forth across

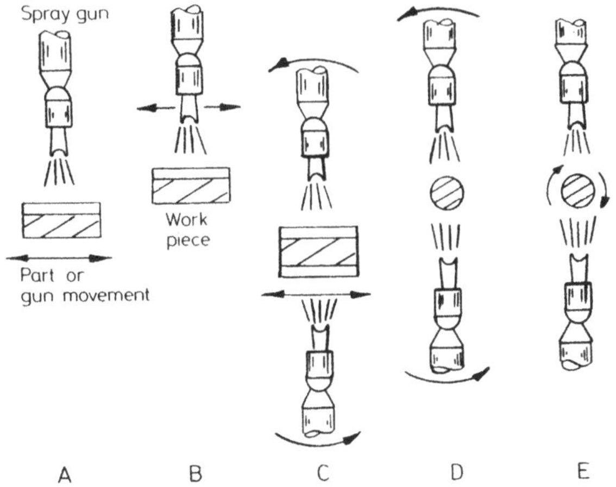

FIGURE 16.3. *Spray gun paint techniques. (Adapted from Ref. [2].)*

a moving part for full surface coverage. (D): Spray guns revolving above and below the part, which may be stationary or revolving. (E): Stationary guns with revolving part. In all cases, the parts are conveyed along at a line speed conducive to the paint coverage required.

The paint system selected—solvent or water based–will determine the attainable degree of adhesion with the type of plastic. Some plastics require a surface treatment or base primer for good adhesion. With other plastics, the solvent system may dissolve the surface of the plastic part to achieve good adhesion. With plastics such as polycarbonate, polystyrene, and acrylics, special care is required to avoid stress cracks. These resins are attacked by solvents and the entire part can, be destroyed.

Water-based systems can be used where solvents attack the resin. But because water-based systems do not chemically bond to the resin, the coating is weaker and not as chemically resistant.

Some polymers and their alloys exude plasticizers during molding and some flame retarded grades may form a surface film on the molded part that impedes adhesion. If this occurs, degreasing or special surface treatment is required. There are special treatments and paint systems that can be used to obtain the adhesion, but they are usually more expensive.

Depending on the plastic and processing conditions used for the part, some may have to be stress relieved before painting to avoid crazing and cracking during curing of the paint. If this problem cannot be easily solved, an alternative paint or cure system may be the only solution. In many cases, the processing conditions can be adjusted to produce stress-free parts or the part and mold modified to eliminate problem areas.

Each paint system will have its own curing method, such as air dry, oven curing, and the newer UV oven. The paint, resin, and cure system must be compatible and selected based on the finish required for the parts and the volume of parts required.

When using solvent-based systems, determine if special handling and procedures are required for air quality and personnel protection. State and federal agencies, such as the Occupational Safety and Health Administration (OSHA), are very strict about which systems can be used safely on the paint line and the air quality exiting the drying station into the environment. Some systems require special operator equipment, with filters and air emission equipment.

Part Cleanliness

Molded parts must be kept clean for both solvent- and water-based paint to avoid defects and poor adhesion. Items to be considered are as follows:

1. Avoid mold releases if possible—especially silicones.
2. Remove dirt, grease, oils, and fingerprints (hand creams). Use clean, white, cotton gloves to handle parts.
3. Use cleaning agents in washtanks that are compatible with the plastic.
4. Use filtration to remove solid contaminants from washtanks.
5. Final wash must be of high quality; hard-water areas must use deionized water for the final rinse.
6. Bag parts and seal them to prevent dust contamination and moisture pickup, if storage before painting is necessary.

The coating quality of the paints, including the cosolvent in water-based systems, is important for adhesion, proper application, and curing. The solvent must be compatible with the plastic substrate to adhere without chemical attack. In addition, the solvent must balance the need for moderate viscosity for easy application and leveling of the paint. It must also have the correct volatility for controlled evaporation during curing.

The solvent flash-off rate must stabilize the coating on the part's surface, but leave the film in a "open" fluid condition. This permits the remaining solvent to pass through without causing surface defects during the cure. Paint lines must maintain a uniform temperature or solvent balancing will be necessary for correct flow and cure.

Water-based paint requires special consideration and handling. Some of these are listed for your consideration and understanding:

1. Use different equipment for applying nonwater-based solvent and water-based paint.
2. Stainless steel or other noncorrosive paint tank components are required to prevent rust.
3. Do not overthin water-based systems. They require higher viscosity for correct application. Use a 10 percent maximum water dilution as the limit for many formulae.

4. Do not shake or agitate pressure-pot or paint during mixing, as water-based systems foam very easily.
5. Keep water-based systems tightly covered, as exposure to air causes skin to develop on the paint surface.
6. Use higher than normal air pressure to compensate for the paint's higher viscosity. This will yield finer atomization and result in faster drying and better adhesion. The finer spray and particle size will allow the resin particles, which are coated with a soap film emulsion for suspension in water, to be more finely laid down and packed on the substrate. As the water evaporates, the emulsion film breaks and exposes the resin to the curing medium. During curing, the paint resin goes through a cross-linking process forming the paint film surface.
7. Flush paint equipment only with water and cleaning concentrates. Lacquer thinners and ketones will soften, but not dissolve, residues.
8. Clean up partially dried paint from nozzle and masks. Use methyl ethyl ketone to soften residue and a soft nonmetallic wire brush to reduce scratches and damage to masks. Fully dried paint requires a stripper.

Part Paint Specifications

The areas on the part to be painted must be clearly noted. If paint masks are required to keep some sections clean, they must be specified. The plug, block, and cap masks are shown in Figure 16.4. The plug mask fills a molded-in depression to keep it free of paint, while the block mask confines paint to a shallow surface area. The cap mask blocks off the surface area, allowing the depressed areas to be painted. Masks are specified using an isometric drawing that shows a view of the top and bottom of the part. The areas to be considered are noted on the drawing in the notes section:

1. Area to be painted—paint, manufacture, texture, and specification per manufacturing procedure.
2. Area not to be painted—masked off.
3. Overspray area—does not matter whether painted or not.

The isometric part drawing is then marked using open, single, or double crosshatching on the respective areas. This ensures that the painter will know the part requirements and will be easily understood. The notes column should specify any special cleaning, surface, and part preparation or primers, as well as any required procedures and application documents.

Hygroscopic parts should be kept as close to dry-as-molded (DAM) conditions as possible to reduce moisture absorption that can cause finish problems. A list of common paint problems and their solutions is discussed in Appendix F. Often, when paint problems occur, the part is scrapped. The paint line must

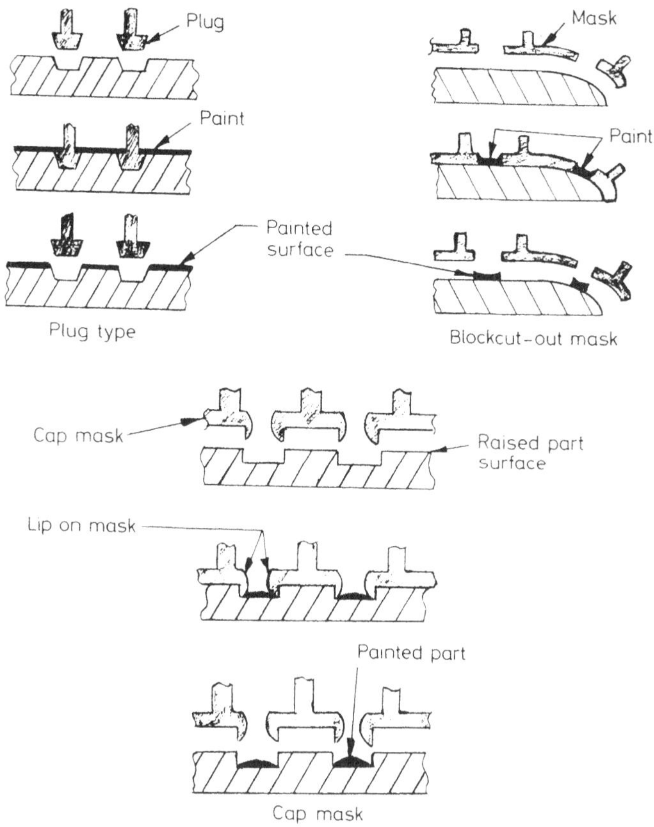

FIGURE 16.4. *Paint masks. (Adapted from Ref. [2].)*

stay in control. The sooner a paint problem is solved, the sooner the line can return to producing a quality decorated painted part.

Graphics

Graphics are used to apply consumer directions and provide product information with contrasting colors. Four common techniques are as follows:

1. Silk-screen printing
2. Pad printing
3. Hot stamping
4. Heat transfer

These systems are easily automated and may utilize one or more of these techniques. The designer can use the comparisons in Appendix F to answer questions early in the program about the decorating system.

Silk Screen

Silk-screen printing is performed directly on a plastic part in single or multiple colors. It can reproduce patterns or lettering on flat and simple curved surfaces from a pattern laid out on a screen. As shown in Figure 16.5, ink or paint is forced through a stenciled fabric on to the part surface. Depending on the complexity and detail required, the process can be automated. Keying the screens exactly to the part's surface is required for uniformity, especially if multiple screens are used.

Pad Printing

Transfer pad printing has been used to decorate plastic parts for more than 25 years. Printing can be on flat, cylindrical, or odd-shaped three-dimensional parts, and it can compensate for such molding irregularities as sink marks. It can be automated with rotary index tables, index shuttles, or linear conveyors. Single or multiple colors can be used, because the inks are capable of wet-on-wet transfer and parts can be handled about 30 seconds after printing.

The transfer pad, which is the most important element in the process, is a mixture of silicone oil, rubber, and other additives. The life of a pad is, on average, 100,000 to 150,000 transfers. The printing plates or cliche, range from 4 square inches up to 14 by 36 inches. They are made from hardened steel, steel foil, plastic, or copper, and material selection is based on the number of impressions required, quality of characters, flexibility, and cost. Plate life can vary from 50,000 impressions for steel foil to over 100,000 for plastic plates to more than 10 million for hardened steel plates. The typical transfer-pad printing process is shown in Figure 16.6.

The ink must have consistent viscosity that is controlled by the printing machine's pump. This reduces rejects and allows unskilled labor to operate the press. Wet-on-wet pad-printing inks allow multicolor, automatic transfer, assembly-line decoration. It uses 12 basic colors that can create more than 400 different shades.

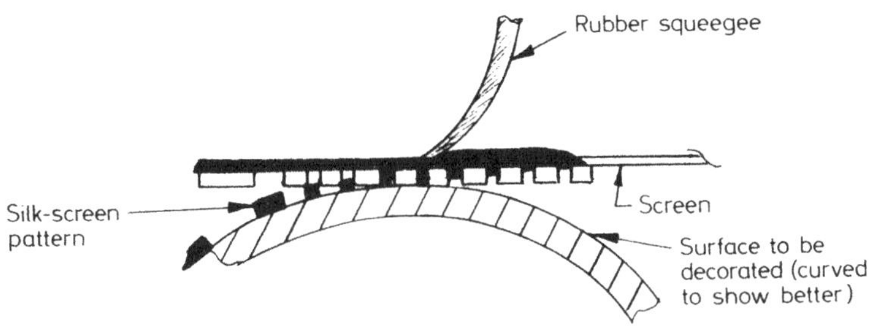

FIGURE 16.5. *Silk-screen decorating. (Adapted from Ref. [2].)*

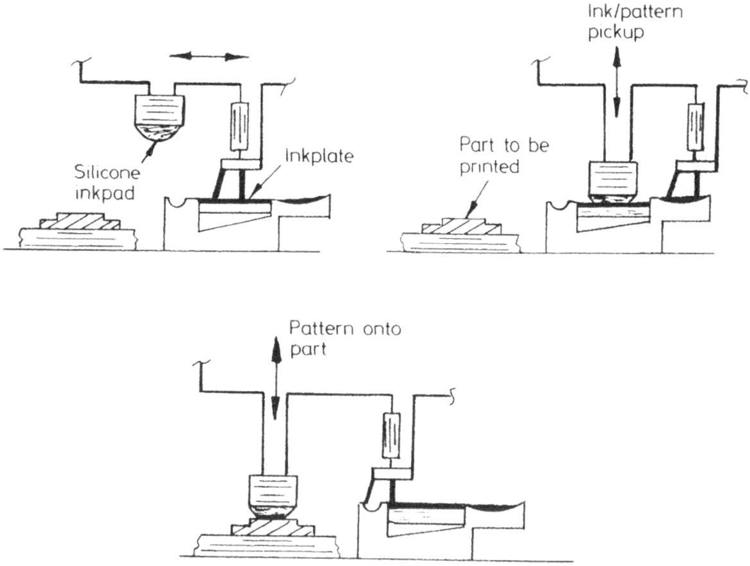

FIGURE 16.6. Transfer pad printing process. (Adapted from Ref. [3].)

Transfer-pad printing is economical because of its long life, particularly when compared to other similar decorating techniques. It has a high quality image in the industry; some machines produce parts with almost zero defects.

A major use for transfer-pad printing is printing keyboard characters for computers, typewriters, phones, and other devices. Entire keyboards can be printed at one time. With the correct inks and finishing operations, character integrity can last for more than 10 million wear cycles. This decorating technique has replaced the more expensive two shot molding for these keyboards.

There is also the rotary-pad printing process, which is faster and more adaptable to the packaging industry and higher-speed applications.

Hot Stamping

Hot stamping uses heated metal or silicone dies to place either heat-activated or adhesive-bonding foil transfers on the part's surface. The transfer is limited to the detail on the die face or mating plastic part. The technique, shown in Figure 16.7, can be automated and parts can be handled immediately after stamping. The heated die can transfer foil directly on to the part's surface in raised, flush, or countersunk areas.

Hot-stamping problems can be avoided if the correct foil, part surface preparation, equipment, setup, and the control are maintained. The part—its material and design—must be compatible with the process. This includes the hot-stamp press, tooling, and procedure for part loading and unloading. The

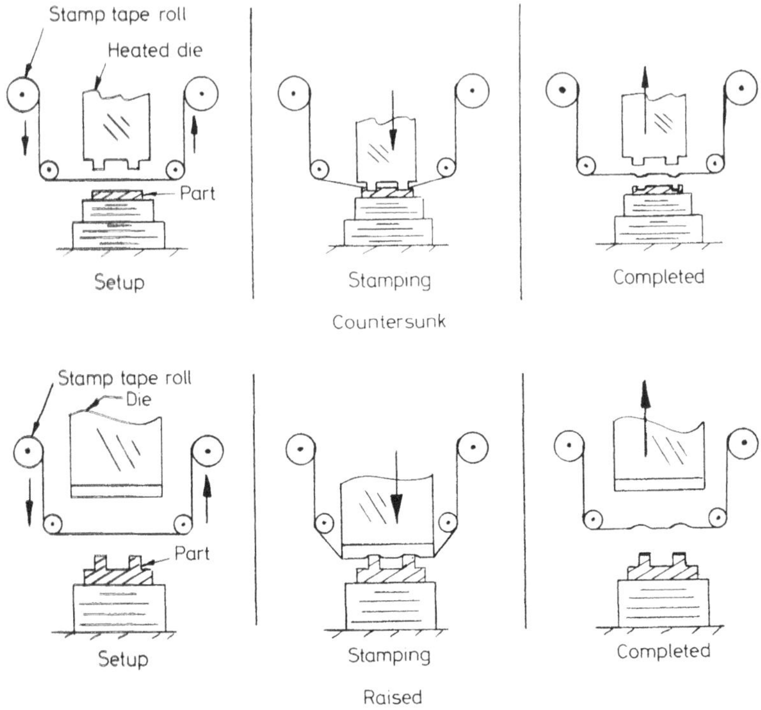

FIGURE 16.7. Decoration by hot stamping. (Adapted from Ref. [2].)

foil must be selected to bond to the part's substrate. If the substrate is coated, the foil must be formulated to adhere to the coating. Surface treatment may be required for polyolefins, acetal, and nylon polymers. The foil should always be tested on the resin to be sure it is correct.

Since foils vary from lot to lot, always save a part of a roll that has worked well. If a problem develops, lot numbers can be compared. If the lot number is different, the operator can change rolls or contact the supplier for an explanation. Foils should always be stored away from heat and moisture, as they are sensitive.

The size of the hot-stamp press is very important. Some industry decorators recommend a 30 percent higher tonnage over the nominal pressure required for a particular job. For a metal die of one square inch of foil-transfer surface, one ton of pressure is required. For a silicone rubber die, 400 to 500 pounds per square inch are recommended. For a silicone rubber roller, with only one line contact between the substrate and roller, only 100 to 150 PSI is required. The equipment and decorating operation are shown in Figure 16.8.

Part decorators should contact their material, foil, and hot-stamp machine suppliers to answer questions before decorating begins. Should decorating problems develop, Appendix F lists solutions to common problems. A bit of

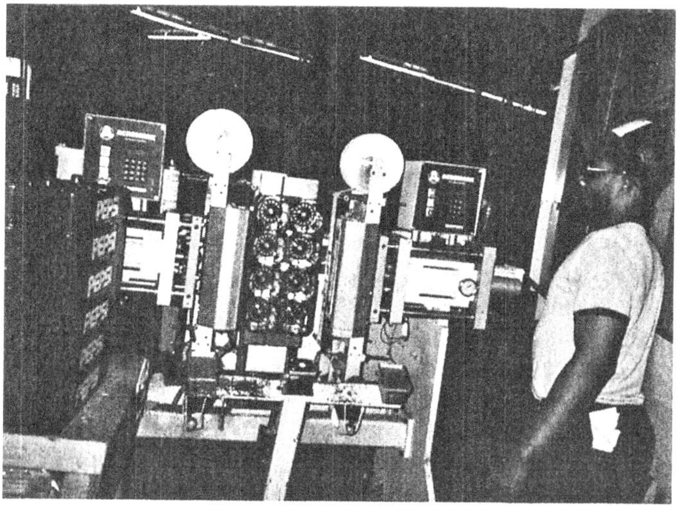

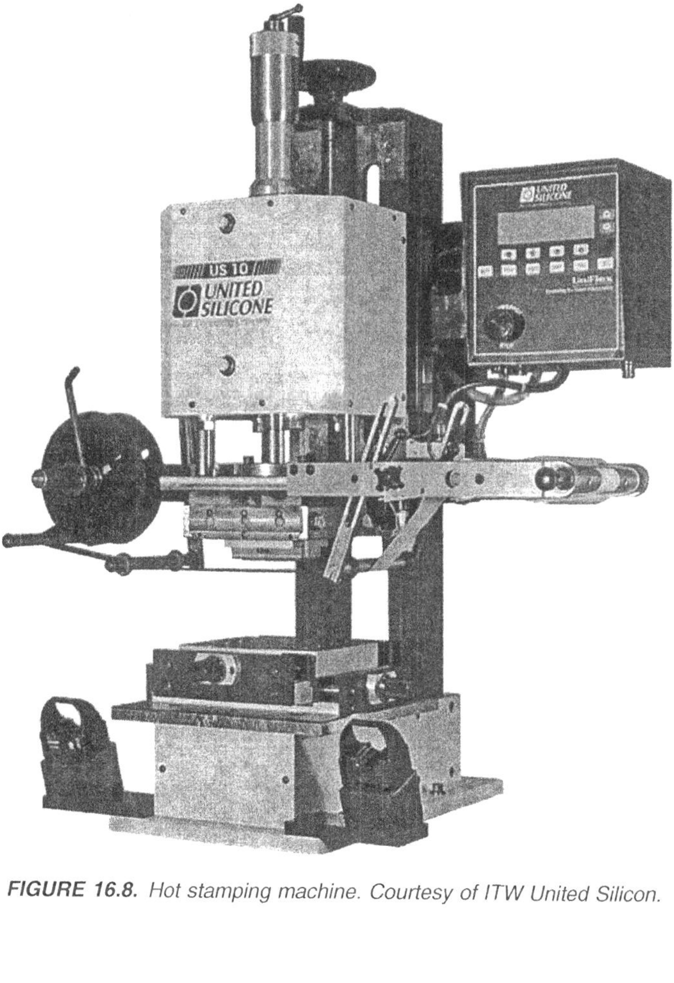

FIGURE 16.8. *Hot stamping machine. Courtesy of ITW United Silicon.*

obvious advice for all decorating techniques is to read the instruction book first.

Heat Transfer

Heat-transfer decorating uses a preprinted and often multicolor foil or other suitable decorating medium. Through a heated die, a bonding system transfers the material to the molded part. A roll-on heat-transfer system reduces the tendency to trap air under the transfer. The system can be automated, with the part and transfer indexing and traveling under the heated-transfer die. Different transfer systems are available depending on requirements. Equipment suppliers should be consulted to answer specific questions.

Two other decorative systems, which are available, are as follows:

1. Spray and wipe for depressed lettering
2. Two-shot molding

SPRAY AND WIPE

The "spray-and-wipe" system is for molded-in depressed lettering or information printed in fine detail. It offers added contrast and protection to the decorated part. Letter or detail-edge contours should be sharp to ensure that they will not be fuzzy as a result of the wiping operation. Detail wiping should leave the filler level as close to the part surface as possible to prevent contamination with foreign materials, grease, and dirt. Contaminants can lodge in the depression during use. This detail more effectively resists abrasion and wear from surface contact than lettering and other decorating methods shown in Figure 16.9.

The spray-and-wipe technique sprays or wipes air dry paints into the part's molded depressions. After drying, the operator uses a low-knap cloth, wet with a suitable solvent, to remove the excess surface paint. This is a work-intensive operation. The operator applies just enough pressure to remove the surface paint but not the paint in the recessed areas. Care must be taken to not have folds in the cloth, which could wipe paint from the depressions.

The dry-wipe, nonsolvent system is similar; the paint filler is sprayed on to the surface. After drying, the filler turns to a dry powder which is carefully removed with a dry cloth. The dry-wipe paint has a lower bonding strength and should not be used on parts subjected to impact or severe vibration.

TWO-SHOT MOLDING

Two-shot molding, as a decorating or information transfer technique, uses a plastic molded-in or over-another plastic preformed part. The encapsulated

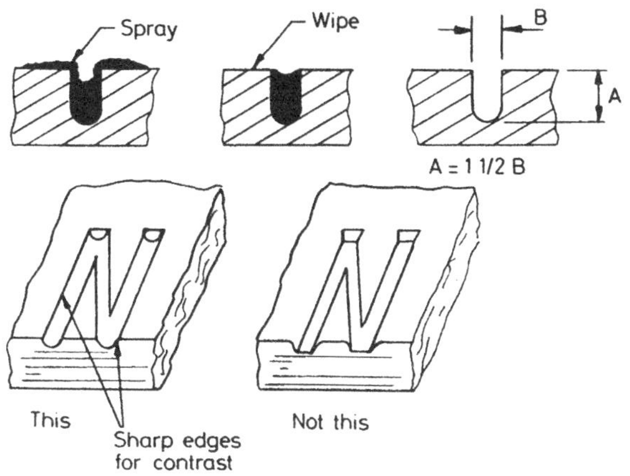

FIGURE 16.9. *Spray and wipe decorating (Adapted from Ref. [2].)*

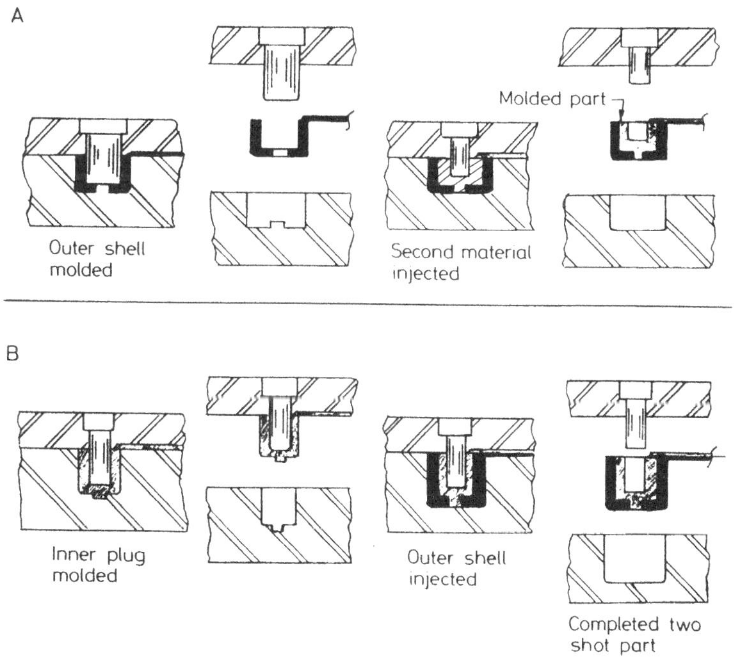

FIGURE 16.10. *Two-shot molding: (A) using two separate molds, and (B) using the mold-indexing technique. (Adapted from Ref. [2].)*

material, which is decorated, is visible. An example is a typewriter key. Two-shot molding is losing popularity. New higher quality printing inks and techniques are less expensive to apply and give equal or better results. Two-shot assemblies were the main products using this technique. Acetal acrylonitrile butadiene styrene (ABS) and polycarbonate/acrylic were the major materials used. The resins require a wide temperature range, so that the second-shot material will not soften or melt the first, thereby causing loss of detail at the insert area. Material shrinkage is important, as the second material must shrink tightly around or fill the part cavity of the first shot without creating a parting line. If two-shot molding or decorating is used, the above areas must be considered in terms of tooling and manufacturing costs, part volume, quality, and customer requirements.

IN-MOLD DECORATING

In-mold decorating (IMD) is a relatively new technique. It uses a continuous ribbon carrier film with heat transfer foils or multicolored decorations. The carrier film is advanced and indexed in front of the open mold cavity. Plastic is injection molded against it after the mold has closed on the film. Under heat and pressure, the decoration releases from the carrier film and bonds to the molded part. The part can have a sophisticated design and shape, including recesses with concave/convex dimensions. Multiple colors and selective metallization are possible. The use of reinforced resins is not recommended, because they can abrade the transfer during the injection stage.

This technique eliminates secondary operations, particularly such "wet" processes as pad printing, silk screening, or spray painting, and exposing the plastic part to potentially damaging solvents. This dry process negates environmental problems. The IMD system can also be used for first- and second-surface decorating, depending on the plastic used. Plastic materials now used are styrene acrylonitrile (SAN), ABS, polycarbonate (PC), polystyrene (PS), thermoplastic polyurethanes, PC/polybutylene terephthalate (PBT) blends, and acrylics.

In mold decorating requires tight control over the positioning of the film in the cavity. Two methods are employed to position the transfer in the mold: a continuous carrier film-feeder machine, which puts a random or multicolor repeat pattern on the part, or a registered film-feeder, which puts a precise decoration or information transfer exactly on the part during each molding cycle. The registered film-feeder requires exact X and Y axis positioning and can hold tight tolerances of graphics on the part to within +/− 0.004 inches. An example of this is an instrument cluster bezel. A film-feeder for an injection molding machine is shown schematically in Figure 16.11, along with the three basic stages of in-mold decorating.

The tooling required for IMD is specially built. A single cavity, three-plate mold using pin-point part gating is preferred. The keys to the process are the specialty film and foils on which the graphics are printed and the film-feeder for exact registration and transfer positioning. The detail and colors are limited by the carrier film. Multicolor and metallic transfers are possible, producing a high-tech, three-dimensional appearance on the special gravure-printed carrier film web.

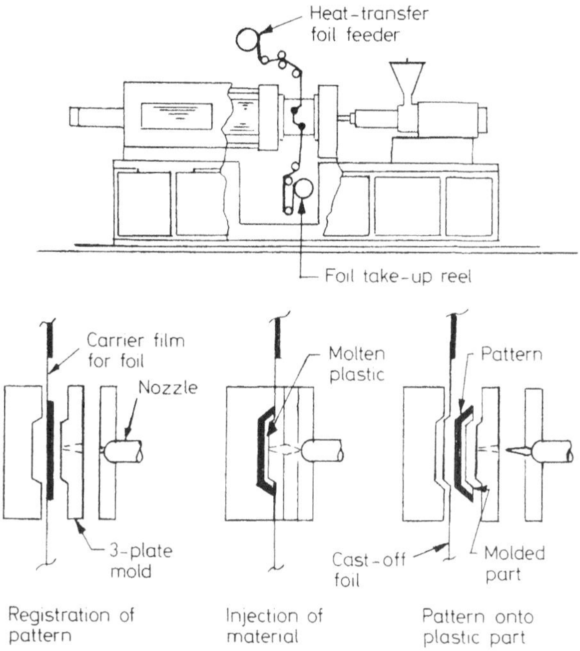

FIGURE 16.11. *IMD film feeder for an injection molding machine. (Adapted from Ref. [10].)*

Design considerations for the part are as follows:

1. Part depth no greater than 6 inches.
2. No sharp corners or edges to cut into the transfer causing an underdecorated surface or graphics (ink) cracking.
3. Raised or recessed areas must be carefully designed to avoid stretching the transfer.
4. Wraparound or recess of graphics is limited because of stretching.
5. Minimum draft must be employed.
6. Maximum depth of part surface to the parting line on the perimeter is 0.25 inches, interior openings 0.15 inches.
7. The overall part size of the registered-transfer decoration in the mold parting-line plane is limited to 15 × 22 inches.

As with all decorating techniques, early consultation with the IMD film supplier and equipment manufacturer is required.

VACUUM METALLIZING

Vacuum metallizing produces a shiny decorated surface, a conductive surface, or a surface protected from electromagnetic interference (EMI). This technique deposits a thin (3 to 5 millionths of an inch) metal deposit or coating on a plastic part in a vacuum chamber. The metal deposited may be aluminum, silver, gold, chromium, or nickel, and is usually in thin foil strips called staples. The procedure uses a line-of-sight process. Molded parts are placed in holding fixtures—either stationary or rotating—in a vacuum chamber. The metal (staple) coating is vaporized at extremely high temperatures (1800°F or higher) by tungsten filaments located at the center of the chamber. The vaporized metal atoms travel in straight lines from the filament and condense on all encountered surfaces. Firing time varies from 15 to 30 seconds to develop the coating thickness specified for the part. A typical vacuum chamber is shown in Figure 16.12, with the parts mounted. The holding fixture must support the part in a noncritical area to avoid surface defects and position it for a uniform coating. Because of the time required to load parts, the chambers have multiple rollin/rollout part holding fixtures to increase productivity.

To obtain good adhesion and reflectivity, a base coat is applied to the plastic substrate. After metallizing, a top coat protects surface from physical or chemical deterioration.

The process for applying the coatings is spray, dipping, or flow coating. Flow coating is preferred when fixtured parts are rotated. The part is sprayed with a nozzle that stops after the coating has flowed evenly over the part. Because the coatings have a 100 percent solid content, UV curing systems are preferred. Drying takes 1 minute, as opposed to 1 hour for convection ovens. In addition, convection ovens can cause further part shrinkage and crazing of the metallized surface. Environmental problems are also eliminated using UV base and top coats, because the solvents are very mild and evaporate rapidly.

Metallizing, as with IMD, can be applied on either the first or the second surface of the plastic part (see Figure 16.13). First surface decorating is usually

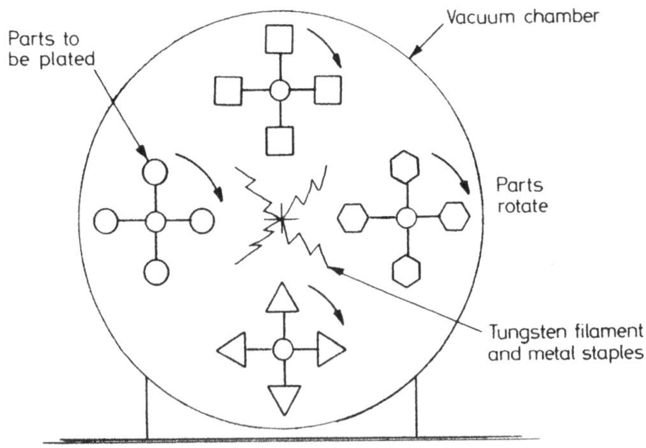

FIGURE 16.12. Chamber for vacuum metallizing. (Adapted from Ref. [2].)

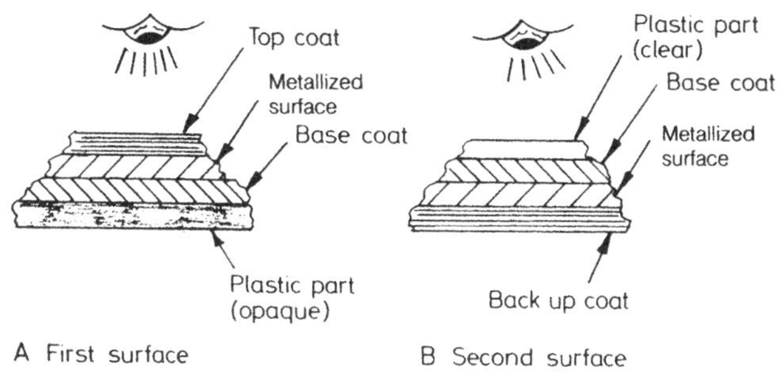

FIGURE 16.13. Metallizing can be applied to either the first or second surface of a part. (Adapted from Ref. [2].)

done on an opaque plastic part, as a control knob, with the reflective surface covering the show surfaces of the part. The metallized surface is then protected by a clear top coat.

Second-surface decorating is done on the reverse surface of a clear, transparent, plastic part, such as a reflective lens, which offers a higher degree of metallizing protection. The metallizing is then viewed through the clear plastic part and base coat.

The metallized coating is rather fragile and not recommended for hard abrasion and wear applications. The durability depends on the wear characteristics of the top and base coat, the environment, and end-use applications.

Plastics that can be vacuum metallized are ABS, PC, acrylic, PS, SAN, nylon, polyesters, modified polyphylene oxide (PPO), and other polymer alloys. Polyolefins can be metalized, but require surface preparation for good adhesion. Plasticized vinyl and cellulosics are not recommended for vacuum metallizing.

Part design is critical, because any defect will be amplified by the metallized surface. Sink marks, weld lines, flow lines, and surface-and-gate imperfections on the metallized surface will show. The base coat does not hide these problems. The parting lines must be exact and the tool surface must be kept clean from a buildup of mold deposits.

Molding conditions and part handling must be tightly controlled after molding to avoid scratches, surface blemishes, and contamination. The areas to be considered are as follows:

1. Minimize bosses—consider sonically welding after molding.
2. Uniform part thickness of 0.100 to 0.150 is best; minimize rib sections.
3. Large flat surface areas should be kept to a minimum—consider texturized surfacing, if acceptable, to hide distortion or weld lines. Use a slight crown of 0.015 in/in in the mold to hide surface distortion.

4. Good cooling/heating layout in tool is required to control warpage, fill, and molded-in part stress.
5. Uniform cavity surface finish.
6. Good venting to avoid burn or flow marks, especially weld lines.
7. Use high-quality tool steels to eliminate corrosion and wear caused by polymer flow.

ELECTROPLATING

For parts exposed to more vigorous handling and use, electroplating is recommended. It produces plating-bond strengths and surface appearances equal to the finishes produced on die castings requiring Class A reflective surfaces.

Electroplating is similar to vacuum metallizing with regard to part design and molding considerations, but the process is entirely different. For electroplating, the plastic surface must be chemically etched to accept the preplate metal base coat. This coat produces the metal-to-plastic adhesion and provides the conductivity for electroplating.

Electroplating is a series of operations. Each is specially developed for the plastic to be plated. These plating operations consist of an etching system, neutralization bath, catalyst bath, accelerator bath, and an electroless nickel or copper bath, followed by the final electroplating operation.

The plating thickness is composed of various metals laid down in a series of steps to provide protection, ensure adhesion, and allow for the differences in thermal expansion of the plastic and metal interface. A leveling and brightening coat provides the final reflective finish. Most final plating consists of a copper/nickel/chrome electroplate, but other, such as bright and antique brass, satin nickel, silver, black chrome, and gold are possible. Plastics that can be electroplated are PC, ABS, nylon, polyesters, PS, polyethylene terephthalate (PET) modified PPO, polyphenylene sulfide (PPS), polyethersulfone, polyetherimide, polyarylether, and blends of these polymers.

To achieve quality plating, three-way interaction between the material supplier, molder, and plater is necessary. Using a "plating grade" of plastic does not always ensure that the plastic parts will be correctly plated. A wide range of factors and ingredients can and will affect the outcome. These include the base polymer, plating process, part design, molding conditions, molded-in stresses, external stress, part complexity, filler types and content, and other additives, such as colorants, flame retardants, heat stabilizers, plasticizers, mold release, and how well they are blended in the injection molding machine.

Most plating operations expose the parts to chemicals that etch out a particular component in the material to achieve adhesion of a plating component. When this is done, molded-in or surface stresses can cause premature part failure. Therefore, the plater needs to know the molding conditions to circumvent plating problems. This allows the plater to adjust the chemistry of the solutions to achieve good adhesion and surface aesthetics.

Certain additives and fillers can cause problems. These include carbon black in polycarbonate and as little as two parts per hundred of titanium dioxide in a resin. Some additives, such as heat stabilizers, plasticizers, and antioxidants, may also leach out of the parts and contaminate the plating solution. Both internal and external mold releases and some processing aids have also been known to cause plating problems. Therefore, the plating information in Table 16.1 should be known by all parties.

The parts should be as stress free as possible. Surface stress will cause overetching in some areas and underetching in others. Either can result in varying degrees of adhesion. With each plastic, adjust molding conditions to achieve as low a stress level as possible. Tests (e.g., dye tests) can be run on molded parts to determine if the conditions used produce stress-free parts. Contacting the material supplier and plater will assist in this area. Because this is a critical area, some platers insist on doing the molding themselves to ensure stress-free parts.

One important quality test is the peel test. A crosshatch tab section is cut on the plated surface down to the substrate. An edge of the plating is raised and a pull/tear value is generated to check conditions and plating adhesion.

TABLE 16.1. Part Plating Information.

Information Needed	Why It Matters
Flame retardant system and concentration level.	Some flame retardants migrate to the surface of the molded part and can cause adhesion problems during plating. Always check with the material supplier for the recommended plating system to be used.
Mold release (internal and external) used.	Certain mold releases can be detrimental to plating adhesion. Avoid silicone oils and waxes.
Fillers used and concentration level.	Even when you use a "plating" resin grade, certain fillers, pigments, and reinforcements can degrade adhesion during plating.
Actual molding conditions.	Platers need to know the injection pressures and melt and mold cavity temperatures used to produce a part; this gives the plater an indication of molded-in stress.
Part stresses.	One major problem platers are confronted with is stress caused by molding conditions or part geometry. Platers need to analyze each part to determine the best possible plating solutions.
Resin used and if an alloy, the components and percentages.	Platers require key information on the resin used in an application. Handling blends and alloys can be particularly difficult. Keep hygroscopic parts DAM prior to plating.

Source: Adapted from Ref. [7].

Since the plating process has many operations, quality checks must be conducted on each step to keep the process in tight control. This is the plater's responsibility.

FLOCKING

Flocking is a new technique for both decorating and sound insulating. It is a finishing operation using synthetic or natural fibers (flocks) as a surface coating. This has a desired visual, tactile, or noise-reduction effect.

Flocking consists of adhesively bonding chopped, short-length fibers to the plastic substrate. The adhesive can be either an epoxy or an oven-cure, urethane, or air-cure type. Because the fibers only adhere where adhesive is placed, patterns and direct placement of fibers are possible by silk-screen application of the adhesive. Fibers are applied using a pnuematic tubular discharge device. The device has a positive charge and the piece to be flocked has a negative charge. This is known as the static electrical field method. In this manner, fibers can be oriented on the part to achieve the desired effect. The common flocking fibers are presented in the following table.

Advantages

Acrylic	Cotton	Nylon	Polyester	Rayon
Easy to dye Inexpensive	Very inexpensive	Durable	Low absorption Durable Dyeable	Low absorption Low price

Disadvantages

Moderate durability	Low durability	High price	High price	Low durability

The part design should avoid sharp corners and deep crevices from both a structural aspect and to ensure more uniform coverage of the part. Most plastics can be flocked, depending on the adhesive and cure system used.

The polyolefins require special surface treatments for good adhesion. Flocking is always finding new applications and, in many cases nonprime resins may be used depending on the end-use requirement. The tooling and molding may be less critical, but must meet the customer's quality requirements.

GRAVURE DECORATING

Gravure printing on simple and complex plastic parts is a new technology for decorating with multicolor patterns, woodgrain and marble patterns, and geometrics. Basically, any pattern or design that can be printed on a water-soluble plastic or organic film can be transferred to a plastic part. This process is trademark protected and known as “Hydro GraFix.” The results are shown in Figure 16.14.

The special ink-patterned film is fed very slowly on to the surface of the water in a rectangular water tank with a controlled laminar-surface flow rate. The inked film floats, the film softens, and dissolves. The ink is left floating on the surface. The rate of film feed into the tank is controlled by the laminar-surface water flow rate of the ink, as it moves toward the parts to be decorated. The parts are located further down the tank’s length.

The parts are mounted on a ferris wheel–type dunking station over the tank. It slowly feeds the parts into the advancing ink flow. The dunking station is speed controlled to match the surface ink flow rate. As the parts enter the surface of the water, the ink pattern is attracted and transferred to the plastic part, wrapping around and contouring itself to the part’s geometry. Orientation of the part on the dipping wheel is such that all surfaces to be coated come in contact with the ink. This is similar to dip-decorating Easter eggs with oil pigments that float on the surface of water.

After the part exits the tank at an ink free area, it is transferred to a drying station where the final top coat is added for surface protection, durability, and luster. Nongeometric style patterns work best—wood grains and marbles—as no two parts are ever the same. Depending on part geometry, some pattern stretching or nonuniformity is possible, as the ink pattern wraps around corners and curved surfaces. Parts with small recesses or blind holes may not be decorated with this technique, as air entrapment does not permit the ink to flow into these areas. Almost all plastics can be decorated, and base coats are used to enhance and increase adhesion of the ink surfaces. Textured surfaces can also be decorated and areas masked off if no decorating is desired.

FIGURE 16.14. Hydro GraFix decorating. (Courtesy of Hydro GraFix Inc., Greensboro, NC).

Part-to-part color and contrast is attained by the pattern's printing quality, but each part will be slightly different. Before this technique, adhesive-backed transfer films or silk screens were used. Complex shape decorating was next to impossible; it was restricted to flat shapes with simple, curved surfaces. Early woodgraining or marbling effects for plastic parts were only possible using a ram machine with a salt-and-pepper resin mix. When injected into the mold cavity, it gave a nonuniform melt flow pattern.

The specific details, requirements, and procedures for all decorating techniques should be obtained from the supplier of the materials and equipment. Special care and quality control at this stage must be exercised. If the decorating is poor, not only is the time and expense of decorating lost but also so is the molded plastic part.

17

Customer and Employee Satisfaction

The use of total quality process control in all of a company's operations will result in increased profits, increased employee morale, fewer missed deliveries, and the ultimate, increased number of satisfied customers.

Continuous training in the methods of operation, verification of results, and the follow through of all employees in performing their daily tasks is necessary for continual improvement. Becoming certified to or operating in compliance with a national standard is now the norm, and companies are requiring their part and components suppliers to improve their product's quality. Noncompliance of a product to meet the customer's requirements and specifications is not permitted.

Major companies are now reducing their supplier base in return for improved quality, pricing, and customer service that is needed to remain competitive in the marketplace. Suppliers need to retain their employees to ensure their systems and products will remain competitive and their company remains a valued supplier.

These ideas and concepts form the basis of the total quality process control methodology. Without a company's commitment to *excellence in manufacturing*, no single system will be successful. This is based on the basics of plastic manufacture, which are as follows:

1. Good part design capable of fulfilling end-use requirements.
2. Material selection based on part requirements, supplier quality, molding ability, and end-use function.

Total Quality Process Control for Injection Molding, by M. Joseph Gordon

3. Mold built to manufacture the part per items (1) and (2), and capable of attaining cycle and tolerance requirements meeting customer specifications.
4. Manufacturing equipment able to produce the products consistently from cycle to cycle within process control limits. An added advantage would be output data showing that the process is in statistical process control during manufacture. This verifies part quality to the customer.
5. Secondary operations, such as assembly and decoration, controlled and capable of adding value to the product.
6. Packaging and shipping on schedule.

Management is working on establishing an employee partnership that is built on trust, partnership, and joint responsibility for producing quality products; all actions are directed toward the awareness of quality in products, service, and personal performance.

Any successful quality program must have motivated people at every level. Quality improvement begins with "you"—the individual worker—with support from management. Unless everyone is committed to quality improvement, the route toward *excellence in quality* will be delayed. Employees and management, working together, can attain sizable reductions in manufacturing costs, as well as improved services to meet "Just-in-Time" delivery and annual growths of 10 to 15 percent by meeting customer product and quality requirements. An implemented quality program clearly results in "continuous improvement" by upgrading of operations, equipment, and personnel training.

QUALITY AWARENESS

To determine your "quality awareness," answer the following questions. Do not be disappointed if your score is only 70 percent positive. You can always learn more about quality procedures and operations.

1. Does your company have quality objectives, and do you understand them?
2. Is your company committed to attaining these goals?
3. Do you offer input to quality/procedure improvements?
4. Can you communicate honestly with your peers, department management, and customers, as part of your job?
5. Do you respect individual differences, whether or not you agree?
6. Can you resolve conflicts with co-workers?
7. Do you seek assistance in solving problems when needed?
8. Do you trust co-worker input, and is there a feeling of cooperation versus competition?

9. Is the welfare of your co-workers of interest to you?
10. Do you want to advance or have you reached your highest level?
11. Do you like your job and the company?
12. Do you want to advance and take courses and training to reach that level?
13. Is there advancement opportunities within the company you can attain?

If you are committed to quality, let it be known and work toward achieving it. Everyone will benefit.

Customers insist on continued quality improvement to increase their competitiveness in the marketplace. If they are successful, the company will profit.

The plastic part supplier must make a firm commitment to these basic concepts, with management and workers acting as a team. All will prosper as a result of major cost savings and improved products through the total quality process control program.

Appendix A

Quality Management System (QMS) Control of Documents Procedure

Purpose

This procedure defines the process for controlling the origination, approval, release, revision, distribution, storage, and disposition of project and organizational QMS documents.

Responsibility/Authority (Perform all steps unless indicated otherwise in a specific step)

[Select: ISO Management Representative or designee
-OR-
Configuration Manager (CM) or designee]

Special Considerations

QMS documents are defined in the Quality System Manual (QSM). All QMS documentation, except QMS records, is covered by this procedure. QMS records are controlled in accordance with the QMS Control of Records Procedure.

This procedure deviates from the standard procedure format, and steps are independent rather than sequential.

Total Quality Process Control for Injection Molding, by M. Joseph Gordon

Inputs

- Potential QMS documents
- Suggestions for new QMS documents
- Requests for change

Entry Criteria

- Potential QMS documents are generated

Outputs

- Approved QMS documents
- Obsolete QMS documents properly marked (if retained)

Exit Criteria

- QMS documents are under CM
- QMS documents are available at appropriate work areas
- Current versions of the internal QMS documents are identified

Activities

Step 1. Ensure organizational and project QMS documents are reviewed and approved prior to release. Review documents for adequacy and compliance with contractual requirements.

NOTE:

— Each organization should identify functional responsibilities that will review and approve various types of documents. *[Direction: Provide the list of functional responsibilities and who will perform them as an appendix to this procedure or include a reference in this step to the document where this information is contained.]*

— Each organization should establish a document numbering scheme and a method of identifying drafts. *[Direction: Provide the document numbering scheme as an appendix to this procedure or include a reference in this step to the document where this information is contained.]*

Step 2. Ensure that the current version of organizational and project QMS documents is identified and readily available for staff members.

NOTE: This can be a list of documents and current version with hard copy access to documents, an electronic database, or a link to the latest documents with appropriate access or another method as determined by CM.

Step 3. Ensure that organizational and project QMS documents are updated as needed. Documentation updates are reviewed and approved prior to release.

NOTE: Each organization should identify functional leads that will review and approve changes to released documents. *[Direction: Provide the list of functional leads and their responsibilities as an appendix to this procedure or include a reference in this step to the document where this information is contained.]*

Step 4. Ensure documents are legible and easily identifiable.

Step 5. Ensure that external QMS documents are identified. Control distribution of external documents and coordinate with the PM.

NOTE:

— This can be an electronic list of documents with appropriate access or another method.

— The distribution of external project documents is usually controlled as the Quality Manager (QM) designates. *[Direction: Describe how external documents will be controlled in an appendix to this procedure or include a reference in this step to the document where this information is contained.]*

— The revision of project external documents to be used is as specified in the project contract or as agreed with the customer.

— The distribution of external non-project documents is usually controlled by the QM or CM.

Step 6. If obsolete (superseded or cancelled) QMS documents are retained, ensure that they are properly marked to prevent unintended use.

Step 7. Ensure QMS documents that are company proprietary are disposed of in accordance with applicable security and/or company directives as well as local instructions and/or customer requirements.

Step 8. When required by a contract or considered appropriate by senior management, make QMS documents available to customers or to customer representatives.

Revisions

Version Number	Date	Description
[Direction: Begin revision history after first, non-draft release.]	*[Sub: date in the format month day, year]*	

Appendix B

Design of Experiments (DOE): Statistical Troubleshooting Process Screening for Reducing the Number of Variables

Once a process is believed to be in control and the optimum manufacturing parameters are controlled, an additional problem develops that products may still not meet customer specifications. Something has changed in the process or environment. The material and/or machine parameters, although not noticeable, may have changed or are different. The plant manufacturing environment as well as the plant-provided services are varying or the tooling has worn sufficiently to now cause a problem. As a result, any number of multiple manufacturing variables are varying but not enough to be easily detected by the operator or system control. This situation is often viewed as the process was in "control" but not totally "capable" of producing the desired product.

What do you do now? The old method was to guess and begin by holding all variables constant but one. Then, through trial and error and typically after a lot of time, the elusive controlling variable is found by running tests on the manufacturing machine and the process. Then, once the controlling variable(s) are determined, they can be adjusted and brought into control to make good parts to the required specification. This technique works, but often the troublesome variable(s) are not easily or quickly detected and the operator has to adjust the system constantly to make acceptable products. This is wasteful,

Total Quality Process Control for Injection Molding, by M. Joseph Gordon

expensive, and often causes a delay in shipping acceptable product to the customer.

There has to be an alternative method available! There is—a rapid screening technique using a statistically designed series of trials in which many variables are changed at the same time in the manufacturing process. This rapid screening technique was developed by Dr. Genichi Taguchi with the name for his method known as Design of Experiments or DOE. The intent of DOE is to rule out nonsignificant variables by modifying an array of variables and combining them in a specific pattern. In this manner, valid statistical information on the impact value of each variable in the process, on the product, can be obtained and evaluated with other variables.

The Taguchi experiments are used once a process is in control and the process parameters are optimized, but the parts still do not meet or drift in and out of specifications. The process can be affected by the material, machine settings, tooling changes, plant environment, or any number of items in the manufacturing system. The manufacturing process is in control but not yet capable. The Taguchi method uses orthogonal arrays—rows of experiments (factors or variables) versus the trial (runs) to be performed. The variables for each factor and run are established using quality control problem analysis techniques. These factors are then tested with each run using a different combination of these variables.

Taguchi's orthogonal arrays capture the most significant variable combination levels for testing. For example, to test independently seven variables at two setting levels, high and low, the complete orthogonal array would be large: 2^7 or 128 separate testing sites. This means to test each variable, at its high and low level, in combination with all the other variables, at each of their levels, would take 128 separate trial runs. There are many combinations of arrays and levels of testing. The tester determines which levels to test and the amount of time the plant can spend to find the elusive variable.

To solve the problem, the Taguchi variable screening analysis is applied. This is a mass-screening technique that uses a statistically designed series of trial runs. Many process variables are changed at the same time in a controlled test environment during the manufacturing process. The goal is to rule out insignificant variables by modifying an array of variables and combining them in specific patterns. In this manner, valid statistical information on the impact of each variable on the process and the product can be obtained. These variables are rapidly evaluated, and concentration is placed on the highly likely causes of the problem. Determining when to use this technique is not always clear. It is based on the customer requirements and on the capability of the manufacturing process, including machinery, tooling, material, and personnel.

A process is in control when the variation in the product, plotted on a bar chart, falls within the three-sigma bell-shaped curve values. This is assuming the process is, itself, stable and within the control limits earlier determined to make the product to customer's specification. At this time, the product may meet customer specifications but not always, and it may drift in and out of tolerance. A process is said to be capable when 99.5% of its products and all of its variations are within the customer's specifications.

The realization that improvement can still be attained or needed in a process usually occurs at this point when the process is in control but not yet capable. At this point, the "screening experiment" is called for to find the contributing factor that leaves the process not yet capable.

SCREENING EXPERIMENT: THE NINE STEPS

The nine steps of the screening experiment are used to improve the process and find the unknown variables needing adjustment to make the process capable and to keep all products within customer requirements.

These nine steps are as follows:

1. Determine what improvements are needed.
2. Brainstorm for ideas on what effects the variable to be improved.
3. Select the factors to be analyzed and the levels to be used.
4. Randomize the experimental runs.
5. Perform the experimental runs.
6. Separate the effects.
7. Test for significance.
8. Analyze the results.
9. Change the process based on the results of the experiment.

STEP 1. DETERMINING THE PROBLEM

When you realize a critical dimension is not where it should be, consider using a DOE to find the missing variable and its required value to manufacture acceptable products. This determination may be based on the continual drift of a product's required dimensions or specifications. Manufacturing processes cannot consistently manufacture the specified products. At this time, the process variables for material, tooling, and machine settings must be evaluated in a minimum duration of time to keep the program on schedule.

This may include, as in this example, an injection molded product, tool cavity dimensions, or die size analysis that is needed to manufacture the product. It is very important that all tooling variables are deemed correct and are stabilized during manufacture. Then, if a change is required, only those variables necessary are changed to make process adjustments and improvements.

At this time, it is important to evaluate the product's manufacturing tooling dimensions that form the product in the manufacturing process. This should be done in all manufacturing processes so it is eliminated as a source of the problem and the operator does not have to keep continually adjusting the process to bring the product into the customer's specifications.

Assuming the tooling is correct, we can begin with the following example to explore the controlling variable(s) in the manufacturing process for the product.

A glass-reinforced, injection-moldable polyester is to be used to make a pin that's length is not meeting customer requirements. The pin must measure 2.00 inches long (+/–) 0.003 inches). But, the average pin length is coming out 1.990 inches, which is 0.010 inches too short. A method must be found to reduce the material shrinkage or pack out the pin in the tool cavity to bring the pin into tolerance.

STEP 2. BRAINSTORMING FOR SOLUTIONS

Different techniques can be used as have been discussed. These may include an abbreviated quality circle, value analysis/evaluation, cause and effect, or fishbone analysis of the process. Also, depending on the seriousness of the problem, a special team of engineers and production personnel can be brought together to solve the problem.

In this situation, the "fishbone" diagram can be used as a useful quality analysis method for this group to determine what variables may be affecting the finished length of the pin. From their discussion, the problem-solving team can draw up a list of possible variables that may be the cause of the problem. Remember, no idea is to be left out no matter how absurd. As an example, the group has listed seven variables that should be investigated as causes for the apparent high material and part shrinkage.

But, effect prior to this stage, the incoming material should have been checked for variability and glass content, which will have a definite effect on mold shrinkage of the material in the tool cavity. Now, assuming the tool and material are within specifications, the variables selected to be investigated are listed from the "fishbone" diagram.

STEP 3. RANKING THE VARIABLES

From the discussion, the variables are ranked in relation to most probable causes, but no discussion on "why" is permitted. All possible variables must be listed and then ranked by the team so as not to miss a possible cause of the problem. The screening process alone will show what variables are significant and their importance of pin length. Often, the aspects of a few ideas may be combined to narrow the list.

Feasibility, practicality, effectiveness, and cost should be considered in ranking the ideas. The group's top seven ideas were as follows:

TABLE B-1. Variable Factors.

Factor	High Level (+)	Low Level (–)
A. Tool temperature	370°F	320°F
B. Cure time	60 sec	30 sec
C. Injection pressure	2500 psi	1500 psi
D. Screw back pressure	200 psi	50 psi
E. Injection time	12 sec	6 sec
F. Flow of material	soft	stiff
G. Secondary hold time	30 sec	15 sec

1. Mold temperature
2. Cure time
3. Injection pressure
4. Screw back pressure
5. Injection time
6. Flow of material in the tool
7. Secondary hold time

At this point, the team decides at what levels the variables will be run during the screening experiment. The variable levels should be selected to show the extreme levels of the equipment's operating range. Therefore, the high and low values should be at the edge of the operation window for both the material and the manufacturing equipment.

The reason for selecting these variable extremes is to point out major changes. Therefore, as a result, if the change does not show a significant effect on the pin variable and only a small statistical result is obtained, then that variable is almost certainly not significant. The selected variables, high and low selected values, are listed in Table B-1 to be used in the screening experiment.

STEP 4. STATISTICAL RANDOMIZING THE RUN

The importance of this step is to select a screening matrix run of (n) times that accommodates (n – 1) variables. In this example, an eight-run design evaluates the seven variables selected. The matrix designs go up by blocks of four; a four-run design evaluates three variables; a twelve-run evaluates eleven variables. For each of the eight trials, the pattern of highs (Hs) and lows (Ls) in the eight-run matrix in Table B-2 dictates whether to use the high or the low level of the variable at the head of each column. As an example, in run #1, variables A, B, C, and E will be run at their high levels, whereas variables D, F, and G will be run at their low levels.

TABLE B-2. Eight-Run Variable Screening Matrix. (Adapted from Ref. [1].)

Run	A	B	C	D	E	F	G
1.	+	+	+	−	+	−	−
2.	−	+	+	+	−	+	−
3.	−	−	+	+	+	−	+
4.	+	−	−	+	+	+	−
5.	−	+	−	−	+	+	+
6.	+	−	+	−	−	+	+
7.	+	+	−	+	−	−	+
8.	−	−	−	−	−	−	−

TABLE B-3. Twelve-Run Screening Experiment Design. (Adapted from Ref. [1].)

Run	A	B	C	D	E	F	G	H	I	J	K
1.	+	+	−	+	+	+	−	−	−	+	−
2.	+	−	+	+	+	−	−	−	+	−	+
3.	−	+	+	+	−	−	−	+	−	+	+
4.	+	+	+	−	−	−	+	−	+	+	−
5.	+	+	−	−	−	+	−	+	+	−	+
6.	+	−	−	−	+	−	+	+	−	+	+
7.	−	−	−	+	−	+	+	−	+	+	+
8.	−	−	+	−	+	+	−	+	+	+	−
9.	−	+	−	+	+	−	+	+	+	−	−
10.	−	+	+	−	+	+	+	−	−	−	+
11.	−	−	−	−	−	−	−	−	−	−	−

The same would be true if only two variables were selected. You use a four-run design and only have columns A and B with the high and low variables dictated by the pattern shown (*n*) Table B-2, columns A and B, for pattern of (Hs) and (Ls). Also, included is a twelve-run variable matrix run in ($n - 1$) or eleven times shown in Table B-3 for a twelve-run screening experiment.

Also, the order of the runs should be randomized to cut down the delay time between high- and low-variable adjustment. This means variables related to temperature should start out either low or high with others more easily adjusted by a variable change within the temperature extreme variable selected. This will normally tend to randomize the runs as required during the experiment.

As an example, the low-temperature trials for tool temperature (2, 3, 5, 8) are run first because the ease in raising temperature rather than lowering it. They are then randomized as (2, 5, 3, 8). The high-temperature trials (1, 4, 6, 7) are then run but randomized as (1, 6, 4, 7). The order of the experiment is, thus, (2, 5, 3, 8, 1, 6, 4, 7).

STEP 5. RUNNING THE EXPERIMENT

The equipment is set up for the first selected set of conditions and allowed to reach steady-state conditions. At this point, select one cavity and start to collect samples. The usual sample size is five; more may be taken but never less than five. The cavity selected depends on the runner system built into the tool. If it is a balance runner system, then any cavity will usually do. But if unbalanced, usually an extreme cavity is selected because of the flow distance and pressure drop anticipated within the tool at this point in the cavity system. This cavity or cavities will usually be the most difficult for obtaining the desired part dimensions because of the reasons mentioned. As a check, you can also select a cavity closer to the sprue to determine whether the variation also occurs within that cavity as the variables are changed.

Once the manufacturing cycle is in equilibrium for the first sample, they are collected with the second and successive set of conditions set on the machine and are allowed to reach equilibrium and samples collected for each sample cycle.

The molded samples are then measured at the critical dimensions after cooling down for a predetermined time period. The same cooling time for each set of samples is always used to rule out differential post-mold shrinkage caused by moisture pickup or other temperature or post-mold shrinkage variables. The average for the sample dimension is obtained for each run and entered in the "Average" column of Table B-4.

The average value (X) is the summation of all sample dimensions (Xn) in the run divided by the number of samples (n) and is the measure of the central tendency.

$$X = \frac{X_1 + X_2 + X_3 + \text{------}Xn}{n} \qquad X = \frac{\Sigma Xn}{n}$$

TABLE B-4. Average Pin Length: Averages and Ranges for an Eight-Run Screening Experiment.

Run	Average Length, in	Range of Length, in
1.	1.992	0.008
2.	2.001	0.005
3.	2.000	0.007
4.	1.990	0.008
5.	1.998	0.009
6.	1.995	0.005
7.	1.998	0.009
8.	1.991	0.006

The range for each run is then calculated for all run samples in a similar manner and entered in the range column of Table B-4. The range value is the difference between the highest and the lowest measured dimension for all samples in the run.

The time for running the trial will vary depending on the variables selected. It may be as short as 1 hour to a few days. But, by changing only one variable at a time, it could take even longer and the variable that has significant effect on the critical pin dimension could be missed.

If the job is run on more than one machine, normal processing can continue on the other machine. A one-manufacturing machine operation production will be lost during the experimental run but can be made up later once the critical variables are determined, with all products then meeting customer specifications. Bad products or rejects will be eliminated and the manufacturer will have a much higher confidence level of being able to accept all the parts produced to reduce inspection time.

STEP 6. SEPARATING THE EFFECTS

Each variable during the trial is run in combination with all other variables to produce test samples for each run. During the analysis, each variable is considered individually whether its low or high level has had an effect on the product's pin length and by how much.

Evaluation begins by the filling in Table B-5 with the values obtained in Table B-4, for the average pin length of each trial run.

TABLE B-5. Matrix for a Seven-Variable Screening Experiment Using Pin Length. (Adapted from Ref. [1].)

Run	A Tool Temp	B Cure Time	C Inject Pressure	D Screw back Pressure	E Inject Time	F Flow Material	G Hold Time
1.	+1.992	+1.992	+1.992	+1.992	+1.992	+1.992	+1.992
2.	–2.001	+2.001	+2.001	+2.001	–2.001	+2.001	–2.001
3.	–2.000	–2.000	+2.000	+2.000	+2.000	–2.000	+2.000
4.	+1.990	–1.990	–1.990	+1.990	+1.990	+1.990	–1.990
5.	–1.998	+1.998	–1.998	–1.998	+1.998	+1.998	+1.998
6.	+1.995	–1.995	+1.995	–1.995	–1.995	+1.995	+1.995
7.	+1.998	+1.998	–1.998	+1.998	–1.998	–1.998	+1.998
8.	–1.991	–1.991	–1.991	–1.991	–1.991	–1.991	–1.991
Sum H	+7.975	+7.989	+7.988	+7.989	+7.980	+7.984	+7.992
Sum L	–7.990	–7.976	–7.977	–7.976	–7.985	–7.981	–7.974
Diff.	–0.015	+0.013	+0.011	+0.013	–0.005	+0.003	+0.017
Effect	–0.004	+0.003	+0.003	+0.003	–0.001	+0.001	+0.004

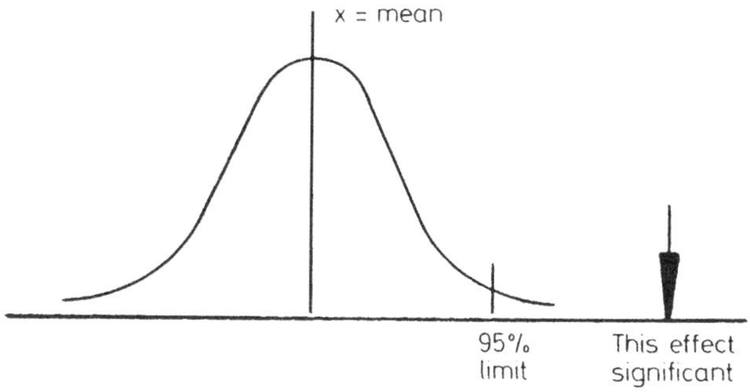

FIGURE B-1. *Bell curve with effects.*

Begin by filling in Table B-5, the data table, for run number 1. The value 1.992 is listed for each variable, and where the "H" or high value is indicated from Figure B-1, mark it (+) plus. Where "L" low, mark it (–) minus. So for run number 1, the highs are listed (+) in columns A, B, C, and E. The lows marked (–) are in columns D, F, and G. This procedure is then continued for each successive run until the completed matrix Table B-5 is completed.

Then, for each variable column, the sum of the highs (+) is written in sum H and, likewise, the sum for the lows (–) in sum L. Subtracting sum H then obtains their difference from sum L. This value is entered as either (+) or (–) depending on the larger of the high or low sum values in the (Diff.) row.

The difference value for each variable is then divided by the number of times the variable was changed from high to low, or four times, and it is entered in the effect row with the same (+) or (–) sign. This will statistically reduce the calculated value to its true value.

By using the plus and minus values, going from a low mold temperature to a high temperature, variable A, decreased the part dimension by 0.004 inches, or the mold shrinkage increased on the pin length.

The lower tool temperature therefore caused a decrease in the part's shrinkage or, more positively, an increase in pin length. Then by reviewing the variable effect line, one can determine how each variable affected the pin length and by what amount. From this data, one can select the significant variable that, if changed, may bring part size within the customer's requirements. The next step in the analysis is very important, it looks at the normal variation within a system and provides the criteria that determine what variables are significant.

For easier identification, use red markings for plus values and blue for minus (or any distinctive color) to identify easily the value changes and the tendency to either plus or minus.

Therefore, by reviewing the variable effect line, one can easily view how each variable affected the pin length when going from low values to high values and by what amount. From these data, one can then select the significant variables that by being changed will improve the process and bring part size within the customer's requirements and specification.

The next step is important in that it looks at the normal variation within a system and provides the criterion that determines what variables are significant.

STEP 7. TESTING FOR SIGNIFICANCE

A variable will be significant if its effect, as calculated by step 1 through step 6, is larger than the system's normal variation. Therefore, a test is required to evaluate normal variation.

The criterion that determines what variables are significant is the normal variation within the system. Based on this, a test is used to determine the normal variation within the manufacturing process producing the pins and affecting their manufactured length.

Any calculated effect in the matrix will be significant if it is larger than the normal variation computed by the following formulas and procedure. Effects smaller than the normal variation will not be significant.

The minimum significant factor effect (MSFE) is a statistic that uses the range values from each run in the calculation. The MSFE is developed by multiplying the standard deviation of an effect (range values) by the Student t value.

Standard Deviation of an Effect (Range pin length/run)

$$R = \frac{\text{Sum of the Ranges}}{\text{Number of trials}} + \frac{\Sigma R}{T} = +\frac{0.057}{8} = 0.007$$

The average range (R) is calculated from values in Table B-1.

An estimate of experimental error (σ EE) is obtained from:

$$\sigma\,\text{EE} = \frac{R}{d_2} = \frac{0.007}{2.326} = 0.003$$

The term d_2 is an estimator used to convert average ranges to standard deviations. It is found in Table B-6 or from statistical tables. The term k is the number of samples collected; in this case it was five per trial run d_2 will vary as the number of samples collected. It is important that the same number of samples are collected for each run, and they should not vary once selected for the other runs.

The standard deviation of an effect (Sd effect) is then calculated:

$$\text{Sd effect} = \frac{2\times\sigma\text{EE}}{\sqrt{N\times k}} = \frac{2\times(0.003)}{\sqrt{8\times 5}} = 0.00095$$

Table B-6. Table of d_2 Estimator Values. (Adapted from Ref. [1].)

Number of Samples N	Estimator (d_2)	Number of Samples N	Estimator (d_2)
2	1.128	16	3.532
3	1.023	17	3.588
4	2.059	18	3.640
5	2.326	19	3.689
6	2.534	20	3.735
7	2.704	21	3.778
8	2.847	22	3.819
9	2.907	23	3.858
10	3.078	24	3.895
11	3.173	25	3.931
12	3.258		
13	3.336		
14	3.407		
15	3.472		

The number of samples measured per run is (k), in this case it is five, and N is the number of runs. It is important that the same number of samples are collected for each run, as d_2 will vary with the number of samples collected.

Next, one must calculate the confidence level desired. This means to estimate/calculate which effect may be significant and where it lies. This would normally be outside the bell curve, three sigma limits.

For a 95.0 percent confidence level, only 2.50 percent of the relative probability remains in each tail of the bell curve. This means that one time in forty an effect will be calculated as statistically significant but will not be significant. Figure B 1 shows this in a form more easily understood.

The Student t distribution is used to approximate a distribution when the ample size is small and where sigma (σ) must be estimated from data that appear normally distributed. Therefore, we consider the sampling distribution of the t statistic.

$$t = \frac{Y - \mu}{\sqrt{(N)n}} = \frac{\text{average} - \text{mean}}{\sqrt{\text{Standard deviation of average}}}$$

Figure B-2 shows how t approaches the normal variant, as the n number of samples becomes larger. The more samples measured, the closer the data will approach the three-sigma distribution curve. Based on the sample size selected, t is then determined by calculating the degrees of freedom df of the experiment. "T" equals the number of runs and k is the sample size measured.

Degrees of freedom:

$$df = T(k-1) = 8(5-1) = 32$$

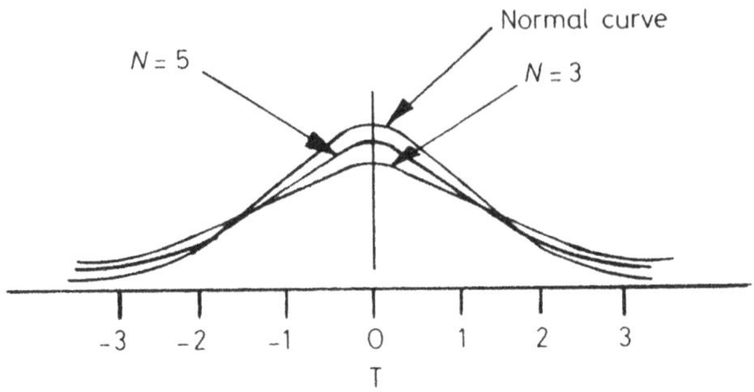

FIGURE B-2. *Comparing normal and* T *distributions.*

Using Table B-7 for a 95.0% confidence level and tracking down the *df* column to 30 yields an approximate or estimated *t* value of 2.042. It can be extrapolated to get the actual value for a *df* of 32, that is 2.0378, but because it is so close, the *df* value of 30 can be used. Therefore, the minimum significant factor effect is calculated by:

Values of *t* are listed in Table B-7.

$$\text{MSFE} = (t \times 0.95df)(\text{Sd Effect}) = (2.042)(0.00095) = 0.002$$

$$\text{MSFE} = 0.002$$

Therefore, any calculated effect equal to or less than the MSFE value, in this example 0.002, will cause minimal or no effect on pin part length. By referring back to Table B-5, one can observe that variables A, B, C, D, and G are greater than the MSFE value, and these variables indicate a positive effect on controlling pin length.

The larger the effect value, the greater the variable will affect the dimension. But all effects above 0.002 should be considered when analyzing the data and reevaluating the process settings.

STEP 8. ANALYZING THE DATA

When a factor is significant, it has a direct effect on the part dimensions. When a factor is significant, it has an effect on the variable of interest that is the part length, for this example. When the effect is positive (+), the variable increases as the factor is increased from the low level to the high level. When the effect is negative (–), the variable decreases as the factor is increased from the low level to the high level.

Therefore, because variable B, C, D, and G were positive, increasing the variable will have a positive effect on increasing part pin length. With variable A, which was negative, decreasing the variable will increase pin length.

TABLE B-7. Values of *t* for Different Degrees of Freedom and Confidence Levels. (Adapted from Ref. [1].)

	Degrees of Freedom				
df	90%	95%	98%	99%	99.9%
1	6.314	21.706	31.821	63.657	6363.619
2	2.920	4.303	6.965	9.925	31.593
3	2.353	3.182	4.541	5.841	12.941
4	2.132	2.776	3.747	4.604	8.610
5	2.015	2.571	3.365	4.032	6.859
6	1.943	2.447	3.143	4.032	5.959
7	1.895	2.365	2.998	3.499	5.405
8	1.860	2.306	2.896	3.355	5.041
9	1.833	2.262	2.821	3.250	4.781
10	1.812	2.228	2.764	3.169	4.587
11	1.796	2.201	2.718	3.106	4.437
12	1.782	2.179	2.681	3.055	4.318
13	1.771	2.160	2.650	3.012	4.221
14	1.761	2.145	2.624	2.977	4.140
15	1.753	2.131	2.602	2.947	4.073
16	1.746	2.120	2.583	2.921	4.015
17	1.740	2.110	2.567	2.898	3.965
18	1.734	2.101	2.552	2.878	3.922
19	1.729	2.093	2.539	2.861	3.883
20	1.725	2.086	2.528	2.845	3.850
21	1.721	2.080	2.518	2.831	3.819
22	1.717	2.074	2.508	2.819	3.792
23	1.714	2.069	2.500	2.807	3.767
24	1.711	2.064	2.492	2.797	3.745
25	1.708	2.060	2.485	2.787	3.725
26	1.706	2.056	2.479	2.779	3.707
27	1.703	2.052	2.473	2.771	3.690
28	1.701	2.048	2.467	2.763	3.674
29	1.699	2.045	2.462	2.756	3.650
30	1.697	2.042	2.457	2.750	3.646
40	1.684	2.021	2.423	2.704	3.551
60	1.671	2.000	2.390	2.660	3.460
120	1.658	1.980	2.358	2.617	3.373
Infinite	1.645	1.960	2.326	2.576	3.291

For example, in variable C, injection pressure, using the higher packing pressure of 2500 psi, versus 1500 psi, increased the pin length. With variable A, tool cavity temperature, increasing the tool cavity temperature caused greater material or pin length shrinkage, thus decreasing the pin length. Therefore, decreasing tool cavity temperature will have a lengthening effect on the pin length.

STEP 9. CHANGING THE PROCESS

Based on the screening experiment, it was discovered that five of the seven variables had an effect on the part pin length. Four variables optimize the length when at their high value, one variable optimizes at its low value, and two variables have little or no effect on part pin length.

Now would be the time to run one more experiment with the significant values at their upper or lower limits to test their effect on part pin length. If in this test the pin length falls within specification limits consistently, the new cycle variable settings would be determined.

This may not often be the case, and fine tuning may be required to obtain the final results. However, the variables and their effects are known and adjustment (fine tuning) can now be accomplished with minimum effort and time. It may in some cases be sufficient to adjust only one of the more meaningful variables to bring part pin size within specifications. But, in a worst-case scenario, the combined effects may not be the right solution. This means combining them together may move the dimension greater or less than the effect of each added together. This interaction of effects is lost in a screening experiment.

If this occurs, a full factorial experiment 2^7, 128 times, is required with each varying in two ways. This experiment is beyond the scope of this presentation.

Any properly designed, performed, and analyzed experiment can yield positive results and is the key solution to tricky part-specification problems. The total process control techniques are then applied to control the manufacturing process and maintain the variables within their processing window.

Appendix C

Checklists

NO. 1 PRODUCT DEVELOPMENT CHECKLIST

Date:

Customer:
Address:
Contact: Phone: Fax: E-mail:
Alternate:

Customer/In-house part development number:
Market established: Benefits to market:
Market size: Estimated volume/Year:
Users of product:

Anticipated sales price: How calculated: How sold:
Manufacture in-house: Outside supplier: Joint:
Outside supplier(s):

Purchased parts required:
Parts:
Suppliers:
Cost:

Total Quality Process Control for Injection Molding, by M. Joseph Gordon

New part: Existing: Redesign REQ'D.:
Metal replacement:
Competition: Who: Sale price:
Market size: Share desired: Estimated sell price:
Patentable: Applied for: Date:
Patent no.:

Program assets available: Required:
Market introduction date anticipated:
Assistance REQ'D.: Type:
Whom: What areas:
Probability of program success:
Estimated completion date:

Project team leader: Alternate:
Project start date:
Decision dates: ______________ ______________ ______________
All REQ'D. Info, available Assets available Start date

Development team members:
Customer if designated:
Sales:
Engineering:
Design:
Production:
Tooling:
Quality:
Purchasing
Finance:
Management:
Suppliers:

Part development (team analysis of product)
Part requirements (general, specific, liability items), be specific:
1.
2.
3.
Benefits to user:
Limitations of current product:
Competitions product evaluation:
Quality requirements:
Improvements possible:
Possible to combine functions:
Possible to change material:

Checklist analysis:
Customer requirements: Date:
Sales/Contract: Date:
Engineering: Date:
Problem: Date:
Design: Date:
Material: Date:
Program scheduling Date:
Manufacturing: Date:
Tooling: Date:
Purchasing: Date:
Suppliers: Date:
Price estimation: Date:
Development: Date:
Assembly: Date:
Decoration: Date:
Packaging: Date:
Shipping: Date:

Agency and code requirements:
Who: What:
Customer requirements:
What:
Program status, continue: Need more information:
What:
Terminate: Reasons:
Continue: Approved by: Date:

Part design & material selection:
Design checklist completed to suit requirements:
Additional information required:
What:
From whom: Required by:
Available: Needs to be developed: By when:
How:
By whom:

Part/Design analysis:
Designer:
Type calculations: Finite Element Analysis:
Stereolithography/Selective Laser Sintering: Model:
Prototype:
Time estimated to complete: Cost:
Completion date:

Material candidates:
A: Why:
B: Why:
C: Why:

Supplier A:
Supplier B:
Supplier C:

Supplier property data available: Requested: Date:
Required data:

If not avaliable, can it be developed: By whom:
By when:
Supplier contact: Phone: Fax:
E-mail:

Prototype: Testable: Type:
Who provides: When:
Specified in contract:
Cost estimate:
Full size: Material:

Prototype testable: Type of tests required:
Actual end use conditions: Simulated:
Conditions:
Requirements to pass:
What identifies failure/pass:
Who determines:
Agency/Code requirements testable:
What:
Testing time:
Test cost:
Number of tests: Samples required:
Supplier test data required:
Procedure defined: Procedure number:
Who does testing:
Contact: Phone: Fax: E-mail:
Who evaluates data
Documentation REQ'D.:
Certification REQ'D.:
Test results in what form:

Pass/Fail:
Comments:

Project status checkpoint:
Continue: Terminate: Need more data:
What:
By whom:

Design finalized:
Customer approved: By whom: Date:
Title:

Material selected:
Supplier: Product code:
Alt. Supplier: Product code:
Can either be subsituted at will:
Decision authorized by only:

Each material must be end-use tested before final approval:
Supplier on certified supplier list:
If not, when:
Who approved:

QA approval status of suppliers:
Supplier certification type required:
Specific lot data:
Typical lot data:
Special requirements:
Approval status:

Purchased parts required: What:
Suppliers:
Certification REQ'D.:
Vendor audited for quality When: Status of audit:

Critical dimensions:
Drawing available for discussion: Drawing no.:
Dimensions attainable: Plastic tolerances:
Number of critical tolerances:
Where 1: 2. 3.

Inserts used: Type:
In mold: At assembly:
Screws used: Type:

Other assembly methods:
SNAP/Press fit:
Sonic:
Thermal:
Solvent/Adhesives:

Quality requirements: (see quality checklist):
What major requirements:
Who determines:
When verified:
By whom:
Test equipment required:
What:
Available:
Supplied by whom:
Cost of testing:
Customer to verify tests: Only data: Both:
Procedure number:
Documentation required:
What:
How to report:

Manufacturing method (see checklist)
Method:
Tooling:
Special REQM'TS.:
Where: By whom:
Contact: Phone: Fax: E-mail:
Capability of equipment evaluated: CP: CpK:
CR:

Personnel training REQ'D.:

Process control used: Closed loop feedback
Other:
Real time process control used:
Manufacturing procedure documented:
Procedure number:
Work instructions:

Mold design (see checklist)
Completed:
All team members approved design:
If not, who disagrees:
Why:
How resolved: By whom: Date:

Special requirements: What:
Mold type: Number of cavities:
Balanced runner system: Replaceable gate block:

Supplier:
Contact: Alternate:
Phone: Fax: E-mail:
Estimated price: Delivered when:
Mold special features: What:
Core pulls/Unscrewing:

Mold flow analysis: By whom: Results:
Mold cool analysis: By whom: Results:

Mold tryout: Where: By whom:
Press size: Ounces: Tons of clamp:
Platten size: Mold fit within:
Process control used for tryout:

Material: Grade: Lot number:
Regrind allowed: Used: Percentage:
How blended:
Trial date: Length of trial:
Good parts produced:
If not, what was problem:
How will it be corrected:
By whom:
When:
Retrial of mold scheduled: When: Where
Mold trial results:

Final mold trial date:
Length of trial: Time: Cycles:
Good parts produced:
Parts meet customer requirements:
If not, what was lacking:
Fixable: Processing: Mold:
Material:
Tool approval: Date: By whom:

Existing tooling: Last molded at:
Maintenance performed to meet product requirements:
Quality assurance approved all dimensions:
Date:
Contact:
Reason for transfer:
Tool drawings available:

Parts list available:
Known problems with tool, documented:
Corrected:
By whom:
Verified: By whom: Title: Date:
Mold approved for production:
By whom: Title: Date:

Assembly method (see checklist)
Completed:
Assembly required: Fixtures REQ'D.: Available:
Must be developed: By when: By whom:
Who pays:
Assembly drawing available: Drawing number:
Type of assembly:
Press fit: SNAP fit: Sonic: Thermal:
Solvents: Adhesives:
Screws: Other or combination of methods:
Repairable: Type allowed:
Sealed unit: Type:
Hand/Machine assembly: Required assembly rate:
Process/Specifications defined: Document number:
Part cleaning required: How: With what:
Must keep part dry as molded: How: With what:
Stored where:

Assembly testing required: What:
Procedure number: Testing specifications:
Specification number:

Assembly equipment required: Is it capable:
Must it be calibrated: To what:
Specifications:

Adhesive/Solvents used: System:
Supplier: Contact: Phone:
Material Safety Date Sheet sheets required with order:
Osha requirements:

Purchased parts in assembly:
What:
Supplier:
Contact: Phone: Fax: E-mail:
Quality rating: Approved supplier:
Inspect before assembly: Specification:
By whom: When: Where:

Decoration method (see checklist)
Completed:

Packaging method (see checklist)
Completed:

Piece-part cost analysis (see price check sheet)
Completed: By whom: Date:
Approved by management: Date:
Approved by production: Date:

Material cost:
Manufacturing cost:
Mold cost:
How paid for:
By whom:
Amortized over production run:

Assembly cost:
Purchased part costs:
Decoration cost:
Packaging cost:

Total cost of program:

Final program analysis:
Continue:
Terminated:
Comments:

Approved by:
Date:
Customer representative:
Date:

Program start date anticipated:

NO. 2 SALES & CONTRACTS CHECKLIST

Date:

Customer:
Address:
Contact: Phone: Fax: E-mail:

Application:
Volume/Year: Anticipated part price:
Release quantity: Frequency:
Part size (sq. in.)
Drawing/Sketch/Prototype available: Type: DWG. No.
Who designs product:

Requirements:
Part development check list used:
Part is new/existing/redesign:
Users:
Agency/Code approval REQ'D.:
Special situation:
Supplier certification REQ'D.: What:

Type of manufacture:
Anticipated material: Type:
Supplier:
Is company capable of supplying:

Purchased parts used: What:
Who furnishes: Inventory REQM'TS.:
Assembly REQ'D.: Type: Checklist used:
Decoration REQ'D.: Type: Checklist used:
Packaging & shipping REQM'TS.:

Mold/Tooling:
Who supplies:
Who designs:
Type: Number of cavities: Balanced:
Special in mold REQM'TS.: What:

Existing mold:
Condition:
Reason for transfer:
Last molder:
Contact: Phone: Fax: E-mail:
In-house trial to access condition:
When: Where:
Mold drawings available: Bill of materials:

Who built current mold:
Contact: Phone: Fax: E-mail:
Special equipment REQ'D. To run mold: What:
Who furnishes:
Process condition records available: Where;
Parts available:
Material used:
Supplier:
Grade: Amount on hand:

New tool:
Who designs:
Who owns:
Anticipated cost:
Customer payment method, direct on approval, with partial payments:
Amortized over production run as parts are delivered:
Who approves tooling:
Who approves first parts off tooling:
Contact: Phone: Fax: E-mail:

Mold check list used:
Maintenance requirements:
Who approves repairs:
Who pays: Approval REQ'D. before repair:
Contact: Phone: Fax: E-mail:

Quality requirements:
Quality checklist used:
Incoming material tests REQ'D.:
Purchased parts tests REQ'D.:
Requirements specified:

Prototype testing required:
Type, model/molded part/SLA/SLS/model, other:
Who furnishes:
Time requirements to provide:
Cost anticipated:
In-process testing REQ'D.: What
Requirements:
Equipment required:
Who furnishes: What:

End-use testing required:
What:
Who performs:
Requirements:
Test equipment required: Who furnishes:

Testing documentation (SPC) required:
Incoming:
Production:
Assembly:
Decoration:
Special documentation REQ'D.:
What:
Customer required documentation prior or at time of shipment:
What:
Problem resolution:
Contact: Phone: Fax: E-mail:

Customer tests at incoming: What testing:
Procedures documented: What:
Who furnishes:
Contact: Phone: Fax: E-mail:
Production:
First article who approves:
Requirements (quality checklist or other used):
Form/Fit/Function:
Aesthetics:
Dimensional:
Color approval:
Special specifications:

Anticipated release quantities per order:
Just-in-time production REQ'D.: Frequency:
Shipment distance:
Inventory requirements: Who pays:
Payment terms/method:
Direct: Amortized: Other:

Contract terms specified:
Release on purchase orders: Timed releases:
Who specifies:
Customer approval to ship: Who:
Shipper specified:
Shipping paid by: Terms:
On release:
Contract terms used: What terms:

Customer quote to:
Address:
Quote due date: Time:
Quote delivered by: How:

Quote items required piece part price:
Scheduling:
Material:
Tooling:
Assembly:
Decoration:
Packing:

Special testing:
Special REQM'TS.:
What:

Quote number:
Quoted by:
Approved by:
Sales contact: Phone: Fax: E-mail:
Submitted date:
Customer response date:

Accepted/Rejected by: Why:
Any special terms required:
What:
Requote allowed if rejected:
Time limits:
Due by date:

Contract submitted date:
Customer approved date:
Customer officer approval:
Contract supplier approved by: Date:

NO. 3 PRODUCT DESIGN AND DEVELOPMENT SCHEDULE CHECKLIST

Date:

Customer:
Address:

Customer contact: Alternate:
Phone: Fax: E-mail:
Program start date: EST. Completion date:
Just-in-time program:
Anticipated quantity/Ship frequency: ______/______

Part name: Job no.:
Drawing available Drawing no.:
Part available: Model/Prototype:

Part description new: Existing: Competitors:
Function of part:

Existing material: Proposed material:
Part weight & specific gravity of material: _______/_______

Number parts in assembly, existing: Proposed:
Function of adjacent parts:
Able to combine functions:
What:

Purchased parts used: Type:
Type:
Type:

Customer incentive for project: Compensation:
Performance improvement: Redesign:
Weight savings: Cost reduction:
Meet new requirements: Alternate supplier needed:
Other considerations:

Production information:
Manufacturing method:
Volume (parts/year):
Supplier, in-house: Outside: Who:

Design considerations (obtain sketch of forces acting on part)
Part function:
Customer liability if failure occurs:

Operating conditions:	Normal	Maximum	Minimum
Temperature:	:	:	:
Service live (hours)	:	:	:
Forces (lbs/torque/etc.)	:	:	:
Durataion of forces time on:	:	:	:
Time off:	:	:	:
Maximum deflection allowed:	:	:	:

Load bearing application:
Buckling considered:

Impact forces: Where on part (sketch): Type:

Repeated: One time: Frequency:
Drop height: Load: Impact energy:

Vibration effects considered:
Vibration input:
Weight of assembly: Weight of internal components:
Mounting of components (method):
Attaches to another assembly or part: Part:
Function of this part:
Operating speed (RPM):
Exciting frequency (CPS):
Displacement (inches/mm):
Acceleration (G forces):

Environmental conditions:
Chemical exposure: Type: Concentration:
Chemical makeup:
Moisture (humidity): Percent: Temperature:
Water exposure type (fresh/salt/boiling/steam):
Temperature:
Radiation: Type: Exposure level: Time:
Sunlight: Exposure type (direct/indirect): Time:
UV protection required: Color fading a factor:
Ambient temperature not operating: Operating:

Electrical requirements:
Maximum current supported:
Maximum voltage:
Insulation properties required: Type:
EMI/EMP protection required: Values required:

Flammability requirements required: Type:
Smoke generation limits REQ'D.: Oxygen level limits:
Must meet UL flame requirements: VO/V1/V2/HB/V5

Agency requirements (UL/CSA/NSF/FDA/ETC.): Agency:
Requirements:
Customer requirements:
Other requirements:

Abrasion & wear required:
Type:
Mating part material:
COEFF. friction REQ'D.: Static: Dynamic:
Lubrication REQ'D.: Type: Chemical composition:
Self lubricating: Internal: External:
Type lubrication allowed:

Safety factor requirements:
Part liability: Severity if failure occurs:
Degree of liability to supplier:
Part failure impact on product/Application:
Consumer/Industry application:
Instructions on product to operate:
Warning labels required:
What information on label:
Who supplies:
Agency requirements/Testing required: What: Who:
Quality requirements:

Critical tolerances:
How many: Where: Tolerance:
Impact on part function if not met:
Part requirements (flash/warpage/sink/porosity):
Specifications established: Within-supplier capability:
Material/Parts incoming testing REQ'D.:
Testing specified:
Testing requirements:
Customer requirements established:
Requirements:

Will customer test incoming product: How: Where:
Tests required to meet:
Test equipment:
Procedure established:
Personnel trained:

Special equipment REQ'D.:
Customer contact: Phone:
Supplier witness tests: Must schedule:
If failure occurs, how are disputes settled:
By whom:

Integration of combining part functions considered:
What operations can be combined:
Material capable:
Assembly methods considered:

Part requires assembly:
SNAP/Press fit:
Mechanical fasteners:
Welding (thermal/sonic):
Adhesives/Solvents:
Other:
Plant has equipment to do assembly: What is required:
Personnel training required: What:
If no, use outside company:
Equipment required:
Cost factor on product:

Decoration requirements:
Colored: Compounded:
Salt & Pepper: Concentrates:
Color: Requirements: Color sample:
Must match mating part: Color/Material/Paint:
Pigment type allowed: Will it affect materials properties.
Testing requirements: What:
Special cleaning required: What:
Affects on material:
Painted: Paint type: Primer:
Number coats:
OSHA requirements involved: What:

Metalized: Type of metal: Thickness:
Plated: Type of metal: Thickness:
Etching of material required: What: Conditions:
In-house: Outside: Who:

Printing: Type: Molded in: On surface:
Surface preparation required: What:
Foils/Decals: Type: Supplier:

Printing:
Information required:
Who furnishes:
Finish on part: Class A: Textured: Depth:
Finish type: High polish: Sample available:
Must match mating part: Sample available:

Testing requirements:
Tests to meet:
Procedure:
Prototype tested:
Production tests:
End-use requirements:

Prototype:
Molded:
Modeled:
SLA/SLS/Other:

Customer information:
Design deadline: Extension time available:
Are all design requirements/End use information available:
If not, what is missing: When available:
End-use test available: Who tests:
Where: How many cycles:
Condition of part: Conditioning required: What:
Contact: Phone:
Who gives final approval of design:

Preliminary cost figures completed:
Anticipated piece part cost:
Quote number:

Have all departments been contacted for their design input:
Sales:
Engineering:
Purchasing:
Manufacturing:
Quality:
Tooling:
Assembly:
Decoration:
Packaging:
Shipping
Material suppliers:
Outside sources required:
Upper management approval:

Additional information required, not listed to assist in understanding completely the function, manufacturing, quality, assembly, and any other abuse or requirements the part must withstand or environmental stresses not listed:

Designer:
Design team signoff:
Company representative: Approval date:

NO. 4 MATERIAL CHECKLIST

Date:

Customer:
Address:

Part name: Job number:
Production start date: Production supervisor:
Material: Product code: Volume:
Alternate supplier: Product code:
Critical parts requiring use of same lot number of material, because of color, dimensions:
Part numbers:
Alternate parts production start date:

Product volume: Product weight:
Material REQ'D, LBS.: Order size, LBS.:
All one lot number or mixed: yes/no Confirmed:

Material certification required*: yes/no Certification to:
Special requirements, material values, color, properties, specification:
Values required:
Certification required with each shipment: yes/no
Prior to receipt of material: yes/no
Test values on material required:
Values required:

Package type: bags-drums-gaylords-bulk:
Price per pound/kilo: Volume discount:
Colored material: yes/no
Method: compounded-salt & pepper-concentrate (type)
Concentrate source:
Color sample required**: yes/no Type: color chip-surface type-resin
Pigment changes permitted: yes/no Must notify if required: yes/no
Notify who:

Special incoming material testing required: yes/no contact:
Tests:

QC contact at receiving:
Production contact:

Material routing on receipt: warehouse-silo-production-outside molder
Hold till testing completed: yes/no Contact:
Disposition if material fails incoming tests:

Notify contacts in: QC:
Production:
Sales:
Purchasing:

Other parts required for product sale:
Product name: Part number:
Supplier: Contact:
Date required:

* See inspection & material flow sheet no.:
** See color match request for verification:

Purchasing representative:

NO. 5 PURCHASING CHECKLIST

Date:

Customer:
Address:
Contact: Phone: Fax: E-mail:

Contract number:
Job number:
Production start date:
Job schedule completed:
Approved by: Date:

Purchasing requirements:
Buyer: Phone:
Buyer: Phone:

Prototypes: Type:
Supplier:
Contact: Phone: Fax: E-mail:
Purchase order REQ'D.: P.O. No.: Date:
Due by: Receiving contact:
Notify department manager(s) when specific materials are received:
Purchasing: Phone:
Production: Phone:
Engineering: Phone:
Assembly: Phone:
Decoration: Phone:
Quality: Phone:
Packaging: Phone:

Material & finished goods and parts:
Prime material supplier:
Grade:
Certification required: What: Specific lot data:
Required prior to or with receiving documents:
Test results required: What: Send to:
MSDS requested with order:
Quantity ordered: Pounds: Package type:
Purchase order number: Ordered date:
Contact: Phone:
Due to arrive on or before:
Shipper pro number required: What is number:
Trucking company:

Receiving notify on receipt:
Purchasing: Phone:
Production: Phone:
Quality: Phone:

Incoming testing required:
Quality contact: Phone:
Production: Phone:

Testing results:
Quality approved by: Date:
Inventory placement:
Where Bar coded to inventory:
Heated area REQ'D.: Temp. REQ'D.:
If rejected, reason: By whom: Date:
Supplier quality contact: Phone: E-mail:
Disposition:
Segregated from current inventory: Where:
By whom:
Special rejection label on packaging:

Reorder required:
Purchasing notified: When: By whom:
New P.O. Number:
Able to meet production start date:
Production notified: Who: When:
Scheduling required to be adjusted:
Sales notified: Who: When:
Customer contacted by sales: When:
Comments:

Tooling:
Supplier:
Contact: Phone: Fax: E-mail:
Contract/P.O. No.: Date entered:
Due by date: Received on date:
Production notified: Whom: Date:
Tooling notified: Whom: Date:
Quality: Whom: Date:
Engineering: Whom: Date:

Purchased parts for molding product:
Parts:
Purchase order number: Entered:
Quantity ordered:

Due in by:
Certification required:
What:
Supplier:
On approval list: Needs approval: Who/When approved:
Contact: Phone: Fax: E-mail:

Incoming testing required:
What:
Quality contact: Alternate:
Accepted/Rejected: By whom: Date:
Reason:
Reorder: P.O. Number: Required by:
Due in:

Purchased parts for assembly:
Parts:
Quantity ordered:
Purchase order number: Date:
Certification required:
What:
Supplier:
On approval list: Needs approval: Who/When approved:
Contact: Phone: Fax: E-mail:

Incoming testing required:
What:
Quality contact: Alternate:
Accepted/Rejected: Date:
Reason:
Reorder: P.O. Number: Required by:
Due by:

Parts:
Quantity ordered: Date:
Purchase order number:
Certification required:
What:
Supplier:
On approval list: Needs approval: Who/When approved:
Contact: Phone: Fax: E-mail:

Incoming testing required:
What:
Quality contact:

Accepted/Rejected: Date:
Reason:
Reorder: P.O. Number: Required by:
Due by:

Purchased parts for decoration:
Parts:
Quantity ordered:
Purchase order: Date:

Packaging:
Purchase order:
Quantity:
Special requirements:
Required by:

Receiving to notify:
Purchasing: Phone:
Production: Phone:

Special manufacturing equipment required:
What:
Supplier:
Contact: Phone: Fax: E-mail:
P.O. Number: Order date: Due date:
Notify:
Purchasing: Phone:
Production: Phone:

NO. 6 QUALITY CHECKLIST

Date:

Customer:
Address:
Contact: Phone: Fax: E-mail:

Part name: Job number:
Manufacturing start date:
Production supervisor:
Material: Product code:
Supplier:

Quality procedures per ISO9000/QS9000/other
Quality inspector:

Customer quality requirements known:
Document: Revision:
Engineering change orders received: What:
Incorporated into production: When: By whom:
Any deviations allowed: What:
Who approved at customer: When: Title:

Part requirements:
Physical: Document: Drawing:
Chemical: Document:
Electrical: Document:
Agency requirements: What:
Code requirements: What:

Part design documented: Drawing No.:
Material documented: Certification received: Results:
Incoming INSP/test results:
Confirmed by: Dept: Title:
Review of procedures by: All current:
Manufacturing: Reviewed by: Date:
Decorating: Reviewed by: Date:
Assembly: Reviewed by: Date:
Final testing: Reviewed by: Date:
Packaging: Reviewed by: Date:
Shipping: Reviewed by: Date:

Material safety data sheets available & current:
Manufacturing equipment maintenance current:
Tooling maintenance current:
Auxiliary equipment maintenance current:

Process control limits established: By whom:
Part quality limits documented: By whom:
Document:
Test & inspection equipment documented: By whom:
Procedure:
Statistical process control data reviewed:
Process control:
Documented for records:
Quality function deployment, analysis completed customer: Date:
Failure mode and effects analysis, analysis completed: Date:
Fishbone analysis completed: Date:
Measurement tool analysis: Date:
Metric requirements completed: Date:
SPC, requirements established: Date:
Six sigma analysis completed: Date:
All measurement items in certification: Date:

Test equipment avaliable:
What is required:
Inspector: Alternate:

Customer on site inspection required: During manufacture:
Final:
Customer inspector: Alternate:
Phone: Fax: E-mail:
Same inspection equipment used by customer:
What if not:
Who supplies:
Agency testing required: Contact:
Address:
Number of parts to send: Sent:
Information due back when:
Information/forms required:

Was real-time process control used during manufacture:
computer output saved and filed:
File name:

Quality records reviewed and signed off for shipment:
By whom: Date:

NO. 7 DESIGN AND DEVELOPMENT SCHEDULE CHECKLIST

Outline program: Start date:

A. Design team:
   1. Primary members: Alternate: Responsibility:
   2. Secondary members: Alternate: Responsibility:

B. Checklist: Start: Finish:
   1. Sales & contract
   2. Part development
   3. Part design
   4. Material
   5. Purchasing
   6. Mold, customer requirements
      Mold, design & materials
   7. Pricing
   8. Vendor survey
   9. Scheduling for manufacture
   10. Manufacturing
   11. Quality
   12. Assembly
   13. Decorating
   14. Packaging & shipping
   15. Problem solving (if required)

C. Design reviews
   1. Preliminary—program review with primary and secondary team input
   2. Design analysis and reviews from the checklists
      a. Part consolidation/function incorporation/value added extras
      b. Material possibilities/selection
      c. Material supplier inputs/data availability
   3. Design layout of part/parts, system
   4. Review of initial design/cost, projections/assembly/decorating, and design team input
   5. Review of mold requirements, type, functions, tolerances, and cost
   6. Manufacturing capability study
      a. Injection molding machine
      b. Mold
      c. Auxiliary equipment
      d. Shipping
      e. Packaging
      f. Material/certify/test/verify
      g. Plant support facilities
      h. Personnel training required
      i. Purchased parts/suppliers
      j. Assembly/decorating

7. Secondary operations
   a. Assembly/type/source/equipment/training
   b. Decoration
   c. Packaging and shipping
8. Quality requirements
   a. In-house—material, mold, process, product, and tests
   b. Supplier requirements/certifications
9. Finalize preliminary design
10. Prototyping, part/system
    a. Method/type/source/schedule
    b. Product requirements
    c. Testing required, code, customer, and agency
    d. Part function/aesthetics review
11. Customer feedback/analysis/conclusions
12. Finalize design
13. Tooling/mold design review with checklist
    a. Schedule, in-house/outside builds
14. Outside support equipment/services required
    a. Define, cost/schedule
15. Evaluate production tooling on manufacturing machines
    a. Process capability requirements
    b. Quality system to be capable and in control
    c. Finalize total quality process control procedure to produce zero defects and monitor in real time
16. Final customer approval/sign off/begin program

NO. 8 PRICE ESTIMATING CHECKLIST

Customer: Date:
Address:
Contact: Phone: Fax: E-mail:
Part name: Job number:
Drawing No.:

PIECE PART COST ESTIMATING PER 1000 PARTS

A. Material	:	:	:	:
B. Resin cost ($/LB)	:	:	:	:
C. Specific gravity (Sg)	:	:	:	:
D. Part weight (lbs)	:	:	:	:
E. Part weight (D × 1000)	:	:	:	:
F. Material cost (B × E)/0.95	:	:	:	:
G. Cycle time (CT)	:	:	:	:
H. Number of cavities (NC)[a]	:	:	:	:
I. Parts/Hour (H/G × 3600)	:	:	:	:
J. Cavity area (projected)[b]	:	:	:	:
K. Clamp force[c] (CF) Tons × (J × Material factor)	:	:	:	:
L. Shot weight (oz) (D × H × W[d] × 16 oz/lb)	:	:	:	:
M. Machine hour cost (Rate × MC[e])	:	:	:	:
N. Processing cost ($/1000 parts) M/I × 1000	:	:	:	:
O. Adjusted processingcosts[f] [N/(0.95)(0.80)]	:	:	:	:
Total cost (Processing per 1000 parts)	:	:	:	:

[a]Assumed three shifts/day, 6 days/week (see footnote f), 1 year's production produced.
[b]Projected cavity area & runner/sprue, mold cavity in square inches × number of cavities, plus runner and sprue area of mold surface in square inches.
[c]80% to 20% maximum shot weight of resin, use material clamp factor to estimate tons of clamp required.
[c]Use reference chart for shot weight.
[e]Adjust for current machine rates.

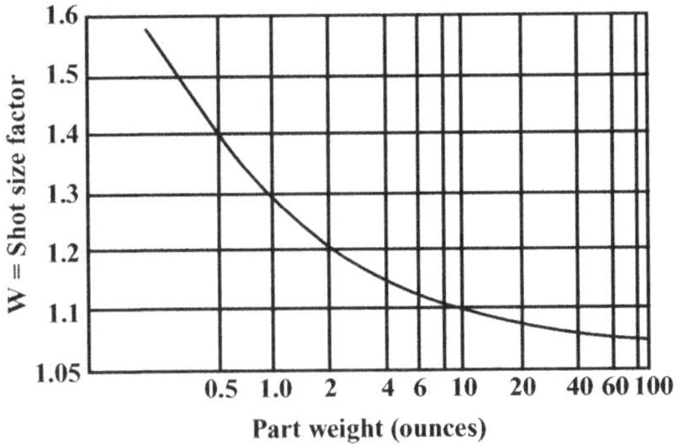

[f]Assumes 95% yield and 80% utility of molding process.

NO. 9 PROGRAM SCHEDULING FOR MANUFACTURE CHECKLIST

Customer: Date:
Address:
Contact: Phone: Fax: E-mail:

Quote number: Completion date: Review date:
Quoted by:
Submitted to customer:
Accept/reject: Reason:
Contract signed: By whom: Title: Date:

Program checklists completed, date:
Sales & contract: Manufacturing:
Part development: Quality:
Part design: Assembly:
Material: Decoration:
Purchasing: Packing & shipping:
Mold: Warranty problem solving:
Pricing:

Manufacturing documents completed:
Job traveler:
Job number for tracking:

Bar coding used: Labels specified: Type:
Information required:

Mold existing: Status:
Repair REQ'D.: What:
By whom: Who pays:
Mold trial date: Where:
By whom: Material:
Accept/reject mold: Reason:
Can mold be modified to make parts: By whom:
When:
Modifications required:
Mold accepted by production: By whom: Date:

New mold: Who designs:
Who pays: How:
Start: Finish:
Schedule determined for manufacture: When:
Mold trial date: Where: By whom:
Material:

Results:
Modifications required:
When completed:
Mold accepted by production: By whom: Date:

Purchasing:
Resin:
P.O./Date: Due in: Package: LBS:
Stored where:
Bar coded internally:
Finished parts: What:
P.O./Date: Due in: Inspection REQ'D.:
What:
Procedure No.:

Stored where:
Special storage required:
What:

Special equipment: What:
P.O./Date: Due in: Inspection REQ'D.:
What:
Procedure No.:

Manufacturing:
Start date: Finish date:
Quantity to produce:
Mold number: To order: Monthly:
Quarterly:
Machine number:
Procedure documented: Document number:

New mold setup: Document data required:
By whom:
Completed: Document number:

Auxiliaries:
Dryer: Chiller: Feeder: :
Conveyor: Part separator: Robot:
Grinder: Weigh scale: Packer:
Blender: Other:

Assembly:
Start date: Finish date:
Special EQM'T.: What:

P.O./Date: Due in:
Special parts:
What:
P.O./Date: Due in: Quantity:
Inspection REQ'D.:

Decoration:
Start date: Finish date:
Special EQM'T.: What:
PO/Date: Due in:
Special parts:
What:
P.O./Date: Due in: Quantity:
Inspection REQ'D.:

Packing:
Start date: Finish date:
Special packaging REQ'D.:
What:
P.O./Date: Due in: Quantity:

Part/material testing requirements:
What:
Where:
When:
By whom:
Test procedure:

Customer to verify:
Certifications required:
What:

Incoming:
Material: Test:
Material: Test:

In process:
Part: Test:
Part: Test:

Assembly:
Part: Test:
Part: Test:

Final:
Test: Type: Inspection:

Special EQM'T. required:
What:
Customer supplied/purchased:
P.O./Date: Due in:

Product certification REQ'D.:
What:
Due to customer: How: When:
To whom:

Invoicing: Date:
Invoice number:
Amount invoiced:
Quantity shipped:
Quantity ordered:
Over/under % allowed: Percent:
Discounts:
Terms:

Reorder anticipated:
When:
Quantity:
Price:

NO. 10 MANUFACTURING CHECKLIST

Date:
Customer:
Address:

Part name: Job number:
Manufacturing start date:
Production supervisor:
Set up time & date: Set up tech.:

Machine number/size Screw type: Check ring:
Nozzle type: Mold insulated from plattens:
Process procedure: Process control chart No.:
Product specification set: Document No.:
Specification limits established: Mean: UCL:
LCL:

Mold number/size:
Ownership:
Sprue bushing fits nozzle: Special bushing required:
Part No.:
Maintenance completed: Signed off by: Date:
Special requirements:
Installation procedure: Number:
Quick change: Mold preheat REQ'D.:
Temperature: Time:
Mold release allowed: Type:

Production equipment:
Dryer: Dry to % moisture: Temperature:
Dry time: Hopper/central/side dryer type: Filters clean:
Desicant good: Clean: Desicant bed dried:
Start before material added: Temp.: Time:

Mold chiller No.: Temperature setting: Flow GP:
Pressure: Cooling medium: Special hoses REQ'D.:
Type: Fittings: Setup procedure available:
Document No.:

Grinder No.: Last inspected:
Filter & unit vacuumed clean:
Blade sharpened: Last material ground:
Screen size in hopper:

Robot No.: Setup procedure: Document No.:
Setup by:
Special instructions required: Document No.:

Part handling:
Operator: Gloves required: Type:
Protect part surface: How: Electro-static:
Conveyor: Machine No.:
Sprue picker: Machine No.:
Mold sweep: Machine no.:

Special operations at press: What:
Equipment REQ'D.:
Special operator training: What:

Package product at press: How:
Special requirements: Packaging supervisor:
Material:
Supplier:
Package type: Hopper loading method:
Hopper capacity:
Drying time required: Sample test prior to start:
% moisture allowed:
Regrind allowed: Percent: How blended at hopper:
Procedure No.: Who adds to hopper: Frequency:

Process control limits established: Document No.:
Quality checks at press: Who approves:
Procedure No.:
Test equipment:
Verify what variables at startup:
Procedure No.:
Who approves saving first production parts:
Samples saved: How many: Quality checks:

Any secondary operations at press:
What:
Special part handling required: What:

Production problems contact:
Quality assurance contact:
Maintenance contact:
Material handling contact:

Parts boxed/palletized/counted/weigh counted/what

Parts to storage/station/holding point:
Parts protected: How & with what:

Any special instructions:

NO. 11 ASSEMBLY CHECKLIST

Date:

Customer:
Address:

Part name: Job number:
Manufacturing start date:
Production supervisor:
Assembly start date: Assembly supervisor:

Assembly drawing No.:
Part numbers:
Type of assembly required:
Procedure available: Procedure number:
Special instructions:
Special equipment required:
Equipment:
Special materials required:
Materials:
Osha requirements:
Operator training required: What: Provided by whom:
Color or texture match required: Type:
Who approves:
Procedure: Procedure number:

Purchased parts required: Purchase order no.:
Received in-house: Where in inventory:
Part numbers:
Quantity required:
Quality approved for assembly:

Total assembly in house: Outside supplier:
Supplier:
Contact: Phone: Fax: E-mail:

Finished testing of assembly required:
Requirements:
Who approves:
Test procedure no.:
Who does testing:
Rejects salvageable: How: Procedure No.:
Who approves:
Disposition of rejects:

Packaging requirements:
Procedure: Procedure No.:

Just-in-time product:
Special instructions:
Documentation required: What:
Documentation to whom:
Shipping contact:

NO. 12 DECORATING CHECKLIST

Date:

Customer:
Address:

Part name: Job number:
Manufacturing start date:
Production supervisor:
Decorating start date: Decoration supervisor:

Decoration required: Type:
Drawing number:
Before or after assembly:
Procedure No.:
Color match/texture required:
Approval by whom:
Procedure No.:
Reject handling procedure:
Salvageable:
Procedure No.:

Part surface preparation required: Type
Part surface testing required:
Test requirements:
Equipment required:
Equipment
Procedure No.:
In-house decoration: Outside: Who:
Training required: What:
Fixtures required: What:
Special: Ordered: Purachase order No.:
Recevied:

Decorating materials ordered:
Supplier:
Materials:
Purchase order No.:
Special requirements:
Certification required: What: By whom:

Osha requirements to be met:
Requirements:
Special equipment required: What: Ordered:
Purchase order No.: Received:

Decorated parts to:
Special handling required:
Parts to storage/station:

Just-in-time product:
Special instructions:
Document:
Packaging contact:

NO. 13 PACKAGING CHECKLIST

Date:

Customer:
Contact: Phone: Fax: E-mail:
Address:

Part Name: Job No.:
Production supervisor:
Manufacturing start date: Quantity:
Packaging required: Type: Special order:
Purchase order issued: P.O. No.: When:
Lead time to order packaging:
Packing due in: Notify whom:

Part protection REQ'D. before packing: What:

Just-in-time manufacture used:
Special packaging required for shipment:
If not, dunnage available:
What type: Who furnishes: Special requirement:
Supplier:
Requirements:
Reusable: How returned to supplier: By whom:

Part packaging performed where: By whom:
Special training REQ'D.:
What:

Number of parts per package:
Number of parts per carton:
Number of cartons per pallet:
Are pallets stackable: How many pallets high allowed:

Storage required before shipping:
Secured area:
QS9000 inspection REQ'D.: What:
Procedure number:
Who pays: How long:

Bar coding required: Who supplies:
Bar code specified:
Special instructions required What:
Lot No.:

Date of manufacture:
Product name:
Product code:
Other information:
Special packaging required: What:
Who supplies:

Packaging prodedure documented:
Document number:

Special trucking required for shipment: What:
Shipper: Contact: Phone:

NO. 14 WARRANTY PROBLEM CHECKLIST

Date:

Customer:
Address:
Contact: Phone: Fax: E-mail:
Industry/market of use:

Problem:

Problem reported by:
Report of problem submitted: Problem.:

Occurred at development: Prototype:
Final design: Production:
Assembly: Decoration: End use: Other:

Failure defined as:

Design-material-purchased parts-assembly-decoration-packaging:
Shipping-other area, describe in detail:

Failure occurrence—once: Several times:
Repeatable: No. times:
Same point or area on part: Variable: Where:
Sketch showing locations:

Sample of failed parts available:
Sent to: When: By whom:

Failure occurred at:
Manufacture: Assembly: Warehouse:
Shipping: End use:

Occurred in winter: Summer: Spring: Fall:
Tropical: Dry area: Other conditions, describe:
Section of country:

Seriousness:
Liability involved: What extent:
Status of failure: Known: Must be investigated:

Molded part failure analysis:
Material: Supplier: Product No.: Lot no.:
Certified by supplier: What certifications:
Incoming test record: Date tested: Test results:
Regrind used: Percentage: Number of passes allowed:

Chemical data available:
Physical data available:
Sample of part retained:

Engineering change order: Number: Date:
Approved by:
Customer approval required: Whom:
Granted by: On date:
All company departments notified of change order and their signature on approval:

Sign-off sheet:
Confirmed by: Title: Date:

Incorporated into part: When:

Agency/code approval granted: By whom: Date:

Part required agency/code certification:
What:
Contact: Phone: Fax: E-mail:

Colored material: Blended where: Concentrate:
Supplier:
Lot/P.O. number:
Contact: Phone: Fax: E-mail:
Any changes in pigment system ingredients during manufacture:
What:
Lot samples available: Test results from supplier:

Molded part: Part number: Date MFG'D.:
Purchased part: Part number: Date MFG'D.:
Supplier:
Mold number:
Cavity number: Consistant with failed parts:
Mold number: Drawing of mold available:
Supplier incoming inspection record: Date: Test no.:

Part failure analysis:

Mechanical failure:
Failure type:
Describe type of failure and if during use:
Customer:
Severity:
Repairable:
How:

Electrical failure:
Used as:
Failure type:
Customer:
Frequency of occurance:
Severity:
Repairable:
How:

Quality assurance:
Any testing showed problems:
Method of testing based on failure:
Any reports of prior failures of this type:
Customer reaction:
Severity:

Material problem:
Molding problem:
End-use application to sever for part/material:
Analysis of failure:

Manufactured in house: Outside molder: Who:
Contact: Phone: Fax: E-mail:
Production date: Lot No.: Shift:
Procedure for manufacture followed:
Problems noted during production: What:
Process control records reviewed: In real time:
By whom:
Molding press No.:
Mold number:
Maintenance performed last:
Maintenance records available:
SPC process data available:

Part analysis/test results:
Visual inspection:
Analytical results:

Physical:
Chemical:
Differential scanning calorimeter:
Thermo gravimetric analysis:
Infared analysis:
Other:
Material supplier analysis/input:

Solution to problem:

Corrective action response assigned to:
Date:
Time estimated to resolve:
Estimated cost to company:

NO. 15 (A) CUSTOMER QUESTIONNAIRE FOR PLASTICS MOLDINGS

SUBJECT: INQUIRY DATED: ______ TENDER TILL: ______

1. Customer: ______________________________ Place: ______________
2. Part designation: ______________________________
3. Drawing no.: ______________________________
4. Project and intended application: ______________________________

____

____

____

PURCHASING QUANTITY:

5. Total purchasing quantity: ____ pieces; annual requirement: _____ pieces
6. Batch size: _____ pieces/day: _____ pieces/week: _____ pieces/month _____

PACKAGING:

7. Packaging materials: ______________________________
8. Quantity per clearing unit: ______________________________
9. Particular specifications: ______________________________

SURFACE QUALITY:

10. Which are the visible surfaces of the molding? Must these have a high-gloss finish? Are the surfaces to be treated? (Painting, hot stamping, plating, printing, etc.) ______________________________

    ____

    ____
11. Where do sprue, gate ejector marks, symbols, company logo, mold cavity marks interfere? Where may they be made? (Mark on drawing) ______

    ____

    ____

DIMENSIONAL ACCURACY:

12. Are the "fine, medium, coarse" tolerances adequate, or must stricter tolerances be agreed on for certain dimensions? (Closer tolerances increase prices) ______________________________

    Tolerances to ______________________________

    Functional dimensions are ______________________________
13. May design changes be made for reasons of production technology?

    Yes ______ No ______

QUESTIONS CONCERNING THE MOLD

14. Mold available: Yes ____ No ____ cavities: size L ____ × W ___ × H ____
15. Mold to be manufactured: cavities; cavities; cavities;
16. Samples in (time applies following final approval of the part drawing)
17. Delivery period for series following approval of samples: ____________
18. Owner of mold: Customer _____ Supplier _____

QUESTIONS CONCERNING THE MATERIAL

19. Raw material types: ________________ Color: ________________
    ________________ _____ Clear _____ Natural
    ________________ _____ Transparent _____ According to sample
    Material spec. no: ________________ (from customer)
20. Which raw material was so far used for the part/project (if applicable) ___
    ____
    ____
21. Of what material is any counterpart made? ________________
    ____
    ____
22. What stresses must the molding withstand? Short ___ Continuous ___
    (a) Mechanical
    (b) Electrical
    (c) Chemical
    (d) Thermal
    (e) Climatic
    (f) Particular light effects

PRICE AND PAYMENT

23. Expected price for the mold: ________________
24. Expected price for the molding: ________________
25. Amortization of mold costs (%):________________
26. Conditions of payment: ________________
    ____
    ____

________________ ________________

Place and Date Customer's stamp and signature

(B) INTERNAL SUPPLEMENTARY QUESTIONNAIRE ON PLASTICS MOLDING

INQUIRY DATED ________ TENDER TILL: __________
CONVERSATION WITH: _________

1. Customer: ______________________ Tel: ______________________
2. Part designation: ____
3. Drawing No.: ____
4. Article design: mold-based technical changes, e.g., position of the parting line, etc., permitting lower piece cost and lower mold cost. For details, refer to drawing. ____________________________________
5. Is the part subject to further processing? Is it joined to other parts? If so, how? Tolerance example, shock loading, etc. ________________
   ____
   ____
6. Further processing: ____ automatic ____ manual ____ stackable ____ nestable
7. Reworking: drilling during assembly fastening of metal parts, etc.
   ____
   ____
8. Ejection tapers: due to surface quality, surface structure
   ____
   ____
9. Type of gating: pin gate, sprue gate, film gate, tunnel gate, etc.; at which surface? May the gate area differ in color from the rest of the surface as a result of machining required? ________________________
   ____
10. Straightness of long parts. Concentricity, flexion, and torsion; contact surfaces; sealing surfaces; permissible sink marks. ________________
    ____
11. Wall thickness ratios dependent on raw material and flow paths; wall thickness reinforcements by means of flow aids ________________
    ____
12. Are there any critical points in the production process? If so, where do they occur and why (material, processing)? ____________________
    ____
13. Total mold output: dependent on article and mold design; guaranteed minimum output without initial mold overhaul ____________________
    ____
14. Questions concerning the mold:
    No. of shots ______ per ______/cycles/min
15. Information concerning the sample: Material ___ Weight ___ grams ___
    Runner weight ___ grams
16. With which particular test specifications must the mold comply?
    (e.g., ASQ, 100% checks, random checks) ________________________
    ____

17. Who supplies the test gauges? ______________________________

18. Handling of complaints; incoming inspection ______________________________

19. Are there other important matters that neither of the two questionnaires has covered? ______________________________

20. Comments: ______________________________

Place and Date ______________________ Participants ______________________

TABLE Mold Design Guidelines.

Part size and volume	Up to 160 square inches of projected area; 1 million or more pieces.	Larger than 160 square inches of projected area; 1 million pieces or less.	Same as previous columns; up to 100,000 pieces (or preproduction).	Prototype (facsimile)
Mold design	Complete details of all part form dimensions in the mold design by moldmaker; all inserts detailed except standard purchased items (ejector pins, bushings, etc.).	Same as previous column.	General concept layout of mold only; details to be supplied by moldmaker optional.	Same as previous column
Cavity/core steel	Hardened AISI H-13, S-7, 420 stainless steel, A-2.	Prehardened steel, P-20, 414 stainless steel.	P-20 cold rolled steel, cast steel.	Aluminum or any of the previous column
Base steel	Prehardened steel, 280–320 BHN, 4140/P-20.	Same as previous column, integral base and/or cavity core.	Any steel.	Any steel or aluminum

TABLE (*Continued*)

Gating	Determined by molder to match part design.	Same as previous column.	Same as previous column.	Same as previous column
Base construction	Standardized A (cavity plate) or B (ejection half) series style bases guided ejection; tapered parting line locks (locators with zero clearance).	A or B series; integral cavity and/or core could be cut in solid base instead of inserts; guided ejection; parting line locks.	Whatever is necessary to hold cavity and core (doweled location on top of plate; universal mold-base system).	Same as previous column
Cooling	In cavity and core block; also in inserts to maintain best possible temperature control (use heat pipes, etc.).	Same as previous column.	Whatever is necessary to simulate production parts (possible sacrifice: longer cycle); typically water in mold base only.	Same as previous column
Area requiring moving components (not in line of draw)	Slides: mechanical or hydraulic; lifters angled or straight; all components to be hardened and made of dissimilar material or hardness to eliminate galling.	Same as previous column.	Possible short cut, but must simulate production parts	Could be loose pieces (manual inserts etc.; which come out with part and are reinserted for next cycle)
Special details	Fragile area of core or cavity inserted for ease of maintenance; bar ejection; blade ejection.	Same as previous column.	Not necessary	Not necessary

Source: Adapted from Ref. [1].

No. 15 (Continued) Checklist for mold designs

DATE ____________
CUSTOMER ______________________________ CONTACT __________
PART ____________ DRAWING NO. _________ REV. NO. _________
MOLD NO. _______ CONTRACT NO. _________
PHONE NO. ______

A. PART DRAWING	MOLD HALF (A) Fired Half	(B) Move Half	COMMENTS
__ Utility of part checked	___	___	___
__ Material selected (Vendor and No.)	___	___	___
__ Shrinkage calculated correct?	___	___	___
__ Gate type and location selected	___	___	___
__ Runner sized for flow distance	___	___	___
__ Matching contours for assembly	___	___	___
__ Tolerances attainable (Major/Minor)	___	___	___
__ Decorating required	___	___	___
__ Inspection equipment noted	___	___	___
__ Cooling fixture required	___	___	___
__ Assembly operations required	___	___	___
__ Draft and part finish noted	___	___	___
__ Quality level requirement	___	___	___
__	___	___	___
__	___	___	___
__	___	___	___
B. INJECTION MOLDING MACHINE			
__ Press No. ___ oz ___ tons ___	___	___	___
__ Screw type ___	___	___	___
__ Process setup sheets available	___	___	___
__ Press available	___	___	___
PRESS DATA			
__ Clamp force required ___	___	___	___
__ Injection press (Max) ___	___	___	___
__ Part surface area ___	___	___	___
__ Nozzle type ___	___	___	___
__ Nozzle angle and diameter ___	___	___	___
__ Sprue bushing matches nozzle ___	___	___	___
__ Tiebar "daylight" (opening) ___	___	___	___
__ Melt capacity of press ___	___	___	___
__ Ejection (hydraulic/mechanical)	___	___	___

	MOLD HALF		COMMENTS
	(A) Fired Half	(B) Move Half	
__ Cycle parameters ________________			
__ Inject/pack pressures ____________			
__ Cycle times ____________________			
__ Barrel temperature profile ________			
__ Back pressure __________________			
__ Shot volume ___________________			
__ Pad length _____________________			
__ Suckback required ______________			
__ Screw RPMs ___________________			
__ Screw recovery time ____________			
__ External connection ____________			
__ Hydraulic ______________________			
__ Air ___________________________			
__ Electrical ______________________			
__ Water _________________________			
__ Locating ring diameter __________			
__ Mold clamping _________________			
__ Core puller and control _________			
__ Ejector release mechanism _______			
__ Ejector coupling _______________			
__ Limit switch ___________________			
__ Heating connections ____________			
__ Regrind allowed _______% ________			
________________________________			
________________________________			
________________________________			
C. AUXILIARY EQUIPMENT			
__ Material drying required			
__ Dryer type __ Temperature __ Time __			
__ Mold temperature control (unit no.) ___			
__ Cool/heat _____________________			
__ Separate circuits in mold _________			
__ Degating of part _______________			
__ In tool ________________________			
__ Hand __________________________			
__ Part separator __________________			
__ Robot _________________________			
__ Special mold support equipment _____			
__ Operation _________ Type _________			
__ Process control limits (estimate) _____			
__ Feed to hopper _________________			

	MOLD HALF		COMMENTS
	(A) Fired Half	(B) Move Half	
D. MOLD (SUMMARY) MOLD NO.			
__ Type ____	____	____	____
__ Dimensions ____	____	____	____
__ Daylight required ____	____	____	____
__ Centering diameter (+/– in.) ____	____	____	____
__ Sprue bushing radius/diameter ____	____	____	____
__ Parting line marked ____	____	____	____
__ Ejector stroke dimensioned ____	____	____	____
__ Heat treatment noted ____	____	____	____
__ Venting noted ____	____	____	____
__ Cavity finish noted ____	____	____	____
__ Draft ____	____	____	____
__ Cooling circuits (individual) ____	____	____	____
__ Special cores for heat removal ____	____	____	____
__ Cores separate cooling circuits ____	____	____	____
__ Ejector system adequate ____	____	____	____
__ Sprue puller (type) ____	____	____	____
__ Balanced cavity layout ____	____	____	____
__ Tolerances anticipated ____	____	____	____
__ Gating into part ____	____	____	____
__ Press for sampling ____	____	____	____
__ Press for production ____	____	____	____
__ Materials for part ____	____	____	____
__ Shrinkage ____	____	____	____
__ Cavities numbered ____	____	____	____
__ Thermocouples in mold for temperature control ____	____	____	____
__ Uniform texture ____	____	____	____
__ Inserts required ____	____	____	____
E. HOT RUNNER TYPE __ VENDOR __			
__ Heat capacity/No. units ____	____	____	____
__ Gate type ____	____	____	____
__ No. of plugs ____	____	____	____
__ Electrical load ____	____	____	____
__ No. thermocouples (type) ____	____	____	____
__ Temperature settings ____	____	____	____
____	____	____	____
____	____	____	____
____	____	____	____

	MOLD HALF (A) Fired Half	MOLD HALF (B) Move Half	COMMENTS
F. UNSCREWING MOLDS			
__ Method/connections dimensioned _____	_____	_____	_____
__ Mechanical _____	_____	_____	_____
__ Electrical _____	_____	_____	_____
__ Hydraulic _____	_____	_____	_____
__ Air _____	_____	_____	_____
__ Internal/external to mold _____	_____	_____	_____
__ Core stripper plate travel _____	_____	_____	_____
G. QUICK MOLD CHANGE			
__ Tool built for QMC _____	_____	_____	_____
__ Service conn. in back plate _____	_____	_____	_____
__ Machine adapted for QMC _____	_____	_____	_____
__ Installation method (equipment available) _____	_____	_____	_____
__ Fittings/type _____	_____	_____	_____
__ Temperature control prior to mounting _____	_____	_____	_____
H. PARTS LIST	_____	_____	_____
__ All items recorded _____	_____	_____	_____
__ Quantities correct _____	_____	_____	_____
__ Spare parts noted _____	_____	_____	_____
__ Vendors listed _____	_____	_____	_____
__ Heat treatment noted & process _____	_____	_____	_____
__ Surface treatment noted and specified _____	_____	_____	_____
_____	_____	_____	_____
_____	_____	_____	_____

COMMENTS

____

____

____

____

Date _______________ Reviewed by _______________ (Design)

_______________ (Purchasing)

_______________ (Production)

_______________ (Quality)

_______________ (GM)

NO. 16 INJECTION MOLD SPECIFICATION SHEET

Date ___________

Customer: ____________________ Contact: ____________________

Part: ____________________ Dwg. No. _____ Rev. No. _______

Mold No. ____________________ Contract No. ____________________

Phone No. ____________________

Material ____________________ Supplier ______ Shrinkage ______

A. MOLD TYPE

- ____ 2 Plate
- ____ 3 Plate
- ____ Multiutility die (MUD)
- ____ Hot runner
- ____ Quick change
- ____ Surface area
- ____ Production life
- ____ (parts/year)

B. CAVITIES

- ____ Number
- ____ Balanced
- ____ Inserted
- ____ Fired half (FH)
- ____ Move half (MH)
- ____ Internal radii
- ____ External radii

RUNNER

- ____ Full round
- ____ Semiround
- ____ Quarter round
- ____ Trapezoidal
- ____ U-shaped
- ____ Cold slug traps

C. GATING

- ____ Sprue gate ____ Outside ____ Inside
- ____ Film
- ____ Submarine ____ Flare ____ Pinpoint
- ____ Tab
- ____ Pin
- ____ Fan
- ____ Hot runner: Type ____________________
  Supplier ____________________

- ____ Standard
- ____ Ring
- ____ Disc
- ____ Full edge
- ____ Spider
- ____ Pinpoint center
- ____ Degating type

D. COOLING (TEMPERATURE)

____

- ____ Optimum cooling
- ____ Cooled cores
- ____ Cooled core pins
- ____ Each cavity separate
- ____ Cooled cavity plates
- ____ Individual circuits
- ____ Surge cooling
- ____ Turbulent flow
- ____ Water line diameter
  FH ___
  MH ___
  Cores ___
- ____ Connections/quick release

E. HEATING (TEMPERATURE)

____

- ____ Hot water
- ____ Hot oil
- ____ Cartridge watts ___
- ____ Coil watts ___
- ____ Band watts ___
- ____ Plugs watts ___
- ____ Cavity
- ____ Core
  FC Temp ___
  MH Temp ___
- ____ Connection quick release

F. MATERIAL

	P-20	H-13	A-2	D-2	420SS	P-6	O-1	S-7	BERYL CU	SIL BRONZ				CASE HRD	HARDN'D	QUENCHED/	TEMPERED		NITRITED	CHROMED		
____ Mold set																						
____ Cavity plates FH																						
____ Inserts FH																						
____ Cavity Plates MH																						
____ Inserts MH																						
____ Slides																						
____ Back-up plates																						
____ Ejector plate																						
____ ____________																						
____ ____________																						
____ ____________																						

G. DEMOLDING

____ Stripper plate
____ Round ejectors
____ Stripper sleeve
____ Two stage strip
____ Flat ejectors
____ Angled

____ Unscrewing timed by ________
____ Hydraulic/air/mechanical
____ Sprue puller plate
____ Slide FH ___
____ Slide MH ___
____ Sprue puller FH/MH Type ____

H. CAVITY SURFACE—DRAFT _____ FH _____ MH _____ CORES

FH	MH	
_____	_____	High polish
_____	_____	Polish
_____	_____	Polish one-direction
_____	_____	Ground
_____	_____	Wet honed

FH	MH	
_____	_____	Sand plasted
_____	_____	EDM
_____	_____	Chrome plate
_____	_____	Textured
		Type ___
		Depth ___
_____	_____	Parting line match
_____	_____	Drawings available

FINISH QUALITY

_____ FH ________
_____ MH ________
_____ Cores _______

I. VENTING

_____ Cavity
_____ Core
_____ Ejector pins
_____ Blind vents

_____ Dead pockets
_____ Dummy pin
_____ Porous vent___to waterline
_____ Vacuum venting
_____ Vent blow back (release)

J. MISCELLANEOUS

____ Temperature sensors
____ Pressure sensors
____ Dial gauge
____ Mold number
____ Cavity numbered
____ Engraving (raised/undercut)
____ Company logo
____ Templates supplied
____ Decorating required
____ Drawings available
____ Insulated plate FH ___ MH ___
____ Sprue
____ Locating ring
____ Standard mold parts
____ Special mold parts
____ ____________________
____ ____________________
____ ____________________

K. MACHINE

____ To fit machine (sample) ___ Production ___
____ Mold dimensions
____ Daylight
____ Quick mold change operation
____ Automatic
____ Semiautomatic
____ Operator assist
____ Robot
____ Anticipated overall cycle time
____ Process control
____ Regrind allowed
____ ____________________

L. MOLDED PART QUALITY

____ To print
____ Critical dimensions number _____ Tolerance ______________
____ How checked ______________________________
____ Attainable w/gating selected
____ Time checked after molding
____ Part end-use environment temperature __________, __________

M. SPECIFIC INFORMATION

____ Receive all detail drawings with mold and parts list
____ Receive sample parts with measured points
____ Receive one set drawings with parts list with mold
____ No corrections for distortion on hardening
____ ______________________________
____ ______________________________

____ Price __________ Cavity
____ Estimated Price __________ Cavity
____ Fixed Price __________ Cavity
____ Delivery time: Weeks ___ Penalty if late ___

____ Mold acceptance based on
____ Parts to print
____ Part (form/fit/function)
____ End-use testing
____ Cavity/mold revisions negotiable
____ Cost
____ By you
____ By us

Design	______	By us	______	By you
Material	______	By us	______	By you
Sampling	______	By us	______	By you

_____ Part drawings available
_____ Mold drawings available
_____ Parts list available

Mold accepted by ____
____
____
____

Date ______________________________ Signature ________________________
____
____
____

NO. 17

Company Logo	CONTRACT BID AND AWARD REVIEW CHECKLIST	
Address	Doc. No.	Revision:
City, State, Zip	Date:	Page: 1 of 7

Contract No.: ____________________ **Name:** ____________________

CONTRACT ADMINISTRATION

Date Submitted: ____________ Date Completed: ____________

Have all parties reviewed their sections for the SOW/proposal/contract? Yes ___ No ___

Customer/agency name: ____________________

Contract number: ____________ Contract type ____________

Date of contract/Modification (MOD): ____________________

Delivery order/task order no.: ____________________

Pricing schedule & total price as proposed: Yes ______ No ______

Price same as proposal: Yes ______ No ______ Why not? ____________

Terms & conditions as proposed: Yes ____ No ____ What: ____________

Sales order and/or business master updates required: Yes ____ No ____

Statement of work same as proposal: Yes ____ No ____ Why not? ________

Contract exceptions: Yes _____ No _____ What: ____________

Contract administration approval sig.: ____________ Date: ____________

Technical proposal items are required and so stated in awarded contract. Yes ______ No ______

Are any Notice of Revision (NOR)'s involved and have they been received Yes ______ No ______

Comments: ____________________

Contracts approval: ____________ Date: ____________

B. ENGINEERING REQUIREMENTS

New equipment Yes ______ No ______Program name: ____________

Lead engineer: ____________ Phone: ____________ E-mail: ____________

Age of system: ___ Replacement items: Yes __ No __ New/old drawings ____

Re-engineering required Yes ___ No ___ What: ____________ Cost allowed:

Drawing package complete: Yes ___ No ___ Missing: ____________

Latest revision of all drawings provided Yes ____ No ____ Bill of Material (BOM) provided Yes ____ No ____

Engineering checklist evaluated: Yes ___ No ___ Results ____________

Software available for all operation, BOM, design, manufacture, drawing format Yes ____ no ____

Company Logo Address City, State, Zip	CONTRACT BID AND AWARD REVIEW CHECKLIST	
	Doc. No.	Revision:
	Date:	Page: 2 of 7

List software available and what additional software is needed with cost and license fee and how many licenses: ________________
Are additional computing hardware/interface requirements needed/acquired: Yes ____ No ____ What (e.g., bar coding, radiofrequency identification): ___
____
Design concurrence: Yes ____ No ____
Design cost concurrence: Yes ____ No ____
Method of evaluation/approval determined Yes ____ No _____ Special Test Equipment reqd. Yes ____ No ____
What: ________________ By whom? ________________
Specifications listed: Yes ____ No ____ What: ________________
Drawing Review Performed Yes ____ No ____ Results ________________
Outside Testing capability required Yes ___ No ___ Who ___ Cost: _____
Data output in what format? ___ Cost in quote? Yes ___ No ___ Cost _____
Defence Contract Management Agency (DCMA) verification required Yes ____ No ____ At site of test Yes ____ No ____ Where ____ First Article Yes/No ____ Advanced Notice required Days ________________
Cost proposal received Yes ____ No ____ What figure ________________
Special Testing required Yes ____ No ____ What ____________ Test Value
Tolerances reqd Y ___ N ___
Do we have the test equipment req'd. Yes ____ No ____ Type: __________
Outside test source needed Yes ____ No ____ Who ____ Cost ____ Test time ____________
Test Data sufficient to price Yes ____ No ____ CME verification required Yes ____ No ____
DCMA verification required Yes ____ No ____ Where/when ____________
Notice required days ________________
Describe 1st Article __________ All Releases __________
Engineering Approval: ________________ Date ________________

C. SPECIAL PROJECTS
Special terms/conditions/requirements
Review labor, rate table (as proposed for effort): Yes ____ No ____
Period of performance acceptable: Yes ____ No ____
Pricing as proposed: Yes ____ No ____
Peachtree updates required: Yes ____ No ____
Contract: MOD value concurrence: Yes ____ No ____

Company Logo Address City, State, Zip	**CONTRACT BID AND AWARD REVIEW CHECKLIST**	
	Doc. No.	Revision:
	Date:	Page: 3 of 7

Cost analyst approval sig: ________________________________ Date: ____
Special projects approval sig.: ______________________________ Date: ____
BOM review and minimum buys identified, Any penalties? Bonus?
Resolution of disputes, terms, Escalators, any hidden charges: ___________
____
Approval by: ___________________________ Date ____________________

D. PROGRAM MANAGEMENT

Program manager approval: __________________________ Date: ________
Delivery schedule approved: Yes ___ No ___ Quantities acceptable Yes ___ No ___ What: ____________________
Delivery schedule included Yes ___ No ___ Quantity/shipment Yes ___ No ___ Qty ________
Shipping method _____________ Shipper specified Yes ___ No ___
Who________________________ Acceptance Location _________________
____
Cost approval: Yes ___ No ___ Price approval: Yes ___ No ___ Who authorizes approval ____
Statement of Work (SOW) Approval: Yes ___ No ___
If government is DD250 electronic form used Yes ___ No ___
Executive program management approval sig.: _____________ Date: _______
Incentives/penalties: Are there any in the SOW or contract: Yes ___ No ___
Type ________ What ____
Status reports required: Yes ___ No ___ Type ______ Format ____________
Frequency ______________________________
Who prepares: _____________________________________ Approval req'd. Yes ___ No ___ Who approves: _______________________________________
Submitted to: ______________________________
CDRL:
Form 1423 items ____
Data item description ____
Data distribution list-item: ____
Personnel copied: ____

E. PURCHASING

Priced bill of material (BOM) include: Yes ___ No ___
Long lead time material identified. Yes ___ No ___ What _______________

Company Logo Address City, State, Zip	**CONTRACT BID AND AWARD REVIEW CHECKLIST**	
	Doc. No.	Revision:
	Date:	Page: 4 of 7

Best and worst delivery times: ____________________/____________________
Validity dates of quotes ____
EOM available Yes ___ No ___
Contract rating ____
Serial numbers or other special markings S# __________ Marking _________
Penalties for early shipment of product Yes ___ No ___ What ____________
Program manager approval sig.: __________________________ Date: _______
Customer supplied items provided Yes ___ No ___ What/by who __________
____
Any classified items Yes ___ No ___ What ______________________________
Software required OTS/Special Yes ___ No ___ Buy/Provided ____________
Who __________________
Budgeted Yes ___ No ___ Estimated cost ________________________________
Source ____
Purchasing approval sig.: ______________________________ Date __________

F. QUALITY ASSURANCE
Quality assurance plan developed for program Yes ___ No ___ By whom ______________________ Date _____________
Requirements/standards to meet ____
Classified items involved Y/N ___
What ______________________________________ Material certification req'd. Y/N ___ What ___
Special data required Y/N ____ What _____________ Data media _________
Tests ________ certificate of compliance, Special Pack/ship, Special materials Yes ___ No ___ What _____________
Supplier specified Y/N ___ What ____________________ Customer property/ equipment ________ Special information ________ Specific shipper ________
Receiving requirements/tests Y/N ___ What __________ Onsite test/approval Y/N ___ What ___
Special first article inspection requirements What ___ Customer notification time ___ Warranty ___
Prototype approval required Yes ___ No ___ By whom __________________
Special documentation Yes ___ No ___ Problem Resolution method ______
Special packaging requirements Y/N What _______________
Test/inspection time/personnel in cost figures Yes ___ No ___ Who Calc. _______ Hours _______

Company Logo Address City, State, Zip	**CONTRACT BID AND AWARD REVIEW CHECKLIST**	
	Doc. No.	Revision:
	Date:	Page: 5 of 7

Quality Requirements what ____________________ Standards as proposed: Yes ___ No ___
SOW? Configuration decided Y/N ___ What ____________________
Technical data package as proposed: Yes ___ No ___
Marking on part defined/legible/dimensioned/fonte type and size provided: Yes ___ No ___ What ____________________
Finish of part specified: Yes ___ No ___ What: ____________________
Quality assurance approval signature.: ____________________ Date: ______

G. MANUFACTURING/OPERATIONS

Special Purchasing requirements, Long lead items identified, Date reqd. ________ Lead time est: ________
Manufacturing costs as proposed: Yes ___ No ___ Why not? ____________
Any special technician training required Yes ___ No ___ What/by whom/ certification regd. ____________________
Special test equipment required/customer supplied, fixtures Y/N ___ What ____________________
Hours of effort (BOE) for all departments estimated and reviewed Y/N ___ What ____________________
Any special items or requirements, Y/N ___ What ____________________
Customer provided material, Y/N ___ What ____________________
First article due: ________ Source inspection, Who, when, ____________ where ____________________
Delivery schedule approved: Yes ___ No ___, Mfg checklist used Yes ___ No ___
Packing materials special/standard Yes ___ No ___ What ____________
Crating/pallets/no. units/pallet stack height ____________________
Manufacturing approval signature.: ________________ Date: ___________

H. COST ANALYST/FINANCE

Correct labor rate table attached: Yes ____ No ____
Accounting classification reference number (ACRN) information included: Yes ____ No ____
POP: ________________________ Dates verified: Yes ___ No ___ Dates of performance ____________________
Contract/MOD value reviewed: Yes ___ No ___ Change: Yes ___ No ___
Contract value same as quote Yes ___ No ___ If not What? ____________

Company Logo Address City, State, Zip	CONTRACT BID AND AWARD REVIEW CHECKLIST	
	Doc. No.	Revision:
	Date:	Page: 6 of 7

Cost analyst approval sig: ______________________ Date: ________
Finance approval sig: ______________________ Date: ________
Payments: sequence of payments (i.e., monthly, quarterly) ______________
Method of payments: ____
Payment address: ____
Special requirements: Yes ___ No ___ What: __________________________
Invoicing: method of invoicing (bill to customer, WAWF FFP, COMBO, Cost Voucher)
Receiving report required: Yes ___ No ___ Give to: ___________________
Review of invoice by USG required: Yes ___ No ___ POC: _____________
E-mail: ____

I. EXPORT / SHIPPING

M. Crews contacted: Export license for shipment outside the U.S. Yes ___ No ___
Military or commercial product ________________ Country _____________
Export license Comm: Dept. of Commerce: _____ Office of defense trade controls _____
Export requirements: ____
Export license military: Export Control Act/International Traffic in Arms Regulations (ITAR)
US State Department License: ______________________________________
Export requirements: ____

J. INVOICING

Method of invoicing (bill to customer, WAWF FFP, COMBO, Cost Voucher)
Receiving report required: Yes ___ No ___ Give to: ___________________
Review of invoice by USG required: Yes ___ No ___ POC: _____________
E-mail: _________________________________

K. SECURITY

Security requirements: Yes ___ No ___
Security Level: (sensitive, non-critical classified, secret, etc.) _____________
Special requirements: ____

L. SPECIAL, CHECKLISTS REQUIRED BY F. MUNRO

Build to print To be developed by ___________________________
Build to specifications To be developed by ___________________________

Company Logo Address City, State, Zip	**CONTRACT BID AND AWARD REVIEW CHECKLIST**	
	Doc. No.	Revision:
	Date:	Page: 7 of 7

R & D type programs To be developed by ______________________
REP
Phase I SBIR. CPFF

M. CONTRACTS MANAGEMENT
Resolution required Y/N ___ Concurrence of cross-functional management (A–I): Yes ___ No ___
Director contracts and procurement approval, Sign: ________________
Date ______
Comments: ____

Checklist Review Completed:
Proposal Lead: ______________________________ **Date** ________

Awarded Contract: ___________________________ **Date** ________

NO. 18 GOVERNMENT PURCHASE ORDER CHECKLIST P.O. _____

The following checklist must be completed and attached to all purchase orders when the initial P.O. total value exceeds $3,000, and any subsequent P.O. amendment/change exceeds $3,000 (the FAR Simplified Acquisition Threshold for fixed price purchasing).

1) Documentation

___ Purchase order checklist
___ Requisition (Copies of traveling requisition cards attached)
___ Federal prime contract number and applicable DPAS rating to apply on purchase order (From Contracts)
___ Company terms and conditions (as appropriate for military or commercial orders)
___ Prime contract-terms and conditions flowdowns (Section I of prime contract from (contracts)
___ Request for quotations/proposals (Check if verbal: ___ Verbal quotes must be documented)
___ Quotations/proposals received and no bids (Must be attached, verbals must be documented by confirming e-mail)
___ Bid Summary (Competitive Bids required if over FAR $ 3,000 micro-purchase threshold for supplies)
___ Purchase Order and P.O. Changes (Include All Revisions)
___ Unique, identification, number (UID) (Govt Orders Mil-Std-130 flowdown)
___ Service contract act applicability/flowdown (Govt orders for services over $100,000)
___ Preference for domestic specialty metals (Govt orders, flowdown is DFARS 252.225-7014)

2) Price Justification

___ Competitive	(Attach quotes)
___ Agreement in place	(Pricing established through negotiation)
___ Inter-company	(Ail Inter-company, transfer pricing established by Pall)
___ Standard catalog pricing	(Catalog Year ___ Page # ___)
___ Price history (Parametric Analysis)	(Refer to PO# ___)
___ Cost analysis	(Required for TINA orders)
___ In house engineering estimate	(Attach estimate)
___ Comparison with same or similar item	(Reference PN)

___ GSA schedule	(Schedule #)
___ Sole source negotiated	(if over $650,000) Subcontractor Cost or Pricing Data required DCAA Field Pricing Assist Audit required FAR 15.404 CME audit and negotiation plan required FAR 15.405–15.406 CME Price Negotiation Memorandum required FAR 15.406 Truth in Negotiations Act (TINA) certification required as of date of agreement on price FAR 15.406

3) Source Justification

___ Competitive	(Attach quotes)
___ Price history	(See above)
___ Customer directed	(Check one: Per blueprint ___ Pei spec ___ Per letter ___) Reference # ___
___ Agreement in place	(Pricing established through negotiation)
___ Inter-company	(Manufactured by Pall)
___ Engineering directed	(Check one: Per blueprint ___ Per spec ___)
___ Program directed	Memo Reference #: ___
___ Only supplier qualified	(Reason: See below
___ Economically justified	(Check one: Tooling ___ Delivery ___ Qualification test ___)
___ Other justifiable reasons	(Reason: ___ See below ___)

4) Certification Requirements (to be on file prior to order placement)

___ Representations and certifications	(Gov't orders over $10,000)
___ Nonsegregated certification attached	(Gov't orders over $25,000)
___ Debarred, ineligible or suspend certification attached	(Gov't orders over $25,000)
___ Clean air and water certification attached	(Gov't orders over $100,000)
___ Anti-lobbying certification attached	(Gov't orders over $100,000)
___ Cost accounting standards	(Gov't orders over $550,000)

N/A Small business plan	(Required if over $550,000)
EEO Preaward certification attached	(Gov't orders over $1,000,000)

Buyers Narrative of Procurement Issues./Concerns/Conclusions

Justification as follows (or reference to attachment Price Negotiation Memorandum).

These words always end the narrative: "Pricing is Fair and Reasonable based on negotiation"

Buyer Name: ______________________ **Date:** ______________________

Appendix D

Supplier Evaluation Survey

Company Logo	SUPPLIER EVALUATION SURVEY	
Address	Doc. No.	Revision:
	Date:	Page 1 of 12

COMPANY: ______________________________
ADDRESS: ______________________________
CITY, STATE & ZIP: ______________________________
PHONE: ( ) _______ **FAX ( )** _______ **E-MAIL** ______________

Date(s) of Evaluation Survey: ____________________
Person(s) Performing Evaluation Survey: ____________________
____

TYPE OF SURVEY:
1.__ PRESURVEY 2. __ INITIAL 3.__ FOLLOW-UP

Company Logo Address	SUPPLIER EVALUATION SURVEY	
	Doc. No.	Revision:
	Date:	Page: 2 of 12

AUDIT RATING: ______ # CLASSIFICATION:
___ ACCEPTABLE
___ CONDITIONAL
___ MARGINAL
___ UNACCEPTABLE

AUDIT LEADER:______________

AUDIT TEAM:________________

____

____

QUALITY ACCREDITATION: _______ ISO 9001 ______ AS 9100 ______

Other: ____

PRODUCT SERVICE:____

APPROXIMATE NUMBER OF PERSONNEL IN WORKFORCE:

MANUFACTURING: ______ ENGINEERING: ______ QUALITY: ______

TOTAL PERSONNEL: _____

UNION ?: ______ NAME: __________________________________

LENGTH OF CONTRACT:__________ EXPIRATION DATE:_________

PRODUCT/SERVICES PROVIDED TO: (INCLUDE APPLICABLE PART NUMBERS AND ACTIVE PURCHASE ORDER NUMBER) ATTACH ADDITIONAL PAGES AS REQUIRED.

____

____

____

SUPPLIER CONTACT:__________ TITLE OR FUNCTION:__________

MEMBERS:____

____

____

DISTANCE TO PLANT MILES/TIME:______________________________

PLANT SQUARE FEET:____

CONDITION OF FACILITIES/EQUIPMENT___________________________

OBSERVATIONS:____

____

Company Logo Address	SUPPLIER EVALUATION SURVEY	
	Doc. No.	Revision:
	Date:	Page: 3 of 12

SCORE	RATING GUIDE (Used to identify the qualities required from their suppliers)
BEST	Considered as the model for Continuous Improvement Processes.
	Continuous improvement, diagnostics in use, process focused.
	Total customer satisfaction is demonstrated.
	Methods, quality deployment applications and results verification are traceable to the quality planning activity.
	Prevention action rather than detection-based.
	Documented system understood with continuous improvement and root-cause corrective action.
	Statistical tools are used appropriately.
	Procedures and work instructions understood by all employees and applied.
BETTER	Proactive and system focused with total customer satisfaction as the goal.
	Decisions and results are traceable to the quality plan.
	Employees are totally involved with the planning, development and quality results.
	System is documented and evidence shows proper use.
	Procedures and work instructions understood by all employees and applied.
GOOD	Existing, little diagnostics, product, service, and standard practice focused.
	Existing, Industry standard practice for the achievement of quality and business objectives.
	Product and some system approach with evaluation performed but little employee involvement
	Acceptable processes and/or procedures, but not understood by all employees involved.
NOT ACCEPTABLE	Nonexistent, short-term thinking.
	No system or processes and/or procedures in place.
	Little or no verifiable evidence.
	No real consistent approach and deployment.
N/A	Not applicable for this supplier

Company Logo	SUPPLIER EVALUATION SURVEY	
	Doc. No.	Revision:
Address	Date:	Page: 4 of 12

A numerical scoring system for individual quality items for supplier/vendor approval criteria is not used. Each supplier is judged based on their capability to provide the goods and services required based on product requirements and specifications, quality, and delivery. Either a Yes or No answer will describe the capability of the supplier. Comments are welcome and identified at each sections list of quality review items.

Quality System and Management **YES—NO**

Quality objectives and responsibilities are clearly stated, widely distributed, and understood throughout the organization. ________

Quality goals and continuous improvement objectives are communicated to all areas of the organization. ________

The quality and reliability goals are aggressive relative to customer expectations and are targeted toward continuous improvement. ________

Management has a proactive "defect prevention" attitude to achieve continuous improvement. ________

Management reviews internal audit results of Manufacturing, Engineering, Purchasing, Materials, and other functional departments. ________5

YES—NO

Management periodically reviews customer complaints with Quality, Manufacturing, Engineering, Purchasing, and Materials. ________

Management identifies resource requirements and provides adequate resources including trained personnel. ________

Company Logo Address	SUPPLIER EVALUATION SURVEY	
	Doc. No.	Revision:
	Date:	Page: 5 of 12

Internal and external quality measurements are available and visible to the organization. __________

Quality system is registered to ISO 9001, ISO/TS16949, or AS 9100. Which:__________ __________

A dated and approved Quality Policy & Procedure manual exists and is available to all areas of the organization. __________10

Customer Purchase Orders are reviewed for accuracy and compliance prior to acceptance to ensure the capability to meet all stated requirements. __________

The Quality organization is independent from manufacturing, engineering, etc. for segregating nonconforming products to assure the product integrity and meet the customer requirements. __________

A documented and active internal audit program is in place. __________

A Supplier Audit/Survey Program is in place. __________

Advanced problem-solving techniques (FMEA, Cpk, DOE, etc.) are used throughout the organization. __________15

A price-of-quality program is in place and reviewed and acted on regularly by senior management. __________

Personnel are trained in the policies and procedures of their operations and departments. __________

Training plans and training records for all employees are clearly documented. __________

A formal method exists for employees to submit suggestions for improvement, and there is a timely feedback mechanism. __________

Company Logo Address	SUPPLIER EVALUATION SURVEY	
	Doc. No.	Revision:
	Date:	Page: 6 of 12

Document Control **YES—NO**

A secure system is in place to ensure that current documents, drawings, and specifications are available and controlled. ______

A documented process is in place to incorporate changes. ______

A system is in place to control records. ______

YES—NO

A secure system is in place for controlling software media used for manufacturing, inspection, and test. ______

Electronic documentation is regularly backed up and placed in a secure location. ______

New product development procedures are in placed and they are followed in the design process. ______

Does the system ensure that the most current customer specifications are available to the Procurement function? ______

Is conformance to customer specifications assured before an order is taken? ______

Purchasing **YES—NO**

A process is in place to assure procured materials are purchased from approved suppliers. ______

Is quality history considered along with price, delivery, and service when making source decisions? ______

Company Logo Address	SUPPLIER EVALUATION SURVEY	
	Doc. No.	Revision:
	Date:	Page: 7 of 12

Does an effective quality monitoring and improvement program exist for their suppliers (including capability/capacity reviews)? ________30

Purchase material requirements are specified. ________

Partnerships or alliances are formed with major suppliers to improve total performance (i.e., quality, cost, and delivery). ________

The company understands the principles of point of use storage / Kanban / Lean Manufacturing and has evidence that systems are in place or concepts are being put into practice. ________

Incoming Material Control **YES—NO**

Receiving inspection facilities and equipment are maintained. ________

A documented receiving inspection process, including testing methods, is in place. ________35

Critical characteristics are clearly communicated.

The results of receiving inspection activity is documented, maintained, and retained. ________

The inspection status of purchased material is clearly identified. ________

Nonconforming purchased material is identified and segregated from conforming material in a secure manner. ________

Company Logo Address	SUPPLIER EVALUATION SURVEY	
	Doc. No.	Revision:
	Date:	Page: 8 of 12

Inventory Control and Production Planning	**YES—NO**
Inventory accuracy is maintained.	________
Suppliers are monitored.	________
There are dedicated Customer Service personnel for a specific customer.	________
Configurations of stocked items are controlled.	________
Order due dates are managed with respect to past due conditions.	________
Supplier deliveries are kept in line with production need dates.	________
Supplier/shop delays are communicated to the customer.	________
A Master Schedule has been developed.	________

Process/Quality Control and Calibration	**YES—NO**
A documented process is in place for verification and control of the quality of in-process material exists.	________
A documented quality plan is in place for assuring the quality of in-process check/final inspection of outgoing product (Cpk, FMEA, etc.).	________
The results of in-process inspection is documented and communicated to senior management on a regular basis.	________
The packaging and storage of in-process products is adequate to prevent damage and theft.	________

Company Logo Address	**SUPPLIER EVALUATION SURVEY**	
	Doc. No.	Revision:
	Date:	Page: 9 of 12

A documented First Article Inspection process exists. ____________

Rejected in-process material is clearly identified, segregated, and dispositioned. ____________

A Lot Control System or equivalent is in effect that provides traceability of the product through the manufacturing process to the raw material or other purchased material. ____________

Statistical process control (SPC) is used throughout the organization for improvement. ____________

Control charts or other statistical methods are being used as needed and are available at the point of application. ____________

Incapable processes or machines are targeted for improvement or replacement. ____________

YES—NO

The corrective action process clearly addresses containment, root-cause failure analysis, feedback to internal and external customers, and follow-up to verify the problem is permanently fixed. ____________

A comprehensive Preventative Maintenance program exists for all equipment. ____________

Housekeeping procedures are documented and followed. ____________

In-process material protected from electrostatic discharge (ESD), deterioration, or damage is in place. ____________

In-process material is identified and controlled. ____________

A written procedure is in place to specify the calibration frequency of all inspection gauges and other measuring devices. ____________

Company Logo Address	SUPPLIER EVALUATION SURVEY	
	Doc. No.	Revision:
	Date:	Page: 10 of 12

Written calibration procedures for all gages and other measuring devices are in place. __________

Calibration standards are up to date and traceable to NIST or other appropriate standards. __________65

Calibration records indicate gauges are calibrated as prescribed by the written procedure. __________

All tools and fixtures that are used as a media of inspection are qualified and identified. __________

Calibration status is clearly indicated on all measurement equipment used to accept or reject material. __________

Manufacturing Capability and Innovation **YES—NO**

This site has the manufacturing capabilities, including personnel, equipment, processes, etc. to meet the manufacture requirements of parts. __________

Gauge R & R (Repeatability and Reproducibility) is used as to determine manufacturing capability. __________70

5S is in place.
- Sort (used daily, used weekly, used monthly)
- Storage (open access, shadow boards)
- Shine (clean, free of clutter)
- Standardize (processes, tools, work instructions)
- Sustain (monitor, measure) __________

New process development procedures exist and are followed. __________

A program is in place to review existing processes to reduce cycle time and cost. __________

Company Logo Address	SUPPLIER EVALUATION SURVEY	
	Doc. No.	Revision:
	Date:	Page: 11 of 12

The company is involved in concurrent engineering or early design input with customers. __________

A tracking system is used to measure improvements (metrics) in manufacturing. __________75

The company works with their suppliers to develop innovation/improvement. __________

Control of Nonconforming Material, Corrective Action & Outgoing Product Quality **YES—NO**

Nonconforming product is effectively controlled and dispositioned. __________

A process for the disposition of nonconforming material, material review board (MRB) is documented. __________

Rejected or suspect material is adequately identified and segregated from production. __________

Suppliers are notified of nonconforming material and corrective action requested. __________80

A process is in place for notifying customers of the shipment of nonconforming material. __________

A process for the disposition of scrap is documented. __________

In-process defects, customer complaints, and supplier-identified defects are analyzed in a timely manner to detect and eliminate potential causes on nonconforming product. __________

Inspection samples are taken on the basis of characteristic classification, i.e., Major/Minor or Critical. __________

Company Logo Address	SUPPLIER EVALUATION SURVEY	
	Doc. No.	Revision:
	Date:	Page: 12 of 12

Results of final inspection are documented, tracked, and trended for improvement. ____________85

Rejected material is clearly identified and segregated. ____________

Packaging and storage of finished products adequate to prevent damage. ____________

Final acceptance facilities and equipment are in place. ____________

Root-cause failure analysis is performed on internal and external failures and appropriate corrective action is implemented. ____________89

Overall Supplier Rating Criteria

Score	**Rating = Total number of YES responses**
	Outstanding and exceeds expectations
	Satisfactory and meets expected performance
	Meets most expectations and needs some improvement.
	Needs significant improvement
	No system or process exists

Appendix E

Mold Problem Solutions

TABLE E-1. Common Molding Problems and Solutions*.

Problem	Cause(s)	Solution(s)
Black streaking	Contaminated plastication system Improper drying	**Machine:** Purge or clean cylinder or nozzle. Check machine cylinder for cracks. Check nozzle seating. Turn on hopper throat water. Clean around top of hopper. Use machine with smaller cylinder. Reduce screw RPM. **Mold:** Check for grease or oil in mold. Reduce hot-runner block temperature. Change to higher temperature mold slide grease. Reduce knockout pin lube level. **Material:** Dry material properly.
Burn marks (on part)	Improper mold venting Injection rate too high Screw RPM too high Back pressure too high Clamp pressure too high	**Machine:** Reduce injection speed backpressure, booster pressure temperature settings. Reduce booster time. Check for resin contamination. **Mold:** Enlarge or clean mold venting. Center mold core if shifted. Change gate location. **Material:** Change grade or flow.
Burn marks (at gate)	Excess shear heat	Same as burn marks on part.
Color fading	Melt temperature too high Improper pigment	**Machine:** Reduce melt temperature. Use machine with smaller cylinder. Speed cycle. **Material:** Use higher temperature pigments. Reduce level of fines in regrind. Change to part painting or dyeing.

Total Quality Process Control for Injection Molding, by M. Joseph Gordon

TABLE E-1. ***(Continued)***

Problem	Cause(s)	Solution(s)
Discoloration	Processing temperatures too high Improperly sized cylinder Contamination	**Machine:** Purge cylinder. Lower melt temperature. Use smaller cylinder machine. Lower rear zone heats, backpressure, screw RPMs. Use smaller shot size machine. **Mold:** Reduce hot runner block temperature. **Material:** Examine material for contamination source.
Splay or silver streaking	Moisture trapped in material Melt temperature too high Cold mold (causes condensation) Poor pressurization Incorrect mold setup	**Machine:** Lower melt and nozzle temperature. Use shorter nozzle. Check nozzle for blockage. Reduce injection speed. Open gates or increase number of gates. Use smaller cylinder machine. Raise backpressure. Check for water source near hopper. Correct excessive melt decompression. Lower screw speed. **Mold:** Check for water leak in mold. Open sprue, runners, gates, and/or vents. **Material:** Dry material before use. Install fresh desiccant bed in dryer. Reduce regrind levels.
Brittleness	Processing temperatures too high Improperly sized cylinder Poor mold design Too much regrind Contaminants/fines present Improper drying	**Machine:** Lower melt temperature. Increase injection speed or, if brittleness worsens, decrease injection speed or increase gate size or number. Use smaller cylinder machine. **Mold:** Lower mold temperature. Increase radii of fillets and corners. **Material:** Decrease amount of regrind. Dry before use. If too dry, reduce drying and lower stock temperatures and pressures. Check for contamination, oversized pieces, excessive fines. Moisturize parts before shipping. Use higher impact grade. Check part relative viscosity.
Distorted parts (warping)	Material is oriented during injection Difference in packing density Molded-in stresses Nonuniform melt-flow into mold Part too hot when ejected Improper knockout system Poor part design	**Machine:** Increase cure time. Decrease hold time. Lower melt temperature. Raise melt temperature for better packing. Reduce drying if resin is too dry. **Mold:** Check for uniformity of knockout action. Lower mold temperature on concave side of warped part and raise it on convex side. Slow mold opening. Redesign part. Open gates, runners, sprues, and nozzle. Relocate gates. **Material:** Switch to nucleated resin.
Delamination (Pits, orange peel, wrinkles)	Processing temperatures too low Improper drying Contaminants present	**Machine:** Raise melt temperature and injection rate and pressure. Speed injection rate. Purge machine cylinder. Raise backpressure. Raise screw speed. Increase hold time. **Mold:** Raise mold temperature. Enlarge vents and gate size. **Material:** Dry material. Search for contamination.

TABLE E-1. (*Continued*)

Problem	Cause(s)	Solution(s)
Flashing (on parting line)	Injection pressure too high Melt temperature too high Mold malfunction	**Machine:** Reduce boost and/or hold pressure. Reduce holding time. Increase clamp force. Move mold to larger press. Reduce boost time, feed settings. Lower melt temperature. Check platen alignment. Make gate open times consistent. Check pyrometer. **Mold:** Align/adjust mold. Examine for contamination on mating surfaces. Check die cavities for mitering. Reduce mold temperature. Improve mold venting. **Material:** Switch to different flow.
Nozzle drool	Nozzle too hot Barrel front zone too hot Pressure build-up	**Machine:** Lower nozzle temperature, melt temperature. Increase decompression time. Switch to smaller orifice nozzle. Use reverse taper nozzle. **Material:** Dry material.
Freeze-off nozzle	Processing temperatures too low Improper nozzle design	**Machine:** Raise nozzle temperature. Increase heater band watt. Move heater bands forward. Check for burnout on nozzle. Decrease cycle time. Use nozzle with larger orifice. Examine nozzle orifice for contamination. Use reverse taper nozzle. **Mold:** Raise mold temperature
Short shots	Insufficient melt volume High-pressure drop in mold Mold malfunction	**Machine:** Increase melt temperature and cycle time. Increase booster pressure and time. Increase feed setting. Check all heater bands with pyrometer. Move mold to larger cylinder press. Check bridge in hopper throat. Turn on hopper throat cooling water. Check machine hydraulics. Increase injection pressure and speed. Repair broken check valve. **Mold:** Raise mold temperature. Open gates, widen runners (or narrow runners if pressure allows). Thicken part design. Improve cold slug well. Clean or enlarge vents. Improve gate locations. Open runners and/or nozzle. Change vent location. **Material:** Use easy flow grade. Reduce drying if too dry. Check for excess lubricant.
Sink marks	Underpacking Mold malfunction Poor part design	**Machine:** Increase holding time and pressure. Raise injection pressure. Reduce melt temperature. Increase cure time. Increase or decrease injection rate by flow control. Increase cushion. Repair check valve. **Mold:** Relocate gates nearer heavy sections. Increase gate size and vents. Lower mold temperature. Open runners, nozzle, and/or sprue. **Material:** Change flow grade.

TABLE E-1. (*Continued*)

Problem	Cause(s)	Solution(s)
Sticking (cavity)	Insufficient part cooling Poor mold design Cores too slender	**Machine:** Decrease injection pressure, hold time, boost time, melt temperature. Increase cure time. **Mold:** Check for undercuts and adequate draft angles. Raise mold temperature. **Material:** Reduce regrind level. Increase lubricant level.
Sticking (sprue)	Insufficient part cooling Poor mold design Cores too slender	**Machine:** Decrease injection pressure, hold time, and boost time. Increase nozzle heat and cure time. **Mold:** Align nozzle seating to sprue bushing. Ensure nozzle orifice is smaller than sprue bushing. Smooth burrs or nicks in sprue bushing. Lower stationary mold half temperature. Increase sprue bushing taper. Use more effective sprue puller. **Material:** Increase lubricant level. Reduce regrind level.
Voids	Slow material set-up in core Moisture in pellets condenses Thick/thin part section	See Sink Marks solutions, except raise mold temperatures. Decrease injection rate if material flows around core pins or makes sharp turns.
Weld lines	Nonuniform material flow Poor mold design Mold malfunction	**Machine:** Increase injection pressure, rate speed, and flow control. Increase melt temperature. Use larger machine. **Mold:** Raise mold temperature. Improve venting in weld area. Change gate location to alter flow pattern. Increase wall thickness. Check for core shift causing unbalanced wall thickness. No mold release spray. **Material:** May be too quick setting. Dry material if moisture or volatiles are trapped at weld. Correct overdrying.
Flow lines	Irregular flow pattern Mold malfunction Moisture in material Mold release agent used	**Machine:** Increase ram speed, melt temperature, backpressure, second stage injection time, barrel/nozzle temperature, and screw RPM. **Mold:** Increase mold temperature. Relocate gates to shorten flow distance. Increase number of gates. Make sure venting is adequate and correctly located. Clean mold cavity surface. Mold surface lube should not be used. **Material:** Dry material before use.

TABLE E-1. (*Continued*)

Problem	Cause(s)	Solution(s)
Excess shrinkage (small parts)	Cooling time too short Processing temperatures too high Poor material selection	**Machine:** Lengthen cooling time. Raise packing pressure. Open nozzle. Increase cushion. Examine check valve. Raise stock temperature to enhance packing; if parts get smaller, lower stock temperature to hasten freezing. Raise injection rate. **Mold:** Lower mold temperature. Open sprue, runners, and gates. **Material:** Change to nucleated resin. Put hot, ejected parts into cold water.
Screw slippage	Equipment wear Wet resin	**Machine:** Check for worn screw, faulty check valve, blockage in bottom of feed hopper. Lower rear barrel temperature controller. **Material:** Check for wet resin, excessive lubricant or fines, poor pellet cut/bulk density.
Jetting	High initial melt surge	**Machine:** Reduce initial injection speed. Increase nozzle temperature, backpressure, and melt temperature. **Mold:** Increase mold temperature, gate size, flare gate. Change gate location.
Part dimensions (too large)	Overpackaging Poor mold design	**Machine:** Verify injection fill time. Decrease hold pressure, overall injection speed, final injection speed backpressure, and front barrel temperature zone. **Mold:** Reduce mold temperature. Adjust mold design to verified shrinkage factor.
Part dimensions (too small)	Underpacking Poor mold design	**Machine:** Verify injection fill time. Increase overall injection speed, hold pressure, hold time, backpressure, and melt temperature. Adjust cushion. **Mold:** Reduce mold temperature. Increase cavity venting and cooling time. Adjust mold design to verified shrinkage factor.
Poor part finish	Mold fill too slow Mold malfunction Poor part design	**Machine:** Increase injection pressure, injection speed, and melt temperature. Decrease nozzle orifice. **Mold:** Make sure venting is adequate. Increase mold temperature. Discontinue using mold lube. Clean and polish cavity surfaces. **Part design:** Reduce abrupt changes in section thicknesses. Avoid overly thick ribbing.
Odor (gassing)	High melt temperature Contamination Resin holdup spots	**Machine:** Observe melt appearance (gassing) and measure melt temperature. Reduce cylinder temperatures if melt temperature is high. Reduce overall cycle to decrease holdup time. Correct hold-up spots in cylinder, adaptor, nozzle, and screw tip. Check valve assembly. Use smaller injection unit. **Material:** Avoid resin contamination.

TABLE E-1. (*Continued*)

Problem	Cause(s)	Solution(s)
Mold deposit	Venting Melt temperature	**Machine:** Decrease injection rate. Decrease melt temperature. Correct hold-up spots in cylinder, screw, and nozzle assembly. **Mold:** Increase gate size, flare gate. Enlarge vents. Change vent location. **Material:** Avoid resin contamination. Use hopper dryer to improve reground resin.
Unmelted particles	Surface defects Weak parts	**Machine:** Increase cylinder temperatures and backpressure. Reduce screw RPM. Use the hopper dryer to preheat resin. Increase overall cycle. Use screw designed for material. Use larger machine or injection unit.
Blush, frost, and folds		**Machine:** Decrease injection rate. Increase mold temperature. Change gate location.
Gate smear		**Machine:** Increase mold temperature. Decrease injection rate. Flare gate. Increase gate size. Change gate location.
Poor surface detail	Improper mold fitting Improper packing	**Machine:** Increase shot size, overall injection speed, backpressure injection fill time, second-stage pressure, and time. **Mold:** Increase mold temperature.

Adapted from Refs. [1, 2, 3, 4, 5].
*This table summarizes information received from equipment, material suppliers, and processors.
RPM, rotations per minute.

TABLE E-2. Process Guide for Clear Plastics.

Problem	Typical Causes	Possible Solutions
Silver streaking, splay (silver-white marks, usually in the flow direction), bubbles	a) Moisture b) Entrained air c) Decomposition products from overheated resin d) Contaminated resin	a) Check for recommended drying conditions; check dryer operation; use desiccant dryer. b) Raise backpressure; reduce screw decompression; add or enlarge vents. c) Reduce barrel temperatures, back pressure, or screw RPM; reduce shear heating by lowering injection rate or injection pressure, enlarge gates or nozzle diameter. d) Check dryer hopper; purge barrel; check vacuum transfer lines.
Voids, sink marks	Material volume shrinks after gate or other melt-path areas have frozen; generally as a result of insufficient part packing from various processing or tooling problems	Increase packing pressure or hold time; reduce injection rate (especially with thick sections); enlarge or relocate gate(s); raise mold temperature for voids, reduce it for sinks; increase shot size; check screw and nonreturn valve for back flow.

TABLE E-2. (*Continued*)

Problem	Typical Causes	Possible Solutions
Dark specks	Caused by flakes of charred resin from machine hangups or improper process control	Purge machine with at least five shots; check for excessive heater temperature; clean nozzle tip and extension; clean nonreturn valve and screw; check for excessively large machine for required shot size.
Haze, cloudiness	Impurities or moisture in resin; process temperatures or pressures too low	Check for contamination from tramp resin; increase resin dryness; increase mold or melt temperature; increase injection or backpressures.
Gate blush	Associated with melt fracture around the gate from stresses caused by process conditions or mold geometry	Raise melt temperature; reduce injection speed; check gate for sharp edges; enlarge gate; check that runner system has cold-slug well.
Brittleness, poor impact properties	Can result from variety of conditions, such as wet resin, molded-in stress, or improper part design	Check resin for dryness; reduce stress potential by raising melt temperature, reducing injection pressure; review part design for stress concentrators like notches, sharp radii, etc.; reduce regrind percentage.
Jetting (ripples in surface)	Associated with uneven spreading of melt into large unrestricted area of cavity at high injection speeds	Cut injection rate; enlarge gate or relocate it away from open area.
Discoloration, burn marks	Indicates overheated resin from many possible causes: a) Overtemperature in barrel or nozzle b) Gate and runner size c) Stagnant material d) Vent problems	a) Check barrel and nozzle heater controls. b) Check for shear heating from inadequate diameters of nozzle orifice, gates, or runners and from sharp edges. c) Make certain barrel, screw, and return valve are free from accumulated resin and look for stagnant areas in hot runner systems. d) Check for blocked vent channels, enlarge or add vents.

RPM, rotations per minute.

TABLE E-3. Troubleshooting Guide of Process Parameters.

Problem	Rapid Response							Slower Response				Possible Causes and/or Solutions
	Packing Pressure	Injection Rate	Screw Forward Time	Back Pressure	Boost Pressure	Melt Decompression	Mold Open Speed	Cycle Time	Melt Temperature	Mold Temperature	Nozzle Temperature	
Splay marks	+	+/–							–		–	Wet material
Flash	–	–							–			Mold needs adjustment, clamp ton low, injection pressure high
Poor weld lines	+	+							+	+		Improve venting, relocate gate, melt temperature
Burn marks		–							–			Improve venting, relocate gate, injection pressure
Surface blemish	+	+/–							+	+		Wet material, mold temperature, gating
Sink marks	+		+		+				+	+		Increase gate size, injection speed
Short shots	+	+	+		+				+	+		Increase gate size, mold temperature
Voids bubbles	+		+	+	+	–			+			Improve venting, increase gate size, drying
Discoloration								–	–		–	Purge machine, clean screw, barrel, and nozzle
Flow lines	+	+	+						+	+		Increase gate size
Delamination		–										Contaminated material
Unmelted pellets				+				+	+			Check heater bands, back pressure
Warpage		+/–	–					+		+		Check cooling line location, mold temperature
Distortion on ejection	–						–					Check mold for draft, undercuts, mold/melt temperature
KO pin penetration	–						–	+				Check for undercuts, hold time
Shot-to-shot variance		+	+									Check for nonreturn valve leakage
Drool						+			–		–	Use reverse taper nozzle, use suckback
Erratic screw retraction				+					+			Check for screw wear, throat temperature, fines
Part sticking	–	–	–					+				Check mold surface, use undercuts in core
Sprue sticking	–		–								+	Check sprue bushing surface, use reverse taper nozzle

IMPORTANT: Make only one adjustment at a time. Allow sufficient cycles or time for process to reach equilibrium before next adjustment.
Plus (+) sign: increase the variable to eliminate the problem.
Minus (–) sign: decrease the variable.
Plus/minus sign (+/–): either adjustment may solve the problem.
Rapid response: by next or following cycle.
Slower response: long time required; process must reach steady state.
Source: Adapted from Ref. [6].

Appendix F

Decoration and Information on Parts

TABLE F-1. Solutions to Common Paint-Line Problems.

Problem	Symptom	Cause	Solution
Blistering	Localized lifting of the paint film from the substrate. Usually because of surface contamination. Most noticeable in accelerated exposure and temperature cycling tests.	Poor wetting caused by oil, grease, mold release, and fingerprints; moisture drawn through coating by hygroscopic cleaning-agent residue from wash tank.	Good housekeeping procedures; more thorough final rinse.
Blushing	Whitish areas in the paint film. Produced by light reflected from condensed moisture droplets trapped in the dried coating.	High relative humidity in the painting area; moisture in air supply or paint hoses. Condensation can be aggravated by cooling from rapid evaporation of solvent, particularly during summer months. Cold parts and materials may cause condensation in winter months.	Apply air conditioning to dehumidify air or use slower solvents during high-humidity months. Check moisture traps in air supply. Bring all materials to room temperature before use. Check for fast solvent evaporation conditions.

Total Quality Process Control for Injection Molding, by M. Joseph Gordon

TABLE F-1. (*Continued*)

Problem	Symptom	Cause	Solution
Cobwebbing	Cotton candy-like filaments of coating resin that form in the air between the gun and part. Produces a lacy pattern in the coating. Most common in lacquers containing volatile solvents like acetone and methyl ethyl ketone and high-molecular-weight resins. Rarely occurs in water-based and two-component systems.	High solvent loss during spraying; too little solvent to begin with.	Check coating system for excessively fast solvent or too high a viscosity. Monitor for excessive paint atomization caused by insufficient pot pressure. Check fast solvent evaporation conditions.
Craters, fisheyes	Dimples or round, craterlike depressions in the paint film. Typical size $\frac{1}{16}$ to ½ in. Usually caused by contaminants on the substrate that prevent localized wetting by the liquid paint.	Mold release residue (especially silicone); machine or compressor oils and lubricants; dirt, grease; hand creams; contaminants from spray washer; overspray from adjacent painting operation; inadequate clean-out of pot and/or paint supply lines. Another possibility: gels in substrate or paint polymer.	Avoid silicones in processing or handling parts to be painted. Check for presence of oil, grease, etc., on problem parts by rinsing or wiping surface before painting with degreasing or alkaline cleaners. Make sure final rinse water is clean.
Dry spray	Rough, sandy feel to coating often with reduced gloss. Paint particles unable to flow properly and blend to form smooth coating.	Paint insufficiently thinned in formulation; high solvent loss during spray.	Reduce coating viscosity; check for fast solvent evaporation conditions.
Low gloss flatness	Reduced sheen or luster of the coating because of reflected light scattering from surface.	Culprits may include the condensed moisture, cleaning agent residues, or migration of internal plastic additives (plasticizers for instance), or dry spray. Also, substrate roughness may show through if paint film too thin.	Correct appropriate condition. If parts made from compounded resins, check for additive exudation, especially if stored in hot areas before painting.

TABLE F-1. (*Continued*)

Problem	Symptom	Cause	Solution
Mottling	Color variations in different areas of the paint. Usually produced by uneven distribution of pigment in coating through separation or settling. More common with metal flake paints, but can occur with nonmetallic systems.	Pigment separations often related to a solvent problem. Problem can occur if solvent evaporates too slowly. Also may occur if coating remains too "fluid" because of excessive solvent, too high coating thickness (from extra touch up pass, for instance), or too slow flash off before oven.	For slow evaporation rate, use more volatile solvent mixture or check for slow solvent evaporation conditions. If too fluid, check for poor atomization (large globules) resulting from low air pressure or high pot pressure.
Orange peel	Pock marked, slightly rough-looking surface resembling an orange skin. Can reflect uneven shrinkage of coating during drying or poor leveling characteristics when applied.	Film shrinkage from fast solvent evaporation; film too thin; failure of coating droplets to knit smoothly because of insufficient fluidity.	Consider less volative solvent; evaluate for thicker film; check possibility of dry spray.
Poor hiding	Show through of background color.	Too much solvent, coating too thin.	Check paint for thorough mixing, use less thinner; reduce atomization air pressure; increase pot pressure; increase coating thickness.
Sags, runs	Thick streaks or areas in the coating produced by gravity on vertical surfaces. May also show up as thickened borders around edges, raised details, and openings in part.	Coating remains runny too long. May be caused by excessive film thickness, or solvent evaporation may be too slow.	Consider faster solvent; check other factors in slow solvent evaporation.
Soak-in, bite	Dull, blushlike, or off-color area, often circular or half moon in shape in dried coating. Seldom discernible in uncoated part.	Attributable to preferential solvent attack in areas where plastic density varies. Typical sites: gate and sprue areas and extremities of long flow molded parts.	Usually a molding problem, reflecting poor packing from gate chilling, cool melt, cold mold, etc.

TABLE F-1. (*Continued*)

Problem	Symptom	Cause	Solution
Solvent pops	Craters, pinholes, or bubbles on coating. May occur during flash-off or in the drying oven.	Associated with coating rupture during solvent escape. If occurs before oven, usually due to high solvent volatility; if in oven, could be too little flash off time, film thickness too heavy, or oven temperature too high (rapid volatilization).	Check solvent volatility and process conditions. For preoven problem, check for factors in fast solvent evaporation. For in-oven problem, consider reducing oven temperature, extending flash time, reducing film thickness.
Wicking	End of fiberglass reinforcement protruding from plastic surface into or through the coating. Affects coating smoothness.	Bent-over fiber ends are released when plastic surface is softened by solvent action or oven heat.	Try less aggressive solvent or lower oven temperature. Adjust molding conditions for a more resin-rich surface.

Source: Adapted from Ref. [1].

TABLE F-2. Comparison of the Four Most Common Decorating Methods.

	Direct Screen Printing	Pad Printing	Hot Stamping	Heat Transfer
Recognition factors	Thicker and more opaque, no clear or adhesive at edges.	Thinner and less opaque, large color areas may look weak.	Colors usually "debossed," can be bright gold or silver.	May have tooling "halo" around design, usually multicolor.
Image size and limitations	Screens can be made any size.	7 × 14-in limit; special machines can print 10 × 20 in.	Limited by pressure of machine and tendency to trap air. Usual range 300 to 500 psi; roll-on solves air entrapment and can be applied to 12 × 24-in area.	Limited by pressure of machine and tendency to trap air; usual range 100 to 300 psi (soft goods as low as 30 to 50 psi); roll-on solves air entrapment and can be applied to 12 × 24-in area.

TABLE F-2. (*Continued*)

	Direct Screen Printing	Pad Printing	Hot Stamping	Heat Transfer
Resolution of detail	Medium.	Fine to medium.	Medium.	Fine—including 133-line four-color process.
Large areas of solid color	Okay, with good equipment and operator.	Not good without multiple prints of color.	Possible, but trapped air can be a problem; roll-on machine will help.	Possible, but trapped air can be a problem; roll-on machine will help.
Opacity	Good.	Poor, with multiple prints only fair.	Good.	Good if screen printed; fair if gravure printed.
Color match	Your responsibility in-house or ordered from outside supplier.	Your responsibility in-house or ordered from outside supplier.	Use closest color of foil available; for long runs, have it customer formulated.	Inks have to be custom formulated when transfer is printed.
Registration of multiple colors	Fair to good depending on equipment, tooling, operator, and size stability of plastic from first to last print.	Fair to very good depending on equipment, whether part stays in same nest through all color prints, and on quality of the tooling.	Fair to good, depending on equipment, tooling, and size stability of plastic.	Very good, four-color process, 133-line screen demands tight registration; so does multicolor and fine detail.
Part shape and limitations	Flat or single curve (cylinder)	Can be irregular or compound curve, but art distortion requires trial and error to correct.	Flat or single curve (cylinder)	Flat, single curve, or slight compound curve; carrier paper wrinkling limits shape and size on compound curves.
Arc limits on cylinder or a cylinder with draft or taper of 1 degree or less	Almost 360 degrees, avoiding ink-to-screen contact on the wrap.	Approximately 100-degree arc for reciprocal machine or 360 degrees for special wrap machines.	Approximately 90-degree arc for reciprocal machine; 360 degree (with a slight overlap preferred) for wrap machines.	Approximately 90-degree arc for reciprocal machine; 360 degrees (with slight overlap preferred) for wrap machine (except if wax release); then 360 minus $^1/_8$ in.

TABLE F-2. (*Continued*)

	Direct Screen Printing	Pad Printing	Hot Stamping	Heat Transfer
Process wet ink or dry	Wet-drying or curing required between and after final color.	Wet-drying or curing required between and after final color; some ink systems can be printed "wet on wet."	Dry—proceed to next process.	Dry—proceed to next process.
Inventory required	Screens of various meshes; inks to make custom colors for particular substrate chemistry; solvents, cleaners, retarders, and squeegees, or order matched color inks as needed.	Pads of various sizes, shapes, and durometers; inks to make custom colors for particular substrate chemistries; solvents, cleaners, and retarders, or order match colors as needed.	Various hot stamping foils in various colors compatible with surfaces to be marked; or order as needed.	Heat transfers for a specific job.
EPA and fire-safety considerations	Flammable materials must be stored and insured accordingly.	Flammable materials must be stored and insured accordingly.	None.	None.
Learning-curve timeframe for system startup	Days to week.	Days to week.	Hours to days.	Hours to days.
New operator leaning curve	Hours to days.	Hours to days.	Minutes to hours.	Minutes to hours.
Setup skill level	Skilled.	Skilled.	Semiskilled.	Semiskilled.
Operator skill level	Semiskilled.	Semiskilled.	Unskilled.	Unskilled.
Part change over new part, new design	Minutes to hours to change tooling (nest), screens, inks, reregister (longer for multiple colors).	Minutes to hours to change tooling (nest), cliches, pads, change inks, reregister (longer for multiple colors).	Seconds to minutes (rarely hours) to change tooling (nest), application head, foil, reregister (longer for multiple colors), and recheck pressure, dwell, and temperature.	Seconds to minutes (rarely hours) to change tooling (nest), application head, roll of heat transfers, and reregister (time same as for single color), recheck pressure, dwell, and temperature.

TABLE F-2. (*Continued*)

	Direct Screen Printing	Pad Printing	Hot Stamping	Heat Transfer
Process variance causing defective "print"	Ink viscosity somewhat critical.	Ink viscosity extremely critical.	Process is quite stable; silicone rubber die "forgives" mild surface blemishes.	Process is quite stable because less pressure is required; softer rubber (50 to 60 durometer) "forgives" many surface blemishes.
Part variance causing defective "print"	Poor surface finish, with such blemishes as sink marks (some defects).	Surface blemishes usually unaffected unless extreme (reduces defects).	Direct ram overcomes thickness changes; toggle machines very sensitive to thickness changes (some defects).	Direct ram overcomes thickness changes; toggle machines very sensitive to thickness changes (some defects).
Cost of equipment	Small to large area (100 sq. in.) not very cost sensitive, but very cost sensitive to multiple colors.	Small to large area cost sensitive; cost sensitive also to multiple colors.	Small to large area cost sensitive (multiple color more than one pass or more than machine).	Cost sensitive to size, not sensitive to multicolor tooling (nest and heads); cost same for one or multiple colors.
Cost of tooling	Low to moderate for single color, moderate to high depending on tolerances for multiple colors.	Low to moderate for single color, moderate to high depending on tolerances required for multiple colors.	Low to moderate for single color, moderate to high depending on tolerances for multiple colors.	Low-to-moderate cost as for single color, even though graphics may be multicolor.
Part input/ output equipment cost	Approximately the same for all processes, but may be more costly for multiple colors.	Approximately the same for all processes, but may be more costly for multiple colors.	Approximately the same for all processes, but may be more costly for multiple colors.	Approximately the same for all processes, but is not more costly for multiple colors.

TABLE F-2. (*Continued*)

	Direct Screen Printing	Pad Printing	Hot Stamping	Heat Transfer
Cost of inks, foil transfers	Inks—not very costly, but sensitive to size.	Inks—not very costly, but sensitive to size.	Foils—cost sensitive to size, costs for multiple colors increase linearly per color area.	Transfers—cost sensitive to size, not as sensitive to additional colors.

EPA, Environmental Protection Agency.

Source: Adapted from Refs. [2] and [3].

TABLE F-3. Solutions to Hot-Stamp Decorating Problems.

Problem	Cause	Solution
Flattened characters when tipping raised letters or beads.	Die too hot, too much pressure on die, or excessive dwell time.	Lower heating setting or head pressure. Reduce dwell timer setting.
Distorted imprint on plastic part.	Skidding of die on contact with foil due to fixture deflection.	Realign die on head slide so that it is directly under the press ram. Modify or redesign fixture.
Blurred image or imprint.	Excessive die heat.	Reduce heater temperature.
Weak impression or no imprint.	Insufficient air pressure.	Check for obstruction in air line or need for larger air line.
Inconsistent transfer of decoration to the parts.	Variation in parts (thickness, warpage, sink marks).	Modify heat, pressure dwell time settings to optimize for parts from all mold cavities.
	Heat variations at die face.	Check that heat control is holding to preset tolerances. Look for air gaps between heater block and dovetail due to die shim, or heat loss caused by the riser block.
Decoration fails to adhere to plastic.	Insufficient cure time or strip delay.	To determine whether stripping time is the problem, manually lay a section of foil on the part, cycle the press, and peel the foil off. If the imprint is good, stripping must be adjusted, either by reducing head upstroke speed, or adjusting stripper bar springs.
	Air trapped under foil.	Check foil feed and die contact with foil to determine cause of entrapment.
	Contamination on part.	Determine what the contaminant is and its source and eliminate it.
	Wrong foil used.	Check compatibility of foil, replace with correct foil.
	Special coating on plastic.	Determine what coating is and change to a foil that is formulated to be compatible.

TABLE F-3. (*Continued*)

Problem	Cause	Solution
Imprint deeper on one end of part than the other.	Machine is not level.	Level machine and mount die directly under arm.
Flaking in decoration, featherly edges, or fill in.	Dwell time too long.	Shorten dwell time by adjusting air flow control valves.
Loss of gloss in foil.	Dwell time too long.	Shorten dwell time.
Inconsistent imprints in multiple part setups.	Part irregularities, i.e., sinks, thickness variations.	Shim the fixtures to compensate for part irregularities.
Inconsistent transfer of decoration to the parts.	Die temperature too low, or inadequate pressure.	If die impresses the plastic, increase die temperature. If imprint does not impress the plastic, boost pressure on die.
	Dwell time too short.	Lengthen dwell time.
	Off-spec foil.	Manually place a section of foil from a roll that has run well on a part and cycle the press. If it prints well, replace the roll on the machine.
Uneven imprint.	Unevenly heated die. Die head cocked off center. Fixture may have shifted.	Check die temperature. Look for cartridge heater outage. Determine why die head is cocked and realign as necessary. Reset fixture as needed.
Repeated void in decoration at same location.	Molded-in part feature, i.e., rib or boss, unsupported by fixture.	Shim or modify fixture.

Source: Adapted from Ref. [4].

Glossary

Abrasion The wearing away of some surface area by its contact with another material.

Absorption *See* Moisture.

Accelerated aging Aging by artificial means to obtain an indication on how a material will behave under normal conditions over a prolonged period.

Accelerated weathering Duplicating or reproducing weather conditions by machine-made means.

Acceptable quality level The minimum quality level at which a product will be accepted or rejected.

Accumulator *(1)* A device for conserving energy in hydraulic systems of molding equipment. *(2)* An auxiliary ram extruder used to provide fast material delivery in a molding machine.

Acetal resins A crystalline thermoplastic material made from formaldehyde. Trade names: Delrin and Celcon.

Acrylics The name given to plastics produced by the polymerization of acrylic acid derivatives, usually including methyl methacrylate. An amorphous thermoplastic material.

Acrylonitrile, Butadiene, Styrene (ABS) A thermoplastic classified as an elastomer-modified styrene.

Adaptor A mechanical reducing mechanism for the barrel to the nozzle; a thermal insulator from the nozzle to the barrel for temperature control.

Total Quality Process Control for Injection Molding, by M. Joseph Gordon

Additive A material added to resin prior to molding or forming to add a desired property or characteristic to the finished product.

Adsorption. *See* Moisture.

Aesthetics The external surface appearance of a plastic part.

Aging The change of a material over time under defined natural or synthetic environmental conditions, leading to improvement or deterioration of properties. *See also* Accelerated aging; Artificial aging.

Air vent A small outlet in a mold used to prevent entrapment of gases.

Alkyd resins The name given to synthetic resins processed from polyhydric alcohols and polybasic acid or anhydrides. Alkyds are thermosetting resins.

Alloy Composite material made by blending polymers or copolymers with other polymers or elastomers under selected conditions, e.g., styrene–acrylonitrile copolymer resins blended with butadiene–acrylonitrile rubbers.

Ambient temperature Temperature of the medium surrounding an object. Used to denote prevailing room temperature.

Amorphous Plastic materials that have no definite order of crystallinity.

Amortized The cost of an item spread out equally over time or a specified number of parts. Frequently used when estimating finished part costs if the cost of a mold or capital equipment is spread out and added into the piece-part cost.

Amplitude The maximum value of a periodically varying value used to describe the energy transmitted from the ultrasonic welding horn to the weld joint.

Analog Refers to a needle on a scale readout that is almost instantaneous from the input signal to the output readout. Determined by the design of the circuitry. Gives an approximate reading based on the detail of the readout scale.

Anneal *(1)* To heat a molded plastic article to a predetermined temperature and slowly cool it to relieve stresses. (Annealing of molded or machined parts may be done dry, as in an oven, or wet, as in a heated tank of mineral oil). Often done with the part in a holding fixture. *(2)* To heat steel to a predetermined temperature above the critical range and slowly cool it to relieve stresses and reduce hardness.

Antioxidant A substance added to a material to inhibit oxidation.

Antistatic agents (antistats) Agents that, when added to the molding material or applied on the surface of the molded part, make it less able to conduct electricity (thus hindering the fixation of dust).

Approved supplier A product supplier who has been rated satisfactory on previous jobs. May involve a detailed analysis of manufacturing and quality capability to be sure they meet customer requirements.

Arc resistance The time required for a given electrical current to render the plastic surface of a material conductive because of carbonization by the arc flame.

Artificial aging The accelerated testing of plastic specimens to determine their changes in properties. Carried out over a short period of time, such tests are indicative of what may be expected of a material under service conditions over extended periods. Typical investigations include those for dimensional stability; the effect of immersion in water, chemicals, and solvents; light stability; and resistance to fatigue.

AS9100 The aerospace equivalent to ISO9000. This standard for aircraft manufacturers and the suppliers is more stringent in their controls for suppliers to supply parts that must have the highest of quality ratings due to safety and reliability.

Ashing The reduction of a polymer by high heat to yield any inorganic fillers or reinforcements. Used to verify the percentage of nonorganic content in the resin.

ASTM American Society for Testing and Materials.

Attribute Unlike a property, it is a quality that is less precisely known and is only ascribed to someone or something.

Automatic mold A mold or die in injection or compression molding that repeatedly goes through the entire cycle without human assistance.

Auxiliary equipment Refers to equipment, other than the injection molding machine and mold, required to insure the manufactured part will be made correctly, including, for example, dryers, chillers, material and part conveyors, and robots.

Back pressure *(1)* A pressure against the free flow of material during extruder running (plasticating) that causes the material to have a high-mixing action. This pressure occurs as the material resists the forward movement of the material in the extruder or an externally controlled hydraulic pressure put against the movement of the extruder in a reciprocation screw machine to create this greater mixing action. *(2)* Resistance of a material, because of its viscosity, to continue flow when mold is closing.

Back taper (Back draft) Reverse draft used in mold to prevent molded articles from drawing freely.

Backing plate In mold construction, a plate used as a support for the cavity blocks, guide pins, bushings, etc.

Balanced mold A mold laid out with runner and cavities spaced and sized for uniform flow, fill, and packing pressure throughout the system.

Ball valve A screw melt sealoff similar to a check ring but designed differently. Uses a round ball to seal off the melt so it does not flow back over the screw flights during the injection cycle.

Bar coding The electronic/optical bar recognition system for identification, storage, printout, and retrieval of specified data and information.

Barrel (Extruder) In injection molding, extrusion or bottle-blowing equipment. It is the hollow tube in which the plastic material is gradually heated and melted and from which it is extruded.

Bezel A grooved rim or flange. An example is a television bezel.

Blanket purchase order (BPO) A purchase order placed with a supplier for materials over a set time period. Customer then releases material as required or as specified.

Bleed *(1)* To give up color when in contact with water or a solvent. *(2)* Undesired movement of certain materials in a plastic (e.g., plasticizers in vinyl) to the surface of the finished article or into an adjacent material. Also called "migration." *(3)* An escape passage at the parting line of a mold, such as a vent, but deeper, which allows material to escape or bleed out.

Blend Any combination of mixtures of a base resin with additives or modifiers. The base resin has been modified.

Blind hole A hole that is not drilled or molded entirely through.

Blister A raised area on the surface of a molded part caused by the pressure of gases or air inside it.

Bloom A visible exudation or efforescence on the surface of a plastic. Bloom can be caused by lubricant, plasticizer, etc.

Blow by Overpaint spray under the edge of a mask.

Blow molding *(1)* A molding process primarily used to produce hollow objects. *(2)* A molding process in which a hollow tube (parison) is forced into a shape of the mold cavity using internal air pressure. The two primary types are injection blow molding and extrusion blow molding.

Blow pin A hollow pin inserted or made to contact the blowing mold so that the blowing media can be introduced into the parison or hollow form and expanded to conform to the mold cavity.

Blowing agents (forming agents) An additive capable of producing a cellular structure in a plastic or rubber mass.

Blueing off Checking the accuracy of mold cutoff surfaces by putting a thin coating of Prussian Blue on one half and checking the blue transfer to the other half.

Blush The tendency of a plastic to turn white or chalky in areas that are highly stressed, such as gate blush.

Boat A tungsten container used to hold staples during vacuum metallizing.

Bolster Spacer or filler in a mold.

Boss Projection on a plastic part designed to add strength, facilitate alignment during assembly, and provide for fastening.

Bottom plate Part of the mold containing the heel radius and push up.

Branched chains In polymer chemistry, side chains attached to the main original chain.

Breaker plate A perforated plate located at the end of an extruder or at the nozzle end of an injection cylinder. It often supports the screens that prevent foreign particles from entering the die. Used to keep unplasticized material out of the nozzle and to improve distribution of color particles.

Breathing (Degassing) The opening and closing of a mold to allow gases to escape early in the molding cycle. When referring to plastic sheeting, "breathing" indicates permeability to air.

Brinell hardness *See* Rockwell hardness.

Bridge Part of a paint spray mask that holds a stem that, in turn, holds a plug or a cap, usually made of heavy wire and silver, soldered to the plate.

Bubble A spherical, internal void; globule of air or other gas trapped within a plastic. *See* Void.

Bubbler A device inserted into a mold face, cavity, or core that allows water to flow deep inside the hold into which it is inserted and to discharge through the open end of the hole. Uniform cooling of the molds and of isolated mold sections can be achieved in this manner.

Burned A carbonized condition showing evidence of thermal decomposition through some discoloration, distortion, or localized destruction of the surface of the plastic. Usually caused by poor venting of the mold cavity.

Burning *(1)* Overheating the resin in the barrel causing discoloration and, if long enough, charring the material. *(2)* Caused by trapped gasses in a poor or nonvented area of the mold. The gasses may ignite, as a result of pressure and temperature, as in a diesel engine, and discolor or char the part.

Butadiene A synthetic rubber used in butadiene–styrene, butadiene–acrylonitrile, and acrylonitrile–butadiene–styrene.

Butt-fusion A method of joining similar forms of thermoplastic materials using heat.

Buttress thread A type of thread used for transmitting power in only one direction. It has the efficiency of the square thread and the strength of the V-thread.

Cadmium A heavy metal element used as a pigment in plastics. Now being replaced because of its hazardous nature.

Cam bar The stationary angled bar or rod used to mechanically operate the slides on a mold for side action core pulls.

Caprolactam A cyclic amide compound containing six carbon atoms. When the ring is opened, caprolactam is polymerizable into a nylon resin known as type-6 nylon or polycaprolactam.

Carbon black A black pigment produced by the incomplete burning of natural gas or oil. It is widely used as a filler, particularly in the rubber industry and wire/cable applications. Because it possesses useful ultraviolet protective properties, it is also used in molding compounds intended for outside weathering applications.

Cartridge heaters Electrical heaters of various outputs enclosed in a jacket that can be selectively located in a mold to heat the surrounding area.

Case harden To harden the surface of a piece of steel to a relatively shallow depth.

Cavity Depression in the mold that usually forms the outer surface of the part. Depending on the number of such depressions, molds are designated as a single cavity mold, a multicavity mold, or a family cavity mold.

Cavity number A sequential number engraved in a mold cavity and reproduced on the molded part for later reference in case a problem ever occurs with the part. Used in multicavity molds of similar parts.

Cavity retainer plate Plates in a mold that hold the cavities. These plates are at the mold parting line and usually contain the guide pins and bushings.

Cellular plastics Foamed plastics.

Cementing A process of joining two similar plastic materials to themselves or to dissimilar materials by means of solvents.

Center gated mold An injection or transfer mold in which the cavity is filled with molding material through a sprue or gate directly into the center of the part.

Chalking Dry, chalklike appearance or deposit on the surface of a plastic. *See* haze; bloom.

Change request A request to modify or alter the dimensions, material, tolerances, or manufacture of a part now in or soon to be in production. Used to insure all interested and involved department personnel are informed and can comment and approve or disapprove of the pending change.

Charpy A type of pendulum test for toughness. *See* Impact test.

Chase An enclosure of any shape, used to *(1)* shrink fit parts of a mold cavity in place; *(2)* prevent spreading or distortion in hobbing; *(3)* enclose an assembly of two or more parts of a split cavity block.

Checkring A material shutoff ring mounted on the front of the screw, behind the screw tip, that allows melt to flow past it when the screw is retracting so that a supply of melt builds up in front of the screw. When the screw moves forward to inject melt into the mold, the check ring moves rearward and seals off the screw flights so that the melt is pushed into the mold.

Chemical resistance Ability of a material to retain utility and appearance following contact with chemical agents.

Chromium plating An electrolytic process that deposits a hard film of chromium metal onto working surfaces of other metals. Used when resistance to corrosion, abrasion, and/or erosion is needed.

Clamping area The largest rate molding area an injection or transfer press can hold closed under full molding pressure.

Clamping force (Clamping pressure) In injection molding, the pressure is applied to the mold to keep it closed despite the fluid pressure of the compressed molding material within the cavity and runner system.

Clamping plate A plate used to fasten the mold to a molding machine.

Clarity Material clearness or lack of haze.

Closed loop System used with microprocessor for control of a machine's cycle. *See* feedback.

Coefficient of expansion The fractional change in a specified dimension (sometimes volume) of a material for a unit change in temperature. Values for plastics range from 0.01 to 0.2 mils/in. °C (ASTM D 696).

Coefficient of friction The value calculated under a known set of conditions, such as pressure, surface, speed, temperature, and material, to develop a number—either static or dynamic—of the resistance of the material to slide or roll. The lower the value, the higher the material's lubricity.

Coining The peening over or compressing of a material to change its original shape or form.

Cold flow A plastic exhibits cold flow when it does not return to its original dimensions after being subjected to stress. *See* Creep.

Cold shot Incomplete parts formed while cycling a molding machine during heatup.

Cold slug The first material to enter an injection mold; so called because in passing through sprue orifice it is cooled below the effective molding temperature.

Cold slug well Space provided directly opposite the sprue opening in an injection mold to trap the cold slug.

Color concentrate A mixture of a measured amount of dye or pigment and a specific plastic material base. A more precise color can be obtained using concentrates than using raw colors. *Note:* Care should be taken to verify that the color concentrate base is compatible with the plastic it is to color. Color concentrate is normally used at 1 to 4 percent of the plastic material to be colored.

Color-fast The ability to resist change in color.

Color standard The exact color a plastic resin or part must match to be acceptable. Resin suppliers often submit color chip samples of the matched resin color to be compared with the molded part. The color chip, or standard, is usually 2 × 3 inches with one polished surface and various textured surfaces on the opposite side. Suppliers use similar standards to verify the color of each lot of resin shipped to their customer.

Combination mold *See* Family mold.

Commodity resin Usually associated with the higher volume lower priced plastics, with low-to-medium physical properties. Examples are polyethylene (PE), polypropylene (PP), polystyrene (PS), acrylic, polyvinyl chloride (PVC), ethylene vinyl acetate (EVA), and acrylonitride butadiene styrene (ABS). Used for less critical applications.

Compound A mixture of polymer(s) with all materials necessary for the finished product.

Compression ratio In the extruder of an injection/blow molder screw, the ratio of volume available in the first flight at the hopper to the last flight at the end of the screw.

Compressive strength Crushing load at he failure of a specimen divided by the original sectional area of the specimen (ASTM D 695).

Concentricity *(1)* The relationship of all circular surfaces with the same center. *(2)* Relationship of all inside dimensions to all outside dimensions. Usually, as with diameter, expressed in thousandths of an inch (F.I.M. = FULL INDICATOR MOVEMENT). Deviation from concentricity is often referred to as runout.

Conditioning The subjection of a material to a stipulated treatment so that it will respond in a uniform way to subsequent testing or processing. The term is frequently used to refer to the treatment given before testing. ASTM standard conditions for a plastic testing laboratory are 23°C + 2°C (73.4°F + 3.6°F) and 50 percent + 5 percent relative humidity.

Conditioning chamber An enclosure used to prepare parts for their next step in the assembly or decorating process. Parts can be stress relieved, humidity or moisture conditioned, or impregnated with another element.

Consigned material Material given over to another supplier for care and use in manufacturing a customer's product.

Contamination Any foreign body in a material that affects or detracts from the part's quality.

Control plan A written plan that lists step-by-step procedures describing how a specific operation will be conducted and followed.

Controllers The instruments, timers, and pressure controls used to control and regulate the molding cycle.

Cooling channels Channels or passageways within the body of a mold through which a cooling medium can be circulated to control temperature on the mold surface. May also be used for heating a mold by circulating steam, hot oil, or other heated fluid through channels as in molding thermoplastic materials.

Cooling fixture A block of steel, wood, or composite material that is similar to the shape of the molded piece. The hot molded part is taken from the mold, placed on it, and allowed to cool, without distorting. Also known as a shrink fixture.

Cooling time The time period required after the gate freezes for the part to solidify and become rigid enough for ejection from the mold cavity.

Copolymer A polymer produced by polymerization of two or more monomers. Can also be done as a secondary compounding operation on an extruder.

Core *(1)* Male element in die that produces a hole or recess in a part. *(2)* Part of a complex mold that molds undercut parts. Cores are usually withdrawn to one side before the main sections of the mold open. *(3)* A channel in a mold for circulation of a heat-transfer medium. *(4)* The central member of a laminate.

Core pin A pin for forming a hole or opening in a plastic mold.

Core pin plate Plate holding core pins.

Coring The removal of excess material from the cross-section of a molded part to attain a more uniform wall thickness.

Corona treatment Exposing a plastic part to a corona discharge increases receptivity to inks, lacquers, paints, and adhesives. *See* Surface treatment.

Corrosion Material that is eaten away by chemical reactions at the surface area.

Crazing Fine cracks that may extend in a network on or under the surface or through a layer of plastic material.

Creep The dimensional change with time of a material under load, following the initial instantaneous elastic deformation. *See* Cold flow (ASTM D 674).

Cross-linking The chemical combination of molecules to form thermally stable bonds within a polymer, not broken by heating.

Crystallinity A state of molecular structure in some resins that denotes uniformity and compactness of the molecular chains forming the polymer. Normally attributed to the formation of solid crystals with a definite geometric form. High crystallinity causes a polymer to be less transparent, or opaque.

Cure That portion of the molding cycle during which the plastic material in the mold becomes sufficiently rigid or hard to permit ejection.

Curing time The time between the end of injection pressure and the opening of the mold.

Cushion The ¼ to ½ inch of resin in front of the screw tip at the end of the injection cycle used to maintain packing pressure on the melt until the cavity gate freezes off.

Cycle The complete, repetitive sequence of operations in a process or part of a process. In molding, the cycle time is the period, or elapsed time, between a certain point in one cycle and the same point in the next.

Daylight opening Clearance between two platens of a press in the open position. Mold daylight describes the opening distance of mold halves for part removal.

Deboss(ed) An indent or cut in design (depressed design) or lettering of a surface.

Decompression The removal of pressure on the melt by an increase in screw flight depth and a positive vent opening in the barrel of a vented barrel extruder or injection molding machine. It allows the melt to expand and degas.

Deflashing Covers the range of finishing techniques used to remove the flash (excess, unwanted material, as filing, sanding, milling, tumbling, and wheelbrating) on a plastic mold.

Degassing *See* Breathing.

Degating The removal of the part from the runner system.

Degradation A deleterious change in the chemical structure, physical properties, and/or appearance of a plastic, usually caused by exposure to heat.

Delaminate To split or separate a laminated plastic material along the plane of its layers.

Density Weight per unit volume of a substance, expressed in grams per cubic centimeter or pounds per cubic foot.

Desiccant Substance that can be used for drying purposes because of its affinity for water.

Design of experiments (DOE) A problem-solving technique developed by Taguchi using a testing process with an orthagonal array to analyze data and determine the main contributing factors in the solution to the problem.

Design stress A long-term stress, including creep factors and safety factors, that is used in designing structural fabrication.

Destaticization Treating plastics materials to minimize their accumulation of static electricity, and subsequently, the amount of dust picked up by the plastics because of such charges. *See* Antistatic agents.

Destructive test Any test performed on a part in an attempt to destroy it; often performed to see how much abuse the part can tolerate without failing.

Deterioration A permanent change in the physical properties of a plastic as evidenced by impairment of these properties.

Diaphragm gate Gate used in molding annular or tubular articles. Gate forms a solid web across the opening of the part.

Die A metal form in making or punching plastic products. It is used interchangeably with mold.

Die drips Carbonized resin drool formed on the face of an extrusion die face during the resin production cycle. If the die face is not kept clean, it can solidify, breakoff, and contaminate the virgin resin.

Dielectric constant Normally the relative dielectric constant. For practical purposes, the ratio of the capacitance of an assembly of two electrodes, separated solely by plastic insulating material, to its capacitance, when the electrodes are separated by air. A relative measure of nonconductance.

Dielectric heating (electronics heating or R.F. heating) The plastic to be heated forms the dielectric of a condenser to which a high frequency (20 to 80 mc.) voltage is applied. Dielectric loss in the material is the basis of the process, which is used for sealing vinyl films.

Dielectric strength The maximum electrical voltage a material can sustain before it is broken down, or "arced through," in volts per mil of thickness.

Dieseling *See* Burning.

Differential shrinkage Nonuniform material shrinkage in a part.

Digital Numberical output device that must index, number by number, from the initial output reading to the final output reading. More accurate than a similar analog device, but slower. Gives an exact reading.

Dimensional stability Ability of a plastic part to retain the precise shape in which it was molded.

Disc gate *See* Diaphragm gate.

Discoloration Any change from the original color, often caused by overheating, light exposure, irradiation, or chemical attack.

Dished Showing a symmetrical distortion of a flat or curved section of a plastic object, so that as normally viewed, it appears concave, or more concave than intended. *See* Warpage.

Dispersion Finely divided particles of material in suspension in another substance.

Domed Showing a symmetrical distortion of a flat or curved section of a plastic object, so that, as normally viewed, it appears convex, or more convex than intended. *See* Warpage.

Double-shot molding A method of producing two-color pieces in thermoplastic materials by successive injection molding operations. The part molded first becomes an insert for the second mold.

Dowel Pin used to maintain alignment between two or more parts of a mold.

Draft A taper or slope in a mold required to facilitate removal of the molded piece. The opposite of this is called back draft.

Drool *(1)* Melt oozing from a nozzle that is not correctly temperature controlled. *(2)* Creation of die drips on the face of an extruder die.

Drop test *See* Impact test.

Dry as molded (DAM) Term used to describe a part immediately after it is removed from a mold and allowed to cool down. All physical, chemical, and electrical property tests are performed on nonconditioned test bars and the results recorded on the data sheets. Parts and test bars in this state (DAM) are felt to be their weakest in some properties as they have not had time to condition or relieve any molded-in stresses.

Dry coloring Method commonly used to color plastic by tumble blending uncolored particles of the plastic material with selected dyes and pigments.

Dryers Auxiliary equipment used to dry resins before processing to ensure that the surface properties are within manufacturer specifications. There are several styles of dryers, including ovens, microwave, hot-air desiccant bed, and refrigeration types.

Ductility The extent to which a solid material can be drawn into a thinner cross-section without breaking.

Duplicate cavity plate Removable plate that retains cavities; used where two-plate operation is necessary for loading inserts.

Durometer hardness The hardness of material as measured by the Shore Durometer.

Dwell *(1)* A pause in the application of pressure to a mold, just before the mold is completely closed, to allow the escape of gas form the thermoset molding material. *(2)* The time between when the injection ram is fully forward holding pressure on the material within the mold and the time the ram retracts.

Dyes Intensely colored synthetics or natural organic chemicals that are soluble in most common solvents and dissolve in the plastic substrate while imparting color. Characterized by good transparency, high-tincturial strength, and low specific gravity.

Ejection The removal of the finished part from the mold cavity by mechanical means.

Ejection time The time in the cycle when the mold opens, the part is ejected, the mold closes, and clamping pressure is applied.

Ejector pin (ejector sleeve) A rod, pin, or sleeve that pushes a molding off of a core or out of a cavity. It is attached to an ejector bar or plate that can be activated by the ejector rod(s) or the press or by auxiliary hydraulic or air cylinders.

Ejector pin retainer plate Retainer plate onto which ejector pins are assembled.

Ejector return pins Projections that push the ejector assembly back as the mold closes. Also called safety pins and position pushbacks.

Ejector rod or bar A bar that activates the ejector assembly when the mold is open.

Elastic deformation A deformation in which a substance returns to its original dimensions on release of the deforming stress.

Elasticity That property of a material by virtue of which it tends to recover its original size and shape after deformation. If the strain is proportional to the applied stress, the material is said to exhibit Hookean or ideal elasticity.

Elastomer A material that at room temperature can be stretched repeatedly under low stress to at least twice its length and snaps back to the original length upon release of stress.

Electric discharge machining (EDM) A metal-working process applicable to mold construction in which controlled sparking is used to erode the work piece.

Electroformed molds A mold made by electroplating metal on the reverse pattern of the cavity. Molten steel may be then sprayed on the back of the mold to increase its strength.

Electronic data interchange (EDI) The exchange of data by customer and supplier computers, usually through a third neutral computer company that safeguards the host computers from unwanted entry. Used for order placement, shipment, receiving, billing, and payment.

Electroplating Deposition of metals on certain plastics and mold for finish.

Elongation The increase in length of a material under test, expressed as a percentage difference between the original length and the length at the moment of the break.

Embossing Techniques used to create depressions of a specific parttern in plastics film and sheeting. Such embossing is in the form of surface patterns on the molded part by photoengraving or a similar process.

Endothermic An action or reaction that absorbs heat.

Enduse The function the part or assembly was originally designed and manufactured to perform.

Engineering resin Associated with plastics having medium to high physical properties used for structural and demanding applications. Examples are nylon, acetal, polybutylene terephthalate (PBT), polyethylene terephthalate (PET), polycarbonate (PC), polyphenylene sulfide (PPS), and liquid crystal polymer (LCP).

Environmental stress cracking (ESC) The susceptibility of a thermoplastic article to crack or craze under the influence of certain chemicals or aging, weather, and stress. Standard ASTM test methods that include requirements for environmental stress cracking are indexed to ASTM standards.

Etch To treat a mold with an acid, leaving parts of the mold in relief to form the desired design.

Ethylene–vinyl acetate A plastic copolymer made from the two monomers, ethylene and vinyl acetate. This copolymer is similar to polyethylene but has considerably increased flexibility.

Exothermic Pertaining to an action or reaction that gives off heat.

Extrusion The plasticizing of a material in an extruder (barrel-and-screw or plunger assembly) and forcing of the molten material or extrudate through a die or into a mold. The initial part of the molding process.

Extrusion plastomer An instrument used to determine melt flow index (MFI). *See* Melt index.

Fabricate To work a material into a finished form by machining, forming, or other operation. In the broadest sense, it means to manufacture.

Failure mode and effects analysis (FMEA) A quality-assurance tool analyzing all business and/or manufacturing operations in a continuous step-by-step manner to determine any variables in an operation that can affect the operation. Once these are determined, develop ways to control the variability and selection of control methods for the control of these variables to product a repeatable good product, cycle to cycle.

Family mold *(1)* A multicavity mold in which each cavity forms a part that often has a direct relationship in usage to the other parts in the mold. Family molds can have more than one cavity making the same part, but they will still always have that same direct relationship to usage. *(2)* A multicavity mold in which each cavity forms one of the component parts of the assembled object. The term often applied to molds in which parts from different customers are grouped together in one mold for economy of production. Sometimes called a combination mold.

Fan gate A shallow gate somewhat wider than the runner from which it extends.

Feathered thread A thread that is thin at its end (comes out sharp) and does not end abruptly. Usually found in screw-machine parts.

Feed throat The section of the hopper mounted on the injection molding machine to feed resin into the feed section of the barrel and screw.

Feedback Information returned to a system or process to maintain the output within specified limits.

Fiber Thin strands of glass used to reinforce both thermoplastic and thermosetting materials. One-inch-long fibers are occasionally used, but the more common lengths are ½ and ¼ inches.

Fill and wipe A decorating process in which depressed letters or a design in a molded plastic part is filled with paint. The excess paint is wiped off, leaving paint only in the depressed areas.

Fill rate The pressure–time relationship used to describe the filling of the mold cavity.

Filler An inert substance added to plastics for the purpose of improving physical properties or processability, or reducing the cost of the material.

Fillet A rounded inside corner of a plastic piece. The rounded outside corner is called a bevel.

Fines Very small particles mixed in with larger particles.

Finish *(1)* To complete the secondary work on a molded part so that it is ready for use. Operations such as filing, deflashing, buffing, drilling, tapping, and degating are commonly called finishing operations. *(2)* The plastic forming the opening of a bottle, shaped to accommodate a specific closure. The ultimate surface structure of a part. *(3) See* Surface finish.

Finite element analysis A stress analysis technique of a part using a computer-generated model that can take finite sections of the part for analysis of the forces and loads the part will experience in service. It generates a part-section analysis that shows the force concentrations in the section and determines if the material selected will be suitable for the part.

First surface The front surface of a plastic part, nearest the eye.

Fishbone diagram A problem analysis technique used to list all the variables and steps in the solution to a problem. All contributing elements are associated with each factor and taken back to their starting point to ensure that all variable elements are considered.

Fissure A narrow opening crack in a material.

Fixture Means of holding a part during a machine or other operation.

Flakes Resin residue formed on the inside of pipes during material transfer. Created by the friction of the pellets against the surface of the transfer piping. With time, they build up, flake off, and can cause feed problems at the throat of the injection molding machine.

Flame retarded A resin modified by flame-inhibiting additives so that exposure to a flame will not burn or will self-extinguish. Some resins will not burn as thermosets; others can be modified to meet agency flame/burning specifications; and others, depending on their base materials, may not be able to be modified.

Flame treatment A type of surface treatment that oxidizes a plastic surface for better reception of paint, inks, and adhesives. *See* Surface treatment.

Flammability Measure of the extent to which a material will support combustion.

Flash Extra plastic attached to a molding along the parting line. Under most conditions, it is objectionable and must be removed before parts are judged acceptable.

Flash gate Usually a long gate extending from a runner parallel to an edge of a molded part along the flash or parting line of the mold.

Flash line A raised line appearing on the surface of a mold and formed at the junction of mold faces. *See* Parting line.

Flash mold A mold in which the faces are perpendicular to the clamping action of the press. The higher the clamping force, the tighter the mold seam.

Flash trap A molded-in lip or blind recess on a part that is used for trapping excess molten material (flash) during an assembly operation. Negates a flash trimming secondary operation.

Flex bar An ASTM specified test bar used to develop physical property data for plastic materials. Usually sized at 4 × ½ × ⅛ = inches thick or thicker, depending on the ASTM specification.

Flexural strength Ability of a material to flex without permanent distortion or breaking (ASTM D 790).

Flock Very short fibers of cotton, wood, glass, etc., used as an inexpensive filler.

Flocking A decorating and/or sound-deadening technique where fibers of different materials are attached to the surface of a plastic part. Fibers can be oriented in specific directions and patterns determined by the techniques used and adhesive patterns laid down on the surface of the part.

Flow *(1)* A qualitative description of the fluidity of a plastic material during the process of molding. *(2)* A quantitative value of fluidity when expressing a melt flow index. *See* Melt index.

Flow chart A line chart that traces a process from start to finish.

Flow-coating A painting process in which the article to be painted is drenched under a curtain of lacquer. The part is withdrawn and rotated until the coating dries.

Flow length The actual distance a material will flow under a set of molding machine conditions. Influenced by the processing and mold design variables, the composition of the polymer, and any additives in the polymer.

Flow line A mark on a molded piece made by the meeting of two flow fronts during molding. Also called weld line or weld mark.

Flow marks Wavy surface appearance on a molded object caused by improper flow of the material into the mold. *See* Splay marks.

Fluidized bed coating Tiny particles of thermoplastic resin are suspended in a gas stream (generally air) and behave like a liquid. A heated article is immersed in this fluidized bed or powder. The thermoplastic resin particles melt and fuse to the heated surface, forming a smooth coating.

Foamed plastics Resins in sponge form. The sponge may be flexible or rigid, the cells closed or interconnected, the density anything from that of the solid parent resin down to, in some cases, 2 lb./cubic foot.

Foil decorating Molding paper, textile, or plastic foils printed with compatible inks directly onto a plastic part, so that the foil is visible below the surface of the part as an integral decoration.

Force *(1)* (physics) That which changes the state of rest or motion in matter, measured by the rate of change of momentum. *(2)* That portion of the mold which forms the inside of the molded part. *See* Core; Plunger.

Force plate The plate that carries the plunger or force plug of a mold and guide pins or bushings.

Freeze-off Refers to the gate area when it solidifies, as well as any area in the resin flow system when the melt becomes too cool to flow and solidifies.

Frequency The number of completed energy transmissions imparted to the welding horn in a vibratory motion.

Friction welding A means of assembling thermoplastic parts by melting them along their line of contact through friction. *See* Spin welding.

Full indicator movement (F.I.M.) A term in current use to identify tolerance with respect to concentricity. "Former practices" terms are Full Indicator Reading (F.I.R.) and Total Indicator Reading (TIR) runout.

Fusion bond *(1)* The joining of two melt fronts that meet and solidify in a mold cavity. *(2)* The bond formed during the assembly operation where the joint line is melted prior to assembly. *See* Hot-plate welding; Induction welding; Ultrasonic sealing.

Galling A surface area that is worn away by another by rubbing against it.

Gardner A type of drop-weight impact test. *See* Impact test.

Gardner test *See* Impact test.

Gas-assisted injection molding (GAM) An injection molding process that introduces a gas (usually nitrogen) into the plasticized material, to form voids in strategic locations.

Gaseous blowing agent A compressed gas, such as compressed air or nitrogen, used to create a cellular structure or controlled voids in a rubber or plastics mass.

Gate In injection and transfer molding, the orifice through which the melt enters the cavity.

Gauges The measuring instruments used to determine whether the part meets customer specifications, including go/no go plugs, micrometers, and vernier calipers.

Gaylord A plastic term used to identify a box of resin versus a bag or drum. Box size and weight of resin can vary depending on the density of the resin and the supplier's box size. Box size usually conforms to the size of a standard pallet on which it is shipped.

Gel permeation chromatograph (GPC) Used to determine molecular weight distribution.

Generic Descriptive of an entire type or class of plastic resins. The base resin is one of a family of polymers, but there may be hundreds of product combinations.

Glass transition point The temperature range indicated by the change from a viscous or rubbery state to a hard, brittle state in a polymer.

Gloss The shine or luster of the surface of a material (ASTM D 673). *See* specular gloss; surface finish.

Graining This refers to wood graining on plastics. This can be done by hand, roller coating, hot stamping, or printing.

Grinder (granulator) Machine with a series of knife blades and a sizing screen to chop up parts, sprue and runners, and other plastic materials for reuse or resale. Available in many sizes, styles, and capacities. Used to make regrind.

Grit blasted A surface treatment of a mold in which steel grit or sand materials are blown onto the walls of the cavity to produce a roughened surface. Air escape from mold is improved and special appearance of the molded article is often obtained by this method.

Guide pins Devices that maintain proper alignment of force plug and cavity as the mold closes. Also called leader pins.

Guideway Usually a T-shaped slot in a mold.

Gusset An angular piece of material used to support or strengthen two adjoining walls.

Hand molds Molds that are removed from the press by the operator, who opens the mold and extracts the part by hand.

Hardness The resistance of a material to compression and indentation. Among the most important methods of testing this property are Brinell hardness, Rockwell hardness, and Shore hardness.

Haze The degree of cloudiness in a plastic material.

Head The end section of the molding machine that consists of the core, die, mandrel, mold, and other parts necessary to form the plastic.

Heading The mechanical, thermal, or ultrasonic deformation of a pin to form a locking attachment to retain whatever is under the deformed material.

Heat-distortion point An arbitrary value of deformation under a given set of test conditions. In ASTM Test D 648, it is defined as a total deflection of 0.010 inches in a rectangular bar supported at both ends under a load of 66 or 264 psi. The temperature is increased 2°C per minute.

Heat sealing A process of joining two or more thermoplastic films or sheets by heat.

Heat stability The resistance of a plastic material to chemical deterioration during processing.

Heat stabilizer An ingredient added to a polymer to improve its processing or end-use resistance to elevated temperatures. The term is used in different contexts depending on the benefit to be derived from the additive. For processing, it retards changes in resin color. For end use, it protects the surface of the part exposed to elevated temperatures. It does not imply that a resin can be used beyond its recommended end-use temperature rating if it is heat stabilized.

Heater bands The only heat source for the barrel and nozzle temperature control divided, usually, into rear, middle, front, and nozzle temperature control sections. They are very accurate resistance heaters with high heat output.

Heating chamber In injection molding, that part of the machine in which the cold feed is reduced to a hot melt. Also called heating cylinder.

Hermetic As in seal, to form a bond that is pressure tight, so that air or gasses cannot enter or escape.

Hiding power The opacity that can be effected with a coating.

Hob A master model in hardened steel used to sink the shape of a mold cavity into a soft steel block.

Hobbing Forming multiple mold cavities by forcing a hob into soft steel cavity blanks. Also called sinking. *See* Hob.

Holding pressure The pressure maintained on the melt after the cavity is filled until the gate is filled and freezes off. *See* packing pressure.

Homopolymer The product of the polymerization of a single monomer (repeating unit).

Hone, honing, honed To impart a precise accuracy to the surface finish of a mold cavity by using a fine-grained whetstone.

Hoop stress The circular stresses referred to in round, usually pressure-type, containers.

Hopper A conical reservoir from which the molding powder or pellets feed into the extruder screw.

Hopper feeder Usually part of the resin drying system, but can be an independent system, to convey material to the machine's feed hopper using vacuum or positive air pressure.

Hot/heated manifold mold A thermoplastic injection mold in which the portion of the mold that contains the runner system has its own heating elements to keep the molding material in a plastic state ready for injection into the cavities, from which the manifold is insulated.

Hot-plate welding The use of a heated tool to cause surface melting of a plastic part at the joint area. It is then removed prior to the joint surfaces being pressed together to form a fusion bond.

Hot-runner mold A thermoplastic injection mold in which the runners are insulated from the chilled cavities and remain hot so that the center of the runner never cools in normal cycle operation. Runners are not usually ejected with the molded pieces. Called insulated runner molds when heating elements are not used in mold. *Note:* A heated manifold mold is a hot-runner mold that is both heated and insulted; an insulated mold is a hot-runner mold that does not contain heaters.

Hot stamping Engraving operation in which roll leaf is dyed or (metallized foil) stamped with heated metal dies onto the face of the plastics. Also called branding.

Hot tip The precise controller and gating mechanism of a hot runner mold.

Hydrolysis Chemical decomposition of a material involving the addition of water.

Hygroscopic Tending to absorb moisture from air.

Impact strength *(1)* The ability of a material to withstand shock loading. *(2)* The work done in fracturing, under shock loading, a specified test specimen in a specified manner. *(3)* The relative susceptibility of plastic articles to fracture under stress applied at high speeds.

Impact test Often associated with the Gardner (ball or falling dart) test, with a known weight falling at a known distance and hitting a part, thereby subjecting it to an instantaneous high load. Could also be a pendulum-type impact test. ASTM impact tests for material properties are the Izod, Charpy, and Tensile Impact tests.

Induction welding The use of radio, magnetic, or electrical energy to form a melt through the application of a foreign medium at the joint line to form a fusion bond.

Inert pigment A pigment that does not react with any components of a paint.

Initiator Any foreign additive mixed in a material to cause a chemical or physical reaction in the melt or liquid stage.

Injection molding A molding procedure whereby a heat-softened plastic material is fed into a cavity (mold), which gives the article the desired shape using a screw or ram. Used with both thermoplastic and thermosetting materials.

Injection pressure The pressure in the mold during the injection of plasticized material into the mold cavity. Expressed in psi, with the hydraulic system pressure being used to indicate changes, when there are no sensors in the mold.

Injection ram. *See* Ram.

Injection time The time it takes for the screw's forward motion to fill the mold cavity with melt.

Inorganic A mineral compound not composed of carbon atoms.

Insert An integral part of plastics molding. It consists of metal or other material, which may be molded into position or may be pressed into the molding after the molding is completed. Also a removable or interchangeable component of the mold.

Ishakawa *See* fishbone diagram.

ISO9000 International Organization of Standardization, the current world-class recognized quality standard for all businesses in the world.

Izod A type of pendulum impact test. *See* impact test.

Izod impact test An impact test in which a notched sample bar is held at one end and broken by a blow. This is a test for shock loading. *See* impact test.

Jetting Turbulent flow of plastic from an undersized gate or thin section into a thicker mold section, as opposed to laminar flow of material progressing radially from a gate to the extremities of the cavity. May also result from shooting material into a mold cavity where there is no core or immediate cavity wall to break up the flow of the material coming through the gate.

Just in time (JIT) A practice developed to minimize customer inventory. The supplier provides the product, at predetermined intervals, so that it can proceed directly to the customer's assembly line. This practice demands excellent quality control and production schedules. Customers who use JIT must demand the same care and treatment from their own suppliers. Suppliers and customers are usually located within a few hours shipping time of each other.

Kaizen A Japanese-developed quality assurance method in which a group of consultants are brought in to a plant with complete control to change, modify, and replace the current manufacturing line. The new methods, procedures, and manufacturing positions are developed for more efficient, economical, and quality proficient operation without management's permission while working with the employees of the company.

Kirksite An alloy of aluminum and zinc used for the construction of prototype molds. It imparts a high degree of heat conductivity to the mold.

Knit line A line on a part where opposing melt fronts meet. Created by material flow around obstructions or multiple gating. *See* Weld line.

Knockout pin A pin that pushes a cured molded article out of a mold. Sometimes called an ejector pin.

Laminar flow Laminar flow of thermoplastic resins in a mold. It is accompanied by solidification of the layer in contact with the mold surface that acts as an insulating tube through which material flows to fill the remainder of the cavity. This type of flow is essential to duplication of the mold surface.

Land *(1)* The horizontal bearing surface of a semipositive or flash mold by which excess material escapes. *(2)* The bearing surface along the top of the flights of a screw in an extruder. *(3)* The surface of an extrusion die parallel to the direction of the melt flow. *(4)* The bearing surfaces of any mold. *(5)* The gate, when entering a part, has either one or two dimensions. There is always one more dimension involved, which is the length of the gate itself. This would be called the land. On a round gate, it is the second dimension. On a rectangular or square gate, it is the third dimension.

Lean The streamlining way of producing products using team methods to remove waste and improve flow, as opposed to batch manufacture leading to potential problems in different lots. Lean manufacturing is a process management uses to have station control over all operations making the product. Toyota Motor Sales developed the system for its reduction of waste in both material and operator motion.

Let-down ratio Quantification of the quantity of one ingredient to be mixed with a base material to obtain the desired results.

Lifters *See* Slides.

Light-resistance The ability of a plastics material to resist fading after exposure to sunlight or ultraviolet light (ASTM D 1501). Light stability is the measure of this resistance.

Limit switch An electromechanical switch positioned to stop or start an action. It is operated by mechanical action on a movable control arm.

Lip A part of paint spray mask going over and down a wall.

Loading tray A device used to load the charge of material or metal inserts simultaneously into each cavity of a multicavity mold by the withdrawal of a sliding bottom from the tray. Also called charging tray.

Locked-in-stresses *See* Residual stress.

Lot number A number assigned to a specific lot of material or parts. Used for traceability and accountability by the supplier and customers on all paperwork for the product.

Lubricants *(1)* A processing aid to assist material flow in the barrel of an injection molding machine. Can be a solid, such as sodium or zinc styrate, or a liquid usually compounded into the base material. *(2)* Internally lubricated resins that use oils, teflon, moleidium disulfide, or other materials to give the molded part a lower coefficient of friction.

Macbeth A lighting system used for checking color.

Manifold A pipe channel, or mold, with several inlets or outlets.

Marriages Poorly cut pellets that are too hot and bond together in strings or clumps that are not trapped and removed by the screen prior to packaging.

Master curve The acceptable or required curve that all subsequent test curves must match.

Matched metal (die) molding Method of molding reinforced plastics between two close-fitting metal molds mounted in a press.

Material review board A panel of representatives from departments of the company who are involved with a product. It decides if the material and/or product meets customer requirements if a question or problem about quality arises.

Matrix Refers to the base resin or material used for a molded part.

Matte finish A type of dull, nonreflective finish. *See* Surface finish.

Melt fracture An elastic strain set up in a molten polymer as the polymer flows through the die. It shows up in irregularities on the surface of the plastic.

Melt front The exposed surface of molten resin as it flows into a mold. The melt front advances as the molten resin is continuously pushed through its center section.

Melt index (MI) or melt flow index (MFI) The amount, in grams, of a thermoplastic resin that can be forced through a 0.0825-inch orifice when subjected to the prescribed force (grams) in 10 minutes at the prescribed temperature (°C) using an extrusion plastometer (ASTM D 1238).

Melt strength The strength of the plastic while in the molten state.

Melt temperature *(1)* The temperature at which a resin melts or softens and begins to have flow tendencies. *(2)* The recommended processing temperature of resin melt for correct processing. *(3)* The temperature of the melt when taken with a pryometer melt probe.

Metal plating The process of plating a plastic part by chemically etching the surface to accept a base metal on which the subsequent layers of metal are deposited. Usually a many stepped process. Not all plastics can be metallized.

Metallizing A general term used to cover all processes by which plastics are coated with metal.

Meter SI length unit equal to 100 centimeters or 39.37 inches.

Metering equipment A machine or system to meter accurately additives or regrind to the machine's hopper or feed throat. Comes in many sizes and types to suit each particular application, including augers, shuttle plates, photoelectric eyes, and positive or negative weight loss belt feeders.

Metering screw An extrusion or injection molding screw that has constant shallow depth and pitch section, usually over the last three to four flights.

Methyl methacrylate An amorphous thermoplastic resin. A common name is acrylic resin.

Microprocessor Computer system that stores, analyzes, and adjusts the controls of a machine based on the parameters established during the operation of the machine it is controlling. Only operates within preset limits. Continuously analyzes output data to adjust and maintain the machine's cycle within programmed limits. Can also store data and output it as directed by programming.

Migration of plasticizer Loss of plasticizer from an elastomeric plastic compound with subsequent absorption by an adjacent medium or lower plasticizer concentration.

Mil English unit of length equal to 0.001 inch or 0.00254 centimeters.

Modifiers Any additive to a resin that improves the processing or end-use properties of the polymer. An example would be plasticizers added to PVC resin to make it soft and pliable and improve its impact strength. All PVC resins use different modifiers to meet desired product requirements. This is true of almost all plastic resins currently manufactured.

Modulus of elasticity The ratio of stress to strain in a material that is elastically deformed (ASTM D 790).

Moisture *(1)* (Absorption) The pickup to moisture from the atmosphere by a material that penetrates the interior. *(2)* (Adsorption) Surface retention of moisture from the atmosphere.

Moisture vapor transmission rate (MVTR) The rate at which water vapor permeates through a plastic film or wall at a specified temperature and relative humidity (ASTM E 96).

Mold (1) (noun) A medium or tool designed to form desired shapes and sizes. (2) (verb) To process a plastic material using a mold.

Mold deposits Material buildup on a cavity's surface due to plate out of resin, usually in a gaseous state. Can also be attributed to additives in a resin adhering to the mold's surface.

Mold height *See* Daylight opening.

Mold open time *See* Ejection time.

Mold release *(1)* A lubricant used to coat a mold cavity to prevent the molded piece from sticking to it, thereby facilitating its removal from the mold. *(2)* Additives put into a material to serve as a mold release. Also called release agent.

Molding A group of plastics processes using molds.

Molding cycle The period of time required to complete a cycle and produce a part/product.

Molding material Plastic material in varying stages of granulation often comprising plastic or resin, filler, pigments, plasticizers, and other ingredients, ready for use in the molding operation. Also called molding compound or powder.

Molding pressure *(1)* The pressure applied directly or indirectly on the compound to allow the complete transformation to a solid dense part. *(2)* The pressure developed by a ram or screw to push molten plastic into a mold cavity. *See* Injection pressure.

Molding shrinkage The difference in dimensions, expressed in inches per inch, between a part and the mold cavity in which it was molded. Both the part and mold cavity are at normal room temperature when measured. Also called mold shrinkage and contraction.

Molecular weight (MW) (average molecular weight) The sum of the atomic masses of the elements forming the molecule, indicating the relative size typical chain length of the polymer molecule.

Molecular weight distribution (MWD) Normally determined using a gel permeatron chromatograph (GPC), MWD is the plot of a fraction of an MW sphere versus the molecular weight.

Monomer A low-molecular-weight-reactive chemical that polymerizes to form a polymer.

Morphology The study of the physical form and structure of material.

Mottle A mixture of colors or shades giving a complicated pattern of specks, spots, or streaks.

Movable platen The moving platen of an injection or compression molding machine to which half of the mold is secured during operation. This platen is moved either by a hydraulic ram or a toggle mechanism.

Multicavity mold A mold having more than one cavity or impression for forming finished items during one machine cycle.

Necking The localized reduction of the cross-sectional area of the object.

Nonpolar Incapable of having a significant dielectric loss. Polystyrene and polyethylene are nonpolar.

Nonreturn valve *See* ball valve; checking.

Nonrigid plastic A plastic that has a modulus of elasticity (either in flexure or in tension) of not over 10,000 psi at 25°C and 50 percent relative humidity (ASTM D 747).

Notch sensitive A plastic material is said to be notch sensitive if it will break when it has been scratched, notched, or cracked. Glass is considered to be highly notch sensitive.

Nozzle The hollow cored metal nose screwed into the extrusion end of *(1)* the heating cylinder of an injection machine or *(2)* a transfer chamber (where this is a separate structure).

Nucleation (nucleator) With crystalline polymers, any foreign additive that assists or acts as a starting site for crystallinity within the resin. These initiators can reduce cycle time by speeding up the crystalline formations, thereby causing the part to solidify faster so its ejection from the mold can occur sooner.

Nylon A generic term for polyamides. A crystalline thermoplastic.

Olefin plastics Plastics produced from olefins (polyolefins). Examples are polyethylene and polypropylene.

Opaque A material that will not transmit light and is not transparent.

Open-hole insert An insert with a hole drilled completely through it.

Optical comparator An inspection machine using optics to compare the outline of a part to its required dimensions on a graphic screen.

Orange peel An unintentional rugged surface that gives an appearance resembling the skin of an orange.

Organic Refers to the chemistry of carbon compounds.

Orientation The alignment of the crystalline structure in polymeric materials so as to produce a highly uniform structure. Can be accomplished by cold drawing or stretching during fabrication.

Orifice An opening in a die or other metal piece used to meter (control the flow of) fluid material.

Out-of-round Nonuniform radius or diameter.

Overflow tab A small, localized extension of a part at a weld-line junction to allow a longer material flow path for the purpose of obtaining a better fusion bond of the meeting melt fronts.

Overlay sheet *See* Foil decorating. Also called surfacing mat.

Oxidation *(1)* Degradation of a material through contact with air. *(2)* A chemical reaction involving a combination with oxygen to form new compounds.

Oxygen index An indication of flammability.

Pack time The amount of time that packing pressure is kept on the screw until the gate freezes off. Occurs immediately after the injection stroke ends.

Packing pressure The pressure applied just before the part cavity fills, which is about 50 percent of the injection pressure required to continue filling the mold without flashing it. Packing pressure is maintained until the gate freezes off.

Pad *See* Cushion.

Paint line The point where two colors meet.

Paint step A break in a smooth surface that allows a mask to rest.

Parallel to the draw The axis of the cored position (hole) or insert parallel to the up-and-down movement of the mold as it opens and closes.

Parallels The support spacers between the mold and press platen or clamping plate. Also called risers or support fillers.

Pareto analysis An analytical and statistical technique used to determine part defect type and quantity. Ranks each type of defect as a percentage of the total number of defects found, based on the quantity of each type of defect.

Part separator A machine or system used to separate parts from the runner system automatically after molding. Separated parts go to their next station and the runner moves to a granulator for reuse if permitted. System may use blades, rigid pins, or a degating station with parts placed by a robot for separation.

Parting agent *See* Mold release.

Parting line *(1)* The point in the mold where two or more metal surfaces meet creating a shut off. *(2)* Mark on a molding or casting where halves of a mold met in closing.

Partitioned mold cooling *See* Bubbler.

Pastel A tint. Mass tone to which white has been added.

Permeability *(1)* The passage of diffusion of a gas, vapor, liquid, or solid through a material without chemically or physically affecting it. *(2)* The rate of the passage in *(1)*.

Perpendicular to the draw 90°from parallel to the draw.

Piece part price The calculated finished part cost based on material, processing, assembly, decorating, and packaging, including productivity and overhead costs.

Pigment Imparts color to plastic while remaining a dispersion of undissolved particles.

Pigmented Color pigments are added to a resin to produce a desired color in the plastic resin after molding. Pigments can be either organic- or inorganic-based materials.

Pinpoint gate A restricted orifice through which molten plastic flows into a mold cavity. Also called restricted gate.

Pitch With respect to extruder or injection molding, the distance from any point on the flight of a screw line to the corresponding point on an adjacent flight, measured parallel to the axis of the screw line or threading.

Plastic *(1)* (noun) One of the high polymeric materials, either natural or synthetic, exclusive of rubbers, which either melt and flow with heat and pressure, as with a thermoplastic, or chemically "set," as with a thermoset material. *(2)* (verb) Capable of flow under pressure or tensile stress.

Plastic deformation The deformation of a material under load that is not recoverable after the load is removed. Opposite of elastic deformation.

Plastic memory A phenomenon of a plastic to return, in some degree, to its original form after heating.

Plasticate To soften by heating or kneading.

Plasticity A property of plastics that allows the material to be deformed continuously and permanently without rupture after the application of a force that exceeds the yield value of the material.

Plasticize To make a material soft and moldable with the addition of heat and/or pressure or a plasticizer.

Plate dispersion plug *See* Breaker plate.

Platens The mounting plates of a press to which the entire mold assembly is bolted.

Plunger The part of a transfer or injection press that applies pressure to the unmelted plastic material to push it into the chamber. This, in turn, forces plastic melt at the front of the chamber out through the nozzle. *See* Ram.

Polyallomers Crystalline thermoplastic polymers made from two or more different monomers, usually ethylene and propylene.

Polyamides A group of crystalline thermoplastics, of which nylon is typical.

Polycarbonate resins An amorphous thermoplastic material. It is transparent and can be injection molded, extruded, thermoformed, and blow molded.

Polyethylene A crystalline type thermoplastic material made by polymerizing ethylene gas.

Polyimide Classified as a thermoplastic, it can not be processed by conventional molding methods. The polymer has rings of four carbon atoms tightly bound together. It has excellent resistance to heat.

Polyliner *(1)* A perforated, longitudinally ribbed sleeve that fits inside the cylinder of an injection molding machine. Used as a replacement for conventional injection cylinder torpedoes (older machines). *(2)* A plastic bag placed inside a carbon or box to prevent material contamination during shipment.

Polymer A high-molecular-weight organic compound—natural or synthetic—whose structure can be represented by a repeated small unit, the MER. Examples are polyethylene, rubber, and cellulose. Synthetic polymers are formed by addition or condensation polymerization of monomers. Some polymers are elastomers; some plastics.

Polymerization A chemical reaction in which the molecules of a monomer are linked together to form large molecules whose molecular weight is a multiple of that of the original substance. When two or more monomers are involved, the process is called copolymerization. Addition and condensation are the two major types of reactions.

Polyphenylene oxide (PPO) An amorphous thermoplastic. This material is noted for its useful temperature range from –275 to 375°F.

Polypropylene A crystalline thermoplastic made by polymerizing propylene gas. It has the lowest density of all plastics, except methylpentene.

Polystyrene An amorphous thermoplastic made by polymerizing styrene.

Polysulfone An amorphous thermoplastic noted for its high strength, high service temperature, low creep, and self-extinguishing properties.

Polyvinyl chloride (PVC) A thermoplastic material made by the polymerization of vinyl chloride with perozide catalysts. The pure polymer is brittle and difficult to process. It yields a flexible material when compounded with plasticizers.

Post annealing Stress relieving of molded parts by external means, hot air or oil, humidity chambers, or submersion in a fluid.

Post mold shrinkage The shrinkage occurring after a part has been removed from the mold. Influenced by the material and chemical properties of the resin and its molding conditions. Also influenced by end-use operating and environmental conditions.

Postforming A process used to impart a shape to a previously molded article.

Potentiometer An electrical control device that senses changes in voltage or a potential difference by comparison to a standard voltage and can transmit a signal to a control switch.

Preplastication Technique of premelting injection molding powders in a separate chamber, then transferring the melt to the injection cylinder. Device used for preplastication is commonly known as a preplasticizer.

Pressfit An interference assembly between two mating parts, with friction holding the parts together. Parts assembled are under considerable stress.

Pressure drop The decrease in pressure on a fluid attributed to the number of turns it has to make and the distance it must flow to fill a cavity.

Pressure gradient lines A hypothetical set of pressure lines in a part created by the material's pressure drop as the part is filled. The further the material flows from the gate, the lower the packout pressure.

Pressure pads Reinforcements distributed around the dead areas in the faces of a mold to help the land absorb the final pressure of closing without collapsing.

Printing on plastics The decoration of plastics by means of various printing processes. Such processes are offset, silk screen, letterpress, electrostatic, and photographic methods.

Process control procedures A separate document, often included as an attachment to the quality control manual, which is a detailed description of the methods to be followed in the manufacture of a product. A copy may be attached to the work order for reference and revision as required should changes in the product occur.

Processing aid A resin additive that improves processing characteristics.

Product certification The certificate or letter stating that the material or product meets or exceeds customer requirements. Values are often listed for the tested or measured results. Signed by a key representative of the company to verify accuracy.

Projected surface area The exposed resin area of a mold on the parting line that transmits the injection pressure on the closed mold halves. Includes part, runner, and sprue surfaces expressed in inches squared of the surface area.

Prototype mold A simplified mold construction often made from a light metal casting alloy or from an epoxy resin in order to obtain information for the final mold and/or part design.

Purging Cleaning one color or type of material from the cylinder of an injection molding machine or extruder by forcing it out with the new color or material to be used in subsequent production. Purging materials are also available.

Pyrometer An electrical thermometer for measuring high temperatures. Unit comes with two probes to measure melt and surface temperatures.

QS16949 Automotive harmonization of the major automotive suppliers of the world with input from the trucking manufacturers. An add on to ISO9000 requiring documentation and verification in greater depth and detail to the automotive suppliers specifications and requirements.

Quality assurance A separate department established to direct the quality function of the business and systems responsibility areas. Major concentration is directed to assisting and auditing the activities of the quality control department in their efforts to ensure that quality products are produced.

Quality circles A quality analysis group consisting of employees with specific departmental knowledge used to provide suggestions and ways to solve a procedural or manufacturing quality problem. If found acceptable, the group's findings and solutions are then passed on to upper management for implementation.

Quality control A department set up to be technically involved in the control of product quality. Involved in the principal inspection and testing of a product, with limited systems responsibility.

Quality control manual A document that states the company's quality objectives, and how they will be implemented, documented, and followed in the manufacture of products.

Quality function deployment Method of obtaining the required information from a customer, supplier, or your own company personnel for solving a problem, improving a product, or providing a required or necessary service to a customer.

Quality rated *See* approved supplier.

Quench A method of rapidly cooling thermoplastic molded parts as soon as they are removed from the mold. This is generally done by submerging the parts in water.

Radio frequency (RF) preheating A method of preheating used to mold materials to facilitate the molding operation and/or reduce the molding cycle. The frequencies most commonly used are between 10 and 100 mc/sec.

Ram The press member that enters the cavity block and exerts pressure on the molding compound designated as the "top force" or "bottom force" by position in the assembly. *See* plunger.

Ram travel Distance ram moves when operating a complete molding cycle.

Real time The present time, or, as an activity is occurring.

Recessed letters Depressed letters.

Reciprocating screw A combination injection and plasticizing unit in which an extrusion device with a reciprocating screw is used to plasticize the material. Injection of material into a mold can take place by direct extrusion into the mold, by reciprocating the screw as an injection plunger, or by a combination of the two. When the screw serves as an injection plunger, this unit acts as a holding, measuring, and injection chamber.

Recycled plastics A plastic material prepared from previously used or processed plastic materials that have been cleaned and reground.

Regrind *(1)* Waste plastics that recovered and processed for reuse. *(2)* Plastics that have been ground or pelletized at least twice.

Reinforced molding compound A material reinforced with special fillers to meet specific requirements, such as rag or glass.

Release agent *See* Mold release.

Relief angle *(1)* The angle of the cutaway portion of the pinch-off blade measured from a line parallel to the pinch-off land. *(2)* In a mold, the angle between the narrow pitch-off land and the cutaway portion adjacent to the pinch-off land.

Residence time The amount of time a resin is subjected to heat in the barrel of an injection molding machine.

Residual stress The stresses remaining in a plastic part as a result of thermal or mechanical treatment.

Resin *(1)* Any of a class of solid or semisolid organic products of natural or synthetic origin, generally of high molecular weight with no definite melting point. *See* Polymer. *(2)* In a broad sense, any polymer that is a basic material for plastics.

Resin pocket An apparent accumulation of excess resin in a small localized section that is visible on cut edges of molded surfaces. Also called resin segregation.

Restricted gate Sometimes referred to as pin-point gate. A small opening between the runner and cavity in an injection or transfer mold.

Retainer plate The plate on which demountable pieces, such as mold cavities, ejector pins, guide pins, and bushings are mounted during molding; usually drilled for steam or water.

Retaining pin A pin on which an insert is placed and located prior to molding.

Rib An object designed into a plastic part to provide lateral, longitudinal, or horizontal support.

Ring gate A gate or annular opening that circles around a core pin or molded part.

Rockwell hardness A common method of testing materials for resistance to indentation in which a diamond or steel ball, under pressure, is used to pierce the test specimen (ASTM D 785).

Runner In an injection or transfer mold, the channel that connects the sprue with the gate to the cavity.

Runner system With plastics, the sprues, runners, and gates that lead the material from the nozzle of an injection machine to the mold cavity.

Salt and pepper blends Resin blends of different concentrate additives, in pellet form, mixed with virgin resin to make a different product. Usually associated with color concentrate blends, that, when melted and mixed by the injection molding machine's screw, yield a uniform colored melt for a part.

SAN Styrene–acrylonitrile copolymers.

Scrap A product or material that is out of specification to the point of being unusable.

Screw The main component of the "reciprocating screw" injection molding machine. Has various sizes, lengths, and compression ratios to feed, compress, melt, and meter for injecting into the mold cavity. Basically divided into three major sections but there can be more. *Feed Section* Deep screw depths to convey the resin into the next screw's section. *Transition Section* Gradually decreasing screw depths when resin is compressed, forced against the barrel's surface, and melts. *Metering Section* The molten melt is further compressed in a shallow, uniform screw depth conveying forward as the screw turns.

Screw plasticating injection molding *See* Injection molding.

Scuff mark An imperfection on a part's show surface caused by dragging the part against the mold's surface during ejection from the mold cavity.

Sealing diameter That portion of a metal insert that is free of knurl and is allowed to enter the mold to prevent the flow of plastic material.

Second-surface decorating A method of decorating a transparent plastic part from the back or reverse side. The decoration is visible through the part, but is not exposed.

Semi-automatic molding machine A molding machine in which only part of the operation is controlled by direct human action. The automatic part of the operation is controlled by the machine according to a predetermined program.

Servomotor An electrical motor or hydraulic piston that supplies power to a feedback system that consists of a sensing element and an amplifier used in the automatic control of a mechanical device.

Shear Stress developed because of the action of the layers in the material attempting to glide against or separate in a parallel direction.

Shear heat The rise in temperature created by the compression and longitudinal pressure on the resin in the barrel by the screw's pumping action.

Shear joint An ultrasonic welding joint design where the welding action is parallel to each part surface. *See* Shear.

Shelf life The time a material, such as a molding compound, can be stored without losing any of its original physical or functional properties.

Shore hardness A method of determining the hardness of a plastic material using a scelroscope. This device consists of a small conical hammer fitted with a diamond point and acting in a glass tube. The hammer is made to strike the material under test and the degree of rebound is noted on a graduated scale. Generally, the harder the material, the greater the rebound (ASTM D 2240).

Short or short shot A molded part produced when the mold has not been filled completely.

Shot The yield from one complete molding cycle, including cull, runner, and flash.

Shot capacity The maximum volume of material that a machine can produce from one forward motion of the plunger or screw.

Shot peening Impacting the surface of the material with hard, small, round beads of materials to disrupt the surface flatness. Used to stress relieve welds and to improve the release of plastic resins on smooth core surfaces.

Shrink fixture *See* Cooling fixture.

Shrinkage In a plastic, the reduction in dimensions after cooling.

Shrinkage allowance The additional dimensions that must be added to a mold to compensate for shrinkage of a plastic material on cooling.

SI units Systems International Units.

Side actions (side coring or side draw pins) *(1)* An action built into a mold that operates at an angle to the normal open-and-close action and facilitates the removal of parts that would not clear a cavity or core on the normal press action. *(2)* Projections used to core a hole in a direction other than the line of closing of a mold, and which must be withdrawn before the part is ejected.

Silicone *(1)* Chemical derived from silica used in molding as a release agent and general lubricant. *(2)* A silicon-based thermoset plastic material.

Silk-screen printing In its basic form, it involves laying a pattern of an insoluble material, in outline, on a finely woven fabric. When ink is drawn across the material, it passes through the screen only in the designed areas. Also called screen process decorating.

Sink mark A depression or dimple on the surface of an injection molded part caused by collapsing of the surface following local internal shrinkage after the gate seals. May also be an incipient short shot.

Six Sigma The new quality term and methodology for identifying a process control technique to control a process within Six Sigma limits that reduces defects to 3.4 defects per million, a reduction of 20,000 times.

Skins *See* Flakes.

Slides Caming sections of a mold cavity that form complex three-dimensional part sections that must operate and move before the part can be ejected from the mold. Used to form openings and sections of parts 90° to the part's release from the mold cavity.

Snap fit An assembly of two mating parts, with one or both parts deflecting until the mating parts are together. They then return to their as-molded condition or nearly so, depending on the design of the attachment. Parts can be under high to low stress after assembly.

Solvent Any substance, but usually a liquid, that dissolves other substances.

Solvent welding (solvent cementing, solvent bonding) A method of bonding thermoplastic articles of like materials to each other by using a solvent capable of softening the surfaces to be bonded. Thermoplastic materials that can be bonded by this method are acrylonitrile butadiene styrene (ABS), acrylics, cellulosics, nylons, polycarbonate, polystyrene, and vinyls.

Specific gravity The density (mass per unit volume) of a liquid or solid material divided by that of water (ASTM D 792).

Specification A written statement that dictates the material, dimensions, and workmanship of a manufactured product.

Specular gloss The relative reflective appearance of a material as judged visually.

Spider gate Multigating of a part through a system of radial runners from the sprue.

Spin welding The process of fusing two objects by forcing them together while one of the pair is spinning, until frictional heat melts the interface. Spinning is then stopped and pressure held until they are frozen together.

Spiral flow test A method of determining the flow properties of a thermoplastic or thermoset material, in which the resin flows along the path of the spiral cavity. The length of the material that flows into the cavity and its weight gives a relative indication of the flow properties of the resin.

Splay marks or splay Marks or lines found on the surface of the part after molding that may be caused by overheating the material, moisture in the material, or flow paths in the part. Usually white, silver, or gold in color. Also called silver streaking.

Split cavity A cavity of a mold that has been made in sections.

Split-ring mold A mold in which a split-cavity block is assembled in a chase to permit the forming of undercuts in a molded piece. These parts are ejected from the mold and then separated from the molded piece.

Spot welding the localized fusion bonding of two adjacent plastic parts. Does not require a molded protrusion or hole in the parts. To be effective, used where two parallel and flat surfaces meet.

Spreader/torpedo A streamlined metal block placed in the path of flow of the plastics material in the heating cylinder of extruders and injection molding machines to spread it into intimate contact with the heating areas.

Sprue Feed opening provided in the injection or transfer mold. Also, a slug formed at this hole. Spur is a shop term for the sprue slug.

Sprue bushing A hardened steel insert in an injection mold that contains the tapered sprue hole and has a suitable seat for the nozzle of the injection cylinder. Sometimes called an adapter.

Sprue gate A passageway through which molten plastic flows from the nozzle to the mold cavity.

Sprue lock or puller In injection molding, a portion of the plastic composition held in the cold slug well by an undercut; used to pull the sprue out of the bushing as the mold is opened. The sprue lock itself is pushed out of the mold by an ejector pin.

Sprue picker *See* Sprue lock or Puller.

Stabilizer An ingredient used in the formulation of some plastics to assist in maintaining the physical and chemical properties of the compounded materials at their initial values throughout the processing and service life of the material.

Staking A term used in fastening. The forming of a head on a protruding stud for the purpose of holding component parts together. Staking may be done by cold staking, hot staking, or ultrasonic heating. *See* Heading.

Starve feeding The controlled metering of resin into the machine's feed section to just fill the screw flights. The hopper is not used to do this; it is performed by an auger, feed belt, or hand.

Stationary platen The plate of an injection or compression molding machine to which the frontplate of the mold is secured during operation. This platen does not move during normal operation.

Stereolithography A three-dimensional printing process that produces copies of solid or surface models in plastic. This process uses a moving laser beam, directed by computer, to print or draw across sections of the model onto the surface of photocurable liquid plastic.

Storage life *See* Shelf life.

Strain The dimensionless numbers (or units of length/length, i.e. inch per inch) that characterize the change of dimensions of a test specimen during controlled deformation. In tensile testing, the elongation divided by the original gage length of the test specimen.

Strength of material Refers to the structural engineering analysis of a part to determine its strength properties.

Stress The force applied to produce a deformation in the material. The ratio of applied load to the original cross-sectional area of a test specimen (psi).

Stress concentration Sections or areas in a part where the molded-in or physical forces are very high or magnified by a force or action. All sharp corners have high-stress concentrations.

Stress crack External or internal cracks in a plastic caused by tensile stresses less than its short-term mechanical strength.

Striation *(1)* A separation of colors resulting in a linear effect or color variation. *(2)* In blow molding, the rippling of thick parisons. *(3)* A longitudinal line in a plastic caused by a disturbance in the melt path.

Stripper plate A plate that strips a molded piece from core pins or cores.

Styrene–acrylonitrile copolymers (SAN) A thermoplastic copolymer of styrene and acrylonitrile. If butadiene is blended with the copolymer, a terpolymer is formed known as acrylonitrile butadiene styrene (ABS).

Styrenic Indicates a group of plastics materials that are polymers, either whole or partially polymerized from styrene monomer.

Submarine gate A type of edge gate where the opening from the runner into the mold is located below the parting line or mold surface, as opposed to conventional edge gating where the opening is machined into the surface of the mold. With submarine gates, the item is broken from the runner system on opening of the mold or ejection from the mold.

Suckback A slight retracting of the screw, usually no more than $\frac{1}{8}$ to $\frac{3}{16}$ inch as the mold opens to suckback any resin that might have drooled out of the nozzle after the sprue was pulled. Correct nozzle type and temperature control can eliminate this if drooling occurs with very fluid resins.

Surface finish Finish of a molded product. Refer to the SPI-SPE Mold Finishes Comparison Kit, available from DME Cop., Detroit, MI.

Surface treatment Any method of treating a material so as to alter the surface and render it receptive to inks, paints, lacquers, and adhesives such as chemical, flame, and electronic treatments.

Surging Unstable pressure buildup in an extruder leading to variable throughput.

Swaging An assembly technique, similar to heading, where the plastic material is deformed to a specific shape to assemble one or more parts.

Tab gated A small removable tab of approximately the same thickness as the mold item, usually located perpendicular to the item. The tab is used as a site for edge gate location, usually on items with large flat areas.

Taguchi *See* Design of experiments.

Tapping Cutting threads in the walls of a circular hole.

Temperature gradient The slope of a graphed temperature curve. An increasing or decreasing temperature profile on the barrel of the molding machine is an example.

Tensile impact test A test whereby the sample is clamped in a fixture attached to a swinging pendulum. The swinging pendulum strikes a stationary anvil causing the test sample to rupture. This is similar to the Izod test. *See* Impact test.

Tensile strength The pulling stress, in psi, at a given point on the material stress–strain curve, usually just before the material tears or breaks. Area used in computing strength is usually the original, rather than the necked-down area (ASTM D 638).

Texturizing The etching or cutting of a pattern on a mold surface to be reproduced on the molded part.

Thermal conductivity Ability of a material to conduct heat.

Thermal expansion The linear rate at which a material expands or contracts due to a rise or fall in temperature. Each material is unique and has its own rate of expansion and contraction. Expressed in in/in °F (mm/mm°C).

Thermal stress cracking (TSC) Crazing and cracking of some thermoplastic resins that results from overexposure to elevated temperatures.

Thermocouple A thermoelectric heat-sensing element mounted in or on machinery and the mold to transmit accurate temperature signals to a control and readout unit.

Thermoplastic (TP) *(1)* (adjective) Capable of being repeatedly softened by heat and hardened by cooling. *(2)* (noun) A material that will repeatedly soften when heated and harden when cooled. Typical of the thermoplastic family are the styrenic polymers and copolymers, acrylics, cellulosics, polyethylene, polypropylene, vinyls, nylons, and the various fluorocarbon materials.

Thermoset (TS) A material that undergoes or has undergone a chemical reaction by the action of heat and pressure, catalysts, ultraviolet light, etc., leading to a relatively infusible state. Typical of the plastics in the thermosetting family are the aminos (melamine and urea), unsaturated polyesters, alkyds, epoxies, and phenolics. A common thermoset goes through three stages. *A-Stage* An early stage when the material is soluble in certain liquids, fusible, and will flow. *B-Stage* An intermediate stage at which the material softens when heated and swells in contact with certain liquids, but does not dissolve or fuse. Molding compound resins are in this stage. *C-Stage* The final stage is the TS reaction when the material is insoluble, infusible, and cured.

Thinner A liquid that can extend a solution but not reduce the power of the solvent.

Thread plug, ring, or core A part of the mold that shapes a thread.

Tie bars Bars that provide structural rigidity to the clamping mechanism of a press and usually guide platen movement.

Timers Analog or digital timers used to control the molding cycle accurately.

T.I.R. (total indicator reading) An abbreviation used to identify tolerances with respect to concentricity. *Note:* The term TIR is a "former practices" term; the more acceptable current term is F.I.M. (full indicator movement).

Toggle or toggle action A mechanism that exerts pressure developed by the application of force on a knee joint. It is used as a method of closing presses and also serves to apply pressure at the same time.

Tolerance A specified allowance for deviation in weighing and measuring or for deviations from the standard dimensions of weight. (SPI Guidelines of Plastics Custom Molders).

Tool *See* Mold.

Torpedo *See* Spreader.

Torsional The twisting or turning motion of a part. Torsional stress is created when one end of a part is twisted in one direction while the other is held rigid or twisted in the other direction.

Translucent The quality of transmitting light without being transparent.

Transparent A material with a high degree of light transmission that can be easily seen through.

TS16949 Automotive harmonization of the major automotive suppliers of the world with input from the trucking manufacturers. An add on to ISO9000 requiring documentation and verification in greater depth and detail to the automotive suppliers specifications and requirements.

Tumbling *(1)* Finishing operation for small plastic article by which gates, flash, and fins are removed and/or surfaces are polished by rotating them in a barrel together with wooden pegs, sawdust, and polishing compounds. *(2)* Adding color to a material through tumble blending.

Tunnel gate *See* Submarine gate.

Ultimate strength Strength (stress in psi) at the break point in tensile test.

Ultrasonic insertion The inserting of metal into a thermoplastic part by the application of vibratory mechanical pressure at ultrasonic frequencies.

Ultrasonic sealing or bonding A method in which sealing is accomplished through the application of vibratory mechanical pressure at ultrasonic frequencies (20 to 40 kc.). Electrical energy is converted to ultrasonic vibrations through the use of either a magnetostrictive or piezoelectric transducer. The vibratory pressures at the interface in the sealing area develop localized heat losses that melt the plastic surfaces effecting the seal.

Unbalanced mold A nonuniform layout of mold cavities and runner system, fill rate, packing pressure, and part quality will vary from cavity to cavity. Used only for noncritical, stand-alone parts.

Undercut *(1)* (adjective) Having a protuberance or indentation that impedes withdrawal from a mold in its normal open/closed movement. Flexible materials can be ejected intact even with slight undercuts. *(2)* (noun) Any such protuberance or indentation; depends also on design of mold.

Unit mold *(1)* Mold designed for quick-changing interchangeable cavity parts. *(2)* A mold that comprises only a single cavity, frequently a pilot for the production set of molds.

Universal testing machine A machine used to determine tensile, flexural, or compressive properties.

UV (ultraviolet) stabilizer Any chemical compound that, when added to thermoplastic material, selectively absorbs UV rays.

Vacuum metallizing Process in which surfaces are thinly coated with metal by exposing them to the vapor of metal that has been evaporated under vacuum (one millionth of normal atmospheric pressure).

Vendor A company or person who sells or supplies a part or service to another for a price.

Vent In a mold, a shallow channel or minute hole cut in the cavity to allow air to escape as the material enters.

Venturi dispersion plug *See* Plate dispersion plug.

Vertical flash ring The clearance between the force plug and the vertical wall of the cavity in a positive or semipositive mold; also the ring of excess material that escapes from the cavity into this clearance space.

Vibration welding *See* Ultrasonic sealing.

Vicat softening temperature The temperature at which a plastic is penetrated to 1 mm depth by a flat-ended circular metal pin, while in a controlled (rate-/-rise) temperature silicone fluid bath (ASTM D 1525).

Vinyl Usually polyvinyl chloride, but may be used to identify other polyvinyl plastics.

Virgin plastics or virgin material Material not previously used or processed and meeting manufacturer's specifications.

Viscosity A measurement of resistance of a material to flow.

Void A void or bubble occurring in the center of a heavy thermoplastic part, usually caused by excessive shrinkage.

Volume Synonym for capacity or displacement.

Volume resistivity The electrical resistance between opposite faces of a 1-cm. cube of insulating material. It is measured under prescribed conditions using a direct current potential after a specified time of electrification. It is commonly expressed in ohm-centimeters. Also called specific insulation resistance (ASTM D 257).

Warpage Dimensional distortion in a plastic object after molding.

Web gate *See* Diaphragm gate.

Weld line *See* Flow line.

Welding Joining thermoplastic pieces by one of several heat-softening processes. Butt fusion, spin welding, ultrasonic, and hot gas are examples of such methods.

Welding horn The sonic-energy transmission and pressure-transmitting tool used for ultrasonic welding. Each welding horn is tuned to specific amplitudes to efficiently perform the welding operation.

Wetting agent An ingredient or solution used to lower the surface tension between two materials, so that good coverage and bonding occur.

Wheelabrating Deflashing molded parts by bombarding with small particles at high velocity.

Witness lines Lines left on a molded part by poor mating and fit of side action cores.

Yield value *(1)* (yield strength) In tensile testing, the stress, usually in psi, at which there is no increase in stress with a corresponding increase in strain: Usually the first peak on the curve. *(2)* (yield point) The specific limiting deviation from the proportional stress–strain curve.

Young's modulus *See* Modulus of elasticity.

Zero defects A quality control method where anyone in the production cycle who discovers a quality problem can stop the assembly line or manufacturing process until it is corrected. The problem associated with this method is that upper management is often never made aware that a problem occurred. This lack of knowledge may prevent a complete repair from being initiated and the problem continues to occur.

Bibliography

CHAPTER 2

1. Feigenbaum, A.V., *Total Quality Control*, New York: McGraw-Hill, 1983, pp. 490–491.
2. Harold, D., "Designing for Six Sigma Capability." *Control Engineering* January 1999: 62–66.
3. "*Best Introductory Book on SPQ*," *DataMyte Handbook* 6th ed., DataMyte Business, Allen-Bradley Comp., Inc., December 1995, pp. 1–15.

CHAPTER 3

1. Union Carbides. RIM Polyurethane Organo Functional Silane A-1.100 Treatment.
2. Tobin, B., "The Five Methods of Quality Control." *Plastics Design Forum* January/February 1986: 74–78.
3. Subject—Quality Policy, G.E. Semiconductor Products Dept., Instruction 3.16 4/26/68 MGF 217, p. 1 of 3.
4. Natarasan, R. et al. "Applying QED to Internal Service System Design," *Quality Progress* February 1999: 65–70.
5. Ohio State University, Quality Dept "(QFD)." 1989.
6. "The Five Pillars of Organieational Excellence!" *Quality Digest*, August 2006: 73, Figure 1.

7. Ford Manufacturing Staff, *Potential Failure, Mode and Effects Analysis for Manufacturing and Assembly Processes, (Process FMEA) Instruction Manual.* December 1983: (Preliminary), Attachment I and Attachment II.

CHAPTER 4

1. Navin, J., "How to Choose the Right Custom Molder." *Plastics Design Forum* September/October 1988: 57–60.
2. *Vendor Audit Survey*, Electrolux Corporation.

CHAPTER 5

1. Subject—Quality Policy, G.E. Semiconductor Products Dept., Instruction 3.16 4/26/68 MGF 217, p. 1 of 3.
2. Nickols, F., "Too Many Types of Quality Problems." *Quality Progress* April 2000: 43–49.
3. Feigenbaum, A.V., *Total Quality Control*, New York: McGraw-Hill, 1983.

CHAPTER 6

1. Feigenbaum, A.V., *Total Quality Control*, New York: McGraw-Hill, 1983.
2. Kirkland, C., ed. "Taguchi Methods Increase Quality and Cut Costs." *Plastics World* February 1988: 44–51.
3. Seader, R., Tobin, B., "Meeting the Impossible Schedule." *Plastics Design Forum* November/December. 1987: 17–24.

CHAPTER 7

1. Beck, R.D., *Plastic Product Design*, New York: Van Nostrand Reinhold, 1970.
2. Fritch, L.W., "Honing Molding Parameters by Measuring Flow Length." *Plastic Engineering* June 1989: 41–44.
3. Miller, B., ed. "Product Quality Problems? How Did You Check Your Resin?" *Plastic World* August 1989: 49–55.

CHAPTER 8

1. Beck, R.D., *Plastic Product Design*, New York: Van Nostrand Reinhold, 1970.
2. *Delrin® Molding Guide*, Wilmington, DE: E.I. Du Pont de Nemours Corp., E-49702.
3. *EDM of Tool Steels*, Totowa, NJ: Uddeholm Corporation, T-101.
4. "Elaboration on Shrinkage." *Plastic Design Forum* September/October 1988: 8.

5. Guscott, M.A., "Effects of EDM on Tool Steels." *PM & E* April 1987: 35–41.
6. Houska, C., "Moldmakimtg—Solving Molding Problems with Beryllium Copper." *Plastics Engineering* January 1990: 32–35.
7. "How to Be Kind to Your Toolmaker (and Keep Your Part Costs Down)." *Plastics Design Forum* September/October 1987: 53–58.
8. *Injection Moulds for Thermoplastics.* Reading, PA: DSM Engineering Plastics North America.
9. Kirkland, C., "Hot-Runner PVC Molding: Give It Another Look." *Plastics World* April 1990: 40–51.
10. Leonard, L., ed. "Solids Modeling: A Quick Route from CAD to CAM." *PM & E* December 1988: 31–32.
11. Manji, J., "Designers Guide to Mold Texturizing." *Plastics Design Forum* November/December 1985: 70–78.
12. Miller, B., ed. "Predicting Part Shrinkage Is a Three-Way Street." *Plastics World* December 1989: 48–52.
13. Mobery, C. "Cycle Time Reduction Through High-Thermal-Conductivity Metals." *PM & E* February: 1988: 42–44.
14. *Modern Plastics Encyclopedia*, New York: McGraw-Hill, 1985–1986.
15. Noller, R., "Understanding Tight Tolerance Design." *Plastics Design Forum* March/April 1990: 61–72.
16. Noller, R., "Tight Tolerance Design." *Plastics Engineering* May 1991: 23–27.
17. Rozena, H., Schmidt, H., "Hot Runner Gate Selection: The Key to Molding New TPs." *Plastics Engineering* October 1988: 41–44.
18. Ruehl, D.E, "What Designers Should Know about Molds." *Plastic Design Forum* November/December 1988: 43–47.
19. Snyder, M., "Software Prompts Mold Maintenance." *PM & E* February 1989: 42–43.
20. Tobin, W.J., "Predicting Tooling Delivery," *PM & E* November 1987: 60–63.
21. Tobin, W.J., "Venting from the Inside," *PM & E* February 1990: 55.
22. "Understanding Tight Tolerance Design." *Plastics Design Forum* March/April 1990: 61–71.
23. Wieder, H.K., "The Science of Keeping Cool." *PM & E* June 1987: 44–46.
24. Wigotsky, V., "Some Mold Design Guidelines." *Plastics Engineering* November 1985: 24–25.
25. *Zytel® Molding Guide*, Wilmington, DE: E.I. Du Pont de Nemours Corp., E-23285.
26. *Standards and Practices of Plastics Molders*, Molders Division of the Society of the Plastics Industry, Inc., 1991.

CHAPTER 9

1. Beek, R.D., "Plastic Product Design." New York: Van Nostrand Reinhold, 1970.
2. Bozzelli, J., Larin, B., "Advanced Injection Screw Is Key to Productivity for Global Competition." *Modern Plastics* December 1990: 66–68.

3. *Delrin® Molding Guide*. Wilmington, DE: E.L. Du Pont de Nemours Corp., E-49702.
4. Miller, B., ed. "Why More Molders Are Using Vented Barrels." *Plastics World* April 1987: 51–53.
5. *Modern Plastics Encyclopedia*. New York: McGraw-Hill, 1985–1986.
6. Rideout, D.K., Kochxjda, M., "Setting and Monitoring Toggle Clamping Force." *Plastics Engineering* February 1985: 47–49.

CHAPTER 10

1. Abbott, S., "Central Water Chillers." *Plastic World* September 1990: 61.
2. Abbott, S., "PW Equipment Profile." *Plastics World* September 1990: 61.
3. "*Autotherm I Liquid Temperature Controllers*." Thermal Care/Mayer, A Division of Midwesco Inc., #2501-057.
4. Benson, T., "Choosing a Central Chilling System." *PM & E* February 1991: 49–50.
5. "C G Beside-the-Press Granulators." *Ball and Jewell Bulletin*, No. ME-65.
6. Cloyd, C.E., "Picking the Perfect Chiller." *PM & E* September 1990: 25.
7. "Dryer Users Monitor Resin Dewpoints Closely." *Plastics World* July 1985: 1990.
8. "Drying Technology." *PM & E* January 1991: 41.
9. "Hopper Dryers." *Plastics World* November 1990: 69.
10. "Keep Your Eye on Your Robot." *Plastics World* December 1990: 28–29.
11. Kirkland, C., ed. "How You Can Afford a Quick-Mold-Change System." *Plastics World* November 1987: 27–41.
12. Miller, B., ed. "PW Troubleshooting." *Plastics World* July 1991: 37–39.
13. Miller, B., ed. "PW Troubleshooting." *Plastics World* March 1991: 38–41.
14. *Modern Plastics Encyclopedia*. New York: McGraw-Hill 1985–1986.
15. *Modern Plastics Encyclopedia*. New York: McGraw-Hill, 1991.
16. *Rynite® Molding Guide*. Wilmington, DE: E.I. Du Pont de Nemours Corp., H-06011 10/88.
17. "Self Cleaning Process Water Filter Eliminates Heat Exchange or Plugging." *Modern Plastics* February 1990.
18. Smith, R.C., "Taking the Mystery Out of Quick Mold Change." *Plastics Engineering* January 1989: 33–35.
19. Smoluk, G.R., "Robotics: Now a Truly Essential Component of Effective Molding Lines." *Modern Plastics* November 1989: 54–58.
20. Sneller, J.A., "Auxiliaries: The Last, Essential Link in CIM." *Modern Plastics* November 1985: 56–58; Snyder, M., ed. "Material Handling." *PM & E* September 1998: 29–30.
21. Snyder, A. Sr., ed. "Robotic Parts—Handling Devices." *PM & E* June 1987: 36–38.
22. "Solving Common Dryer Desiccant Problems." *Plastics World* June 1990: 9–10.

23. "Sprue/Runner System Produces Uniform Granulate." *Plastics World* February 1991: 67.
24. Stoughton, P., "*Troubleshooting Tips*," Wexford, PA: Concur Group.
25. "The Shortest Distance Between Two Points." *The Conair Line*. Wexford, PA: Conair Group.
26. Toensmeier, P.A., "Granulators Moves Deepest into the Processing and Recycling Mainstream," *Modern Plastics* May 1990: 49–51.
27. Toensmeier, P.A., "Granulators Are Coming into the Mainstream—at Affordable Prices." *Modern Plastics* 1987: 81–84.

CHAPTER 11

1. Berenter, J.R., Werner, S.A., "Ford Molds a Better Bumper Via Statistical Process Control." *Modern Plastics* February 1986: 68–74.
2. Caren, S., "CAD/CAE for Plastic Protyping." *Plastics Design Forum* March/April 1991: 61–64.
3. "Design of Experiments: The Mystery of Taguchi Explained." *Quality Management*, Section 2, Bureau of Business Practice Inc., Arlington, VA, 1988, 3–6.
4. Feigenbaum, A.V., *Total Quality. Control.* New York, McGraw-Hill, 1983.
5. Gerbig, S.D., "A Guide to Processing Glass-Reinforced Resins." *Plastics Engineering* February 1985: 37–39.
6. Hassel, A. von., "Integrated System Combines Mold Production, Part Design." *Plastics World* November 1991: 12–13.
7. Hertzer, R.A., Green, M.W., "Closed-Loop Servohydraulics: Boon to Automation." *PM & E* November 1988: 38–39.
8. Leonard, L., ed. "The Big Sell Is Automation." *PM & E* February 1981: 25–26.
9. Leonard, L., ed. "Plug Those Auxiliaries into the SPC Loop." *PM & E* October 1989: 45–48.
10. Leonard, L., ed. "Will Rapid Prototyping Be Part of Your Future?" *Plastics Design Forum* January/February 1991: 15–22.
11. Lindsay, K.F., "Rapid Prototyping Shapes Up as Low Cost Modeling Alternative." *Modern Plastics* August 1990: 40–43.
12. McCarthy, L.R., ed. "The ABCs of Flash in Injection Molding." *Plastics World* July 1989: 32–35.
13. Miller, B., ed. "Fast Prototyping Makes Models Sooner and Better." *Plastics World* February 1991: 44–47.
14. Miller, B., ed. "Cavity Pressure Control Keeps Molding Quality on Target." *Plastics World* November 1989: 62–66.
15. Olmsted, B.A., "Solving the Shot-to-Shot Variation Dilemma." *PM & E* December 1987: 38.
16. Tobin, W.J., "Making It Work." *PM & E* May 1987: 28–30.
17. Wigotsky, V., "Engineering Resins: Producing Precision Parts." *Plastics Engineering* September 1985: 33–46.

CHAPTER 12

1. Hall, W.J., "More Realistic Testing of Injection-Molded Parts." *Plastics Engineering* January 1985: 43–45.
2. Progelhof, R.C., Throne, J.L., Patel, U., "Drop-Weight Testing: What It Tells Designers." *Modern Plastics* February 1988: 106–110.
3. *Quality Control Tests*. Phillips Chemical, No. TSM-291, January 1982.
4. Wigotsky, V., "Real-Life Plastic Tests Aim to Tell It Like It Is." *Plastics Engineering* April 1987: 27–35.

CHAPTER 13

1. Leonard, L., ed. "Conveyors: Pathways to Productivity." *PM & E* February 1990: 49–50.
2. Snyder, M., ed. "Robotic Parts-Handling Devices." *PM & E* June 1987: 36–38.
3. Wilder, R.V., "Robots, Sure, But There's a Lot More to Injection Molding Takeoff." *Modern Plastics* November 1988: 47–50.

CHAPTER 14

1. *A Guide for Designing with Engineering Plastics*. Wilmington, DE: E.I. Du Pont de Nemours Corp., 1990.
2. Beck, R.D., *Plastic Product Design*. New York: Van Nostrand Reinhold, 1970.
3. "Designing for Impact Resistance." *Plastics Design Forum* November/December 1989: 104.
4. *Designing with Akulon and Arnite*. Heerlen, The Netherlands: DSM.
5. *Designing with Plastics: The Fundamentals*. Engineering Plastics Division, Holchst Celanese, 93-46 15M/490.
6. *General Design Principles—Module 1*. Wilmington, DE: E.J. Du Pont de Nemours Corp., E-62617.
7. Kelly, D., "An Argument for a New Approach to Product Development." *Plastics Design Forum* January/February 1987: 80.
8. Lewis, G., "Designing to Reduce Assembly Costs." *Plastics Design Forum* January/February 1986: 54–60.
9. Mehta, K.S., "Identifying and Correcting Part-Design Problems." *Plastics Design Forum* November/December 1986: 35–46.
10. "Nominal Wall Thickness." *Plastics Design Forum* March/April 1991: 96.
11. "Planning for Screw-Holding Bosses." *Plastics World* April 1990: 39.
12. "Rigid-Vinyl Design Considerations." *Plastics Design Forum* November/December 1990: 82.
13. Spier, I., "The Most Common Mistakes Made by Design Engineers Working in Plastics." *Plastics Design Forum* March/April 1986: 24–30.
14. "The Living Hinge." *Plastics Design Forum* May/June 1989: 96.

15. Tobin, W., "Design by the Numbers." *Plastics Design Forum* March/April 1988: 59–64.
16. Tobin, W., "Why Plastics Fail." *Plastics Design Forum* January/February 1987: 45–49.

CHAPTER 15

1. Beck, R.D., *Plastics Product Design*. New York: Van Nostrand Reinhold, 1970.
2. *Designing with Plastics: The Fundamentals*. Engineering Plastics Division, Holchst Celanese 93-46 15M/490.
3. *Focused Infrared Melt Fusion*. Branson Ultrasonics Corp., EM89-175.
4. *General Design Principles—Module 1*. Wilmington, DE: E.I. Du Pont de Nemours Corp., E-62617.
5. Gollagan, S.T., "Laying the Groundwork for Ultrasonic Welding." *Plastics Engineering* August 1985: 35–37.
6. Kirkland, C. Sr., ed. "Boost Part Repeatability with Insert Molding." *Plastics World* August 1999: 34–37.
7. Klein, A.J., ed. "Joining Plastics." *Plastics Design Forum* September/October 1988: 39–52.
8. Klein, A.J., ed. "Update on Adhesives." *Plastics Design Forum* May/June 1989: 59–65.
9. McCarthy, L., "Part Design Is Critical for Good Welds." *Plastics World* June 1999: 62–67.
10. Malloy, R.A., Orroth, S.A., "Turning the Screw on Boss Design." *Plastics Engineering* April 1985: 43–45.
11. Miller, B., ed. "Adhesives Toughen Up, But Stay User-Friendly." *Plastics World* May 1987: 39–43.
12. *Modern Plastics Encyclopedia*, New York: McGraw-Hill, October 1991.
13. Snyder, M., ed. "Automation, Part IV: Assembly and Inspection." *PM & E* November 1988: 35–36.
14. *Ultrasonic Staking and Spot Welding of Thermoplastic Assemblies*, Newtown, CT: Ultra Sonic Seal, 1985.
15. *Ultrasonic Weldability-Compatibility Chart for Thermoplastics*. Dukane Corp., Form No. 10348-K-84.

CHAPTER 16

1. Ballway, B., "Finishing Plastics." *Plastics Design Forum* September/October 1988: 77–80.
2. Beck, R.D., *Plastic Product Design*. New York: Van Nostrand Reinhold, 1970.
3. Grob, H., "Pad Printing Impresses the Plastic Industry." *Plastics Engineering* June 1987: 39–41.

4. Hydro Grafix, Greensboro, NC.
5. Kirkland, C. Sr., ed. "Shop Wisely for Heavy-Metal Free Colorants," *Plastics World* October 1990: 49–54.
6. "Lack of Communication Poses Problems with Plating." *Plastics World* May 1991: 46–47.
7. Lodge, C. Sr., ed. "New Label Materials Boost In-Mold Appeal." *Plastics World* May 1989: 39–42.
8. McConnell, V.P., ed. "In-Mold Heat Transfer Decoration." *Plastics Design Forum* May/June 1990: 55–62.
9. Smoluk, G., "Vacuum Metalizing: Now More Than Just a Decorating Process." *Modern Plastics* May 1987: 38–40.
10. Tobin, W., "Painting Plastics." *Plastics Design Forum* September/October 1987: 77–84.
11. Ward, W., "Flocking Decorative and Useful Fiber Finishes." *Plastics Engineering* July 1987: 31–33.

CHAPTER 17

1. "Molder Halves Defect and Part-Inspection Costs: Quality Control." *Modern Plastics* July 1991: 17.
2. Rakstis, T.J., "Quality Starts with You." *Quality First*. Chicago, IL: Dartnell Corporation, 1991.

APPENDIX B

1. Schleckser, J., "Troubleshooting Technique Shortens Path to Quality." *Plastics Engineering* July 1987: 35–38.
2. Schlechser, J., *Troubleshooting Techniques*, No. 0267-017-0.5D, Rogers Corporation, 1986.

APPENDIX C

1. Wigotsky, V., "Some Mold Design Guidelines." *Plastic Engineering* November 1985: 24–25.

APPENDIX E

1. "Common Molding Problems and How to solve Them." *Plastics World Directory* 1988: 414–418.
2. "In Molding Clear Plastics, Success Starts with Clean." *Plastics World* January 1991: 32–33.

3. Leonard, L., Sr., ed. "SPC/SQC Products Proliferate." *PM & E* April 1991: 31–34.
4. Schleckser, J., "Troubleshooting Technique Shortens Path to Quality." *Plastics Engineering* July 1987: 35–38.
5. Schleckoer, J., *Troubleshooting Techniques* No. 0267-017-0.5D, Table 4, Rogers Corporation, Carol Stream, IL, 1986.
6. Leonard, L., ed. "IMM Controls: On the Brink of Brilliance." *PM & E* August 1989: 32–34.

APPENDIX F

1. Miller, B., ed. "How to Track Down Paint-Line Problems," *Plastics World* July 1989: 39–42.
2. Janco, R.A., "Graphics for Decorating Plastics." *Plastics Design Forum* September/October 1987: 86–88.
3. Snyder, M., ed. "Decorating: Evolution in Progress." *PM & E* November 1987: 33–38.
4. Lodge, Charles "Foil Choice, Press Size Are Key to Hot Stamp Decorating." *Plastics World* November 1991: 32–35.

Index

Total Quality Process Control for Injection Molding, by M. Joseph Gordon

Printed and bound by CPI Group (UK) Ltd, Croydon, CR0 4YY

18/09/2026

14964974-0002